Horticulture and Agriculture

Horticulture and Agriculture

Edited by **Thelma Bosso**

SYRAWOOD
PUBLISHING HOUSE
New York

Published by Syrawood Publishing House,
750 Third Avenue, 9th Floor,
New York, NY 10017, USA
www.syrawoodpublishinghouse.com

Horticulture and Agriculture
Edited by Thelma Bosso

International Standard Book Number: 978-1-68286-231-5 (Hardback)

Printed in China.

Contents

	Preface	**IX**
Chapter 1	***In vitro* Induction and Proliferation of Callus from Immature Cotyledons and Embryos of *Juglans regia* cv. 'Xiangling'** Xiao-Dong Cai, Gui-Yuan Wang and Wen-Juan Cao	**1**
Chapter 2	**Potential of Second Crops of Oil Radish and White Mustard as Fall Grazing Pasture for German Mutton Merino Sheep** Péter Póti, Ferenc Pajor, Ákos Bodnár, Károly Penksza, János Tőzsér and Csaba Gyuricza	**8**
Chapter 3	**The Effect of Water Deficit and Nitrogen on the Antioxidant Enzymes' Activity and Quantum Yield of Barley (*Hordeum vulgare* L.)** Arezu Movludi, Ali Ebadi, Soodabeh Jahanbakhsh, M. Davari and Ghasem Parmoon	**11**
Chapter 4	**Micropropagation of the Endangered and Decorative Specie *Dianthus serotinus* Waldst. et Kit.** Marija Marković, Mihailo Grbić and Matilda Djukić	**18**
Chapter 5	**Effects of Saline and Alkaline Stresses on Growth and Physiological Changes in Oat (*Avena sativa* L.) Seedlings** Zhanwu Gao, Jiayu Han, Chunsheng Mu, Jixiang Lin, Xiaoyu Li, Lidong Lin and Shengnan Sun	**26**
Chapter 6	**Multicriteria Analysis of the Effects of Field Burning Crop Residues** Vasilica Stan, Gina Fîntîneru and Mircea Mihalache	**32**
Chapter 7	**Genetic Diversity and Relationships in Local Varieties of Eggplant from Different Cultivar Groups as Assessed by Genomic SSR Markers** Santiago Vilanova, Maria Hurtado, Adriana Cardona, Mariola Plazas, Pietro Gramazio, Francisco J. Herraiz, Isabel Andújar and Jaime Prohens	**40**
Chapter 8	**Phenolic Profile of the Kernel of Selected Persian Walnut (*Juglans regia* L.) Cultivars** Géza Bujdosó, György Végvári, Veronika Hajnal, Gitta Ficzek and Magdolna Tóth	**47**
Chapter 9	**Regulation of Root Length and Lateral Root Number in Trifoliate Orange Applied by Peroxide Hydrogen and Arbuscular Mycorrhizal Fungi** Chun-Yan Liu, Yong-Ming Huang, Ying-Ning Zou and Qiang-Sheng Wu	**53**
Chapter 10	**Kernel Quality of Some Sweet Corn Varieties in Relation to Processing** Ozlem Alan, Gulcan Kinaci, Engin Kinaci, Zekiye Budak Basciftci, Kenan Sonmez, Yasemin Evrenosoglu and Imren Kutlu	**58**

Chapter 11 **Exogenous Proline and Betaine-induced Upregulation of Glutathione Transferase and Glyoxalase I in Lentil (*Lens culinaris*) under Drought Stress** 64
Md. Rezwan Molla, M. Rawshan Ali, Mirza Hasanuzzaman, Mahamud Hossain Al-Mamun, Asgar Ahmed, M. A. N. Nazim-Ud-Dowla and Md. Motiar Rohman

Chapter 12 **Genetic Similarity Assessment among Selected Naked Oat Cultivars and Breeding Lines Using ISSR Markers** 72
Edyta Paczos-Grzeda, Piotr Tomasz Bednarek, Aneta Koroluk, Zygmunt Nita, Zofia Banaszak, Andrzej Bichonski, Marek Chmiel, Agnieszka Gradzielewska, Katarzyna Nowaczyk, Aleksandra Szolkowska, Krystyna Werwinska and Patrycja Wieczorek

Chapter 13 **Phenotypic and Molecular Screening of Apple Genotypes to Woolly Apple Aphid Resistance** 79
Saeid Abu-Romman and Mazen Ateyyat

Chapter 14 **Grassland Degradation Decrease the Diversity of Arbuscular Mycorrhizal Fungi Species in Tibet Plateau** 84
Xiao-Bu Cai, Yue-Lin Peng, Min-Na Yang, Tao Zhang and Qiang Zhang

Chapter 15 **Nitrogen and Zinc Interaction Improves Yield and Quality of Submerged Basmati Rice (*Oryza sativa* L.)** 91
Hakoomat Ali, Zuhair Hasnain, Ahmad N. Shahzad, Naeem Sarwar, Muhammad K. Qureshi, Shazia Khaliq and Muhammad F. Qayyum

Chapter 16 **Microsporogenesis of Apricot Cultivars in Hungary** 99
Veronika Hajnal, Zarif Omid, Márta Ladányi, Magdolna Tóth and László Szalay

Chapter 17 **Growth and Photosynthetic Response of Two Persimmon Rootstocks (*Diospyros kaki* and *D. virginiana*) under Different Salinity Levels** 105
Meral Incesu, Berken Cimen, Turgut Yesiloglu and Bilge Yilmaz

Chapter 18 **Nutrition Intensity in Ternary Diagrams Interpretation for Some Ornamental Species Cultivated on Organic Substrate with Increased Biological Activity** 111
Roxana Maria Madjar, Gina Vasile Scăețeanu, Mircea Mihalache, Catalina Calin, Valentin Sebastian Dan and Adrian Peticila

Chapter 19 **Quality Changes of Different Sweet Cherry Cultivars at Various Stages of the Supply Chain** 117
Fatih Sen, Rustu Efe Oksar, Mina Golkarian and Sevde Yaldiz

Chapter 20 **Nutritional and Nutraceutical Components of Commercial Eggplant Types Grown in Sinaloa, Mexico** 123
Guillermo Niño-Medina, Dolores Muy-Rangel, Alfonso Gardea-Béjar, Gustavo González-Aguilar, Basilio Heredia, Manuel Báez-Sañudo, Jorge Siller-Cepeda and Rosabel Vélez De La Rocha

Chapter 21 **Assessment of the Potential of 1-Methylcyclopropene Treatments to Maintain Fruit Quality of the Common Fig (*Ficus carica* L. cv. 'Bursa Siyahi') during Refrigerated Storage** 130
Okan Ozkaya, Songül Çömlekçioğlu and Hatice Demircioğlu

Chapter 22 **Calcium and Calmodulin Involve in Mycorrhizal and Root Development in Trifoliate Orange Colonized by *Rhizophagus intraradices*** 137
Qiang-Sheng Wu, Qiu-Dan Ni, Yan-Cong Que and Wei Huang

Chapter 23 **Cutting Type and IBA Treatment Duration Affect *Teucrium fruticans* Adventitious Root Quality** **143**
Leo Sabatino, Fabio D'anna and Giovanni Iapichino

Chapter 24 **The Effect of Different Irrigation Water Levels on Grain Yield, Yield Components and Some Quality Parameters of Silage Maize (*Zea mays indentata* Sturt.) in Marmara Region of Turkey** **147**
Abdullah Karasu, Hayrettin Kuşcu, Mehmet Öz and Gamze Bayram

Chapter 25 **Phenolics Content, Antioxidant Activity and Color of Green Walnut Extracts for Preparing Walnut Liquor** **155**
Sina Cosmulescu, Ion Trandafir, Violeta Nour, Mira Ionica and Felicia Tutulescu

Chapter 26 **Phenolic Compounds and Antifungal Activity of *Hedera helix* L. (Ivy) Flowers and Fruits** **160**
Marcel Pârvu, Laurian Vlase, Alina E. Pârvu, Oana Rosca-Casian, Ana-Maria Gheldiu and Ovidiu Pârvu

Chapter 27 **Impact of Climate on Vegetation Change in a Mountain Grassland – Succession and Fluctuation** **166**
Florin Păcurar, Ioan Rotar, Albert Reif, Roxana Vidican, Vlad Stoian, Stefanie M. Gärtner and Robert B. Allen

Chapter 28 **Effect of Light Intensity and Temperature on Growth and Quality Parameters of Grafted Vines** **176**
Bülent Köse

Chapter 29 **Molecular Role of Nitric Oxide in Secondary Products Production in *Ginkgo biloba* Cell Suspension Culture** **185**
Hossam Saad El-Beltagi, Osama K. Ahmed and Adel E. Hegazy

Chapter 30 **Revegetation of Abandoned Sulphur Flotation Waste in Keçiborlu, Isparta/Turkey: Heavy Metals Concentrations of Growing Media and *Agropyron elongatum* Grass** **192**
Huseyin Kalkan, Mustafa Kaplan and Şule Orman

Chapter 31 **Spatial Distribution of Grapes Sugar Content and its Correlations with Climate Characteristics and Climate Suitability in the Huşi (Romania) Wine Growing Region** **199**
Liviu Mihai Irimia, Cristian Valeriu Patriche, Georgeta Mihaela Bucur and Hervé Quénol and Valeriu V. Cotea

Chapter 32 ***In Vitro* and *in Planta* Activity of Some Essential Oils against *Venturia inaequalis* (Cooke) G. Winter** **208**
Géza Nagy, Tamás Hochbaum, Szilvia Sárosi and Márta Ladányi

Chapter 33 **Effects of Salt and Water Stress on Plant Growth and on Accumulation of Osmolytes and Antioxidant Compounds in Cherry Tomato** **214**
Mohamad Al Hassan, Marina Martínez Fuertes, Francisco José Ramos Sánchez, Oscar Vicente and Monica Boscaiu

Chapter 34 **Molecular Cloning and Expression Analysis of a MADS-Box Gene (*GbMADS2*) from *Ginkgo biloba*** **225**
Xiaohui Wang, Junhuan Cheng, Feng Xu, Xingxiang Li, Weiwei Zhang, Yongling Liao, Shuiyuan Cheng and Xianbin Li

Chapter 35 **Effect of Planting Systems and Growing Media on Earliness, Yield and Quality of Strawberry Cultivation under Soilless Culture** **231**
Nafiye Adak and Hamide Gubbuk

Chapter 36 **Morphological and Molecular Characterization and Relationships of Turkish Local Eggplant Heirlooms** **237**
Hatice Filiz Boyaci, Volkan Topcu, Akin Tepe, Isilay Karasahin Yildirim, Mehmet Oten and Aytekin Aktas

Permissions

List of Contributors

Preface

Horticulture is a sub-field of agriculture and deals with all the aspects of cultivating a vegetable garden. In the present scenario, horticulture has immense relevance as it plays a significant role in increasing yield and building pest resistance. This book provides scientific and technical insights into various concepts of horticulture, like plant conservation, landscape restoration, landscape and garden design, etc. It will be an essential companion to the students of botany, agricultural sciences, horticulture, and associated fields of study. The text contains detailed explanations of new techniques and their applications in a multidisciplinary approach and will cater to the needs of research scholars as well.

This book is a comprehensive compilation of works of different researchers from varied parts of the world. It includes valuable experiences of the researchers with the sole objective of providing the readers (learners) with a proper knowledge of the concerned field. This book will be beneficial in evoking inspiration and enhancing the knowledge of the interested readers.

In the end, I would like to extend my heartiest thanks to the authors who worked with great determination on their chapters. I also appreciate the publisher's support in the course of the book. I would also like to deeply acknowledge my family who stood by me as a source of inspiration during the project.

Editor

In vitro Induction and Proliferation of Callus from Immature Cotyledons and Embryos of *Juglans regia* cv. 'Xiangling'

Xiao-Dong CAI*, Gui-Yuan WANG, Wen-Juan CAO

*Yangtze University, College of Horticulture and Gardening, Jingzhou, Hubei, P.R. China; caixiao.dong@163.com (*corresponding author)*

Abstract

Callus holds great potential for biotechnology applications in plants. In this study, several experiments were performed to establish a protocol for *in vitro* induction of good quality callus by using immature cotyledons and embryos of *Juglans regia* cv. 'Xiangling' as the explants. DKW basal media supplemented with different concentrations of plant growth regulators were applied to determine the induction rate, texture and color of the recovered callus. Results showed that the optimum culture medium for the callus induction from these explants was a DKW basal medium consisting of 1 mg/L BA, 2 mg/L KT 250 mg/L glutamine, 500 mg/L casein hydrolysate, 200 mg/L Vc, 50 g/L sucrose and 2.8 g/L Gelrite (i.e., the T8 medium). The average rate of callus induction on the medium reached 80.7% for the immature cotyledons and 66.7% for the immature embryos, and all the callus was characterized by a friable, creamy or yellow appearance. *In vitro* proliferation of callus was not very successful as a result of the fact that the friable callus with creamy or yellow color grew very slowly, whereas the fast-growing callus was easy to turn brown and died eventually. Even though further investigation will be required to overcome the problem of browning and assess the regeneration ability of the callus, this approach for the production of friable callus may have valuable applications for walnut and other woody trees.

Keywords: browning, callus induction rate, somatic embryogenesis, tissue culture, walnut

Introduction

The common walnut (*Juglans regia*), one of the ancestral species of *Juglans*, is a popular fruit crop with a potential for nut production, as it is abundant in nutritious substances and dietary minerals (Vahdati *et al.*, 2008, 2009). Walnut has also been considered as a valuable tree species owning to its use for forestry in the world. In general practice, walnut is usually propagated by seeds and can also be vegetatively propagated via graft and cutting method. Propagation by seeds is undesirable because of high heterozygosis and long growth period of the seedlings. Grafting the scion cultivars onto a rootstock is another general way for walnut clonal propagation, but it was reported to be labor-intensive, time-consuming and costly (Rezaee *et al.*, 2008). Cutting technique was also extensively employed in walnut, but the main difficulty of this method was related to low rooting ability of the nodal segments from mature trees (Payghamzadeh and Kazemitabar, 2011). Since the first report on *in vitro* stem culture of black walnut (*J. nigra*) (Cummins and Ashby, 1969), the potential of *in vitro* propagation of walnut initiated from apical buds, nodal segments, leaves, petioles, cotyledons or embryos has been widely explored during the past four decades. Most of these researches focused on browning phenomenon, shoot or bud elongation, root induction, embryo germination or somatic embryogenesis (Fernández *et al.*, 2000; Vahdati *et al.*, 2006, 2008, 2009; Amiri, 2006; Fatima *et al.*, 2006; Leal *et al.*, 2007; Avilés *et al.*, 2009; Toosi and Dilmagani, 2010; Payghamzadeh and Kazemitabar, 2010, 2011; Amiri and Gharati, 2012; Gotea *et al.*, 2012). In spite of its recalcitrant behavior under *in vitro* conditions, plant tissue culture provides an alternative approach for propagation of walnut species.

Advanced biotechnology has attracted considerable attention in diverse plants. A basic prerequisite of biotechnology application is the availability of a reliable *in vitro* regeneration protocol. As it is well known, plants can be regenerated from callus cultures via somatic embryogenesis or adventitious shoot organogenesis. Somatic embryogenesis is used as a model system for large-scale vegetative propagation of a wide range of plants species (von Arnold *et al.*, 2002) including some forest trees (Lelu-Walter *et al.*, 2013). In walnut, previous reports suggested that somatic embryogenesis could be induced from immature cotyledons, embryos, embryonic axes and endosperm on an appropriate medium (Tulecke *et al.*, 1988; Tang *et al.*, 2000; Fernández *et al.*, 2000; Vahdati *et al.*, 2006; Fatima *et al.*, 2006; Toosi and Dilmagani, 2010; Payghamzadeh and Kazemitabar, 2011). Callus was always used as initial material to produce somatic embryoids (Aftab, 2012) and adventitious shoots (Tang and Newton, 2005) in many woody plants, but this has not been documented in walnut. A lack of good quality callus may be a major bottleneck to successful regeneration of somatic embryoids or shoots from callus in walnut.

Callus is an attractive target explant for studying germplasm and improvement of plant cultivars, such as somaclonal variations, somatic hybridization and genetic transformation, etc. Therefore, in order to apply advanced biotechnology in these research fields, it is essential to develop an efficient callus induction system of walnut. The success of callus induction is to a large extent determined by the genotype, the developmental stage and types of explants, and the nutrients and plant growth regulators provided by the culture medium. Generally speaking, callus may arise from proliferating cells of the parent tissue by placing explants of appropriate size on a growth-supporting medium under sterile conditions. In walnut, callus has been induced from young plant tissue like endosperm (Tulecke *et al.*, 1988), immature cotyledons (Neuman *et al.*, 1993; Long *et al.*, 1995; San and Dumanoglu, 2006), embryos (Breton *et al.*, 2004; Toosi and Dilmagani, 2010) and embryonic axes (Tang *et al.*, 2000; Fernández *et al.*, 2000) and even from leaf explants (Farsi *et al.*, 2012). Culture conditions such as basal medium and auxin/cytokinin ratio are also important factors to be responsible for callus formation. For example, Neuman *et al.* (1993) found that the concentration of 2, 4-D and TDZ played an important role in callus production from walnut cotyledons. Fernández *et al.* (2000) emphasized that 2, 4-D was less effective at inducing callus than NAA or IBA, and they believed that auxins facilitated callus formation in walnut. According to the report of Payghamzadeh and Kazemitabar (2010), callus fresh weight of walnut depended on the concentration of BAP and IBA, and the 0.01 and 0.1 mg/L IBA as well as 1.5 mg/L BAP favored callus induction from immature embryonic bodies. Although efforts have been made to improve the protocol of callus induction of walnut, so far it is still a quite difficult task and far from being solved. Thus, further investigation will be needed to establish a perfect induction system of walnut callus. Herein, *in vitro* callus induction from the immature cotyledons and embryos of *J.regia* cv. 'Xiangling' was studied in response to the DKW basal media supplemented by different concentrations and types of plant growth regulators. Most efforts were directed towards establishing a feasible protocol for the induction and multiplication of walnut callus by numerical investigation and morphological characterization of the callus.

Materials and methods

Plant materials

The open pollinated fruits of *J.regia* cv. 'Xiangling' were harvested at around 8 weeks post-anthesis and prechilled at 4 °C for 20 days before the experiment was conducted, as pretreatment with low temperature may improve the efficiency of callus induction.

Disinfection of explants

Firstly the fruit exocarps were removed and the remaining parts of the fruits were rinsed thoroughly under running tap water for 60 min to eliminate phenolic compounds. Then the remains of the fruits were surface-sterilized by immersion in 70% of ethanol for 5 min followed by disinfection with 2% hypochlorite sodium solution plus 0.01% Tween 20 for 15 min. Thereafter the samples were rinsed 5 times with sterile distilled water. Prior to inoculation, immature cotyledons and embryos were carefully dissected out from the sterile fruits through the suture at the pedicel end using a sharp-pointed forceps and a scalpel blade.

Callus induction

In the present study, the explants were cultured on 13 different media (as shown in Tab.1), i.e., solidified DKW basal media (Driver and Kuniyuki, 1984) supplemented

Tab. 1. Effects of different concentrations of plant growth regulators on induction rate, texture and color of callus from immature cotyledons of *J. regia* cv. 'Xiangling'

Treatment	Plant growth regulators (mg/L)					Callus induction (%)	Callus	
	NAA	BA	2,4-D	KT	TDZ		Texture	Color
CK	0	0	0	0	0	0 f	/	/
T1	1	0	0	0	0	32.6±3.4 d	Compact	Brown
T2	2	0	0	0	0	49.6±5.6 bc	Friable	Creamy
T3	0	1	0	0	0	57.88±6. 7 b	Compact	Green
T4	0	2	0	0	0	55.6±6.7 b	Compact	Brown
T5	0	0	1	0	0	43.7±5.6 c	Friable	Creamy
T6	0	0	2	0	0	71.9±9.0 a	Semi-compact	Green
T7	1	1.5	0	0	0	73.3±10.1 a	Compact	Green
T8	0	1	0	2	0	80.7±8.9 a	Friable	Creamy
T9	0	2	0	3	0	67.9±.6.8 a	Friable	Creamy
T10	0	1	2	1	0	23.0±4.6 e	Semi-compact	Brown
T11	0	1	0	0	1	56.3±6.8 b	Semi-compact	Brown
T12	0	0	1	0	1	29.6±4.6 de	Semi-compact	Brown

Note: Mean values with the same letters are not significant differences (LSD test, $p<0.05$)

with different concentrations of plant growth regulators (NAA, BA, 2, 4-D, KT or TDZ) , designed especially to facilitate callus induction of walnut according to previous reports (Long *et al.*, 1995; Tang *et al.*, 2000; Fernández *et al.*, 2000; Breton *et al.*, 2004; San and Dumanoglu, 2006; Toosi and Dilmagani, 2010; Farsi *et al.*, 2012). All media contained 250 mg/L glutamine, 500 mg/L casein hydrolysate, 200 mg/L Vc, 50 g/L sucrose and 2.8 g/L Gelrite (Sigma), pH 5.8. For the immature cotyledons, the explants were cut into pieces around 0.5 cm^2 and placed horizontally on the medium. The experimental design was completely randomized, and five 100-ml Erlenmeyer flasks containing 40 ml medium were used for each treatment with four explants per flask. For the culture of immature embryos, the explants were aseptically removed from the fruits and oriented with their apical ends protruding from the medium (only one explant per flask), and two replicates each with nine or ten embryos were used for each treatment. All cultures were incubated at 25°C and 12 h/12 h (light/dark) cycles under cool white fluorescent lamps (2500 Lux). 30 days later, the appropriate induction media were selected based on the results of the induction rate, texture and color of the recovered callus. The regenerated cultures were transferred to fresh medium for further induction of callus.

Callus proliferation

In order to obtain much more callus, callus proliferation was also performed in this study. About 0.3 g callus of good quality was picked up carefully and transferred to fresh media. Five flasks were used for each treatment, and the experiment was replicated three times. The callus was harvested at 20 days after culture and fresh weight was measured.

Experimentation and data analysis

All experiments were carried out as a complete randomized design. DATA analysis was performed using the SPSS 16.0. Prior to analysis, data of callus induction rate were arcsin-square root transformed. Variance analysis and means comparison were done using the LSD test (ANOVA). P-values <0.05 were consid¬ered statistically significant.

Results and discussion

Callus induction from the immature cotyledons

After 3 days of culture, the excised immature cotyledons gradually changed color from white to green on all media, and the initiation of callus was observed around the cut edge of the explants after another one week of culture on all media except medium CK, the control. The induction rate, texture and color of the induced callus were different with respect to different media tested 30 days later (Tab.1). In spite that a portion of regenerated cultures slowly became brown in color and even died at last, callus could be observed in all treatments with the exception of the control (Fig.1A). However, most of the calls was compact or semi-compact with a green or brown color (Fig.1B, C and D), and friable callus was only observed on the media T2, T5, T8 (Fig.1F) and T9 (Fig.1E). Somatic embryoids were also occurred on the T9 medium (arrow indicated in Fig.1E), although the induction rate was very low in the present research. The mean rate of callus induction on the T8 medium reached 80.7%, which was the highest percentage among all the treatments. The statistical analysis indicated that the mean callus induction rates were not significantly different in response to the treatments T6, T7, T8 and T9, but all of them had significantly higher rates of callus induction than the rest medium(P<0.005) . Taking into account the induction rate and the quality of the callus, both the T8 and T9 medium were the appropriate media for callus induction from the immature cotyledons of *J. regia* cv. 'Xiangling'.

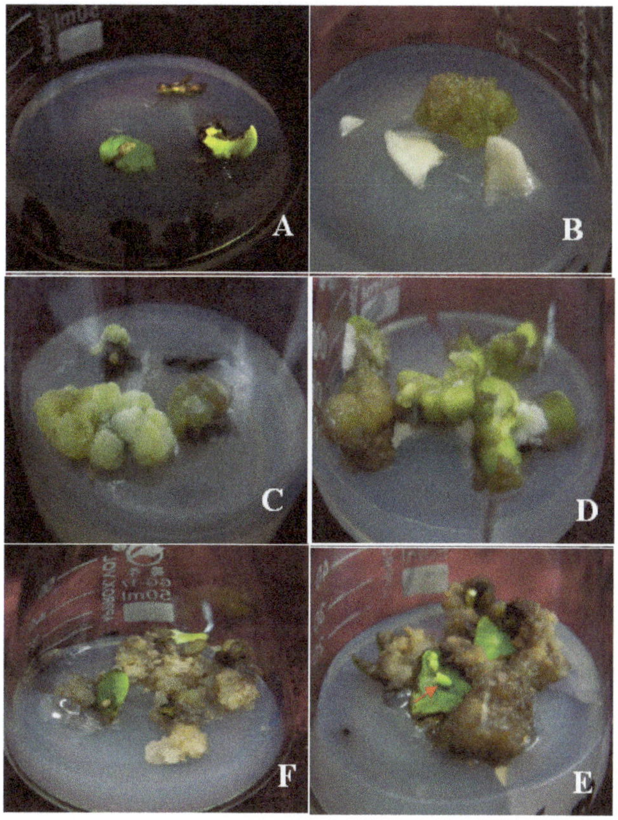

Fig. 1. *In vitro* induction of callus from immature cotyledons of *J. regia* cv. 'Xiangling' on solidified DKW basal media supplemented with 250 mg/L glutamine, 500 mg/L casein hydrolysate, 200 mg/L Vc, 50 g/L sucrose, 3 g/L Gelrite (Sigma) and different concentrations of plant growth regulators. (A) Explants cultured on the CK medium. (B) Semi-compact callus with a green color induced on the T6 medium containing 2 mg/L 2, 4-D; (C) Compact and light green callus induced on T3 medium containing 1 mg/L BA; (D) Compact and green callus induced on T7 medium containing 1 mg/L NAA and 1.5 mg/L BA; (E) Embryos (arrow indicated) and friable callus regenerated on T9 medium containing 2 mg/L BA and 3 mg/L KT; (F) Friable and creamy callus regenerated on T8 medium containing 1 mg/L BA and 2 mg/L KT

Callus induction from the immature embryos

Four media (T2, T5, T8 and T9) for effective induction of friable callus and the CK (as a control) were chosen to regenerate callus from immature walnut embryos. Generally, the immature embryos turned green after 4 days of culture (Fig.2A). Almost all immature embryos were germinated on all the media after 30 days of culture.

Fig. 2. *In vitro* culture of immature embryos of *J. regia* cv. 'Xiangling' on solidified DKW basal media supplemented with 250 mg/L glutamine, 500 mg/L casein hydrolysate, 200 mg/L Vc, 50 g/L sucrose, 3 g/L Gelrite (Sigma) and different concentrations of plant growth regulators. (A) An embryo after 4 days of culture; (B) Regenerated plants after 18 days of culture on CK medium (C) Developed roots but no shoot and callus on T2 medium containing 2 mg/L NAA; (D) Regenerated callus and a green shoot induced on T5 medium containing 1 mg/L 2,4-D; (E) Regenerated friable callus with a yellow color and malformed shoots on T8 medium containing 1 mg/L BA and 2 mg/L KT; (F) Regenerated friable callus with a dark yellow color and plants without a root system on T9 medium containing 2 mg/L BA and 3 mg/L KT

However, different culture media had profound effects on the development of the embryos. The regenerated products had roots and shoots when cultured on the control medium (Fig.2B), and had several roots and no shoot on the T2 medium (Fig.2C), whereas shoots without a root was observed on the rest three media (Fig.2D, E and F). The entire callus appeared creamy in color at the beginning of culture, but later gradually became yellow and even dark yellow. As shown in tab.2, callus formation could be observed on all media except the control, and the maximum percentage (66.7%) of callus induction was achieved on the T8 medium. One-way ANOVA analysis revealed that the mean rate of callus induction on T8 medium was significant higher ($P<0.005$) than those on others.

Callus induction from the regenerated compact/semi-compact callus

Friable callus has been proven to be crucial for the establishment of cell suspension cultures, which played an important role in successful somatic hybridization (Davey *et al.*, 2005) and genetic transformation (Chumakov and Moiseeva, 2012) in many plants. In this study, the compact callus or semi-compact callus was separated from the cultures carefully and were cutted into appropriate pieces with a diameter of 3–5 mm with a sharp-pointed forceps and a scalpel blade, and then placed on the tested media, T2, T5, T8 and T9. After 20 days of culture, it was possible to obtain fresh callus from most of the compact callus or semi-compact callus under all the culture conditions, and its texture and color were variable depending on the concentration and the kind of plant growth regulators. When cultured on the T2 medium, the callus had a semi-compact aspect with a yellowish color (Fig.3A). The creamy callus with a soft texture was observed both on the medium T9 (Fig.3B) and T8 (Fig.3C). Unfortunately, the callus on the culture medium T5 showed a compact appearance with a yellow and green color (Fig.3D).

Proliferation of induced walnut callus

To screen the optimum medium for callus multiplication, about 0.3 g friable callus with creamy color was picked up carefully and transferred to fresh T2, T5, T8 and T9 medium and the CK, respectively. As showed in Tab.3, the mean weight of the fresh callus ranged from 0.64 g

Tab. 2. Effects of different concentrations of plant growth regulators on the induction rate, texture and color of callus from immature embryos of *J. regia* cv. 'Xiangling

Treatment	Plant growth regulators (mg/L)				Callus induction (%)	Callus	
	NAA	BA	2,4-D	KT		Texture	Color
CK	0	0	0	0	0 c	/	/
T2	2	0	0	0	40.6 ± 4.4^{b}	Friable	Dark yellow
T5	0	0	1	0	33.4 ± 7.9^{b}	Friable	Yellow
T8	0	1	0	2	66.7 ± 7.9^{a}	Friable	Yellow
T9	0	2	0	3	46.9 ± 4.4^{b}	Friable	Dark yellow

Note: Mean values with the same letters are not significant differences (LSD test, $p<0.05$)

Tab. 3. Effects of different media on proliferation of the induced callus of *J. regia* cv. 'Xiangling'

Treatment	Plant growth regulators (mg/L)				Callus fresh weight (g /flask)	Callus	
	NAA	BA	2,4-D	KT		Texture	Color
CK	0	0	0	0	0.75 ± 0.08^{c}	friable	Creamy
T2	2	0	0	0	0.64 ± 0.06^{c}	friable	Creamy
T5	0	0	1	0	2.03 ± 0.14^{a}	friable	Brown
T8	0	1	0	2	1.18 ± 0.05^{b}	friable	yellow
T9	0	2	0	3	1.25 ± 0.12^{b}	Water-soaked	yellow

Note: Mean values with the same letters are not significant differences (LSD test, p<0.05)

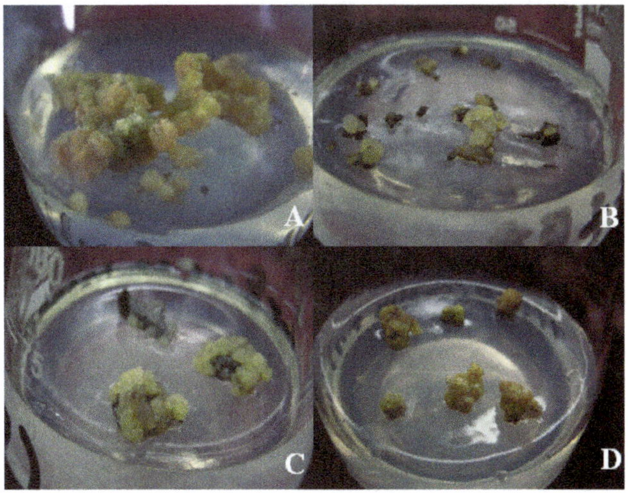

Fig. 3. Induction of callus from the compact callus or semi-compact callus of *J. regia* cv. 'Xiangling'on DKW basal media supplemented with 250 mg/L glutamine, 500 mg/L casein hydrolysate, 200 mg/L Vc, 50 g/L sucrose, 3 g/L Gelrite (Sigma) and different concentrations of plant growth regulators. (A) Regenerated semi-compact callus with a yellowish color on T2 medium containing 2 mg/L NAA; (B) Creamy callus with a soft texture induced on T8 medium containing 1 mg/L BA and 2 mg/L KT; (C) Creamy callus with a soft texture induced on T9 medium containing 2 mg/L BA and 3 mg/L KT; (D) Compact callus induced on T5 medium containing 1 mg/L 2, 4-D

per flask to 2.03 g per flask. The T9 medium containing 1 mg/L 2, 4-D seemed to be the most efficient on multiplication of the callus, and the mean weight of the fresh callus increased over 10 folds and reached 2.093 g per flask, with a significant higher level than those of other treatments (P<0.005). However, the callus on the T5 medium was water-soaked and presented a greater degree of browning (Fig. 4A). On the T8 (Fig. 4B) and T9 (Fig. 4C) medium, most of the callus grew quickly, with a friable and yellow appearance, and a portion of callus died 20 days after culture initiation as a result of browning. The callus cultured on the T2 (Fig. 4D) and the control medium (Fig. 4E) was not as abundant as that on the above-mentioned media, but it had a friable texture with less browning. All of these indicated that the adopted media were not optimum for callus proliferation owning to the fact that rapid-growing callus turned brown easily, and callus of good quality developed very slowly in the study.

Callus can probably be obtained in all plant species provided that the appropriate explant types, culture media compositions and cultural environmental conditions are employed. Walnut is extensively distributed throughout the world. Although efforts have been made to induce callus in walnut as described above, it is far from being completely comprehended. Therefore, this research focused on callus induction from the immature cotyledons, immature embryos and the compact or semi-compact callus (regenerated from the former two types of explants) in *J. regia* cv. 'Xiangling'. According to the above results, friable callus

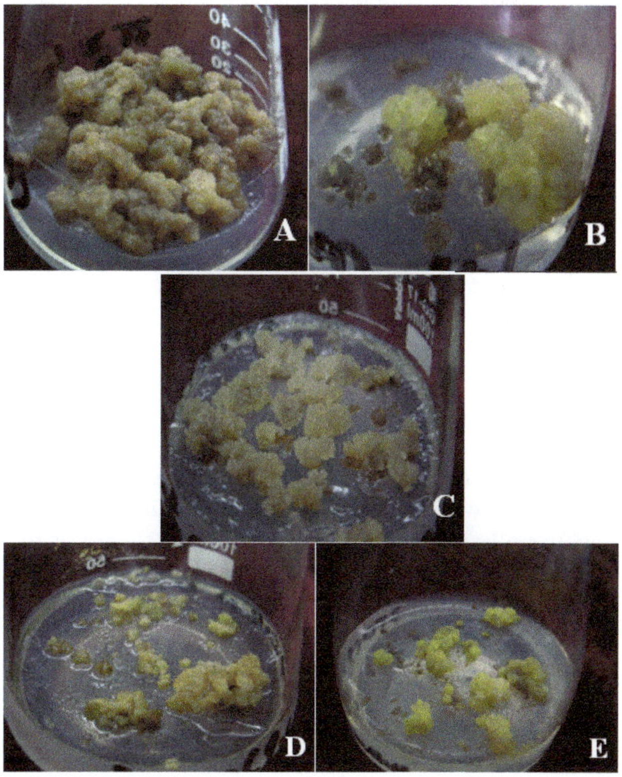

Fig. 4. Proliferation of callus of *J. regia* cv. 'Xiangling' on DKW basal media supplemented with 250 mg/L glutamine, 500 mg/L casein hydrolysate, 200 mg/L Vc, 50 g/L sucrose, 3 g/L Gelrite (Sigma) and different concentrations of plant growth regulators. (A) Water-soaked and browning callus on T5 medium containing 1 mg/L 2, 4-D; (B&C) Friable and yellow callus on T8 medium containing 1 mg/L BA and 2 mg/L KT (B) and T9 medium containing 2 mg/L BA and 3 mg/L KT (C); (D&E) Friable and creamy callus on T2 medium containing 2 mg/L NAA (D) and CK medium (E)

was achieved when these explants was cultured on the T2, T5, T8 and T9 medium in this study. However, as shown in Tab. 1, 2 and 3, most of induced callus was easy to turn brown, and browning also occurred during the course of callus proliferation. Browning is one of the major obstacles in walnut tissue culture, and it is attributed to the production of high concentrations of phenolic compounds which can be easily oxidized (Payghamzadeh and Kazemitabar, 2011). In the present study, although 200 mg/L Vc was used as an antioxidant to alleviate or avoid browning, the phenomenon occurred all the same. Previous reports suggested that frequent subculturing or incubating cultures in total darkness after inoculation could overcome the problem of browning in plant tissue cultures (Long *et al.*, 1995; Payghamzadeh and Kazemitabar, 2011). Some other measures should be adopted to solve the problem of browning in future research.

Callus has the potential to differentiate into a plant by a direct or indirect way, and good quality callus is of great value for research on germplasm and improvement in plants. In general, embryogenic callus has been considered as an ideal initial tissue for biotechnological researches such as induced mutation *in vitro*, genetic improvement and protoplast manipulation. Embryogenic callus could be obtained from the youngest anthers (Prado *et al.*, 2010), embryos (Tang and Newton, 2005) and ovules (Cardoso *et al.*, 2012) in a number of woody trees. In the present paper, we obtained callus from the immature cotyledons, immature embryos and the compact or semi-comapct callus of walnut. However, it is a fact that walnut is one of the most recalcitrant species *in vitro*. Many problems like easy browning, poor proliferation, bad elongation and difficult rooting, are always obstacles for walnut propagation by tissue culture. In this study, the responses of the callus to all adopted media during the course of multiplication were not very satisfying as a result of the fact that the color of rapid-growing callus was easy to turn brown, whereas the callus in good quality developed very slowly. Farsi *et al.* (2012) suggested that somatic embryoids induction from callus was more difficult than from immature cotyledons, zygote embryos and endosperm in *J. regia*. Therefore, further investigation is needed to examine the regeneration capacity of the callus induced in this study.

Conclusion

In conclusion, friable callus with a creamy or yellow color was induced from the immature cotyledons, the immature embryos and the compact or semi-compact callus (regenerated from the former two types of explants) in *J. regia* cv. 'Xiangling'. The T8 medium, i.e. solidified DKW basal media plus 1 mg/L BA, 2 mg/L KT, 250 mg/L glutamine, 500 mg/L casein hydrolysate, 200 mg/L Vc, 50 g/L sucrose and 2.8 g/L Gelrite, was optimum for callus induction from these explants in this study. The adopted media were not very appropriate for *in vitro* multiplication of the callus because callus of good quality grew very slowly, whereas the fast-growing callus was easy to turn brown and even died eventually. Further investigation is required to overcome the problem of browning and evaluate the regeneration ability of the callus. However, this protocol for the induction of friable callus may have valuable applications for walnut and other woody trees.

Acknowledgements

This research was financially supported by Hubei Key Laboratory of Economic Forest Germplasm Improvement and Resources Comprehensive Utilization (20011BLKF239) and partly by the National Natural Science Foundation of China (No. 30900979).

References

Aftab F (2012). Progress and prospects for efficient micropropagation of woody plants, p. 363-377. In: Ashraf M, Öztürk M, Ahmad MSA, Aksoy A (Eds.). Crop Production for Agricultural Improvement. Springer, Netherlands.

Amiri ME (2006). Effect of mineral concentration on tissue-cultured walnut (*Juglans regia* var. Zeiabadi) growth. Acta Hortic 705:383-386.

Amiri ME, Gharati S (2012). Influence of medium composition on multiplication of walnut (*Juglans regia* L.) growth. J Med Plants Res 6(8):1482-1485.

Avilés F, Ríos D, González R, Sánchez-Olate M (2009). Effect of culture medium in callogenesis from adult walnut leaves (*Juglans regia* L.) Chilean J Agric Res 69(3): 460-467.

Breton C, Cornu D, Chriqui D, Sauvanet A, Capelli P, Germain E, Jay-Allemand C (2004). Somatic embryogenesis, micropropagation and plant regeneration of "Early Mature" walnut trees (*Juglans regia*) that flower *in vitro*. Tree physiol 24(4): 425-435.

Cardoso JC, Martinelli AP, Latado RR (2012). Somatic embryogenesis from ovaries of sweet orange cv. Tobias. Plant Cell Tiss Organ Cult 109(1):171-177.

Chumakov MI, Moiseeva EM. Technologies of agrobacterium plant transformation in planta (2012). Appl Biochem Biotech 48(8):657-666.

Cummins JN, Ashby WC (1969). Aseptic culture of *Juglans nigra* stem tissues. For Sci 15:102-103.

Davey MR, Anthony P, Power JB, Lowe KC (2005). Plant protoplasts: status and biotechnological perspectives. Biotechnol Adv 23(2):131-171.

Driver JA, Kuniyuki, AH (1984). In vitro propagation of Paradox walnut *Juglans hindsii* × *Juglans regia* rootstock. Hort Sci 19:507-509.

Farsi M, Vahdati K, Lotfi M, Mirmasoumi M. (2012) Open the windows toward somatic embryogenesis of leaf explants of persian walnut (*Juglans regia* L.). Int J Nuts Related Sci 3(1):13-22.

Fatima A, Kamili AN, Shah AM (2006). Plantlet regeneration

from excised embryonal axes, shoot apices, and nodal segments of *Juglans regia* L. Acta Hortic 705:387-392.

Fernández H, Pérez C, Sánchez-Tamés, R (2000). Modulation of the morphogenic potential of the embryonic axis of *Juglans regia* by cultural conditions. Plant Growth Regul 30(2):125-131.

Gotea R, Gotea I, Sestras RE, Vahdati K (2012). *In vitro* propagation of several walnut cultivars. Bulletin UASVM Horticulture 69(1):167-171.

Leal DR, Sánchez-Olate M, Avilés F, Materan ME, Uribe M, Hasbún R, Rodríguez R (2007). Micropropagation of *Juglans regia* L, p. 381-390. In: Jain SM, Häggman H (Eds.) Protocols for micropropagation of woody trees and fruits. Springer, Heidelberg.

Lelu-Walter MA, Thompson D, Harvengt L, Sanchez L, Toribio M, Pâques LE (2013). Somatic embryogenesis in forestry with a focus on Europe: state-of-the-art, benefits, challenges and future direction. Tree Genet Genomes DOI 10.1007/s11295-013-0620-1

Long LM, Preece JE, Van Sambeek JW (1995). Adventitious regeneration of *Juglans nigra* L. (eastern black walnut). Plant Cell Rep 14(12):799-803.

Neuman MC, Preece JE, Van Sambeek JW, Gaffney GR (1993). Somatic embryogenesis and callus production from cotyledon explants of Eastern black walnut. Plant Cell Tiss Organ Cult 32(1):9-18.

Payghamzadeh K, Kazemitabar SK (2010). The effects of BAP and IBA and genotypes on *in vitro* germination of immature walnut embryos. Inter J Plant Produc 4(4):309-322.

Payghamzadeh K, Kazemitabar SK (2011). *In vitro* propagation of walnut-A review. Afr J Biotechnol 10(3):290-311.

Prado MJ, Grueiro MP, González MV, Testillano PS, Domínguez C, López M, Rey M (2010). Efficient plant regeneration through somatic embryogenesis from anthers and ovaries of six autochthonous grapevine cultivars from Galicia (Spain). Scientia Hortic 125(3):342-352.

Rezaee R, Vahdati K, Grigoorian W, Valizadeh M (2008). Walnut grafting success and bleeding rate as affected by different grafting methods and seedling vigor. J Hortic Sci Biotechnol 83:94-99

San B, Dumanoglu H (2006). Somatic embryogenesis from immature cotyledons of apomictic and non-apomictic seeds in walnut (*Juglans regia* L.). Turk J Agric For 30(2):111-117.

Tang H, Ren Z, Krczal G (2000). Improvement of English walnut somatic embryo germination and conversion by desiccation treatment and plantlet development by lower medium salts. *In Vitro* Cell Dev Biol Plant 36:47-50.

Tang W, Newton RJ (2005). Plant regeneration from callus cultures derived from mature zygotic embryos in white pine (*Pinus strobus* L.). Plant Cell Rep 24(1):1-9.

Toosi S, Dilmagani K (2010). Proliferation of *Juglans regia* L. by *in vitro* embryo culture. J Cell Biol Genet 1(1):12-19.

Tulecke W, McGranahan G, Ahmadi H (1988). Regeneration by somatic embryogenesis of triploid plants from endosperm of walnut, *Juglans regia* L. cv Manregian. Plant Cell Rep 7(5):301-304.

Vahdati K, Bayat S, Ebrahimzadeh H, Jariteh M, Mirmasoumi M (2008). Effect of exogenous ABA on somatic embryo maturation and germination in Persian walnut (*Juglans regia* L.). Plant Cell Tiss Organ Cult 93(2):163-171.

Vahdati K, Jariteh M, Niknam V, Mirmasoumi M, Ebrahimzadeh H (2006). Somatic embryogenesis and embryo maturation in Persian walnut. Acta Hortic 705:199-205.

Vahdati K, Razaee R, Mirmasoomi M (2009). Micropropagation of some dwarf and early mature walnut genotypes. Biotechnol 8:171-175.

von Arnold S, Sabala I, Bozhkov P, Dyachok J, Filonova L (2002). Developmental pathways of somatic embryogenesis. Plant Cell Tiss Organ Cult 69(3):233-249.

2

Potential of Second Crops of Oil Radish and White Mustard as Fall Grazing Pasture for German Mutton Merino Sheep

Péter PÓTI[1*], Ferenc PAJOR[1], Ákos BODNÁR[1], Károly PENKSZA[2], János TŐZSÉR[1], Csaba GYURICZA[3]

[1]*Szent István University, Institute of Animal Husbandry, Páter Karoly 1, H-2100 Gödöllő, Hungary; poti.peter@mkk.szie.hu (*corresponding author)*
[2]*Szent István University, Institute of Botany and Plant Physiology, Páter Karoly 1, H-2100 Gödöllő, Hungary*
[3]*Szent István University, Institute of Crop Production, Páter Karoly 1, H-2100 Gödöllő, Hungary*

Abstract

Second crops of oil radish (*Raphanus sativus*) and white mustard (*Sinapis alba*) being in the bloom stage were evaluated for nutritive value for wether lambs (n=5 per crop, BW=35.6 kg) and *in situ* grazability by ewes (n=15 per crop, BW=50 kg). Animals received no supplemental feed. The two forage crops differed in dry matter (DM) (135 vs. 154 g/kg) as well as in crude protein (148 vs. 206 g), crude fibre (141 vs. 194 g) and N-free extract (550 vs. 482 g) per kg DM. In the feeding trial, total daily DM intake of lambs from oil radish and white mustard differed (1063 vs. 1203 g/head) but was similar in percent bodyweight (3.3-3%). Apparent digestibility of the two forage crops were similar for crude protein (69-70%) and N-free extract (70-71%), but differed ($P<0.001$) for crude fibre (63 vs. 54%); the values were similar for TDN (61.6-62.9%), DE (11.36-11.61 MJ), ME (9.32-9.52 MJ) and NE_m (5.70-5.88 MJ) per kg DM. In the 4-day grazing trial was used rotational grazing, based on results, the oil radish provided a carrying capacity twice as high as white mustard's notably, 1700 vs. 740 ewe grazing days per ha - equal to 60 ewes grazing one hectare for 28 and 12 days, respectively. Thus, oil radish seemed better to be used as fall grazing pasture for mutton sheep.

Keywords: oil radish, white mustard, nutritive value, fall grazing sheep

Abbreviations: dry matter (DM), total digestible nutrients (TDN), digestible energy (DE), metabolisable energy (ME), net energy for maintenance (NE_m)

Introduction

Fast-growing cold-tolerant brassicaceous plants grown as a second-crop can provide good fall pastures for grazing sheep. Such as are, for example, oil radish (*Raphanus sativus* L. *var. oleiferus*) and white mustard (*Sinapis alba* L.). Planted into cereals' stubble in early-August, both are ready to graze by sheep, goats or games from mid-September under conditions of Hungary (Antal, 2005). The catch crops play an important role in the improving the soil fertilising as green manure. These crops improve the soil's water retaining capacity (Stivers-Young, 1998) and structure (Unger and Vigil, 1998).

In spite of a long history of usage, little information is available on the feed value of oil radish and white mustard, average value of dry matter and crude protein was 130-160 g/kg forage and 120-150 g/kg dry matter (DM) for oil radish as well as 140-160 g/kg forage and 180-200 g/kg DM for white mustard, the digestibility of crude protein and fibre was 69-65 % and 70-54% for oil radish and white mustard (Kakuk and Schmidt, 1988; Schmidt *et al*., 2000). Moreover, even less on their *in situ* grazability by sheep (Antal *et al*., 1976; Antal, 2005; McLean, 2007). To date, no comparative feeding or grazing trials have been carried out with these two annuals by sheep. Forage nutritive value is a function of the feed intake and the efficiency of extraction of nutrients from the feed during digestion and the feed intake is affected by the amount of forage available, composition of forages (fibre, protein and minerals in the DM), feeding frequency, the animal's gut capacity and the passage of undigested DM through the animal (Adesogan *et al*., 1993).

In recent times, interest in the use of second crops as green manure has grown Europe-wide. This has prompted us to evaluate the potential of oil radish and white mustard as fall grazing pasture for sheep in the frame of a green manure study.

Materials and methods

A feeding and a grazing trial were conducted with German Mutton Merino sheep at the Experimental farm of the Szent István University, Gödöllő, Hungary in fall 2009 and 2010.

The oil radish (*Raphanus sativus* L. *var. oleiferus*) and white mustard (*Sinapis alba* L.) without fertiliser were grown at the Crop Production and Biomass Utilisation Demonstration Centre of the Szent István University, Gödöllő, Hungary for animal feeding in 2009 and 2010. The experiments with green plants grown as secondary crops were used. The forage crops used were oil radish and white mustard being in the bloom stage, grown on 0.5 ha plots each.

In the feeding trial (2009), 10 wether lambs (about 9 months old; initial body weight 35.6 kg) were used. They were housed in separate barn and received daily 2 kg meadow hay per head for 7 days before the trial began. Lambs were then kept in individual metabolism crates. Five being fed whole shoots of either forage, mown by hand ca. 5 cm above ground level. Periods included 7 days for adaptation to forages, followed by 7 days of digestibility trial while 10 kg per forage were provided. The daily feed allotment in both periods was given in two equal meals at 8 am and 4 pm for 2 hours allowing refusals no higher than 20%. Drinking water, salt blocks were available at all times. Orts were collected; weighted and daily intakes were recorded per animal. Faeces samples were collected by lamb at 24-hr intervals over the last 5 days, and composite samples (50 g) were prepared and stored frozen until analysed. Fresh forage and faecal samples were analysed for dry matter, crude protein (N x 6.25), crude fibre, ether extracts and ash according to Weender analysis; the digestibility of crude nutrients was estimated by the indicator method hydrochloric acid insoluble ash (Hungarian Feed Codex, 2004). Total digestible nutrients (TDN), digestible energy (DE), metabolisable energy (ME) and net energy for maintenance (NE_m) were calculated (Schmidt *et al.*, 2000).

In the grazing trial (2010), 15 dry ewes were allocated to graze either forage stand *in situ*. No concentrate supplement was given except salt licks. Drinking water was available at all times. Animals were gradually introduced to forage: 1-2 hrs per day initially and building up unrestricted access in 5 days. White mustard and oil radish pastures were rotationally grazed for 4 days using electric fences. Ewes were allowed 12 hrs of grazing from 8 am to 8 pm and kept overnight in a shed nearby the plots. Daily feeding allocation was calculated for a 50 kg dry ewe on the basis of the average maintenance requirement (DM: 850-1150 g; NE_m 5.82 MJ kg DM), equal to 1 kg of pasture dry weight (herbage mass) of average quality. Herbage DM yield was estimated by the clipping method using a 2 m square frame. Five random samples were taken (3 cm above ground) from each plot. The fresh weight per m^2 was multiplied by 10,000, then multiplied by the expected crop DM percentage to estimate DM yields per ha. Crop residuals were cut, weighted and DM yields calculated per ha as before. Paddock size was calculated by the pre-grazing herbage mass per m^2 and the forage DM requirement of the flock. Ewe carrying capacity of pastures was calculated by the formula given below:

Carrying capacity (hd/ha) = [Pre-grazing DM yield - Post-grazing DM yield (kg/ha)] / DM required (kg/hd)

All variables were subjected to analysis of variance (ANOVA) using the statistical package of SPSS 14.0. Means were compared by F-test and Student's t-test. Differences were considered significant with $P < 0.05$.

Results and discussions

Forage quality varies considerably with the plant species/genotype, maturity, season, and the growing conditions. Accordingly, the two brassicaceous species (of similar maturity and grown at the same site) differed to some extent in compositional quality (Tab. 1).

Tab. 1. Nutrient composition of the two forage crops fed to wether lambs

Item	Oil radish	White mustard
Dry matter (g/kg fresh forage)	135	154
Per kg DM		
Crude protein (g)	148	206
Ether extract (g)	31	29
Crude fibre (g)	141	194
Ash (g)	130	89
Nitrogen-free extracts (g)	550	482

White mustard and oil radish composition values are concordance with previous data (Kakuk and Schmidt, 1988; Schmidt *et al.*, 2000). White mustard was higher in dry matter, crude protein and crude fibre but lower in ash and N-free-extracts relative to oil radish. The difference in DM and crude fibre content might be related to a lower leaf/stem ratio present in mustard than radish.

In the feeding trial (Tab. 2), wether lambs consumed per head oil radish or white mustard shoots nearly in the same daily amounts (7.9-7.8 kg); their total DM intakes slightly differed (by 140 g) per day and per $BW^{0.75}$ (by 10 g), as well as in percent bodyweight (by 0.39 %). It was due partly to the small between-species differences in DM content (135 g/kg forage for oil radish, 154 g/kg forage for white mustard).

Tab. 2. Daily intake, digestibility and energy content of the two forage crops fed to wether lambs of lambs

Item	Oil radish	White mustard	SEM
Daily intake (kg)	7.88	7.82	0.06
Daily DM intake (g)	1063.01^a	1203.50^b	24.95
DM intake / $BW^{0.75}$ (g)	72.95^a	82.59^b	1.71
DM intake / BW (%)	2.99^a	3.38^b	0.07
Crude protein (%)	69	70	0.76
Crude fat (%)	61^a	52^b	1.57
Crude fibre (%)	63^a	54^b	1.69
Nitrogen-free extracts (%)	70	71	1.11
TDN (% DM)	61.6	62.9	6.36
DE (MJ kg DM)	11.36	11.61	0.12
ME (MJ kg DM)	9.32	9.52	0.10
NE_m (MJ kg DM)	5.70	5.88	0.09

^{ab}=P<0.05 - different letters in a row denote significant differences

Both forage species proved fairly digestible in lambs. Values of investigated parameters are within the ranges for ruminants reported by authors (Kakuk and Schmidt, 1988;

Schmidt *et al.*, 2000). The digestibility coefficients for crude protein and N-free extracts were similar, but the values for crude fibre and crude fat in white mustard significantly ($P<0.05$) lower compared with oil radish, respectively.

In the grazing trial, the two ewe flocks consumed voluntarily the test forages in different amounts due to different preference for the plant parts (Tab. 3).

Tab. 3. Daily feed allocation and carrying capacity of the two forage crops for ewes

Item	Oil radish	White mustard
Number of ewes	15	15
Daily forage DM requirement (kg/ewe)	1.0	1.0
Total DM requirement for 15 ewes for 4 days (kg)	60	60
Estimated herbage DM yield (kg/m^2)	0.236	0.180
Total grazing area required (m^2)	254	333
Paddock size required (m^2/day)	63	83
Pre-grazing DM yield (kg/ha)	2360	1800
Utilised DM yield (%)	72	41
Carrying capacity (ewes/ha)	1700	740

Notably, one utilised 72% of the oil radish biomass without preference for leaves over stems, whereas the other utilised 41% of the white mustard biomass leaving the stems nibbled off the leaves. As a general rule, grazing animals usually prefer the younger plants and leaves rather than the stems. Their first preference is new green leaves, if not present, they will choice older green leaves, followed by green stems, then dry leaves, and finally dry stems (Vallentine, 2001). Sheep in previous trials grazed down the entire above-ground biomass of oil radish without trampling loss, when they were herded on the radish stand from the nearby pasture for 1 hr am and 1 hr pm (Antal *et al.*, 1976). Lower voluntary intakes of forage crops at DM contents above 12% may reflect the animal's difficulty in harvesting the crops *in situ* (Barry *et al.*, 1971). In the present trial, the estimated herbage DM content was ≈12% for oil radish and ≈16% for white mustard.

For a grazing situation, an average carrying capacity of a good *Brassica* pasture would be around 3830 ewe grazing days per ha. Fodder radish produced reportedly 1588-1974 lamb grazing days per ha (Koch *et al.*, 2002). No data available on the carrying capacity of white mustard grazed by lambs in the U.K. (McLean, 2007). Under rotational grazing in the present trial, oil radish provided a carrying capacity twice as high as white mustard's notably, 1700 vs. 740 ewe grazing days per ha equal to 60 ewes grazing one hectare for 28 and 12 days, respectively. These results demonstrate that "leafy" oil radish is superior to "stemmy" white mustard to be used as fall grazing pasture for sheep.

Conclusions

In the two trials reported here each of the plant species were used as forage for the sheep nutrition. In terms of the crude fibre and crude protein digestibility, particularly in terms of the carrying capacity the oil radish was found to be more favourable than white mustard. However in practical grazing situations, sheep on low dry matter content forage crops must have access to a source of roughage (straw, hay) to provide at least 25% of the daily dry matter ration adequate for maintenance of proper rumen activity.

Based on our results, recommended the oil radish and white mustard plants in the experiment was suitable for use as fall grazing by sheep.

Acknowledgements

This work was supported by Kutató Kari Kiválósági Támogatás - Research Centre of Excellence - 17586-4/2013/TUDPOL project.

References

Adesogan AT, Sollenberger LE, Newman YC, Vendramini JMB (1993). Factors affecting forage quality. SS-AGR-93/AG 161, In: Newman Y (Ed.) Florida Forage Handbook, 1-5.

Antal J, Jenei E, Szántó Iné (1976). Elaboration of cultural practices for oil radish (*Raphanus sativus var. oleiferus* (in Hungarian). Növénytermelés 25:375-382.

Antal J (2005). Plant-breeding 2 (in Hungarian). Mezőgazda Kiadó, Budapest.

Barry NT, Drew KR, Duncan SJ (1971). The digestion, voluntary intake, and utilisation of energy by Romney hoggets fed five winter forage crops. New Zealand Journal of Agricultural Research 14:835-846.

Hungarian Feed Codex (2004). II/2 OMMI, Budapest.

Kakuk T, Schmidt J (1988). Animal nutrition (in Hungarian). Mezőgazdasági Kiadó, Budapest.

Koch DW, Kercher C, Jones R (2002). Fall and winter grazing of Brassicas - a value-added opportunity for lamb producers. Sheep and Goat Research Journal 17(2):1-13.

McLean B (2007). Alternative forage crops for finishing lambs. ADAS CYMHU Wales Saskatchewan Ministry of Agriculture: Grazing management for sheep production. www.agriculture.gov.sk.ca.

Schmidt J, Várhegyi Jné, Várhegyi J, Túriné CÉ (2000). The evaluation of ruminant feedstuffs for energy and protein content (in Hungarian). Mezőgazda Kiadó, Budapest.

Stivers-Young L (1998). Growth, nitrogen accumulation, and weed suppression by fall cover crops following early harvest of vegetables. Hort Sci 33:60-63.

Unger PE, Vigil MF (1998). Cover crops effects on soil water relationships. J Soil Water Conserv 53(3):200-207.

Vallentine JF (2001). Grazing Management. 2nd Edition. Academic Press, London.

The Effect of Water Deficit and Nitrogen on the Antioxidant Enzymes' Activity and Quantum Yield of Barley (*Hordeum vulgare* L.)

Arezu MOVLUDI[1], Ali EBADI[2], Soodabeh JAHANBAKHSH[3], M. DAVARI[4], Ghasem PARMOON[5]*

[1]*University of Mohaghegh Ardabili, Faculty of Agriculture, Department of Agronomy, Daneshgah Street, Ardabil, Iran; arezumovludi@yahoo.com*
[2]*University of Mohaghegh Ardabili, Faculty of Agriculture, Department of Agronomy, Daneshgah Street, Ardabil, Iran; ebadi@uma.ac.ir*
[3]*University of Mohaghegh Ardabili, Faculty of Agriculture, Department of Agronomy, Daneshgah Street, Ardabil, Iran; soodabehjahanbakhsh@yahoo.com*
[4]*University of Mohaghegh Ardabili, Faculty of Agriculture, Department of Plant Protection, Daneshgah Street, Ardabil, Iran; mdavari@uma.ac.ir*
[5]*University of Mohaghegh Ardabili, Faculty of Agriculture, Department of Agronomy, Daneshgah Street, Ardabil, Iran; ghasem.parmoon@gmail.com (*corresponding author)*

Abstract

The effects of water deficit and nitrogen fertilizer were studied on antioxidant enzymes activity and quantum yield of barley. An experiment carried out in greenhouse in factorial subject based on a completely randomized design with three replications. Irrigation schedules imposed at three levels of 85%, 60% and 35% field capacity (FC), and nitrogen were applied in quantities of 40, 80 and 120 kg N ha^{-1}. We determined Catalase (CAT), Peroxidase (POX) Polyphenol oxidase (PPO) activities, proline, chlorophyll and carotenoid content, quantum yield and grain yield. The results showed that severe stress (35% FC) increased the activities of CAT, POX and PPO enzymes and proline content, whereas the carotenoids, chlorophyll *a* and chlorophyll *b* decreased. Water deficiency caused the reduction in the quantum yield and the grain yield by 34%. Application of 80 kg of N during stress treatments resulted in higher enzyme activity and proline content. High amount of nitrogen reduced carotenoids, chlorophyll *a*, chlorophyll *b*, and in contrast, enhanced quantum yield. Application of 120 kg N ha^{-1} increased the yield up to 37% under mild stress (35% FC). Correlation coefficient and path coefficient showed that, grain yield was affected directly by amount of carotenoids and quantum yield.

Keywords: catalase, grain yield, peroxidase, polyphenol oxidase, proline

Introduction

Barley (*Hordeum vulgare* L.) is a resistant plant to salinity and water deficit and has a high application in human nutrition and malt industry. The water shortage is the main factor of crop yield reduction in semi-arid areas (Goodroad and Jellum, 1988). Water deficiency stress led to increase in amounts of reactive oxygen species (ROS) such as superoxide anion ($O_2^{\bullet -}$), hydrogen peroxide (H_2O_2), hydroxyl radical ($HO^{\bullet}$) and singlet oxygen (O_2) in plant cell. These active molecules, as a single molecule, cause severe damage to macro molecules and the cell structure or induce some defensive responses at the different molecular, cellular and/or physiological levels (Ohe *et al.*, 2005; Ariano *et al.*, 2005). Drought reduces cells division and inhibits the cells growth due to the reduction of turgor potential and it, also decreases their water content. Also this stress can damage pigments and plastids, reduce chlorophyll *a*, chlorophyll *b* and other carotenoids, hydrolyze proteins and prevalent photochemical reactions in most plants (Reddy *et al.*, 2004; Valladares and Pearcy, 1997).

The enzymatic and non-enzymatic mechanisms are involved in plant defensive strategies to face reactive oxygen species. The enzymes are including superoxide dismutase, glutathione reductase, peroxidase, and catalase and ascorbate peroxidase. In general, the adaptation to drought conditions depends on the amount of reactive oxygen species that are neutralized by the antioxidant system to be kept on relatively low level (Mascher *et al.*, 2005). The cooperation of these components forms the important cycles such as ascorbate-glutathione, and xanthophylls (Mittler *et al.*, 2004). The components of these cycles as a defensive mechanism enable the cells to inhibit the production of ROS or scavenge them and reduce their harmful effects (Asada, 2000). There are some studies on antioxidant defensive systems facing aging and environmental adverse conditions, but the results are very different. While some antioxidant activity in some species can be reduced, the antioxidant activity in other species is increased or remained unchanged (Hodges and Forney, 2000). Ohe *et al.* (2005) reported that the aging reduced oxidizing enzymes including superoxide dismutase, ascorbate peroxidase and catalase in spinach and tobacco leaves. Proline accumulation is the most important contributor osmolytes involved in osmo regulation (Cattivelli *et al.*, 2008). It increases during the stress in the cytoplasm, and protects the intercellular macro molecules structure, cause osmotic adjustment and maintained the cell turgor potential (Bokhari and Trent, 1985). The aim of this study was to investigate the effects of nitrogen consumption on some physiochemical traits of barley under water deficit condition.

Materials and methods

This study was conducted as a factorial experiment was carried out based on a completely randomized design (CRD) with three replications under greenhouse condition in 2012. Irrigation schedules imposed at three levels (85%, 60% and 35% FC), and nitrogen applied at 40, 80 and 120 kg N ha^{-1}.

Quantum yield

The quantum yield was measured by the uppermost fool expanded leaf using a fluorometer (chlorophyll fluorometer; Optic Science-OS-30 USA). For this purpose, the plants adapted to darkness for 20 minutes by using one special clamp then the fluorescence amounts were measured in 1000 (μM photon m^2s), and calculation was performed using following formula (Arnon, 1949):

$$Ø_{PSII} = (F_m - F_0)/ F_m$$

$Ø_{PSII}$; quantum yield amount of photosystem II, F_m or maximum fluorescence after a saturated light pulse on plants adapted to darkness and F_0, the minimal fluorescence in the light adapted, which was determined by illumination with far-red light.

Photosynthetic pigment content

Chlorophyll content measured in 0.2 g fresh leaf tissue, which gradually worn with 80% acetone and the solution volume was brought 20 ml using acetone 80%. Then it was centrifuged for 10 minutes at 400 rpm and the absorbance at 645, 663 and 470 nm was recorded by a spectrophotometer. Chlorophyll and carotenoids were obtained based on the following equations (Arnon, 1949):

$$\text{Chlorophyll } a = (19.3 \times A663 - 0.86 \times A645)\ V/100\ W$$
$$\text{Chlorophyll } b = (19.3 \times A645 - 3.6 \times A663)\ V/100\ W$$
$$\text{Carotenoid} = (1000\ A470 - 1.82\ C_a - 85.02\ C_b)\ /198$$

Determination of enzyme activities

Enzyme activity was measured three times during the day (pre-dawn, midday and evening), and every time, 0.2 g of fresh tissue was used for the enzyme activity. In order to extract the protein, 0.2 g of fresh leaf was pulverized in a mortar using liquid nitrogen and 1 ml of Tris –HCl (0.05 M, pH = 7.5) buffer was added. The mixture centrifuged for 20 minutes at 13.000 rpm, at 4 °C and the supernatant were used for determining the enzyme activity (Sudhakar *et al.*, 2001). CAT activity was assayed according to Karo and Mishra (1976). The 60 μl protein extraction was added to Tris buffer (50 mM, pH =7) and 0.3 ml H_2O_2 5 mM in the ice bath, then the absorbance curve was considered at 240 nm. The enzyme activity was obtained for (OD mg^{-1} protein min^{-1}) from fresh tissue. POX activity was measured as described by Karo and Mishra (1976). The 50 μl protein extracts was added to 2.5 ml extraction buffer, containing 100 μM Tris buffer 100 mM and H_2O_2 5 mM and 10 mM pirogalol in the ice bath and absorbance was read at 425 nm. Furthermore, PPO enzyme activity was measured by Karo and Mishra (1976) method, as it follows:100 μl protein extract was dissolved in 1.5 ml Tris 0.2 M and 0.3 ml pirogalol 0.02 M and the resulting composition was placed in the bain-marie bath at 25 °C for 5 minutes and then the absorbance at 420 nm was recorded. Also, the evaluation of protein carried out by Bradford (1976) method, 0.2 g of plant tissue was squashed with 0.6 ml extraction buffer and was centrifuged at 11.500 rpm for 20 minutes at 4 °C. The supernatant was transferred to the new tubes and centrifuged for 20 minutes at 4.000 rpm. To measure the protein amount, 10 μl of obtained extract was added to 5 μl Bradford solution and 290 μl extraction buffer and the absorbance rate was read at 595 nm.

Determination of proline content

In order to measure proline, 0.5 g of plant fresh tissue was crushed in 10 ml sulpho acetic acid solution to obtain a homogeneous mixture. Then, the solution was smoothed using whit-man no. 2 and 2 ml dimenhydrinate reagent and 2 ml glacial acetic acid were added. The extract was mixed and stirred on bain-marie at 100 °C for one hour and then 4 ml toluene added and the extract was vortexed to form two separate phases. The supernatant was read at 520 nm by a spectrophotometer (Batesetal, 1973).

Grain Yield

Grain yield was calculated by weighting the grain per plant (g plant-1).

Statistical Analysis

An experimental design was arranged in factorial in greenhouse, based on a completely randomized design (CRD) with three replications. Means were compared using LSD at Pvalue=5%. Correlation coefficients and path coefficient for chlorophyll *a* and *b*, carotenoid, quantum yield, proline accumulation, CAT, PPO, POX activity and grain yield were analyzed by the method of Dewey and Lu (1959).

Results and discussion

The water limitation and nitrogen fertilizer significantly affected the photosynthetic pigment content. The highest content of chlorophyll *a* (2.14 mg g^{-1} Fw) and chlorophyll *b* (0.95 mg g^{-1} Fw) was obtained in non-stress condition (85% FC or control) and 120 kg N ha^{-1} respectively. The lowest chlorophyll *a* (1.57 mg g^{-1} Fw) and chlorophyll *b* (0.40 mg g^{-1} Fw) were determined within severe stress (35% FC) and minimum N (40 kg N ha^{-1}). Furthermore, additional nitrogen enhanced the chlorophyll content under stress condition, therefore 120 kg N ha^{-1} significantly increased the chlorophyll content (80% chlorophyll *a* and 33% chlorophyll *b* during 35% FC) in comparison with 40 kg N ha^{-1}. Water limitation caused the reduction in carotenoids and PSII quantum yield, while the application of nitrogen increased them. The highest content of carotenoids (9.98 mg g^{-1} Fw) and quantum yield (0.83) was observed at 85% (FC) and 120 kg N ha^{-1}. The lowest amount of carotenoids (7.11 mg g^{-1} Fw) and quantum yield (0.75) was achieved under severe stress treatment (35% FC) and 40 kg N ha^{-1} (Tab. 1). Water stress reduced chlorophyll and carotenoids content. Photosynthetic pigments were lost because water shortage and it might due to the reduction in essential compounds synthesis such as chlorophyll pigment (Arnon, 1949). Degradation of pigment protein complex and oxidative complex causes damage to the chloroplast lipids, pigments and proteins (Tambussi *et al.*, 2005). Application of nitrogen increased the chlorophyll and carotenoids contents, which indicates the nitrogen impact in relieving stress effect. Moran *et al.* (2000) indicated that nitrogen application enhances chlorophyll *a* and *b* and total chlorophyll content. Such results were reported by Singh *et al.* (2012).

Tab. 1. The effect of water deficit and nitrogen on carotenoid, chlorophyll content and quantum yield of barley leaves (data are means of three replications ±standard error)

Treatment		Carotenoid	Chlorophyll *a*	Chlorophyll *b*	$Ø_{PSII}$
WD*	N**	(mg g^{-1} Fw)	(mg g^{-1} Fw)	(mg g^{-1} Fw)	
85%	40	7.78±0.055	2.05±0.025	0.68±0.035	0.81±0.001
	80	8.24±0.229	2.13±0.017	0.89±0.003	0.82±0.009
	120	9.98±0.084	2.14±0.052	0.95±0.028	0.83±0.009
60%	40	7.11±0.136	1.95±0.038	0.65±0.023	0.77±0.001
	80	7.58±0.153	2.02±0.020	0.75±0.027	0.79±0.004
	120	8.67±0.27	2.13±0.033	0.87±0.017	0.80±0.004
35%	40	7.11±0.142	1.57±0.037	0.40±0.044	0.75±0.002
	80	7.25±0.173	1.59±0.020	0.48±0.015	0.80±0.001
	120	8.46±0.136	1.84±0.007	0.58±0.029	0.80±0.003
$LSD_{0.05}$		0.49	0.09	0.11	0.013

*WD; water deficit (FC %), **N; nitrogen (kg ha^{-1}). Difference between mean difference treatments significant differences (LSD test, $P < 0.05$)

Nitrogen as a crucial element in plant growth plays a paramount role in the grain protein content, chlorophyll content, cell protoplasm and leaf photosynthetic activity (Delfin *et al.*, 2005). Application of nitrogen can adjust stomatal movements and it increases the green area duration (Yang and Zhang, 2006). Gene's expression of ABA synthesis and its activity was affected by nitrogen (Yang *et al.*, 2001). Probably, chlorophyll, exopeptidases and endopeptidases enzymes involved in chlorophyll degradation, influenced by ABA and the reduction of their amount caused the increasing of chlorophyll and carotenoids contents via nitrogen consumption (Hortensteiner and Feller, 2002). Our results showed that, chlorophyll and carotenoids were influenced indirectly, by proline accumulation. Proline maintains turgor potential and protects the intercellular macromolecules structure and thereby is presence affected the amount of chlorophyll and carotenoids. Water stress reduced the quantum yield, which indicates photosystem II efficiencies. The loss of this trait represents the occurrence of light inhibitory effect by environmental stresses (Kocheva *et al.*, 2004). During electron transport to photosystem I, drought stresses may cause interferences in Hill's reaction and thereby quantum yield of photosystem II can be reduced (Zelatev and Yordanov, 2004). Nitrogen increases fluorescence quantum efficiency through the development of leaf area, assimilates distribution and keeps up the photosynthetic efficiency (Arduiniet *et al.*, 2006). Similar results have also been reported by Huang *et al.* (2004).

The antioxidant enzymes activity was altered by the lack of water and nitrogen application. CAT activity was affected by water deficiency and nitrogen used at different times during the day. The highest CAT activity (0.109 OD mg protein min^{-1}) was recorded in severe stress (35% FC) and 80 kg N in the afternoon, and the lowest value (0.014 OD mg protein min^{-1}) at (85% FC) and 120 kg N at midday. The enzyme activity changed at different times during the day, so that the maximum and minimum activity was observed at PM and AM respectively. The CAT activity decrease reached to 54.4% during the different daytimes, under non-stress condition (85% FC) and low N access (40 kg N ha^{-1}). In contrast, by adding 80 kg of N, the enzyme activity upraised to 90% at PM, while using up to 120 kg ha^{-1} N diminished it to 9.5%. Increase stress intensity and N deficiency showed less effect on the enzyme activity at different daytimes. The enzyme activity's was lower during midday than in the morning and PM (Tab. 2). PPO activity significantly changed under water deficit treatment and nitrogen application during different times of the day. The highest activity (0.426 OD mg protein min^{-1}) was determined under severe stress treatment (35% FC) and 40 kg N ha^{-1} in the sunrise and the lowest activity (0.036 OD mg protein min^{-1}) under mild stress (60% FC) and 40 kg N ha^{-1} at midday. The results showed that PPO enzyme activity exhibited no clear relationship with stress despite the significant enzyme activity. As the enzyme activity was significantly lower than control during mild stress (60% FC), its increasing trend was observed under severe stress (35% FC). The water stress and nitrogen application increased PPO activity. Highest (0.35 OD mg protein min^{-1}) and lowest (0.09 OD mg protein min^{-1}) enzyme activity, were observed under severe stress (35%) with 120 kg N ha^{-1} and non-stress (85% FC) and consumption of 40 kg N ha^{-1} respectively, which represented 74% variation in the enzyme activity (Tab. 2).

POX activity changed during the daytime. Its activity was observable (0.18 OD mg protein min^{-1}) in the morning, but, with time, it reached to 0.17 OD mg protein min^{-1} at noon (5% reduction) and in the evening it raised again to 23.5% (from 0.17 to 0.21 OD mg protein min^{-1}) (Tab. 3). The water deficient stress and nitrogen had an impact on the POX activity and it increased their activity. Highest (0.35 OD mg protein min^{-1}) and lowest (0.09 OD mg protein min^{-1}) enzyme activity, were observed under severe stress (35%) with 120 kg N ha^{-1} and non-stress (85% FC) and consumption of 40 kg N ha^{-1} respectively, which represents 74% variation in the enzyme activity (Tab. 3).

Tab. 2. Effect of water deficit and nitrogen application on CAT, and PPO activity of barley

Treatment		CAT (OD mg protein min^{-1})			PPO (OD mg protein min^{-1})		
WD*	N**	Daytime			Daytime		
		Morning	Noon	Afternoon	Morning	Noon	Afternoon
85%	40	0.046±0.014	0.037±0.009	0.021±0.004	0.043±0.008	0.079±0.012	0.075±0.012
	80	0.026±0.006	0.014±0.001	0.040±0.018	0.096±0.009	0.113±0.008	0.127±0.024
	120	0.015±0.002	0.014±0.001	0.019±0.004	0.150±0.055	0.128±0.021	0.161±0.042
60%	40	0.058±0.009	0.024±0.002	0.039±0.003	0.068±0.004	0.036±0.008	0.078±0.008
	80	0.042±0.010	0.039±0.006	0.033±0.003	0.095±0.015	0.090±0.036	0.190±0.016
	120	0.033±0.006	0.030±0.007	0.034±0.003	0.184±0.040	0.105±0.048	0.181±0.017
35%	40	0.061±0.016	0.067±0.004	0.083±0.025	0.426±0.165	0.181±0.058	0.365±0.065
	80	0.070±0.016	0.051±0.008	0. 109±0.016	0.168±0.044	0.139±0.013	0.21±0.034
	120	0.059±0.011	0.050±0.002	0.067±0.010	0.210±0.057	0.190±0.015	0.241±0.062
$LSD_{0.05}$			0.016			0.126	

*WD; water deficit (FC %) **N; nitrogen (kg ha^{-1}), CAT; catalase, and PPO; poly phenol oxidase. Difference between mean difference treatments significant differences (LSD test, P < 0.05).

Tab. 3. Effect of nitrogen on proline content, POA activity and grain yield of barley underwater deficit

Treatment		POX	Proline	Grain yield
WD*	N**	(OD mg protein min^{-1})	(µg g)	(g $plant^{-1}$)
85%	40	0.09±0.005	0.11±0.004	1.00±0.146
	80	0.12±0.007	0.20±0.036	1.16±0.085
	120	0.14±0.006	0.14±0.006	1.54±0.036
60%	40	0.15±0.009	0.32±0.030	0.63±0.022
	80	0.16±0.010	0.62±0.044	0.68±0.030
	120	0.17±0.008	0.41±0.080	0.77±0.055
35%	40	0.21±0.017	1.76±0.063	0.34±0.031
	80	0.28±0.017	2.54±0.239	0.40±0.040
	120	0.35±0.022	2.11±0.031	0.47±0.024
LSD 0.05		0.029	0.265	0.191

*WD; water deficit (FC %), **N; nitrogen (kg ha^{-1}), POX; Peroxidase; Difference between mean difference treatments significant differences (LSD test, P < 0.05).

The results showed that the POX enzyme has the most indirect effects on the quantum yield. This enzyme acts as a catalyst for a quinone from phenols in the presence of oxygen (Yilmaz *et al.*, 2003). Perhaps its effect on the photosynthetic cycle and electron transport chain is observable in the photosynthesis process, during the photosynthetic light cycle when it reduces the quantum yield. The antioxidant enzyme activity (CAT and POX) increased during water deficit stress. Simirnoff and Colombe (2000) reported that CAT, POX and SOD enzymes in barley increased under drought stress, these results are consistent with Bai and Sui (2006) results on maize. Indeed, the increasing of CAT and POX activity during the stress is a defensive mechanism, in plant, against free oxygen radical, causing detoxification and decomposition of hydrogen peroxide produced by the cells and it prevents the breakdown of plant proteins (Ariano *et al.*, 2005). Huang *et al.* (2004) showed that nitrogen deficiency decreased CAT and POX activity in rice over time. We observed that CAT activity is modified by changing the time of the day so that the highest activities were observed at PM. Drought could have been exacerbated during mid-day with high radiation, which increased the evapotranspiration of plants. In addition, POX exhibited the most direct impact on CAT, which affected indirectly the quantum yield. PPO activity was increased under water deficit treatments and nitrogen application, which caused irregular changes on PPO. PPO in most of the higher plants exists and acts as a catalyst for quinine from phenols in the presence of O_2. The enzyme activity depends on plant genotype and growth conditions and plays a role in adventitious root formation, root development organization (Yilmaz *et al.*, 2003) cell division and primary differentiation (Huystee and Cairns, 1982). Proline accumulation was increased with nitrogen application and water deficit as one of the physiological responses to tolerance to the biotic and abiotic stress (Geravandi *et al.*, 2011).

Proline has significantly changed during water deficient stress and nitrogen consumption. By increasing the stress intensity, proline content increased, whereas in all the water treatments, mild applications of N (80 kg) lead to the highest value from this view points. The highest content of proline (2.54 μg g) was observed under severe stress (35% FC) and 80 kg ha^{-1} N (it indicates of 220 fold increasing compared with control). Applying 120 kg ha^{-1} N in comparison with 80 kg ha^{-1} reduced the proline content (Tab. 3). During the drought stress, water potential is reduced and the reduction of turgor pressure causes proline accumulation in the cytoplasm (Misra and Gupta, 2006). Proline protects the intercellular macro molecules structure and proteins, neutralizes free radicals, reduces cytoplasmic pH and maintain the proper ratio of $NADP^+/NADPH$ in metabolism and increase different enzymes activities (Szabados and Savoure, 2009) and it is also used as a combined storage of organic nitrogen, during reconstruction (Sairam and Tyagi, 2004).

The grain yield also was affected by irrigation schedules and nitrogen application. Water stress reduced yield to 66% at 35% FC. Nitrogen increased the yield during stress. Results showed that the highest grain yield (1.535 g $plant^{-1}$) was obtained at 85% FC and 120 kg N ha^{-1} and the lowest (0.341 g $plant^{-1}$) at 35% FC and 40 kg ha^{-1} (Tab. 3). The grain yield was correlated with all traits. Chlorophyll *a*, chlorophyll *b*, carotenoids and quantum yield, were all positively correlated with yield, while the antioxidant enzyme activities and proline content showed a negative correlation. Furthermore, chlorophyll *a* and *b* had a negative correlation with the enzymes and proline (Tab. 4).

Tab. 4. Correlation coefficient of water deficit and nitrogen

Chlorophyll *a*	Chlorophyll *b*	Carotenoid	$Ø_{PSII}$	Proline	CAT	PPO	POX	Grain yield
Chlorophyll a 1.00	0.848**	0.499**	0.127	-0.870**	-0.723**	-0.508**	-0.636**	0.666**
Chlorophyll *b*	1.000	0.572**	0.224	-0.838**	-0.776**	-0.408*	-0.635**	0.679**
Carotenoid		1.000	-0.004	-0.357	-0.370	0.037	-0.263	0.592**
$Ø_{PSII}$			1.000	-0.406*	-0.475*	-0.319	-0.652**	0.668**
Proline				1	.815**	.687**	0.859**	-0.751**
CAT					1	.534**	0.797**	-0.728**
PPO						1	0.738**	-0.423*
POX							1	-0.791**
Grain yield								1

* Significant at the 5% level, ** Significant at 1% level

Tab. 5. Path analysis coefficient for grain yield under water deficit and nitrogen fertilizer

Trait	Direct Effects	Indirect Effects							
		Chlorophyll *a*	Chlorophyll *b*	Carotenoid	$Ø_{PSII}$	Proline	CAT	PPO	POX
Chlorophyll *a*	0.313	-	0.236	-0.093	0.176	0.835	0.145	0.991	-0.869
Chlorophyll *b*	-0.123	-0.252	-	0.426	0.237	0.454	-0.372	-0.219	0.199
Carotenoid	0.426	0.010	0.015	-	0.141	0.342	-0.079	0.157	-0.226
$Ø_{PSII}$	0.529	0.330	0.099	0.205	-	-1.089	-0.594	-0.588	1.209
Proline	-0.115	0.469	-0.223	-0.463	-0.109	-	0.738	0.111	-0.466
CAT	-0.002	-0.149	-0.202	-0.100	0.122	0.038	-	0.209	0.200
PPO	-0.011	0.132	-0.055	-0.115	-0.447	-0.201	-0.001	-	-0.267
POX	-0.069	-0.379	-0.212	-0.440	0.066	-1.081	1.233	0.349	-

The path coefficient results showed that the carotenoids content and quantum yield had the most pronounced direct effects and antioxidants activity, while CAT, POX and PPO had the less noticeable direct effects on the grain yield, so it could be mentioned that in one-unit increment of carotenoids and quantum yield, the grain yield increased with 0.42 and 0.529 units respectively. While increasing each unit in CAT, POX and PPO activity the grain yield was reduced 0.002, 0.09 and 0.011 respectively. Carotenoids content was negatively correlated with chlorophyll *a* and *b*, also changing each unit of them indirectly altered 0.010 and 0.015 of carotenoids (Tab. 5). Quantum yield was negatively correlated with POX and CAT activity and proline content. It seems that each trait has indirect effects on other traits ultimately and may affect grain yield. Chlorophyll *a* and *b* both affect carotenoid content (0.010 and 0.015) respectively and this affects yield with 0.42. Therefore, chlorophyll *a* and chlorophyll *b* had an indirect impact on the yield via carotenoids 0.004 and 0.00. Likewise, other traits influenced the yield by interfering and affecting each other (Tab. 5).

Peroxidase activity and quantum yield had the highest indirect effects on proline amount. The reduction of photosynthesis and plant production due to the decrease in quantum yield can affect the production of proline. Grain yield reduction under drought stress may be attributed to accelerated phonological stages that permit the plants to escape from drought effects or loss of reservoir capacity (Turner *et al.*, 2001). Gadallah (2000) noted that the reasons of grain yield reduction under severe water stress are due to higher sensitivity of plants at the reproductive growth than vegetative growth, stage which results in the reduction of assimilate allocation to grain. It seems that the yield reduction at different levels of drought stress is the result of differences between quantum yield and decreased membrane stability as well. Stress prevents the plant from producing desired substrates and energy use via disturbances in the electron transfer pathway and destruction of tissue associated with photosynthesis (Paknejad *et al.*, 2007). Also, damages from reactive oxygen species on the membrane lipids and the peroxidation of membrane polyunsaturated fatty acids were reported as the main reason of the decrease in grain yield during stress (Blokhin *et al.*, 2003).

Conclusions

The results showed that the water deficit stress reduced photosynthesis and chlorophyll content of the plants, which ultimately reduced grain yield. Our results suggested that plants apply defensive mechanisms, such as antioxidant enzymes and proline to overcome and mask negative effects of stress. Nitrogen application was determined as an optimum strategy for most desirable traits that reduce stresses effect. Among the measured traits, quantum yield had the greatest impact on the grain yield.

Acknowledgment

The authors wish to thank Dr. Saber Zahri, Dr. Hossain Haidari Sharif Abad for providing necessary funds and research facilities required for this investigation.

References

Arduini I, Masoni A, Ercoli L, Mariotti M (2006). Grain yield, and dry matter and nitrogen accumulation and remobilization in durum wheat as affected by variety and seeding rate. Eur J Agron 25:309-318.

Ariano S, Bartolomeo D, Cristos X, Andras M (2005). Antioxidant defences in Olive trees during drought stress: changes in activity of some antioxidant enzymes. Functional Plant Biology 32:45-53.

Arnon DI (1949). Copper enzymes in isolated chloroplast polyphenol oxidase in *Beta vulgaris*. Plant Physiol 24(1):1-15.

Asada K (2000). The water-water cycle as alternative photon and electron sinks. Philosophical Transactions of the Royal Society of London. Series B: Biological Sci 355(1402):1419-1431.

Bai LP, Sui FG (2006). Effect of soil drought stress on leaf water status, membrane permeability and enzymatic antioxidant system of maize. Pedosphere J 16: 326-332.

Bates L, Waldren SRP, Teare ID (1973). Rapid determination of free proline for water stress studies. Plant Soil 39:205-207.

Blokhin O, Virolainen E, Fagerstedt KV (2003). Antioxidants, oxidative damage and oxygen deprivation stress: a Review. Ann Bot 91:179-194.

Bokhari UG, Trent JD (1985). Proline concentrations in water stressed grasses. J Range Manag 38:37-38.

Bradford MM (1976). A rapid and sensitive for the quantitation of microgram quantities of protein utilizing the principle of protein-dye binding. Analytical Biochemistry 72:248-254.

Cattivelli L, Rizza F, Badeck FW, Mazzucotelli E, Mastrangelo AM, Francia E, Mare C, Tondelliand A, Stanca M (2008). Drought tolerance improvement in crop plants: An integrated view from breeding to genomics. Field Crops Res 105:1-14.

Delfin S, Tognetti R, Dsiderio E, Alvino A (2005). Effect of foliar application of N and umic acids on growth and yield of durum wheat. Agron Sustainable Developm 25:183-191.

Dewey DR, Lu KI (1959). A correlation and path coefficient analysis of components of crested wheat grass seed production. Agronom J 51:515-518.

Gadallah MA (2000). Effect of indole-3-acetic acid and zinc on the growth, osmotic potential and soluble carbon and nitrogen components of soybean plants growing under water deficit. J Arid Environ 44:451-467.

Geravandi M, Farshadfar E, Kahrizi D (2011). Evaluation of some physiological traits as indicators of drought tolerance in bread wheat genotypes. Russian J Plant Physiol 58(1):69-75.

Goodroad LL, Jellum MD (1988). Effect of nitrogen fertilizer rate and soil pH on Nitrogen use efficiency in crop. Plant and Soil 106:85-89.

Hodges DM, Forney CF (2000). The effects of ethylene, depressed oxygen and elevated carbon dioxide on antioxidant profiles of senescing spinach leaves. J Expt Bot 51:645- 655.

Hortensteiner S, Feller U (2002). Nitrogen metabolism and remobilization during senescence. J Exp Bot 53:927-937.

Huang ZA, Jiang DA, Yang Y, Sun JW, Jin SH (2004). Effects of nitrogen deficiency on gas exchange, quantum yield, and antioxidant enzymes in leaves of rice plants. Photosynthetic 42 (3):357-364.

Huystee RB, Cairns WL (1982). Progreno and prospects in the use of peroxidase to study cell development. Phyto Chemis 21:1843-1847.

Karo M, Mishra D (1976). Catalase, peroxidase and polyphenol oxidase activity during rice leaf senescence. Plant Physiol 57:315-319.

Kocheva K, Lambrev P, Georgiev G, Goltsev V, Karabaliev M (2004). Evaluation of chlorophyll fluorescence and membrane injury in the leaves of barley cultivars under osmotic stress. Bioelectrochemistry 63(1):121-124.

Mascher R, Nagy E, Lippmann B, Hornlein S, Fischer S, Scheiding W, Neagoe A, Bergmann H (2005). Improvement of tolerance to paraquat and drought in barley (*Hordeum vulgare* L) by exogenous 2-aminoethanol: effects on superoxide dismutase activity and chloroplast ultrastructure. Plant Sci 168:691-698.

Mittler R, Vanderauwera S, Gollery M, Breusegem FV (2004). Reactive oxygen gene network of plants. Tren Plant Sci 9:490-498.

Moran JA, Mitchell AK, Goodmanson G, Stockburger KA (2000). Differentiation among effects of nitrogen fertilization treatments on conifer seedlings by foliar reflectance: a comparison of methods. Tree Physiol 20:1113-1120.

Neelam M, Gupta A K (2006). Effect of salinity and different nitrogen sources on the activity of antioxidant enzymes and indole alkaloid content in *Catharanthus roseus* seedlings. J Plant Physiol 163:11-18.

Ohe M, Rapolu M, Mieda T, Miyagawa Y, Yabuta Y, Yoshimura K, Shigeoka S (2005). Decline in leaf photo oxidative-stress tolerance with age in tobacco. Plant Sci 168:1487-1493.

Paknejad F, Nasri M, Tohidi Moghadam HR, Zahedi H, Jami Alahmad M (2007) Effects of drought stress on chlorophyll fluorescence parameters chlorophyll content and grain yield of wheat cultivars. J Biological Sci 7(6):841-847 (In Persian).

Reddy AR, Chaitany KV, Vivekanandan M (2004). Drought-induced responses of photosynthesis and antioxidant metabolism in higher plants. J Plant Physiol 161:1189-1202.

Sairam RK, Tyagi A (2004). Physiology and molecular biology of salinity stress tolerance in plants. Current Sci 86(3):407-421.

Simirnoff N, Colombe SV (2000). Drought influences the activity of enzymes of the chloroplast hydrogen peroxide scavenging system. J Experim Bot 39:1097-1108.

Singh G, Agnihotri RK, Reshma RS, Ahmad M (2012). Effect of lead and nickel toxicity on chlorophyll and proline content of Urd (*Vignamungo* L.) seedlings. Intern J Plant Physiol Biochem 4(6):136-141.

Sudhakar C, Lakshmi A, Giridarakumar S (2001). Changes in the antioxidant enzyme efficacy in two high yielding genotypes of mulberry (*Morus alba* L.) under NaCl salinity. Plant Sci 167:613-619.

Szabados L, Savoure A (2009). Proline: a multifunctional amino acid. Trends in Plant Sci 15(2):89-97.

Tambussi EA, Nogues S, Araus JL (2005). Ear of durum wheat under water stress: water relations and photosynthetic metabolism. Planta 221:446-458.

Turner NC, Wright GC, Siddique KHM (2001). Adaptation of grain legumes (Pulses) to water limited environments. Advances in Argon 71:193-231.

Valladares F, Pearcy RW (1997). Interactions between water stress, sun-shade acclimation, heat tolerance and photo inhibition in the Sclerophyll *Heteromeles arbutifolia* M. Plant Cell Environ 20:25-36.

Yang J, Zhang J (2006). Grain filling of cereals under soil drying. New Phytol 169:223-236.

Yang J, Zhang J, Wang Z, Zhu Q, Liu L (2001). Water deficit-induced senescence and its relationship to the remobilization of pre-stored carbon in wheat during grain filling. Agron J 93:196-206.

Yilmaz H, Taskin T, Otludil B (2003). Polyphenol oxidase activity during rooting in cuttings of grape (*Vitis vinifera* L.) varieties. Turkish J Botany 27:495-498.

Zlatev ZS, Yordanov IT (2004). Effects of soil drought on photosynthesis and chlorophyll fluorescence in bean plants. Bulg J Plant Physiol 30(3-4):3-18.

4

Micropropagation of the Endangered and Decorative Specie *Dianthus serotinus* Waldst. et Kit.

Marija MARKOVIĆ*, Mihailo GRBIĆ, Matilda DJUKIĆ

*Belgrade University, Faculty of Forestry, Ulica kneza Višeslava, 1, 11000 Belgrad, Serbia; marija.markovic@sfb.bg.ac.rs (*corresponding author); mihailo.grbic@sfb.bg.ac.rs; matilda.djukic@sfb.bg.ac.rs*

Abstract

During past decades, great attention has been paid to propagation of endangered plant taxa in order to preserve biodiversity. The aim of this study was to optimize a protocol for *in vitro* propagation of the critically endangered and decorative species *Dianthus serotinus* Waldst. et Kit. The effects of different concentration of MS salt (Murashige and Skoog) of the culture, medium pH and different carbohydrates (sucrose, glucose, and fructose) on shoot multiplication were examined. The best results were obtained on half-strength MS (Murashige and Skoog) medium, whose pH was 5.8, with sucrose supplied at a concentration of 3%, when shoots with 1-2 nodes or shoot tips (with terminal buds only) were used as explants. The shoots were rooted (76.7%) on half-strength MS medium containing 0.5 $mg{\cdot}L^{-1}$ NAA (1-naphthaleneacetic acid). The obtained plantlets were successfully acclimatized (89%) in a 4:1 mixture of peat and sand and they flowered the following year. Presented protocol enables successful *in vitro* propagation of *D. serotinus*.

Keywords: acclimatization, fructose, glucose, *in vitro* culture, pH value, sucrose

Introduction

Dianthus serotinus Waldst. et Kit. (Caryophyllaceae) is an endemic Pannonian species included in category V (vulnerable taxa) of the IUCN Red List of endangered species. It is under legal protection and has the status of extremely endangered species in Serbia (Law on Environmental Protection, Rulebook on protected species, 2010). This species is found in small numbers only locally in the area of Subotica - Horgoš Sands, where it is threatened by the expansion of agricultural land and afforestation (Boža, 1999). *D. serotinus* is a decorative plant with bluish-green leaves and white flowers, which have a mild scent and a threadlike-rimmed corolla. This endangered species blooms all summer long and grows on poor sandy soils. It is drought tolerant, which makes it suitable for use as an ornamental in dry-stone constructions, rock and cottage gardens (Gajić, 1986; Boža, 1999).

To enable its ex situ and in situ conservation, we decided to propagate *D. serotinus* using the micropropagation method, which is a convenient method for rapid and efficient propagation of endangered species (Pence, 1999). Marković *et al.* (2007) established a sterile *D. serotinus* culture and investigated the effect of different concentrations of phytohormones on the multiplication and rooting of shoots of this species. However, their work examined the effect of phytohormones in a small number of treatments, and the acclimatization of the plantlets obtained was not performed. Considering that *D. serotinus* grows on humus-poor sand with pH ranging from 7.0 to 8.1 in natural habitat (Boža, 1999) and that pH medium in previous research was 5.8 (Marković *et al.*, 2007) we needed to improve protocol for regeneration and multiplication of this plant species.

Beside different concentrations of phytohormones that affect the development of in vitro cultures, important components are carbohydrates as sources of energy and carbon and osmotic agents that regulate the uptake of other culture constituents. In addition, the concentration and type of a carbohydrate in a medium can affect organogenesis and the development of an *in vitro* culture (Mohamed and Alsadon, 2010; Nowak *et al.*, 2004; Todorović *et al.*, 2006; Thorpe *et al.*, 2008). Although sucrose is the most widely used carbohydrate and the cheapest one (Thorpe *et al.*, 2008), other carbohydrates, mainly glucose and fructose, had a better effect on the development of in vitro cultures of some species, i.e. seven species of the genus *Alnus* (Tremblay and Lalonde, 1984; Barghchi, 1988), *Bougainvillea* 'San Diego Red' (Steffen *et al.*, 1988), *Castanea sativa* Mill., *C. crenata* Siebold et Zucc. (Chauvin and Salesses, 1988), *Nepeta rtanjensis* Diklić et Milojević (Mišić *et al.*, 2005b), *Corylus avelana* L. (Yu and Reed, 1993), *Fagus* spp. (Cuenca and Vieitez, 2000). Therefore, variations of carbohydrate components in a growth medium are inevitably taken into consideration in the optimization of protocols for *in vitro* propagation of certain plant species.

A large number of studies have been published on the micropropagation of endangered and/or endemic *Dianthus* species in the Balkans. They include the studies of endemic endangered species of the Czech Republic *D. arenarius* L. ssp. *bohemicus* (Novák) O. Schwarz (Kovác, 1995) and *D. superbus* L. ssp. *superbus* Domin (Mikulík,

1999), the endemic species of the Balkan Peninsula *D. petraeus* Waldst. & Kit. ssp. *noeanus* (Boiss.) Tutin (Radojević *et al.*, 1997), *D. giganteus* d'Urv. ssp. *croaticus* (Borbás) Tutin and *D. ciliatus* Guss. ssp. *dalmaticus* (Čelak.) Hayek (Radojević *et al.*, 2006, Radojević *et al.*, 2010), the endemic and endangered species of Romania *D. spiculifolius* Schur (Butiuc-Keul *et al.*, 2001), *D. petraeus* Waldst. & Kit. ssp. *simonkaianus* (Péterfi) Tutin (Miclăuş *et al.*, 2003), *D. callizonus* Schott & Kotschy (Holobiuc and Blindu, 2006), *D. nardiformis* Janka (Holobiuc *et al.*, 2009, 2010a), *D. pratensis* M. Bieb. subsp. *racovitzae* (Prodan) Tutin (Cristea, 2010), *D. henteri* Heuff. ex Griseb. & Schenk (Cristea *et al.*, 2010), *D. glacialis* Haenke ssp. *gelidus* (Schott, Nym. et Kotschy) Tutin (Holobiuc *et al.*, 2010b), *D. giganteus* d'Urv. subsp. *banaticus* (Heuff. ex Griseb. & Schenk) Tutin (Pop and Pamfil, 2011), in Greece *D. fruticosus* L. (Papafotiou and Stragas, 2009) and many other studies. Also, to date, many studies regarding the optimization of the in vitro protocol of important decorative *Dianthus* taxa, including *D. chinensis* L., *D. barbatus* L. had been published (Jethwani and Kothari, 1993; Jethwani *et al.*, 1994; Pareek *et al.*, 2004). All mentioned studies investigated the optimal concentration of phytohormones added to MS culture media or less often half-strength MS media that differed depending on the species (Murashige and Skoog, 1962). The pH level was adjusted to 5.8 before autoclaving and 2% or even more often 3% sucrose was used as a source of carbohydrates. However, none of these papers investigated the effect of pH or different concentrations of different carbohydrates on the in vitro development of *Dianthus* spp. cultures. Taking all this into account, the aim of our research was to perform a detailed and complete study of the improvement of a protocol for the micropropagation and acclimatization of *D. serotinus*. This investigation included examining of the impact of culture medium pH, various carbohydrates and the development of in vitro cultures on the media with reduced concentrations of MS salts.

Materials and methods

The *in vitro* culture was established from seeds collected at the locality of Subotica - Horgoš Sands, in the vicinity of Palić. The seeds were surface disinfected by immersion in 4% NaOCl (sodium hypochlorite) supplemented with 2-3 drops of the Tween 20 for 20 minutes before three rinses in sterile, distilled water according to the procedure described by Marković *et al.* (2007). Three weeks later, nodal and shoot tip cuttings (with terminal buds only) were cut from the seedlings. These cuttings were placed on a multiplication culture medium containing Murashige and Skoog (MS) salts and vitamins (Murashige and Skoog, 1962), 3% sucrose, 0.8% agar and 0.5 mg · L^{-1} BAP (6-benzylaminopurine) and 0.5 mg · L^{-1} NAA (1-naphtalenacetic acid), as recommended by Marković *et al.* (2007) to obtain the material of sufficient quantity to set up the next experiment. Before the experiment setup, the shoots were cultured for 2 weeks on MS basal culture medium without phytohormones.

In all experiments three different types of explants (shoot tip cuttings with terminal buds only, single node cuttings with axillary buds only and shoots with 1-2 nodes) were placed on either solid MS or half-strength MS culture medium containing 0.8% agar, supplemented with 0.5 mg · L^{-1} BAP and 0.5 mg · L^{-1} NAA. The effect of culture medium pH supplemented with 3% sucrose on shoot induction and multiplication of *D. serotinus* was investigated. In this experiment the pH was adjusted before autoclaving the culture medium (20 min at 120 °C) to the four graded values (5.8, 6.8, 7.8 and 8.8). In the second experiment we investigated the effect of graded concentration (1%, 3%, 5% and 7%) of three different carbohydrate source (sucrose, glucose and fructose) on shoot induction and multiplication of *D. serotinus*. Considering that high concentration of sugar decreased culture medium pH (Thorpe *et al.*, 2008), in this experiment the pH value of all tested media was set to 6.3 before autoclaving. During the multiplication stage, the culture medium was poured into glass vessels of 5 × 5 × 13 cm dimensions. The vessels contained 25 mL of the culture medium and 5 explants of the same type were placed in each of them. The experiments were repeated 3 times with 20 explants per treatment. After 25 days of cultivation, the following parameters were determined: the number of shoots per explant, the number of nodes and shoot length.

During the rooting stage, the shoots (with 1-3 nodes) were placed on basal medium (MS or half-strength MS) supplemented with NAA (0.0-0.5 mg · L^{-1}). The shoots were placed in glass containers of 8 x 8 x 15 cm dimensions, with 100 mL of medium and 15 shoots were placed in each of them. The experiments were repeated 3 times with 30 explants per treatment. After 15 days, the percentage of rooted plants, the number of roots and the length of the longest root per explant were determined.

During the acclimatization stage, the rooted plantlets were planted in three different mixtures of culture substrates, including (1:1) and (4:1) mixtures of peat and sand, and a (2:2:2:1) mixture of peat, sand, garden soil and burned farmyard manure. Before use, the soil was treated with a 1.5% solution of Previcur-N fungicide. Each treatment consisted of three replications with 30 plantlets. The planted plants were covered with perforated plastic wrap to maintain high relative humidity during the first 15 days of acclimatization. Ventilation was applied once daily for 5-10 minutes, after which the plastic wrap was removed. The plants were grown for 10 more days, and after that, their survival rate was recorded.

Plant cultures for all *in vitro* experiments were grown in a growth room under a 16/8-h (day/night) photoperiod and a photon flux rate of 50 μmol · m^{-2} · s^{-1} at 25±2°C in the light of fluorescent white pipes.

Statistical data processing was performed using an appropriate application for statistical analyses (Statgraphics, version 5.0). The significance of differences between the means was determined by the analysis of variance (ANOVA, $p < 0.05$) and the least significant difference (LSD) test.

Results and discussion

Effect of the medium pH on shoot development

The *in vitro* culture was successfully established according to the procedure described by Marković *et al.* (2007). The concentration of MS salts (MS or half-strength MS) and culture medium pH, affected the frequency of shoot regeneration on the media of different pH levels, whereas this frequency was unaffected by explant type (Tab. 1). Contrary to expectations, regardless of the concentration of mineral salts, the shoot regeneration on explants was high on nutrition media with low pH. The difference between parameters obtained on the media of 5.8 and 6.8 pH levels was generally not statistically significant. On the media with a pH level of 7.8, the frequency of regeneration significantly decreased, whereas almost none of the explants developed on alkaline media with a pH level of 8.8 so the growth parameters couldn't be measured. In addition, the frequency of regeneration was higher on half-strength MS media (from 60.5 to 98.9%) compared to MS media (from 35.3 to 73.5%). However, the shoots on MS media were better developed, containing a higher average number of shoots per explant (Tab. 2). A high frequency of regeneration on half-strength MS media was achieved in the species *D. nardiformis* (Holobiuc *et al.*, 2010a). However, it is not clear whether the lower concentrations of salts were more favourable for this species because the MS and half-strength MS media were supplemented with different concentrations and types of hormones.

Tab. 1. Frequency of shoot regeneration on different explants cultured on graded pH media after 25 days *in vitro*

pH value	Medium	Single node cuttings (%)	Shoot tips (%)	Shoots (%)
5.8	MS_0	58.9 ± 9.1 ^{abc}	65.5 ± 8.8 ^{c}	73.5 ± 9.1 ^{b}
6.8	MS_0	50.2 ± 8.0 ^{bc}	68.3 ± 9.1 ^{c}	72.1 ± 8.2 ^{b}
7.8	MS_0	35.3 ± 7.6 ^{c}	30.2 ± 7.2 ^{d}	38.2 ±8.6 ^{c}
8.8	MS_0	0.0 ± 0.0 ^{d}	3.3 ± 1.7 ^{e}	1.7 ± 1.7 ^{d}
5.8	1/2 MS_0	70.3 ± 8.3 ^{a}	98.9 ± 1.1 ^{a}	97.3 ± 2.7 ^{a}
6.8	1/2 MS_0	65.2 ±7.5 ^{ab}	90.8 ± 8.3 ^{ab}	92.5 ± 6.7 ^{a}
7.8	1/2 MS_0	60.5 ± 7.5 ^{ab}	81.5 ± 6.9 ^{b}	73.8 ±7.1 ^{b}

MS_0 - basal medium containing MS mineral salts; 1/2 MS_0 - basal medium containing half strength MS mineral salts;
Note: Values followed by different letters are significantly different at the P < 0.05 level according to the least significant difference test

In determining the average number of nodes per explant, only the nodes that could be used for the next subculture, i.e. those whose axillary buds did not develop into new shoots were taken into account as potential nodal cuttings for the next subculture, and their number varied, although it could be observed that their number was generally high on media with low pH (Tab. 2).

Tab. 2. The average number of shoots and shoot nodes regenerated on different explants cultured on graded pH media after 25 days *in vitro*

pH value	Single node cuttings MS_0	Single node cuttings 1/2MS_0	Shoot tips MS_0	Shoot tips 1/2MS_0	Shoots MS_0	Shoots 1/2MS_0
No. of shoots per explant						
5.8	10.1 ± 1.1 ^{a}	4.5 ± 0.7 ^{a}	8.8 ± 0.7 ^{ab}	7.8 ± 0.6 ^{b}	10.7 ± 0.6 ^{ab}	7.5 ± 0.3 ^{ab}
6.8	5.8 ± 0.8 ^{b}	5.3 ± 0.7 ^{ab}	11.5 ± 0.6 ^{b}	7.2 ± 0.5 ^{b}	11.5 ± 0.3 ^{a}	5.8 ± 0.4 ^{b}
7.8	6.0 ± 1.2 ^{b}	5.8 ± 1.0 ^{b}	6.9 ± 0.6 ^{a}	4.6 ± 0.3 ^{a}	9.3 ± 0.3 ^{b}	8.2 ± 0.3 ^{a}
No. of nodes per explant						
5.8	11.1 ± 0.8 ^{a}	7.1 ± 0.7 ^{ab}	12.3 ± 1.2 ^{ab}	15.1 ± 1.3 ^{a}	11.2 ± 1.2 ^{ab}	13.5 ± 1.0 ^{a}
6.8	4.6 ± 0.4 ^{b}	7.9 ± 0.7 ^{ab}	14.9 ± 1.0 ^{a}	11.9 ± 0.9 ^{ab}	14.5 ± 1.2 ^{a}	10.5 ± 0.8 ^{ab}
7.8	6.1 ± 1.1 ^{ab}	8.9 ± 0.3 ^{a}	7.1 ± 0.9 ^{b}	9.2 ± 1.1 ^{b}	8.1 ± 0.6 ^{b}	9.3 ± 0.7 ^{b}

MS_0 - basal medium containing MS mineral salts; 1/2 MS_0 - basal medium containing half strength MS mineral salts;
Note: Values followed by different letters are significantly different at the P < 0.05 level according to the least significant difference test.

Because of the high level of variability in shoot length, the shoots were placed in different length categories (less than 10 mm, 10-20 mm and more than 20 mm) and the number of shoots belonging to a certain length category was expressed as a percentage of the total number of shoots. Generally, the shoots were longer on MS basal media. In some cases (nodal explants on media with 5.8 and 6.8 pH levels), over 40% of the shoots were longer than 20 mm, while in the case of half-strength MS media, the percentage of shoots longer than 20 mm did not exceed 20% on all media, regardless of the culture medium pH and explant type (Fig. 1). The importance of medium pH for in vitro development of plants is already known (Leifert *et al.*, 1992; Harbage and Stimart, 1996). Nevertheless, in *Maranta leuconeura* cv. Kerchoviana different pH levels of a culture medium had no significant effect on the development of cultures, and there were no statistically significant differences in either rooting percentage or the average number of leaves per shoot and shoot length. Only the differences in the number of shoots per explant were observed (Ebrahim and Ibrahim, 2000). However, Bhatia and Ashwath (2005) showed that a change in medium pH did not significantly affect the number of shoots formed per explant of *Lycopersicum esculentum* cv. Red Coat, although it had an effect on their length. In addition, Ostrolucka *et al.* (2004) revealed that a lower pH level of 5.0 was more favourable than pH 5.5 for acidophilic *Vaccinium corymbosum* 'Duke'. Additionally, a research on the effect of pH on axillary shoot proliferation in *Vaccinium vitis-idaea* L. showed that the optimum pH ranged from 4.0 to 5.5, depending on the cultivar (Ostrolucka *et al.*, 2010). On the other hand, the research of Mišić *et al.* (2005a) showed that the high pH levels of 7.0 - 7.2 were the most suitable for the proliferation of lateral buds and the rooting of shoots of the calciphilic species *Nepeta rtanjensis*. However, contrary to expectations, our research showed that pH levels ranging from 5.8 - 6.8 were the most suitable for the species *D. serotinus*, while at higher pH levels the development of shoots in in vitro culture significantly decreased. A possible explanation for this phenomenon may lie in the specific conditions of in vitro culture and the fact that the adoption of mineral salts from culture medium, and thus the growth of plants, is influenced by not only the concentration of salts, but also pH, temperature, and the biochemical or physiological status of the plant tissues (George and de Klerk, 2008). Also, specific conditions of in vitro culture have a different influence on growth and development of the species belonging to distinct plant families with different metabolism and the response of *D. serotinus* perhaps was depending on interaction of mentioned factors, in different way than other studied species. Therefore, we can assume that a higher pH value would be better in case of a different composition of mineral components of the culture medium. This was confirmed by the data listed in Tab. 1 showing that better results were obtained on half-strength MS media with pH 7.8 than on MS media with pH values of 5.8 and 6.8 for the same type of explant.

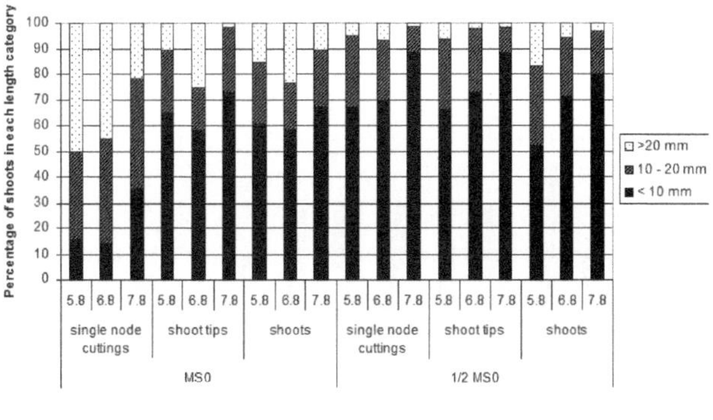

Fig. 1. The effect of graded medium pH on length of *D. serotinus* shoots regenerated on different explants on full-strength MS basal medium (MS_0) and half-strength MS basal medium ($1/2MS_0$)

The effect of carbon source on shoot regeneration

Generally, sucrose had the most favourable effect on the regeneration of explants. Glucose had the second most favourable effect, while the lowest frequency of regeneration was recorded on media containing fructose (Tab. 3). When explants were cultured on half-strength basal media, a low frequency of regeneration occurred only on nutrition media with high concentrations of glucose or fructose (7%). However, the frequency of regeneration was significantly lower on MS basal media. It was also affected by explant type, showing a higher regeneration rate of shoots than in the case of shoot tips and single node cuttings. Similarly, the highest average number of shoots was formed on MS basal medium with sucrose (from 6.2-nodal cuttings to 9.5-shoot tip cuttings). The average number of shoots was lower on media containing glucose and fructose, and generally, the best results were obtained if the concentration of these sugars was 3% (Fig. 2). Similar results were obtained by observing the average number of nodes that can be used for the next subculture, the MS basal media with sucrose supplied at a concentration of 3%, were the best, while the media with glucose and fructose, and half-strength MS medium with sucrose yielded poorer results (Fig. 3). As far as shoot length is concerned, regardless of the type and concentration of sugar, significantly longer shoots were formed on MS basal media. The percentage of shoots longer than 20 mm was generally 50% or higher on media containing glucose and sucrose, whereas on half-strength MS basal media it usually did not exceed 10%, reaching 32% only on the medium with 1% glucose (Fig. 4). In the examination of the effect of concentration and type of carbohydrate on the *in vitro* growth and organogenesis of the species *Rindera umbellata*, the highest frequency of adventitious buds formation on the stem explants was reached on medium with 0.06M ($\approx$ 2%) sucrose, but the highest values of fresh and dry weight were obtained on medium with 0.1M ($\approx$ 3.4%) sucrose (Perić *et al.*, 2012). In contrast, glucose provided optimum growth and morphogenesis of the shoots of *Nepeta rtanjensis* (Mišić *et al.*, 2005b). During the shoot multiplication of the species *Fagus sylvatica* and *F. orientalis* on media containing (3-4%) glucose, the number of shoots per explant was higher than on media containing sucrose or fructose. However, the effect of different sugars on shoot length varied depending on the clone of *F. orientalis* cultivated (Cuenca and Vieitez, 2000). Similarly, during the micropropagation of different cultivars and clones of *Vaccinium vitis-idaea*, optimum effects of carbohydrate type and concentration differed (Debnath, 2005). Considering that sucrose is the most commonly used carbohydrate (Thorpe *et al.*, 2008), the results obtained in our research could be expected ones. Nevertheless, taking into account above-mentioned reports of better effect of other carbohydrates (e.g. glycose) on in vitro growth and development, and different response of different plants species and even the clones to variation of type and concentration of carbohydrates, we could assume that optimum results during in vitro culture of some other *Dianthus* species possibly would be achieved with carbohydrate other than sucrose.

Tab. 3. Frequency of shoot regeneration on different explants cultured on media supplemented with different carbohydrate source and graded concentration after 25 days *in vitro*

Type of sugar	Concentration	Single node cuttings (%)		Shoot tips (%)		Shoots (%)	
		MS_0	1/2 MS_0	MS_0	1/2 MS_0	MS_0	1/2 MS_0
sucrose	1%	85.0 ± 11.2 ^{a}	91.6 ± 5.7 ^{a}	68.4 ± 7.2 ^{b}	96.7 ± 2.1 ^{a}	88.3 ± 6.4 ^{a}	95.0 ± 1.8 ^{a}
sucrose	3%	73.3 ± 6.5 ^{abc}	90.0 ± 5.6 ^{a}	90.0 ± 5.6 ^{a}	98.3 ± 1.7 ^{a}	90.0 ± 4.1 ^{a}	96.7 ± 2.0 ^{a}
sucrose	5%	70.0 ± 6.8 ^{abc}	90.0 ± 5.6 ^{a}	81.7 ± 8.3 ^{ab}	96.7 ± 1.1 ^{a}	90.0 ± 2.3 ^{a}	96.7 ± 2.7 ^{a}
sucrose	7%	83.3 ± 5.8 ^{ab}	88.3 ± 8.1 ^{ab}	73.3 ± 8.3 ^{ab}	91.7 ± 5.3 ^{a}	86.7 ± 5.1 ^{a}	95.0 ± 2.4 ^{a}
glucose	1%	21.7 ± 4.3 ^{c}	98.3 ± 1.7 ^{a}	36.6 ± 10.5 ^{bc}	98.3 ± 1.7 ^{a}	60.0 ± 5.1 ^{ab}	98.3 ± 1.7 ^{a}
glucose	3%	18.3 ± 5.0 ^{c}	98.3 ± 1.7 ^{a}	65.0 ± 9.1 ^{b}	96.7 ± 2.4 ^{a}	85.0 ± 4.8 ^{ab}	98.3 ± 0.8 ^{a}
glucose	5%	16.7 ± 5.2 ^{cd}	95.0 ± 3.3 ^{a}	51.6 ± 8.9 ^{bc}	96.7 ± 2.4 ^{a}	56.7 ± 8.2 ^{b}	98.3 ± 0.8 ^{a}
glucose	7%	6.7 ± 3.3 ^{de}	91.7 ± 4.1 ^{a}	11.7 ± 4.6 ^{cd}	80.0 ± 5.6 ^{a}	15.0 ± 3.3 ^{bc}	95.0 ± 2.4 ^{a}
fructose	1%	11.7 ± 3.7 ^{cd}	96.7 ± 3.3 ^{a}	35.0 ± 5.2 ^{cd}	98.3 ± 0.8 ^{a}	18.3 ± 2.6 ^{bc}	96.7 ± 1.8 ^{a}
fructose	3%	48.3 ± 7.1 ^{bc}	96.7 ± 3.3 ^{a}	48.3 ± 5.7 ^{c}	95.0 ± 2.3 ^{a}	41.7 ± 5.8 ^{bc}	93.3 ± 2.3 ^{a}
fructose	5%	25.0 ± 6.2 ^{c}	80.0 ± 8.8 ^{ab}	20.0 ± 5.1 ^{cd}	70.0 ± 5.4 ^{b}	28.3 ± 5.1 ^{bc}	80.0 ± 7.1 ^{ab}
fructose	7%	3.3 ± 1.7 ^{de}	65.0 ± 7.4 ^{b}	5.0 ± 2.7 ^{d}	61.7 ± 6.7 ^{b}	23.4 ± 4.5 ^{bc}	76.7 ± 7.3 ^{b}

MS_0- basal medium containing MS mineral salts; 1/2 MS_0 - basal medium containing half strength MS mineral salts;
Note: Values followed by different letters are significantly different at the $P < 0.05$ level according to the least significant difference test

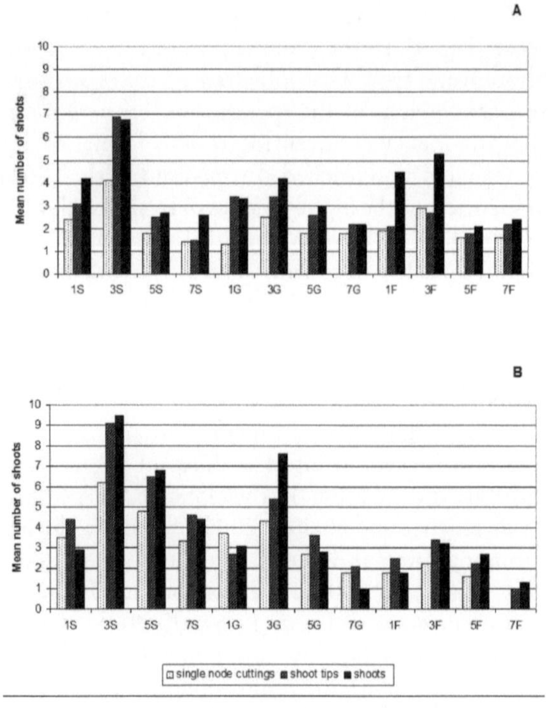

Fig. 2. The effect of graded concentration of different carbohydrate source on number of regenerated *D. serotinus* shoots on full-strength MS basal medium (B) and half-strength MS basal medium (A).

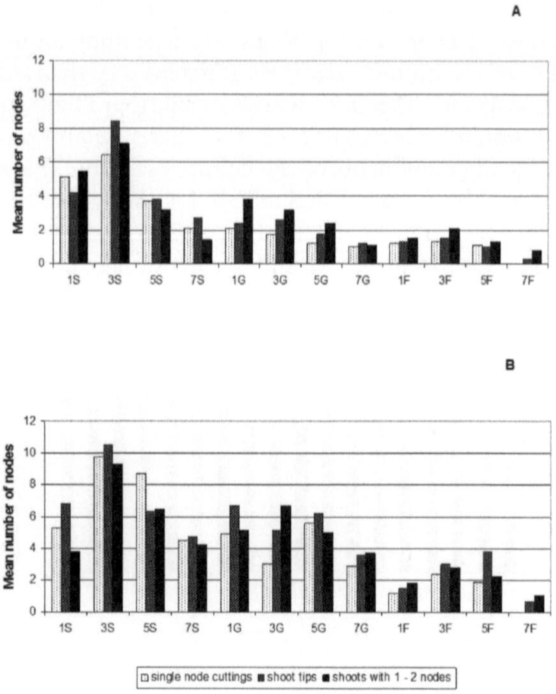

Fig. 3. The effect of graded concentration of different carbohydrate source on number of *D. serotinus* shoot nodes regenerated on full-strength MS basal medium (B) and half-strength MS basal medium (A)

(Figs. 2 and 3: the letters indicate different carbohydrate source and numbers indicate graded concentrations of different carbohydrate source; S-sucrose, G-glucose, F-fructose; 1 – 1%, 3 – 3%, 5 – 5%, 7 – 7%)

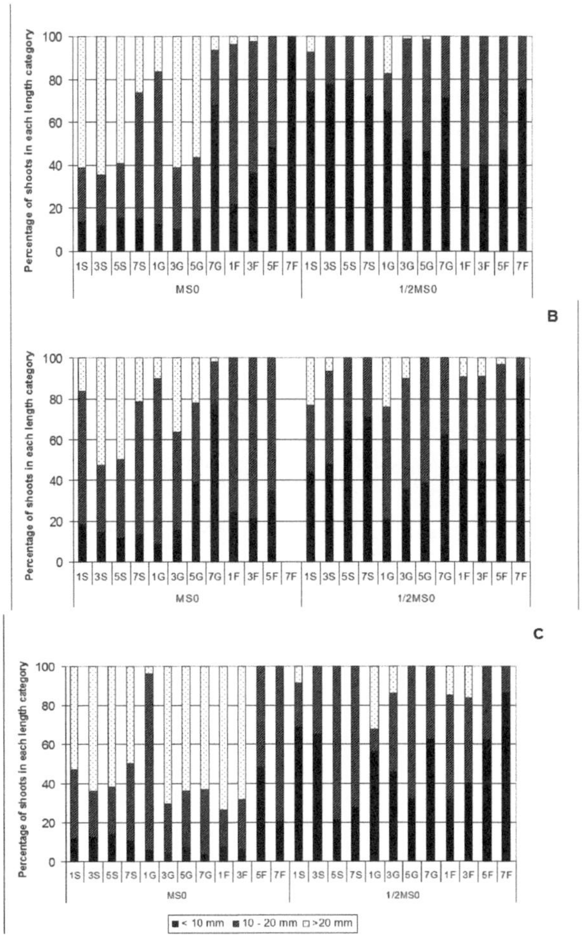

Fig. 4. The effect of graded concentration of different carbohydrate source on length of *D. serotinus* shoots regenerated on shoot tips (A), single node cuttings (B) and shoots (C)

(The letters indicate different carbohydrate source and numbers indicate graded concentrations of different carbohydrate source; S-sucrose, G-glucose, F-fructose; 1 – 1%, 3 – 3%, 5 – 5%, 7 – 7%; MS_0-full-strength MS basal medium, 1/2MS_0 - half-strength MS basal medium)

Rooting and acclimatization

The rooting of shoots was the most successful on half-strength MS medium with 0.5 mg·L^{-1} NAA (76.7%), while the percentage of rooted shoots on MS medium with the same concentration of NAA was slightly lower (68.3%), which corresponds to the results previously described by Marković *et al.* (2007). The concentration of MS salts (MS or half-strength MS) had no significant effect on the percentage of rooting when media with the same concentration of NAA were compared. Similarly, during the propagation of *D. nardiformis* the rooting on MS media with half-strength MS or even 1/4 MS did not significantly affect the quality of rooting (Holobiuc *et al.*, 2010a).

Significantly poorer results were obtained on other media (Tab. 4), characterized by a lower concentration of NAA or a lack of hormones. The rooting percentage was lower than 50%, and the average number of roots per explant was almost halved (Tab. 4). The acclimatization of plantlets was the most successful (89%) in a 4:1 mixture of peat and sand, the least successful acclimatization of plantlets was recorded in a mixture of peat, sand, garden soil and burned farmyard manure (80%), and the acclimatized plants flowered the following year. When compared with other carnation species the rooting (68.3%) and acclimatization (89%) of *D. serotinus* obtained in this study can be considered satisfactory. Considerably better results in terms of rooting (100%) and acclimatization (97%) were obtained in the species *D. deltoides* L. (Markovic *et al.*, 2013), and the rooting of *D. superbus* ssp. *superbus* was successful (100%), although the acclimatization was relatively low (61.5%) (Mikulík, 1999). However, in some studies of other carnation species the rooting percentage was significantly lower (*D. henteri* - 20%, *D. spiculifolius* - 30%, *D. giganteus banaticus* - 40%) (Pop and Pamfil, 2011).

Tab. 4. Percentage of rooted shoots, average number of roots and the mean length of the longest root per explant

NAA (mg·L^{-1})	Medium	Rooting percentage (%)	No. of roots per explant	Mean length of the longest root (mm)
0	MS_0	28.3 ± 6.1 ^{cd}	6.7 ± 1.2 ^{bc}	16.8 ± 2.6 ^{a}
0.05	MS_0	23.3 ± 6.5 ^{d}	7.2 ± 1.0 ^{bc}	17.1 ± 2.0 ^{a}
0.1	MS_0	38.3 ± 5.5 ^{c}	13.0 ± 3.5 ^{b}	14.8 ± 2.1 ^{a}
0.5	MS_0	68.3 ± 7.2 ^{ab}	21.1 ± 3.1 ^{a}	14.5 ± 1.7 ^{a}
0	1/2 MS_0	35.0 ± 7.1 ^{c}	5.6 ± 1.3 ^{c}	15.5 ± 2.7 ^{a}
0.05	1/2 MS_0	36.7 ± 6.8 ^{c}	5.8 ± 1.8 ^{c}	16.2 ± 2.3 ^{a}
0.1	1/2 MS_0	48.4 ± 4.9 ^{bc}	6.3 ± 1.2 ^{bc}	11.3 ± 1.8 ^{ab}
0.5	1/2 MS_0	76.7 ± 7.2 ^{a}	10.5 ± 1.8 ^{b}	12.8 ± 1.5 ^{ab}

MS_0- basal medium containing MS mineral salts; 1/2 MS_0 - basal medium containing half strength MS mineral salts;
Note: Values followed by different letters are significantly different at the $P < 0.05$ level according to the least significant difference test

Conclusion

Taking the obtained results into account, it can be concluded that micropropagation of the endangered species *D. serotinus* should be performed on half-strength MS medium with a pH level ranging from 5.8 to 6.8. The most favourable sugar as carbon source is sucrose supplied at a concentration of 3%, and shoots with 1-2 nodes should be used as explants, although satisfactory results have also been achieved by using shoot tips. This research provide complete protocol for successful in vitro propagation of the critically endangered species *D. serotinus*, and its results could be useful for both activities aimed at biodiversity protection in sandy regions of Pannonian Basin and floricultural production of this species.

Acknowledgement
This work was supported by the Ministry of Education and Science of Serbia, grant No 43007.

References

Barghchi M (1988). Micropropagation of *Alnus cordata* (Loisel.) Loisel. Plant Cell Tiss Org 15:233-244.

Bhatia P, Ashwath N (2005). Effect of Medium pH on Shoot Regeneration from Cotyledonary Explants of Tomato. Biotechnology 4(1):7-10.

Boža P (1999). *Dianthus serotinus* Waldst. & Kit., p. 252-254. In: Stevanović V (ed.). Crvena knjiga flore Srbije. Ministarstvo za životnu sredinu Republike Srbije, Biološki fakultet Univerziteta u Beogradu, Zavod za zaštitu prirode Republike Srbije, Beograd.

Butiuc-Keul A, Suteu A, Munteanu-Deliu C, Deliu C (2001). Study on the in vitro preservation of *Dianthus spiculifolius* Schur. Contrib Bot 36:137-145.

Chauvin JE, Salesses G (1988). Advances in chestnut micropropagation (*Castanea* sp.). Acta Hort 227:340-345.

Cristea V (2010). Photoautotrophic *in vitro* culture of endemic and endangered *Dianthus* species from Romania. Todesco, Cluj-Napoca, 227 p. (in Romanian).

Cristea V, Brummer AT, Jarda L, Miclăuş M (2010). *In vitro* culture initiation and phytohormonal influence on *Dianthus henteri* – a Romanian endemic species. Rom Biotech Lett 15(1):25-33.

Cuenca B, Vieitez AM (2000). Influence of carbon source on shoot multiplication and adventitious bud regeneration in in vitro beech cultures. Plant Growth Regul 32:1-12.

Debnath SC (2005). Effects of carbon source and concentration on development of lingonberry (*Vaccinium vitis-idaea* L.) shoots cultivated in in vitro from nodal explants. *In Vitro* Cell Dev Pl 41:145-150.

Ebrahim MKH, Ibrahim IA (2000). Influence of medium solidification and pH value on *in vitro* propagation of *Maranta leuconeura* cv. Kerchoviana. Sci Hortic - Amsterdam 86:211-221.

Gajić M (1986). Flora i vegetacija Suboticko-horgoške peščare, Šumarski fakultet Univerzitet u Beogradu, Beograd, Šumsko gazdinstvo, Subotica 163-164 p. (in Serbian).

George EF, de Klerk GJ (2008). The Components of Plant Tissue Culture Media I: Macro- and Micro-Nutrients, p. 65-114. In: George EF, Hall M A, De Klerk DJ (Eds.). Plant Propagation by Tissue Culture, 3rd Edition, Springer, AA Dordrecht, Netherlands.

Harbage JF, Stimart DP (1996). Effect of pH and 1-indole-3-butyric acid (IBA) on ro oting of apple microcuttings. J Am Soc Hort Sci 121:1049-1053.

Holobiuc I, Blindu R (2006). Improvement of the micropropagation and *in vitro* medium-term preservation of some rare *Dianthus* species. Contrib Bot 41:143-151.

Holobiuc I, Blindu R, Cristea V (2009). Researches Concerning *In Vitro* Conservation of the Rare Plant Species *Dianthus nardiformis* Janka. Biotechnol Biotech Eq 23:221-224.

Holobiuc I, Catana R, Cristea V (2010a). Researches concerning *in vitro* cultures optimization of the vulnerable species *Dianthus nardiformis* Janka. Analele Universitatii din Oradea - Fascicula Biologie 17(1):116-121.

Holobiuc I, Mitoi M, Blindu R, Helepciuc F (2010b). The establishment of an *in vitro* gene bank in *Dianthus spiculifolius* Schur. and *D. glacialis* ssp. *gelidus* (Schott Nym. et Kotschy) Tutin: II. Medium-term cultures characterization in minimal growth conditions. Rom Biotech Lett 15(2):5111-5119.

Jethwani V, Kothari SL (1993). Micropropagation of *Dianthus barbatus* and *D. chinensis* through cotyledonary node culture. Plant Tiss Cult 2:91-96.

Jethwani V, Sharma VK, Kothari SL (1994). Micropropagation of *Dianthus chinensis* and *Dianthus barbatus* through shoot tip culture. J Indian Bot Soc 73:357-358.

Kovác J (1995). Micropropagation of *Dianthus arenarius* subsp. bohemicus - an endangered endemic from the Czech Republic. Botanic Gardens Micropropagation News 8:106-108.

Law on Environmental Protection, Rulebook on protected species (2010). Official Gazette of the Republic of Serbia, No. 5/10, Belgrade, Serbia.

Leifert C, Pryce S, Lumsden PJ, Waites WM (1992). Effect of medium acidity on growth and rooting of different plant species growing in vitro. Plant Cell Tiss Org 30(3):171-179.

Marković M, Grbić M, Skočajić D, Đunisijević-Bojović D (2007). Uticaj balansa fitohormona na multiplikaciju izdanaka i ožiljavanje vrste *Dianthus serotinus* Waldst. & Kit. Glasnik Šumarskog Fakulteta 95:83-94.

Marković M, Popović M, Vilotić D (2013). Micropropagation of *Dianthus deltoides* L. through shoot tip and nodal cuttings culture. Arch Biol Sci 65(1):17-22.

Miclăuş M, Cristea V, Deliu C (2003). Micropropagation on *Dianthus petraeus* W. et K. ssp. simonkaianus (Péterfi) Tutin. Contrib Bot 38:77-84.

Mikulík J (1999). Propagation of endangered plant species by tissue cultures. Acta Universitatis Palackianae Olomucensis, Biologica 37:27-33.

Mišić D, Ghalawenji N, Grubišić D, Konjević R (2005a). Micropropagation and Reintroduction of *Nepeta rtanjensis*, an Endemic and Critically Endangered Perennial of Serbia. Phyton 45:9-20.

Mišić D, Maksimović V, Todorović S, Grubisić D, Konjević R (2005b). Influence of carbohydrate source on *Nepeta rtanjensis* growth, morphogenesis and nepetalactone production *in vitro*. Israel J Plant Sci 53:103-108.

Mohamed MAH, Alsadon AA (2010). Influence of ventilation and sucrose on growth and leaf anatomy of micropropagated potato plantlets. Sci Hortic 123:295-300.

Murashige T, Skoog F (1962). A revised medium for rapid

growth and bioassays with tobacco tissue cultures. Physiol Plantarum 15:473-497.

Nowak B, Miczynski K, Hudy L (2004). Sugar uptake and utilization during adventitious bud differentiation on *in vitro* leaf explants of 'Wegierka Zwykta' plum (*Prunus domestica*). Plant Cell Tiss Org 76:255-260.

Ostrolucka MG, Gajdosova A, Ondruskova E, Lateekova A, Libiakova G (2010). Effect of medium pH on Axillary Shoot Proliferation of Selected *Vaccinium vitis-idaea* L. Cultivars. Acta Biol Cracov Bot 52(2):92-96.

Ostrolucka MG, Libiakova G, Ondruskova E, Gajdosova A (2004). *In vitro* propagation of *Vaccinium* species. Acta Universitatis Latviensis, Biology 676:207-212.

Papafotiou M, Stragas J (2009). Seed germination and *in vitro* propagation of *Dianthus fruticosus* L. ISHS Acta Hort 813:481-484.

Pareek A, Kantia A, Kothari SL (2004). *In vitro* cloning of ornamental species of *Dianthus*. Indian J Biotech 3:263-266.

Pence VC (1999). The application of biotechnology for the conservation of endangered plants, p. 227-241. In: Benson EE (Ed.). Plant Conservation Biotechnology, Chapter 15, Taylor and Francis, London.

Perić M, Dmitrović S, Živković S, Filipović B, Skorić M, Simonović A, Todorović S (2012). *In Vitro* Growth, Morphogenesis, and Acclimatization of Endangered *Rindera umbellata* (Waldst. & Kit.) Bunge. Hort Science 47:1123-1128.

Pop T, Pamfil D (2011). *In vitro* Preservation of Three Species of *Dianthus* from Romania. Bulletin UASVM Hort 68(1):414-422.

Radojević LJ, Ćalić-Dragosavac D, Špirić J, Stevanović B, Stevanović V (2010). *In vitro* umnožavanje u kulturi segmenata stabla of *Dianthus ciliatus* ssp. *dalmaticus* i *D. giganteus* ssp. *croaticus* (Caryophyllaceae). Botanica Serbica 34(2):153-161.

Radojević LJ, Marinković N, Jevremović S (1997). Vegetativno razmnožavanje u kulturi meristema i segmenata stabla *Dianthus petraeus* Waldst. et Kit. subsp. *noeanus*. Glasnik instituta za botaniku i Botaničke bašte Univerziteta u Beogradu 31:73-77.

Radojević LJ, Špirić J, Stevanović B, Stevanović V (2006). *In vitro* culture of several endemic species Dianthus in Balkan Peninsula. Proceedings of XXII International Symposium Section Ornamentals, 11-15 September 2006, 36. Sanremo, Italy.

Steffen JD, Sachs RM, Hackett WP (1988). Growth and development of reproductive and vegetative tissues of *Bougainvillea* cultured *in vitro* as a function of carbohydrate. Am J Bot 75:1219-1224.

Thorpe T, Stasolla C, Yeung EC, de Klerk GJ, Roberts A, George EF (2008). The Components of Plant Tissue Culture Media II: Organic Additions, Osmotic and pH Effects, and Support Systems, p. 115-174. In: George EF, Hall MA, De Klerk DJ (Eds.). Plant Propagation by Tissue Culture, 3rd Edition, Springer, Dordrecht, Netherlands.

Todorović S, Grubisić D, Giba Z, Mišić D, Konjević R (2006). Sucrose effects on *in vitro* fruiting and seed production of *Centaurium pulchellum*. Biol Plantarum 50:771-774.

Tremblay FM, Lalonde M (1984). Requirements for *in vitro* propagation of seven nitrogen-fixing *Alnus* species. Plant Cell Tiss Org 3:189-199.

Yu X, Reed BM (1993). Improved shoot multiplication of mature hazelnut (*Corylus avelana* L.) *in vitro* using glucose as carbon source. Plant Cell Rep 12:256-259.

5

Effects of Saline and Alkaline Stresses on Growth and Physiological Changes in Oat (*Avena sativa* L.) Seedlings

Zhanwu GAO[1,2], Jiayu HAN[1], Chunsheng MU[1*], Jixiang LIN[3*], Xiaoyu LI[4], Lidong LIN[5], Shengnan SUN[1]

[1]*Key Laboratory of Vegetation Ecology of Ministry of Education, Institute of Grassland Science, Northeast Normal University, Changchun, 130024, China; mucs821@gmail.com (*corresponding author)*
[2]*Department of Geography, Baicheng Normal College, Baicheng, 137000, China*
[3]*Alkali Soil Natural Environmental Science Center, Northeast Forestry University, Key Laboratory of Saline-Alkali Vegetation Ecology Restoration in Oil Field, Ministry of Education, Harbin, 150040, China; jixiang851012@gmail.com (*corresponding author)*
[4]*Key Laboratory of Wetland Ecology and Environment, Northeast Institute of Geography and Agroecology, Chinese Academy of Science, Changchun 130012, China*
[5]*Dongtou Fisheries Science and Technology Research Institute, Dongtou, 325700, China*

Abstract

Two neutral salts (NaCl and Na_2SO_4) and alkaline salts ($NaHCO_3$ and Na_2CO_3) were both mixed in 2:1 ratio, and the effects of saline and alkaline stresses on growth and physiological changes in oat seedlings were explored. Result showed that biomass, water content and chlorophyll content decreased while cell membrane permeability significantly increased under alkaline stress. Saline stress did not have obvious effect on pH value in tissue fluids of shoot and root, but alkaline stress increased pH value in root tissue fluid. The contents of Na^+, Na^+/K^+, SO_4^{2-} increased more, and K^+, NO_3^-, $H_2PO_4^-$ decreased more under alkaline stress, the Cl^- content increased obviously under saline stress but had little change under alkaline stress. The increments of proline and organic acid were both greater under alkaline stress, but organic acid content kept the same level under saline stress. Alkaline stress caused more harmful effects on growth and physiological changes in oat seedlings especially broke the pH stability in root tissue fluid. Physiological adaptive mechanisms of oat seedlings under saline stress and alkaline stress were different, which mainly took the way of accumulating organic acid under alkali stress but accumulating Cl^- under saline stress.

Keywords: alkali stress, chlorophyll, organic acid, proline, salt stress

Introduction

As saline soil widely distributes in more than 100 countries, and covers about 10% of arable area over the world, soil salinization becomes one of the environmental problems over the world. Cations of soluble salts in saline-alkaline soil mainly include Na^+, Ca^{2+}, Mg^{2+} and K^+, anions mainly include Cl^-, SO_4^{2-}, HCO_3^-, CO_3^{2-} and NO_3^- (Tannic 1990). The characteristic of saline-alkaline soil in Northeast China is that Na^+ is the main cations and Cl^-, SO_4^{2-}, HCO_3^- and CO_3^{2-} are the main anions (Yang *et al.*, 2007, 2008). Once soil contains alkaline salt such as $NaHCO_3$ and Na_2CO_3, pH value of the soil will increase. At this time, plants suffer from not only saline stress but also alkaline stress caused by high pH. The damage effect of alkaline stress caused by alkali salts such as $NaHCO_3$ and Na_2CO_3 on plant is more obvious than that of saline stress caused by neutral salts such as NaCl and Na_2SO_4 (Yang *et al.*, 2009). Soil salinization and alkalization frequently co-occur such as in the west Songnen plain of Northeast China, soil salinization and alkalization in grassland and farmland becomes more seriously, which cause losses in agricultural productivity and pose an ecological crisis for humans.

Oat (*Avena sativa* L.) is an annual crop, which belongs to the family of Poaceae. It is a world-wide cultivation crop and mainly distribute in temperate latitudes of the Northern Hemisphere. Oat has no strict requirements for cultivation soil, so it can be usually planted in various kinds of soil. Due to the good saline-alkaline tolerance, oat grew better than other crops when planted in semiarid and saline-alkaline soils, such as wheat. Up to now, most researches focus on crops e.g. sunflower (Shi and Sheng, 2005), and grasses such as *Leymus chinensis* and *Puccinellia distans* (Shi and Sheng, 1998; Li *et al.*, 2004) and make progresses on the growth, physiological response mechanisms and adaptive strategies of these crops/grasses to saline and alkaline stresses. However, there was little research on the growth and saline-alkaline tolerance of oat to saline-alkaline stresses. In order to study the saline or alkaline tolerance of oat seedlings, two neutral salts (NaCl and Na_2SO_4) and two alkaline salts ($NaHCO_3$ and Na_2CO_3) were mixed to simulate different saline-alkaline stress conditions to treat on oat seedlings, and compared the effect of saline and alkaline stresses on growth and physiological changes of this species.

Materials and methods

Plant materials

Oat adopted in this experiment is new cultivar 'Baiyan 2' which cultivated by Baicheng Agricultural Academy in Jilin Province of China. Uniform and full seeds were sown in 20-cm diameter plastic opened pots containing washed sands. After germinated, each pot was fully irrigated by Hoagland's nutrient solution daily. Each pot contained 20 seedlings before treatment. All pots were placed outdoors and protected from rain.

Saline and alkaline stresses treatment

Two neutral salts NaCl and Na_2SO_4 and two alkaline salts $NaHCO_3$ and Na_2CO_3 were both mixed in a 2:1 molar ratio, respectively. Six concentration treatments were applied to both groups: 48, 72, 96, 120, 144 and 168 mM, which simulated 12 different saline and alkaline stresses gradients (pH values were from 7.03 to 7.29 in salt stress, and were from 9.84 to 10.69 in alkaline stress).

When oat seedlings were 4 weeks old, 39 pots with uniform growth were selected and divided into 13 groups, with 3 pots per group. Each pot was a single replicate with three replicated per group. One group was for comparison and the other 12 groups were used for different treatments. Stress treatments were applied daily at 17:00 with 500 mL of nutrient solutions containing the appropriate stress salts. The comparison group was fully irrigated only by Hoagland's nutrient solution. All treatments were continuously treated for 9 days.

Measurement of growth and physiological indexes

Livability: livability=n/N ('n' represented the number of viable plants after the 9-day treatment, and 'N' was the total number of tested plants).

Biomass: Samples were harvested at 7:00 the next morning of the last treatment. All plants in each pot were taken out carefully and cleaned by distilled water. After absorbing the water on the surface of seedlings by absorbent paper, 10 seedlings per pot were taken for measuring plant height and fresh weight (FW); and then were oven-dried at 105 °C for 15 min and then dried at 80 °C to a constant weight, the dry weight (DW) was recorded. The rest materials were dried for measuring other indexes.

Tissue pH: 5 g fresh sample of shoot and root were taken and fully cleaned by distilled water for 3 times, after absorbing surface water by filter paper, the tissue fluid was extruded by injector, instantly measuring pH value by digital pH meter (PHSI-4A type).

Water content: water content=FW-DW (FW was the fresh weight of plants; DW was dry weight of plants).

Chlorophyll content: After 9 days of stress treatment, fresh functional leaves were collected from the plants and cut into small segments to determine the concentration of chlorophyll *a* and *b*. Chlorophyll was extracted from 0.5 g fresh samples with 80% acetone solution. The extracted solution was centrifuged. The optical density (OD value) of the diluted solution was measured at 663 and 645 nm. The concentrations of chlorophyll *a* (C_a) and *b* (C_b) were calculated using formula (1) and (2), respectively.

$$Ca=0.0127OD_{663}-0.00269OD_{645} \quad (1)$$
$$Cb=0.0229OD_{645}-0.00468OD_{663} \quad (2)$$

Leaf conductivity: membrane permeability is reflected by the relative electrical conductivity, ELR was determined as described by Lutts *et al.* (1996), which was defined as the ratio of the electrical conductivity of leaves with intact membranes (S1) to those with membranes destroyed by boiling water treatment (S2). A fresh leaf (1.0 g) was cut into 1 cm long segments, divided equally, and placed in 2 vials with 20 mL deionized water each. One vial was incubated at 25 °C on a rotary shaker for 2 h, and the electrical conductivity of the solution (S1) was then determined using a conductivity gauge. The other vial was placed in boiling water for 5 min, and the electrical conductivity of the solution (S2) was measured after equilibration to 25 °C.

$$ELR=S_1/S_2 \quad (3)$$

Content of dissociative proline: Free proline was assayed by the acid-ninhydrin method. The dried shoot tissue samples (0.1 g) were homogenized in 10 ml of 3% sulfosalicylic acid and then filtered. The filtrate (2 mL) was mixed with 2 ml acid-ninhydrin and 2 mL of glacial acetic acid in a test tube. The mixture was placed in a water bath for 1 h at 100 °C. The reaction mixture was extracted with 4 ml toluene and the chromophore containing toluene was aspirated, cooled to room temperature, and the absorbance was measured at 520 nm with Spectrometer.

Organic acid: Organic acid was measured by DX-300 type ionic chromatographic system (DIONEX, Sunnyvale, USA), and the measurement conditions were as follows: ICE-AS6 analytical column, CDM-II electric conductivity detector, AMMS-ICE II interference suppressor, mobile phase was 0.4 mM perfluorobutyric acid, flow speed was 1.0 mL/min, column temperature was 20 °C, sample size was 50 μL (Zhang and Mu, 2009). The total content of organic acid in the experiment was summation of all organic acid components which could be measured, including citric acid, malic acid and methanoic acid, lactic acid, acetic acid, succinic acid and oxalic acid.

Inorganic ion: Three cations, Na^+, K^+ and dissociative Ca^{2+} were determined by atomic absorption spectrophotometer (TAS-990, Purkinje General, Beijing). Three anions, NO_3^-, Cl^-, SO_4^{2-} were determined by DX-300 ionic chromatographic system (DIONEX, Sunnyvale, USA), and the measurement conditions were as follows: AS4A-SCion exchange column, CD M-II electric conductivity detector, mobile phase was $Na_2CO_3/NaHCO_3$ = 1.7/1.8 mM. $H_2PO_4^-$ was determined by molybdenum blue staining method.

Data analysis

Statistical tests were performed using SPSS 13.0 and by one-way ANOVAS (SPSS Inc, Chicago, IL, USA), the significance level was set at $P \leq 0.05$.

Results

Effects on the growth of oat seedlings

Under saline stress, the livability of seedlings at all salinity was 100%. It did not reach seedling lethal degree even though at the highest salinity (168 mM, pH was 7.29). Under alkaline stress, the livability decreased obviously when the salinity was more than 96 mM and pH was over 9.96 ($P<0.01$). When salinity was more than 144 mM and pH was over 10.39, most seedlings were dead, and the livability was only 18% ($P<0.01$, Fig. 1 A).

Biomass of the seedlings did not change when salinity was less than 48 mM and pH was less than 7.03, but gradually and visibly decreased when the salinity was more than 72 mM and pH was more than 7.12. Under alkaline stress, the greater stress concentration was, the lower biomass was, and the decrements were much larger than that in saline stress ($P<0.05$, Fig. 1 B).

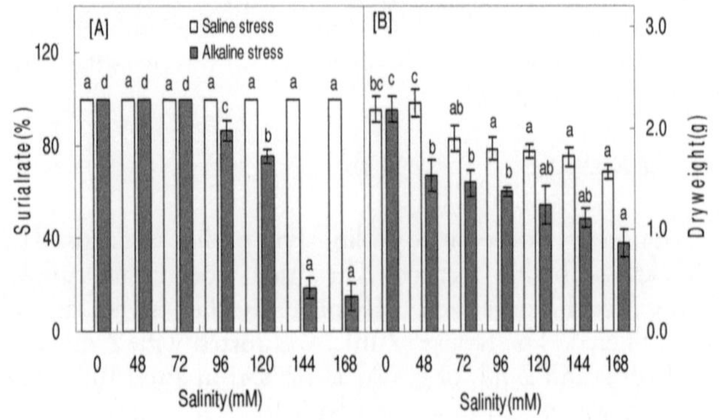

Fig. 1. Effect of salt and alkali stresses on survival/livability (A) and biomass (B) of oat seedlings, the 4 week-old oat seedlings were treated with salt stress ($NaCl:Na_2SO_4=2:1$; pH 7.03-7.29) and alkali stress ($NaHCO_3:Na_2CO_3=2:1$; pH 9.84-10.61) for 9 days, the values are means (±SE) of triplicate samples

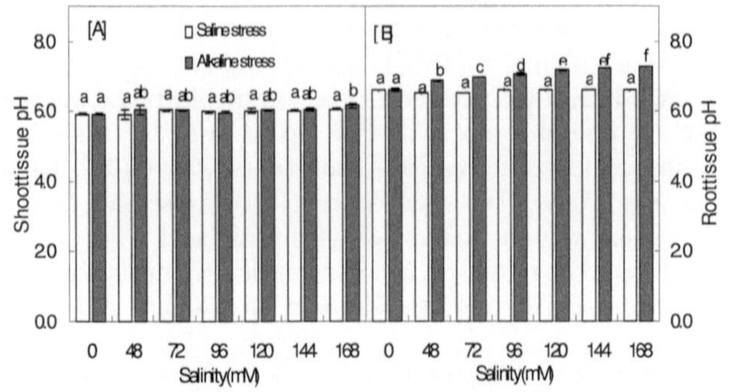

Fig. 2. Effect of salt and alkali stresses on shoot (A) and root (B) pH values of oat seedlings, the 4 week-old oat seedlings were treated with salt stress ($NaCl:Na_2SO_4=2:1$; pH 7.03-7.29) and alkali stress ($NaHCO_3:Na_2CO_3=2:1$; pH 9.84-10.61) for 9 days, the values are means (±SE) of triplicate samples

Effects on tissue pH, chlorophyll content, cell membrane permeability and water content

The pH value of tissue fluid from aboveground tissues was not affected by saline stress and alkaline stress (Fig. 2 A). The pH value of root tissue fluid kept stably under saline stress, but increased obviously with the increasing alkaline stress concentration ($P<0.05$, Fig. 2 B).

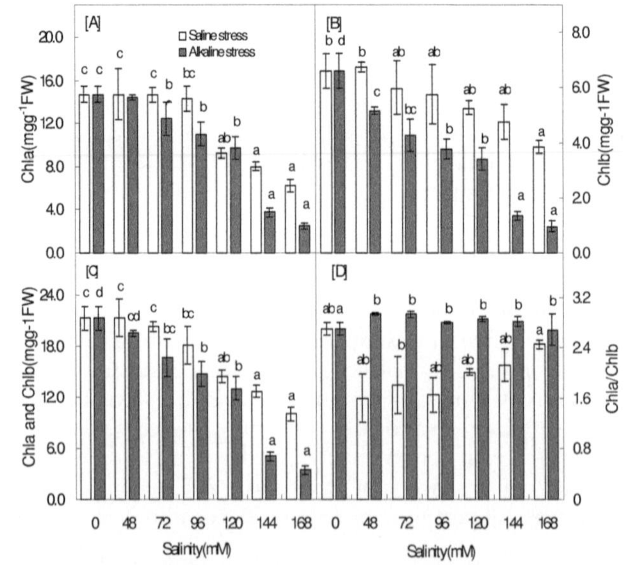

Fig. 3. Effect of salt and alkali stresses on chlorophyll contents of oat seedlings, the 4 week-old oat seedlings were treated with salt stress ($NaCl:Na_2SO_4=2:1$; pH 7.03-7.29) and alkali stress ($NaHCO_3:Na_2CO_3=2:1$; pH 9.84-10.61) for 9 days, the values are means (±SE) of triplicate samples

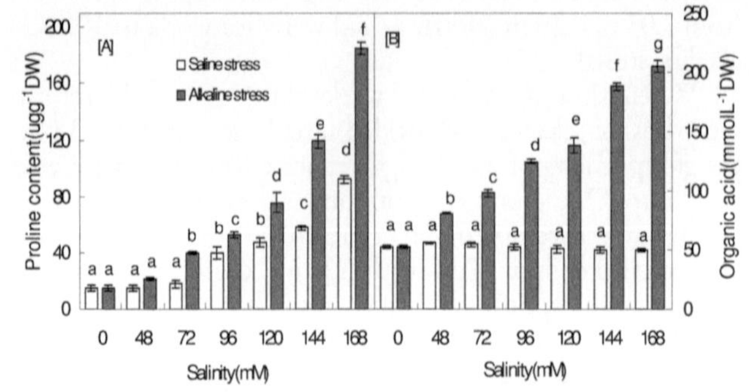

Fig. 4. Effect of salt and alkali stresses on proline (A) and organic acids (B) of oat seedlings, the 4 week-old oat seedlings were treated with salt stress ($NaCl:Na_2SO_4=2:1$; pH 7.03-7.29) and alkali stress ($NaHCO_3:Na_2CO_3=2:1$; pH 9.84-10.61) for 9 days, the values are means (±SE) of triplicate samples

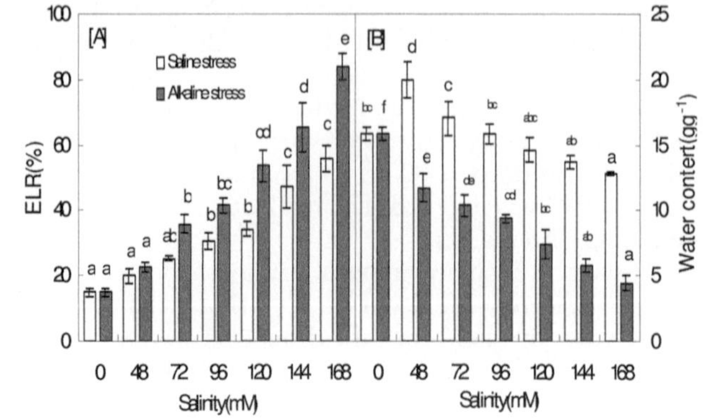

Fig. 5. Effect of salt and alkali stresses on bio-membrane permeability (A) and water content (B) of oat seedlings, the 4 week-old oat seedlings were treated with salt stress ($NaCl:Na_2SO_4=2:1$; pH 7.03-7.29) and alkali stress ($NaHCO_3:Na_2CO_3=2:1$; pH 9.84-10.61) for 9 days, the values are means (±SE) of triplicate samples

With the increasing salinity under saline and alkaline stress, the total amount of chlorophyll *a*, chlorophyll *b* and chlorophyll was visibly decreased, and the decrement was higher under alkaline stress than that under saline stress ($P<0.05$, Fig. 3 A, B, C); the cell membrane permeability was visibly increased, and the increment was higher under alkaline stress than that under saline stress ($P<0.05$, Fig. 5 A).

The water content was significantly higher than the comparison group when the salinity was lower than 72 mM and pH was less than 7.06 and then visibly decreased when the stress concentration was higher than 96 mM and pH was more than 7.14 ($P<0.05$, Fig. 5 B).

Effects on internal inorganic ions and organic solutes

With the increasing salinity, the Na^+ contents were obviously increased under both stresses. The accumulation amount of Na^+ under alkaline stress was 46.5 times than that of the comparison group at the highest salinity, which was 1.8 times higher than that under saline stress (Fig. 6A). The K^+ contents was visibly decreased ($P<0.05$, Fig. 6B), but the amount of Ca^{2+} was increased significantly, and the increment of Ca^{2+} was greater under saline stress than that under alkaline stress (Fig. 6C). The ratio of Na^+/K^+ was visibly increased with the increasing salinity and alkalinity, which was 4.7 times under alkaline stress than that under saline stress at the highest stress concentration (Fig. 6D).

With increasing salinity, the Cl^- contents was obviously increased under saline stress ($P<0.01$), but changed a little under alkaline stress. Compared with controls, Cl^- content was 5.5 times higher at the highest salinity than that of the control (Fig. 7A). SO_4^{2-} contents increased in both saline and alkaline stresses, and the increment in alkaline stress was greater than that in saline stress (Fig. 7B).

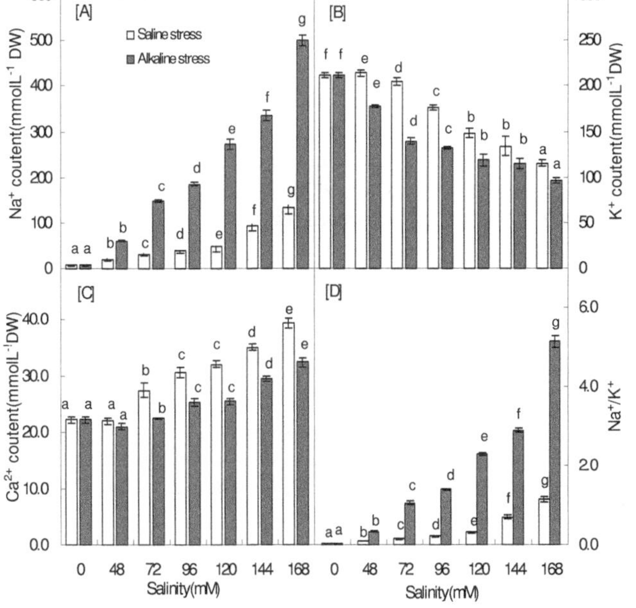

Fig. 6. Effect of salt and alkali stresses on Na+(A), K+(B), Ca₂+(C), Na+ /K+(D) in shoots of oat seedlings, the 4 week-old oat seedlings were treated with salt stress ($NaCl$:Na_2SO_4=2:1; pH 7.03-7.29) and alkali stress ($NaHCO_3$:Na_2CO_3=2:1; pH 9.84-10.61) for 9 days, the values are means (±SE) of triplicate samples

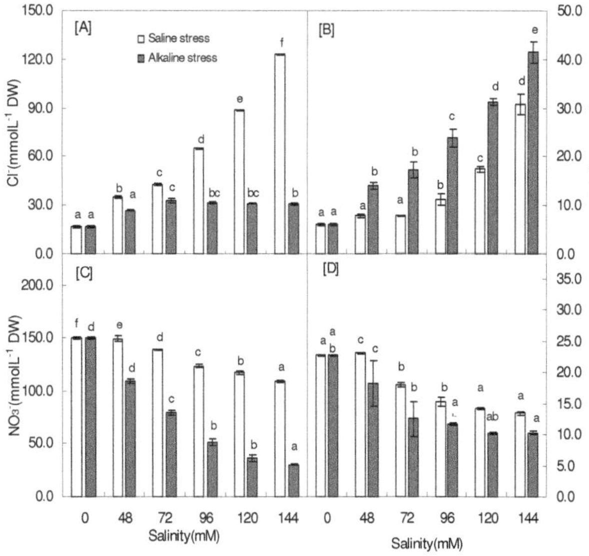

Fig. 7. Effect of salt and alkali stresses on Cl^-(A), SO_4^{2-} (B), NO_3^- (C), $H_2PO_4^-$ (D) in shoots of oat seedlings, the 4 week-old oat seedlings were treated with salt stress ($NaCl$:Na_2SO_4=2:1; pH 7.03-7.29) and alkali stress ($NaHCO_3$:Na_2CO_3=2:1; pH 9.84-10.61) for 9 days, the values are means (±SE) of triplicate samples

Both of NO_3^-, $H_2PO_4^-$ contents decreased in saline and alkaline stresses, and the decrement in alkaline stress was greater than that in saline stress ($P<0.05$, Fig. 7 CD).

With the increasing salinity and alkalinity, proline content was obviously increased, and the increment was greater in alkaline stress than that in saline stress ($P<0.01$, Fig. 4A). The organic acid content kept unchanged in saline stress, but was obviously increased in alkaline stress ($P<0.05$), which was 3.9 times than that of the control at the highest alkalinity (Fig. 4B).

Discussion

Inhibiting effect of saline and alkaline stresses on oat seedling growth

Biomass is comprehensive expression of plants responding to saline and alkaline stresses and also an accurate index of saline and alkaline resistance. Our results represented that low strength saline and alkaline treatments could promote growth of oat seedlings but high pH value had obvious inhibiting effect on growth ($P<0.01$).

Generally, livability is representing for the selection of seedlings under environment. Alkaline salt inhibited the growth of oat seedlings, and can endanger the existence of plants. Different plant also has different saline or alkaline resistances. For same plant, its resistance to saline or alkaline is different in growth periods. Our result represented that livability decreased along with the increasing salinity and pH value in saline and alkaline stress, and the increasing alkaline strength had more visible effect on livability (Fig 1).

Effects of saline and alkaline stresses on physiological changes of oat seedlings

Plasma membrane is the first protective screen for salt to enter cells, and is also the first part of plant organism to be suffered from saline or alkaline stress. As one of osmotic resistance indexes, the permeability of plasma membrane can directly represent the stability of plant cell for intracellular environment and adaptability and resisting ability for changes from extra-cellular environment. In this experiment, the conductivity increased with the increasing salinity, which was similar with the features of single alkaline salt stress (Shi and Yin, 1993, Shi and Zhao, 1997; Wang *et al.*, 2009), and mixed saline-alkaline stress on *Puccinellia tenuiflora*, broomcom, wheat and *Triticum aestivum-Agropyron* (Shi and Wang 2005; Shi and Sheng, 2005). The above stated that the greater the salinity was, the bigger the injury degree of membrane was. In the same salinity, the higher the pH value was, the more obvious the injury degree for membrane was ($P<0.05$).

Under saline stress, the physiology of plants is disturbed because of insufficient water, absent mineral nutrition and inadequate energy. Cell membrane system and membrane structure were relative to photosynthesis, their broken directly or indirectly affected chlorophyll content, which causes decreasing of photosynthesis. In the end, the growth of plants is inhibited by not gaining adequate material and energy from photosynthesis. Generally, chlorophyll *a* is helpful for absorb long wave light and chlorophyll *b* is good for absorb short wave light. In the experiment, the total amount of chlorophyll was decreased under saline and alkaline stress, which inhibited growth of plants leading to decrease biomass (Yang *et al.*, 2008).

Proline was an effective osmotic adjusting material of plants responding to stress situation (Song *et al.*, 2006; Yang *et al.*, 2008). Larger accumulation of proline represents stronger osmotic adjusting ability and resistivity for bad environment of plants (Yang *et al.*, 2008). For example, the content of dissociative proline gradually increased by testing guineagrass and sunflower. The unique and chief factor of alkaline salt was high pH value, such as the stress factor of Na_2CO_3 for plants was high pH value but not osmotic stress and ion toxicity (Shi and Sheng, 1998; Shi *et al.*, 1998). Under Na_2CO_3 stress, the content of dissociative proline in *Puccinellia tenuiflora* seedlings increased with the increasing salinity (Sun *et al.*, 2000). The content of proline of the treated oat groups in this experiment was uniformly higher than that of the comparison group. With the increasing of salinity and pH value, the content of proline increased and the differences became remarkable (Fig. 4A), which stated that the osmotic ability and resistance for bad environment of *Avena sativa* were strengthen (Yang *et al*, 2009).

Organic acid normally exist in plant in organic anions formation, which not only neutralize excess cations (Yang *et al.*, 2009), but also compensate loss of negative charge caused by nutrition deficiency and heavy metal stress (Zhang and Mu, 2009). To balance excess Na^+ accumulation and compensate inorganic anion loss under alkaline stress (Fig. 45), oat seedlings adopted the response system of largely accumulating organic acid. The large accumulation of organic acid under alkaline stress and no accumulation under saline stress proved that organic acid might be a special response for high pH stress (Fig. 4B).

Alkaline stress might disturb selective adsorption of K^+-Na^+ that leads to unbalance of K^+-Na^+ in cells (Yang *et al.*, 2008). Normally, saline stress could inhibit plants for absorbing cations such as Ca^{2+}. But in our experiment, the content of Ca^{2+} slightly increased, which might be the special response of oat to saline and alkaline stresses. Halophyte often accumulates inorganic anions such as Cl^-, SO_4^{2-} in saline stress and reduces NO_3^-, $H_2PO_4^-$ to keep the balance of cations and anions (Yang *et al.*, 2007, 2008). Oat seedlings accumulated Cl^- and SO_4^{2-} largely, which represented that the salt resistance mechanism of oat was nearly the same with that of halophyte. Alkaline stress visibly inhibited the accumulation of Cl^- but irritated accumulation of SO_4^{2-} in high pH alkaline stress, the absorption of anions such as NO_3^-, H_2PO_4 of halophyte was inhibited (Yang *et al.*, 2008), which stated that the inhibitions for anion types of *Avena sativa* and halophyte in alkaline stress were different and the mechanism thereof needed future research.

Alkaline stress visibly irritated accumulation of Cl^- and SO_4^{2-}, and similar with halophyte that high pH also inhibited the absorption of NO_3^-, $H_2PO_4^-$ (Yang *et al.*, 2008). This result stated that there was a different anion adjustment mechanism between *Avena sativa* and halophyte under alkaline stress that need further investigation.

Alkaline resistant mechanism of oat seedlings

The key point of why the effect of alkaline stress was higher than saline stress was that high pH. One way for plants to adapt to high pH was to accumulate micromolecule metabolite with buffering function (such as organic acid, proline etc.) for adjusting internal pH value. Adjusting pH value was a process of consuming energy, which lowered intracellular pH and inhibited growth of plants simultaneously. Plants adjusted intracellular pH by adjusting micro-environment pH around root to keep relative balance of metabolism and ions. Plants can adjust pH in micro-environment outside roots by excreting H^+, organic acid and discharging CO_2 through respiration (Wang *et al.*, 2009). Intracellular pH can be adjusted mainly by accumulating organic acid largely in vacuoles (Shi and Sheng, 2005). The large accumulation of organic acid in cells of oat seedlings under alkaline stress can compensate water and adjust intracellular pH. Therefore, it was stated that accumulating organic acid in large amount was a physiological response mechanisms of oat to the alkaline stress.

The serious damage of alkaline stress on photosynthetic pigments and membrane system represented that alkaline stress might affect cytoplasm, which might break structure and functions of chloroplast and reduce synthesis of chlorophyll (Yang *et al.*, 2007). Alkaline stress disturbed balance of inorganic ions, promoted absorption of Ca^{2+} and Na^+ and inhibited accumulation of K^+, and decreased contents of NO_3^-, $H_2PO_4^-$ and accumulated SO_4^{2-} largely. The above two points seemed to indicate that alkaline stress might seriously disturb metabolism of SO_4^{2-}, NO_3^-, $H_2PO_4^-$, and the accumulation of both proline and organic acid under alkaline stress might be irritability response for the disturbance.

Acknowledgments

This work was supported by the Fundamental Research Funds for the central universities (DL12BA32, 2572014EA04), the National '973' Programme (2015CB150801), and was also funded by National Natural Science Foundation of China (31172259, 31370432).

References

Li CL, Zhou H, Zhang GF, Zhang YJ, Meng L (2004) Effects of saline stress on germination of leafiness Alfalfa seeds. Chinese J Grassland 26:21-25.

Lutts S, Kiner, Bouharmont J (1996). NaCl-induced senescence in leaves of rice (*Oryza sativa* L.) cultivars differing in salinity resistance. Ann Bot 78:389-398.

Shi D, Sheng Y (2005). Effects of various salt-alkali mixed stress conditions on sunflower seedings and analysis of their stress factors. Environ Exp Bot 54:8-21.

Shi D, Wang DL (2005). Effects of various salt-alkaline mixed stresses on *Aneurolepidium chinense* (Trin.) Kitag. Plant and Soil 271:15-26.

Shi DC, Sheng YM (1998). Manual simulation of compound saline-alkaline ecological condition and effect on growth of guineagrass. Acta Botanica Sinica 40 (12):1136-1142.

Shi DC, Sheng YM, Zhao KF (1998). Simulation of complex Alkali-saline conditions and their effects on the growth of Chinese seedlings. Acta Prataculct Sinica 1:36-41.

Shi DC, Yin LJ (1993). Difference between salt (NaCl) and alkaline (Na_2CO_3) stresses on *Puccinellia tenuiflora* (Griseb.) Scribn. et Merr. Plants. Acta Bot Sin 35:144-149 (in Chinese with English abstract).

Shi DC, Zhao KF (1997). Effects of sodium chloride and carbonate on growth of *Puccinellia* and on present state of mineral elements in nutrient solution. Acta Prataculct Sin 6:51-61 (in Chinese).

Song J, Feng G, Tian C.Y, Zhang FS (2006). Osmotic adjustment traits of *Suaeda physophora*, *Haloxylon ammodendron* and *Haloxylon persicum* in field or controlled conditions. Plant Science 170:113-119.

Sun GR, Yan XF (2000). Effect of Na_2CO_3 stress on content of free amino acid of *Puccinellia tenuiflora* seedling. Plant Research 20:71-72.

Tanic KK (1990). Agricultural salinity assessment and management. Society of Civil Engineers, New York 3-5.

Wang YX, Zhang B, Wang T (2009). Effects of saline stress on chlorophyll, bataine content and cell membrane permeability of alfalfa. Practa Science 26:53-56.

Yang CW, Chong J, Kim C, Li CY, Shi DC, Wang DL (2007). Osmotic Adjustment and ion balance traits of an alkali resistant halophyte *Kochia sieversiana* during adaptation to salt and alkali conditions. Plant Soil 294:263-276.

Yang CW, Shi DC, Wang DL (2008). Comparative effects of salt stress and alkali stress on growth, osmotic adjustment and ionic balance of an alkali-resistant halophyte *Suaeda glauca* (Bge.). Plant Growth Regul 56:179-190.

Yang CW, Zhang ML, Liu J, Shi DC, Wang DL (2009). Effects of buffer capacity on growth, photosynthesis, and solute accumulation of a glycophyte (wheat) and a halophyte (*Chloris virgata*). Photosynthetica 47:55-60.

Zhang JT, Mu CS (2009). Effects of saline and alkaline stresses on the germination, growth, photosynthesis, ionic balance and anti-oxidant system in an alkali-tolerant leguminous forage *Lathyrus quinquenervius*. Soil Science and Plant Nutrition 55:685-697.

6

Multicriteria Analysis of the Effects of Field Burning Crop Residues

Vasilica STAN*, Gina FÎNTÎNERU, Mircea MIHALACHE

*University of Agronomic Sciences and Veterinary Medicine, 59 Mărăşti Bd., Sector 1, 011464, Bucharest, Romania; vstan@agro-bucuresti.ro (*corresponding author)*

Abstract

Burning crop residues is frequently used by Romanian land users to clean agricultural fields after crop harvest for ease in post-harvest soil tillage. Huge amounts of crop residues biomass, on very large areas, were burned in Romania in the last twenty years, as compared to other countries. There are several reasons (e.g. the lack of equipment to gather the crop residues and to transport and store them, the diminishing of the livestock after 1990, the absence of other alternatives, especially in the 1990s, but also the lack of information regarding the good practices) that are evocated to support the use of this method. However, this method is not a sustainable one since it can cause many environmental damages, especially related to soil properties (physical, chemical and biological), greenhouse gas emission and crop yields. Contrary to the above stated, crop residues' addition to the soil may restore damaged soil structure, improve aggregate stability, soil water retention, soil fertility, increase total organic carbon (TOC) and total nitrogen (TN) etc. The purpose of this paper is to make a multicriteria analyze of the effects of crop residue management on the soil, agricultural productivity and environment. At the same time, the use of crop residues biomass as a source of energy is presented as an alternative, given its potential ability to offset fossil fuels and reduce CO_2 emissions.

Keywords: crop residues, soil properties, crop yields, greenhouse gases, sustainability

Introduction

Open field burning of crop residues is a simple method for land users to clean agricultural land after crop harvest to facilitate soil tillage. Crop residue burning has proved to be efficient in controlling insects, diseases, and the emergence of invasive weed species (Pintilie *et al.*, 1985; Kutcher and Malhi, 2010; Estrellan and Iino, 2010; Gonçalves *et al.*, 2011). But, as a general practice, its use has been questioned for several reasons, such as: economic loss (Liu *et al.*, 2008; Kludze *et al.*, 2013), environmental degradation (Campbell *et al.*, 1998; Liu *et al.*, 2008; Zang *et al.*, 2008; Awasthi *et al.*, 2010; Estrellan and Iino, 2010; Viana *et al.*, 2013), adverse health impact (Viana *et al.*, 2008; Awasthi *et al.*, 2010; Agarwal *et al.*, 2012) and loss of soil organic matter (Lal, 2008; Granged *et al.*, 2011) which may contribute to the overall deterioration of agricultural soil productivity.

High soil quality equates to high productivity without significant soil or environmental degradation. Soil quality describes the status or specific condition of the soil as a result of its management (Karlen *et al.*, 2003). Evaluation of soil quality is based on physical, chemical and biological characteristics, which vary as a function of the applied management, such as tillage, crop rotation, and the handling of crop residues (Carter, 1994). Management practices that simultaneously improve soil properties and yield are crucial to sustain high crop production and minimize detrimental impact on the environment. Crop residues are important in the maintenance and protection of soil quality (Kludze *et al.*, 2013), and the returning crop residue associated with some other agronomic practices, such as no tillage and crop rotation may improve soil quality and productivity, and may also be useful for the environment (Malhi *et al.*, 2006; Fuentes *et al.*, 2009).

Concerns regarding the effects of field burning crop residues have arisen in Romania in the past twenty years. Therefore, the Romanian government, as well as researchers in the field of agronomy, the organizations fighting for environment quality and protection, and the civil society are acting against the practice of field burning crop residues that has taken place with regaining ownership of agricultural land and with the agricultural land privatization. However, in Romania, in the last twenty years, crop residues were burned on the largest areas, as compared not only, but also with European countries, even considering that before 1990 such practice was very rare and required very well-grounded scientific and economic approaches. The environmental consequences of field burning residues are not seen only in terms of air pollution, but also in terms of intake decreasing of organic matter in the soil, that is important to increase soil fertility and to improve its physical, chemical and biological properties which consequently affect the agricultural productivity.

The purpose of this paper is to analyze, from an agronomic, economic and environmental perspective, the effects of crop residues management on the soil, agricultural productivity and environment, to reveal the need to limit the practice of field burning residues to strict needs and to illustrate the potential of crop residues to induce sustainable changes in dynamic soil quality (Franzluebbers, 2002). Also, the use of crop residues biomass as a source of energy is presented as an alternative. This is only the first step in the scientific approach of this problem. Several projects are ready for future research in this area.

Field burning crop residues: a Romanian study-case

Burning crop residues after harvest is becoming a very common practice in Romania (Fig. 1) since 1990. There are some reasons which have always been evocated by the land users to justify the field burning crop residues: (1) the lack of equipment to gather the crop residues and to transport and store them, especially for the small and medium farmers; (2) the diminishing of the livestock after 1990 (NIS, 2013) when Romania started the transition to the market economy; (3) the lake of information regarding the good practices especially for the small farmers; (4) the lake of interest for an alternative use of crop residues, especially during the period 1990-2000 or afterwards; (5) the argument brought up by most farmers (small or big farmers) that the soil tillage practice is of good quality if the crop residues are burned before. All these reasons are not sustainable, so the land managers need to know the effect of their management on soil ecosystem functioning, in addition to the effect of environmental factors beyond their control (Franzluebbers, 2002).

Fig. 1. Evening open field burning of crop residues in Giurgiu county, Romania. Photo source: Vasilica Stan and Françoise Picard-Bonnaud, 26th August 2013

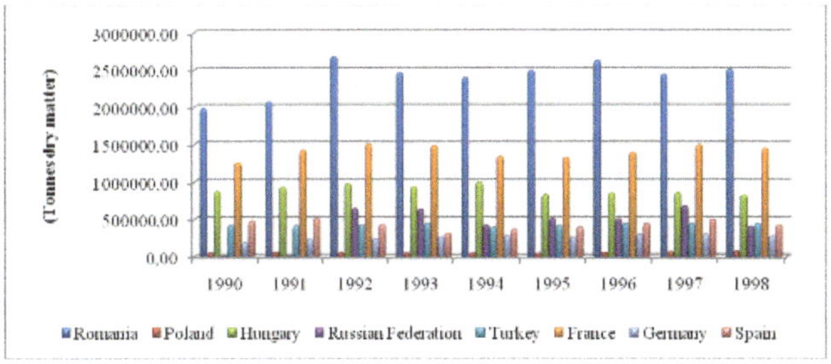

Fig. 2. a.The quantities of biomass burned in Romania compared with different European countries (1990-1998), FAO (2013), all GHG Crops + (Total)

In Fig. 2.a. and 2.b., and in Fig. 3 the annual quantities (tons of dry matter) of crop residues, according to FAO (2013), are given, which represent aggregated data that may include official, semi-official or estimated data between 1990 and 2010. Figs. 2.a. and 2b. (1990-2006), presents the situation before the accession to the European Union (EU), and Romania is compared with old European countries such as France, Germany, Spain where agriculture is well developed, with new European countries, such as Hungary or Poland, which are much closer to Romania from an economic point of view, and with non-EU countries, such as the Russian Federation and Turkey which also have a developed agriculture. In the 1990-2006 period, in Romania, large amounts of crop residues were burned (between 2 million and more than 2,5 million tons biomass dry matter) as compared to the above European countries. After 2007 and until 2010 (Fig. 3), the amount of crop residue burned in Romania still remained high in comparison to other countries, but a decrease of these quantities (below 2,0 million tons biomass dry matter) and a continuous downward trend occur, due to the new strategies and environmental policies.

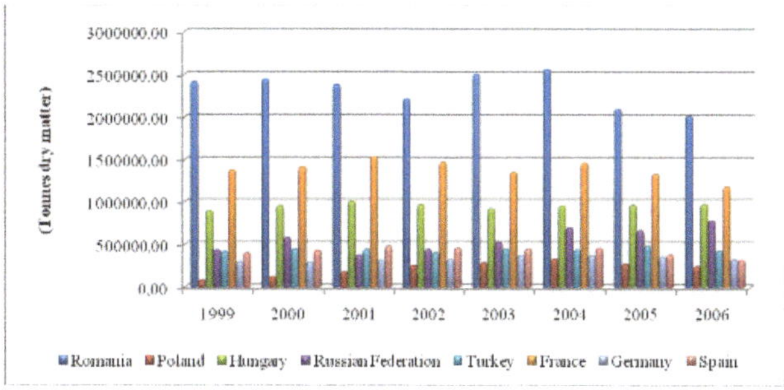

Fig. 2.b. - The quantities of biomass burned in Romania compared with different European countries (1999-2006), FAO (2013), all GHG Crops + (Total)

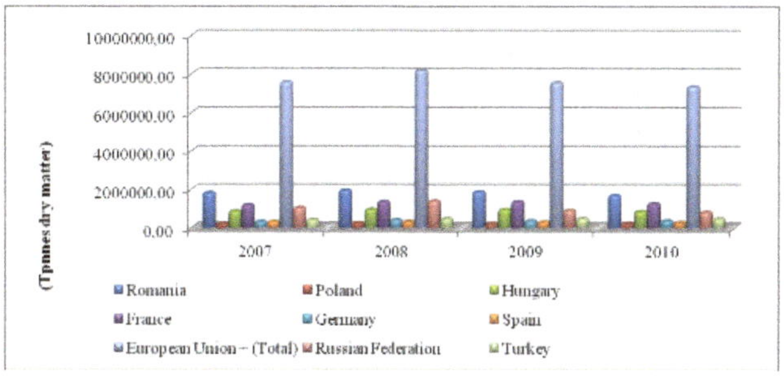

Fig. 3. The quantities of biomass burned in Romania after EU accession (2007-2010) compared with different European countries, FAO (2013), all GHG Crops + (Total)

The Soil Quality Monitoring of Romania (Tab. 1) shows that 62.24% of agricultural soils are characterized by low humus content, 21.34% by a medium content, 13.28% by either extremely-low and very low humus content and only 3.14% of soils are characterized by a high and an excessively high content. Generally, the reduced organic matter content is due to the slope processes and to the neglection of organic fertilization in the last 30-40 years, the agriculture being mostly based on chemical fertilization, and this, only at low rates. The content of humus that characterize the soil classes is presented in Tab. 2 (Canarache et al., 1987). Based on these considerations, a multicriteria analysis of the effects of field burning crop residues can be considered.

Tab. 1. The distribution of agricultural monitoring sample plots, level I by class of soil humus content (Dumitru *et al.*, 2000)

	No. Plots	Extremely low	Very low	Low	Medium	High	Very high	Extremely high	Excessive
					(%)				
Humus content	670	1.64	11.64	62.24	21.34	1.94	0.45	0.60	0.15

Tab. 2. Classes of soil humus content (Canarache *et al.*, 1987)

	Humus content for different texture classes (%)		
	Sand	Loamy	Clay
Extremely low	≤0.2	≤0.6	≤1.0
Very low	0.3-0.5	0.7-1.3	1.1-2.0
Low	0.6-1.0	1.4-3.0	2.1-5.0
Medium	1.1-2.0	3.1-6.5	5.1-10.0
High	2.1-5.0	6.6-10.5	10.1-16.0
Very high	5.1-8.7	10.6-13.9	16.1-21.0
Extremely high	8.8-20.0	14.0-26.5	21.1-35.0
Excessive	≥20.1	≥26.6	≥35.1

A multicriteria analysis of crop residues field burning effects

Concerns related to intensive agriculture and its effects on the environment have led many researchers, everywhere in the world, in the last 30-40 years, to study the effects of some agricultural practices, such as field burning crop residues, on soil or in the atmosphere (Malhi *et al.*, 2006; Singh *et al.*, 2007; Zhang *et al.*, 2008; Fuentes *et al.*, 2009; Soon and Lupwayi, 2012 etc.). Therefore, a multicriteria analysis of crop residues field burning effects will refer to soil properties (physical, chemical and biological), greenhouse gas emission and crop yields. In the same time, the energetic value of crop residues will be considered a very important criterion, as this is one of the most important alternatives to field burning residues.

The effects of residue removal on soil properties and environmental quality need to be investigated more vigorously, as it is essential for these effects to be carefully weighed on a site-specific basis when making a determination about residues removal (Kludze *et al.*, 2013).

Crop residues and soil properties

The vast majority of soil organic matter (SOM) originates from plant inputs, although this material may pass through several trophic levels prior to acquiring the characteristics of stable SOM (Johnson *et al.*, 2007). Soil C (soil carbon) sequestration benefits refers to: soil quality improvement, the increase of soil productivity, the reduction of soil erosion and sedimentation, the decrease of eutrophication and water contamination (Lal, 2007). Therefore, the soil deprivation of C and N (soil nitrogen) with removal of straw suggests that the practice of removing straw from fields for on-farm and industrial uses in the long run may result in soil degradation (Campbell *et al.*, 1998).

Many authors (Malhi *et al.*, 2006; Fuentes *et al.*, 2009; Verhulst *et al.*, 2011; Soon and Lupwayi, 2012; Li *et al.*, 2013) have reported the effects of crop residues when they were retained on the top soil, especially when this practice was combined with crop rotation and soil tillage (minimum tillage or no tillage/zero tillage). Moreover, incorporation of crop residues, such as straw, into the soil, builds up soil carbon (C) as well as soil nitrogen (N) and returns valuable nutrients to the ecosystem. Such build-up of soil C and N will not take place if the straw is removed from the soil (Nguyen *et al.*, 2013).

Management that causes a decline in soil quality reduces its functional abilities, whereas stewardship preserves these abilities. Given the resilience of nature, appropriate soil management techniques can be expected to restore the ecosystem functions once degraded (Franzluebbers, 2002). Retaining instead of burning residues provides several potential benefits (Hartemink, 2008; Rayment, 2003) such as improving soil physical, chemical and biological properties, and reducing fertiliser requirements through recycling nutrients in the residues, sequestering C etc. Successful integration of crop residue management strategies into cropping systems requires understanding of how crop residues influence cycling of nutrients from soil and fertilizers, as well as their effects on soil chemical, physical and biological properties, and crop production (Malhi *et al.*, 2006).

Effects of organic inputs over time on soil physical properties (soil aggregate stability, soil bulk density, water retention etc.) was reported by several authors (Zhang and Peng, 2006; Singh *et al.*, 2007; Fuentes *et al.*, 2009; Yao *et al.*, 2009). The organic products additions to the soil increase aggregate stability by a factor of 1.1-10.0, and this increase is related to the decomposition dynamics of the inputs (Abiven *et al.*, 2009). Soil aggregate stability is very important to avoid soil erosion. Malhi *et al.* (2006) reported that during a 4-year experiment, the addition of straw increased the proportion of larger aggregates by 3% for >38 mm and by 1% for 12.7-38.0 mm size and decreased the proportion of wind erodible aggregates by 1% for 0.42-0.83 mm and by 3% for <0.42 mm size. They also noted that the effects of straw retention (SR) on soil aggregate stability were improved when straw retention was combined with no tillage (NT) practice, which resulted in a lower proportion of wind-erodible aggregates (34%) and a higher proportion of large aggregates (37%), whereas conventional tillage (CT) + not straw retention (NSR) combination resulted in a great proportion of wind-erodible aggregates (50%) and a low proportion of large aggregates (18%). Therefore, an appropriate management of organic matter additions to soils may increase aggregate stability and thus, reduce crusting and erosion problems (Abiven *et al.*, 2009).

It was also shown that the increase of soil carbon content by using crop residues may restore damaged soil structure in the case of puddling in rice, and may increase the size and stability of aggregates, water retention and infiltration and decrease bulk density, dispersion ratio and soil strength (Singh *et al.*, 2007). Moreover, the results of a tillage/crop residue management study (Sparrow *et al.*, 2006) showed that wet aggregate stability was significantly higher with crop residues (16.2%) than without crop residues (13.8%) averaged across all tillage treatments.

Fuentes *et al.* (2009) reported that one of the benefits of retaining residues in the plots subjected to zero and conventional tillage was the reduction in both moisture spatial variability and soil mechanical resistance, but as no residues remained on the soil surface water flowed more easily and the surface was sealed because of the decreased aggregate stability. So, straw retention treatment combined with no tillage kept high soil moisture, rather than no straw retention treatment in the top soil (0-15 cm depth) as it was reported by Malhi and Lemke (2007), which is very important, especially for the arid areas.

Not only crop residues are a primary substrate for the replenishment of soil organic matter, but they also serve as an important source of plant nutrients (Lal, 1995). Crop residues have traditionally been incorporated into the soil as a source of nutrients and organic matter (Redy *et al.*, 2001; Soon and Lupwayi, 2012). Soil organic matter is an ecosystem component with agronomic and environmental functions and is affected by soil management (Franzluebbers, 2002; Diekow *et al.*, 2005). Crop residues are a renewable and vital organic resource for maintaining soil productivity. Intensive cropping with no return of crop residues and other organic materials results in loss of soil organic matter and that is not sustainable (Singh *et al.*, 2007). As it was already mentioned, build-up of organic matter in soil is a slow process and it takes many years to accumulate significant amounts of organic matter in soil (Malhi *et al.*, 2006) that are able to improve soil properties.

Crop residues, such as wheat or soybean residues, added either alone or in combination with P fertilizers, can improve soil P fertility by decreasing P sorption capacity and favouring a building-up in labile inorganic P, and moderately labile organic P (Reddy *et al.*, 2001). There are differences between residue types regarding nutrient content. As it was reported by Li *et al.* (2013), soybean residues have the highest total nitrogen concentration and the lowest C/N ratio, whereas maize residues show the opposite trend. The residue type and placement significantly affect cumulative C mineralization. The cumulative mineral N was higher in soils amended with soybean residues than in those amended with maize residues. For these reasons, crop rotation should be associated with the type of crop residues retention and no tillage, as another very important element for a sustainable soil management technique.

A significant increase of top soil organic carbon content, when crop residues were retained, was also reported by different authors: (Malhi *et al.*, 2006, Singh *et al.*, 2007; Cayuela *et al.*, 2009; Fuentes *et al.*, 2009; Thorburn *et al.*, 2012; Navarro-Noya *et al.*, 2013). Crop residue retention had a more pronounced effect on the soil bacterial communities when no tillage was applied. As it was reported by Navarro-Noya *et al.* (2013), retaining the crop residues significantly increased total organic carbon (TOC) and total N (TN) in soil, but decreased the pH. Also, regarding the soil pH, Fuentes *et al.* (2009) showed that when soil was cultivated with maize and wheat, the pH was significantly affected by treatments, but only in the first 5 cm layer. In conventional tillage with residue retention (CT+R), conventional tillage without residue retention (CT–R) and zero tillage (ZT) with residue retention (+R) treatments, pH ranged from 6.0 to 6.5.

Campbell *et al.* (1998) reported that 6 years of no tillage on a thin Black Chernozem in Saskatchewan, Canada, in a fallow-wheat-wheat rotation tended to reduce TOC and TN when straws were removed. At the end of four growing seasons, the mass of TOC and TN in the 0-15 cm soil depth tended to be higher, whereas light fraction organic matter (LFOM), light fraction carbon (LFC), and light fraction nitrogen (LFN) were significantly higher under straw retained (S) than under straw removed (NS) treatments. Compared to NS, the S treatment increased TOC by 3%, TN by 6%, LFOM by 23%, light fraction organic carbon (LFOC) by 37%, and LFN by 36% (Malhi *et al.*, 2006). After eight crop seasons, tillage and straws management had no significant effect on TOC and total organic N (TON) in the top 15 cm of soil, but the LFOC and LFON were higher with S than NS, and also higher under NT than under CT. Compared to NS, the S treatment increased LFOC by 24% and LFON by 10%. Increase in light fraction of organic C and N fractions due to straw retention was closely associated with greater input of C and N to soil through straw and chaff in the S treatments compared to NS treatments (Malhi and Lemke, 2007). Sparrow *et al.* (2006) also reported that carbon (C) biomass in the 0-10 cm depth increased with decreasing tillage intensity. Soil microbial biomass C and N concentrations were much higher at the surface 10 cm than in the 10–20 cm depth (519 mg C kg^{-1} and 39 mg N kg^{-1} in the 0-10 cm depth versus 229 mg C kg^{-1} soil and 13 mg N kg^{-1} soil in 10-20 cm depth). Microbial biomass C:N ratio in the 0-10 cm depth was also the highest in NT and the lowest in disked twice. Therefore, crop residues returned to crop lands can sustain soil organic carbon (SOC) content and improve soil fertility and biological activity (Cayuela *et al.*, 2009).

Field burning crop residues and the greenhouse gas emissions

Generally, agricultural activities are significant producers of CH_4 and N_2O (IPCC, 2007). Of the three main gases that are influenced by land management and that are responsible for the potential greenhouse effect, CO_2 has the greatest climate forcing potential (57%), while CH_4 and N_2O account for 27% and 16%, respectively (CAST, 1992). Modern agriculture contributes to the production of atmospheric greenhouse gases (GHG) with about 14% of global net CO_2 emissions (IPCC, 2007).

Field burning of crop residue converts a great deal of nutrients to gaseous form, which is then lost from the site

(Haider, 2013). It has been reported that field burning of crop residues represents a major source of greenhouse gase (CO_2, CO, CH_4, H2, CH_3Cl, NO, HCN, CH_3CN, COS, etc.) and aerosols (Crutzen *et al.*, 1979; Johnson *et al.*, 2007; Sahai *et al.*, 2007; Zhang *et al.*, 2008; Vadrevu *et al.*, 2013). In Asia, the annual contribution from open field burning of biomass is estimated to emitting 0.37 Tg of SO_2, 2.8 Tg of NO_x, 1100 Tg of CO_2, 67 Tg CO and 3.1 Tg methane (CH_4) (Streets *et al.*, 2003).

There are different alternatives to field burning crop residues (e.g. soil amendment, biomass feedstock for energy production etc.). As it was already argued, sequestration of C from plant biomass into soil organic matter (SOM) is a key sequestration pathway in agriculture. There are multiple benefits to sequestering C in forest and agricultural soils, beyond the obvious benefit of offsetting CO_2 emissions (Johnson *et al.*, 2007).

Crops residues and their effects on crop yields

From an economic point of view, crop residues, such as straw, are valuable, as they may contribute to improve soil organic matter and soil properties which may result in agricultural productivity (Kludze *et al.*, 2013). Crop residues can have significant effects on crop production, especially when their use is in association with other agricultural practices such as crop rotation, reducing the number of tillage or zero tillage. There are many ways in which crop residues may become benefic for the crop yields (e.g. soil moisture conservation, soil properties improving nutrient content etc.). In a Gray Luvisol, sandy clay loam soil and a second 4 years cycle rotation experiments, Malhy and Lemke (2007) reported that the beneficial effect of zero tillage (ZT) and straw return, on crop yield, is accomplished mainly through conservation of soil moisture in a dry year or in a dry region. Soil moisture content was higher in ZT than in conventional tillage (CT) by 1.5% in the 0-15 cm and by 1.9% in the 15-30 cm depth and also higher in retained straw (R) than in not retained straw (NR) by 1.9% in the 0-15 cm and by 2.6% in the 15-30 cm depth. Also, zero tillage practiced for 14 years, with crop residues retained in the field, resulted in a clayey Cumulic Phaeozem soil with a better quality and, in addition, producing higher wheat and maize yields than the plots subjected to conventional tillage (Fuentes *et al.*, 2009). Residues retention, on average, increased the wheat grain yield by 1.31 times (Bakht *et al.*, 2009). Regarding plant growth, zero tillage with residue retention was characterized by a slower initial growth than conventional tillage practices, but this was compensated by increased crop performance in the later stages, with a crucial influence on final grain yield (Verhulst *et al.*, 2011).

Crop residues and their energetic use

In 2010, about 62% of renewable energy in the European Union (EU) was generated from biomass (Szabó *et al.*, 2011). Agricultural biomass continues to gain attention as a source of alternative energy, given its potential ability to offset fossil fuels and reduce CO_2 emissions, while simultaneously providing an added source of income to farmers (Kludze *et al.*, 2013). The use of biomass as raw materials for bioenergy and biochemical production is encouraged by the need for a secure energy supply, a reduction of fossil CO_2 emissions and a revitalization of rural areas (Cherubini and Ulgiati, 2010).

Our strong dependence on fossil fuels results from the intensive use and consumption of petroleum derivatives which, combined with diminishing oil resources, cause environmental and political concerns (Cherubini and Ulgiati, 2010) and many problems such as energy crisis and global warming since the Industrial Revolution (Liu *et al.*, 2008). An alternative can be represented by lignocellulosic materials. In fact, lignocellulosic feedstocks can be supplied either from dedicated crops or as residues from agricultural, forestry and wood industry (Cherubini and Ulgiati, 2010). Biomass fuels (e.g. wood waste, crop residues, energy crops), in contrast, are considered renewable and carbon neutral (Nguyen *et al.*, 2013). This feedstock is made of three main components (cellulose, hemicellulose and lignin) which can be refined into different final products using a set of jointly applied technological processes. Among the different possible feedstocks, agricultural residues are a widespread lignocellulosic biomass source available in many countries (Cherubini and Ulgiati, 2010).

In areas where plants are cultivated on large surfaces, such as Romanian plain, there is a potential for removing straws or other crop residues for alternative uses, given the recent increasing interest for production of biofuels as a result of the EU Directive on the promotion of the use of energy from renewable sources (2009/28/EC), which sets an overall target of 20% renewable energy to be reached by 2020. In the framework of targets fixed by this directive, each member state had to adopt a National Renewable Energy Action Plan (NREAP) detailing the mandatory national targets in terms of sources and uses of renewable energy (Monforti *et al.*, 2013). The agricultural crop residues contribution to primary energy production has to be evaluated in the NREAPs template (European Commission, 2009).

Given the concentrated crop production regions in the Romanian South Plain or the Romanian West Plain, crop residues from corn (*Zea mays* L.), wheat (*Triticum aestivum* L.), rape (*Brassica rapa* L.), sunflower (*Hellianhus annuus* L.) and soybean (*Glycine max* L. Merr.) are considered to be particularly promising sources of biomass feedstock to replace conventional fuel. Employment of agricultural areas for the production of biomass crops, at the expense of food crops production and the burning of huge amounts of crop residues proves to be unprofitable.

Conclusions

In Romania, large amounts of crop residues are burned in open field. However, this is not a sustainable way to remove them from agricultural fields since it may affect environment (soil properties, air quality etc.), and agricultural productivity. In this study, based on current estimates of the quantity of crop residues burned in the open field, a multicriteria analysis of the negative effects on the environment has been presented and the benefits of incorporating crop residues in agricultural soils or their use for energy, have been given as alternatives. These alternatives can contribute to the increase of the sustainability of crop residue disposal and can reduce the risks posed by the open field burning of crop residues.

Increasing atmospheric emissions of greenhouse gases cause the biggest concerns on the effects of burning crop residues on the environment. This burning is associated also with the impoverishment of soil organic matter and consequently, with the degradation of the physical, chemical and biological properties of soil and its fertility reduction. Therefore, carbon sequestration in soil is a key solution to limit burning crop residues in the open field. Besides, using alternatives for agricultural crops residues, such as fuel production, is able to deliver new amounts of energy, but also to support farmers for land release and protect the environment.

To better control the effects of burning crop residues, as well as for the adoption of effective solutions, an up-to-date data of crop residue field burning and special investigations are necessary. Also, both the short and long-term effects of crop residues burning on soil properties and environmental quality need to be investigated more rigorously.

References

Abiven S, Menasseri S, Chenu C (2009). The effects of organic inputs over time on soil aggregate stability-A literature analysis. Soil Biology and Biochemistry 41:1-12.

Agarwal R, Awasthi A, Singh N, Gupta PK, Mittal SK (2012). Effects of exposure to rice-crop residue burning smoke on pulmonary functions and Oxygen Saturation level of human beings in Patiala (India). Science of the Total Environment 429:161-166.

Awasthi A, Singh N, Mittal S, Gupta PK, Agarwal R (2010). Effects of agriculture crop residue burning on children and young on PFTs in North West India. Science of the Total Environment 408:4440-4445.

Bakht J, Shafi M, Jan MT, Shah Z (2009). Influence of crop residue management, cropping system and N fertilizer on soil N and C dynamics and sustainable wheat (*Triticum aestivum* L.) production. Soil and Tillage Research 104:233-240.

Cayuela L, Sinicco T, Mondini C (2009). Mineralization dynamics and biochemical properties during initial decomposition of plant and animal residues in soil. In: Li LJ, Han X-Z, You M-Y, Yuan Y-R, Ding X-L, Qiao Y-F (2013). Carbon and nitrogen mineralization patterns of two contrasting crop residues in a Mollisol: Effects of residue type and placement in soils. European Journal of Soil Biology 54:1-6.

Campbell CA, Selles F, Lafond GP, McConkey BG, Hahn D (1998). Effect of crop management on C and N in long-term croprotations after adopting no-tillage management: comparison ofsoil sampling strategies. In: Malhi SS, Lemke R, Wang ZH, Chhabra BS (2006). Tillage, nitrogen and crop residue effects on crop yield, nutrient uptake, soil quality, and greenhouse gas emissions. Soil and Tillage Research 90:171-183.

Canarche A, Florea N, Râşnoveanu A, Latiş L (1987). Soil Survey Methodology Elaboration, 3rd Part, Pedological Indicators, Bucharest 106 p.

Carter MR (1994). Strategies to overcome impediments to adoption of conservationtillage. In: Carter MR (Ed.), Conservation Tillage in Temperate Agroecosystems. In: Fuentes M, Govaerts B, De Leónc F, Hidalgo C, Dendooven L, Sayre KD, Etchevers J (2009). Fourteen years of applying zero and conventional tillage, crop rotation andresidue management systems and its effect on physical and chemicalsoil quality. European Journal of Agronomy 30:228-237.

CAST (1992). Preparing U.S. Agriculture for Global Climate Change. Task Force Report No. 119. Council for Agricultural Science and Technology. In: Johnson JM-F, Franzluebbers AJ, Weyers SL, Reicosky DC (2007). Agricultural opportunities to mitigate greenhouse gas emissions. Environmental Pollution 150:107-124.

Cherubini F, Ulgiati S (2010). Crop residues as raw materials for biorefinery systems – A LCA case study. Applied Energy 87:47-57.

Commission Regulation (EC) No 1122/2009 of 30 November 2009 laying down detailed rules for the implementation of Council Regulation (EC) No 73/2009 as regards cross-compliance, modulation and the integrated administration and control system, under the direct support schemes for farmers provided for that Regulation, as well as for the implementation of Council Regulation (EC) No 1234/2007 as regards cross-compliance under the support scheme provided for the wine sector.

Council Regulation (EC) No 73/2009 of 19 January 2009 establishing common rules for direct support schemes for farmers under the common agricultural policy and establishing certain support schemes for farmers, amending Regulations (EC) No 1290/2005, (EC) No 247/2006, (EC) No 378/2007 and repealing Regulation (EC) No 1782/2003.

Crutzen PJ, Heidt LE, Krasnec JP, Pollock WH, Seiler W (1979). Biomass burning as a source of the atmospheric gases CO, H2, NwO, NO, CH3C1 and COS. In: Vadrevu KP, Giglio L, Justice C (2013). Satellite based analysis of fire-carbon monoxide relationships from forest and agricultural residue burning (2003-2011). Atmospheric Environment 64:179-191.

Diekow J, Mielniczuk J, Knicker H, Bayer C, Dick DP, Kögel-Knabner I (2005). Soil C and N stocks as affected by cropping systems and nitrogen fertilisation in a southern Brazil Acrisolmanaged under no-tillage for 17 years. Soil and Tillage Research 81:87-95.

Dumitru M, Dumitru E, Gament E, Cârstea S, Ciobanu C, Cojocaru G, Plaxienco D, Manea A, Calciu I, Motelică DM, Enache R, Radnea C, Vrânceanu N, Mashali AM (2000). Soil Quality Monitoring in Romania. GNP Publishing, Bucharest, 53p.

Estrellan CR, Iino F (2010). Toxic emissions from open burning. Chemosphere 80:193-207.

European Commission (EC). Commission decision of 30.6.2009 establishing a template for National Renewable Energy Action

Plans under Directive 2009/28/EC.

FAO (2013). FAOSTAT. Burning Crop Residues metadata, http://faostat.fao.org/site/738/

Franzluebbers AJ (2002). Soil organic matter stratification ratio as an indicator of soil quality. Soil and Tillage Research 66:95-106.

Fuentes M, Govaerts B, De Leónc F, Hidalgo C, Dendooven L, Sayre KD, Etchevers J (2009). Fourteen years of applying zero and conventional tillage, crop rotation and residue management systems and its effect on physical and chemical soil quality. European Journal of Agronomy 30:228-237.

Gonçalves C, Evtyugina M, Alves C, Monteiro C, Pio C, Tomé M (2011). Organic particulate emissions from field burning of garden and agriculture residues. Atmospheric Research 101:666-680.

Granged AJP, Zavala LM, Jordán A, Bárcenas-Moreno G (2011). Post-fire evolution of soil properties and vegetation cover in a Mediterranean heathland after experimental burning: A 3-year study. Geoderma 164:85-94.

Haider MZ (2013). Determinants of rice residue burning in the field. Journal of Environmental Management 128:15-21.

Hartemink AE (2008). Sugarcane for bioethanol – soil and environmental issues. In: Thorburn PJ, Meier EA, Collins K, Robertson FA (2012). Changes in soil carbon sequestration, fractionation and soil fertility in response to sugarcane residue retention are site-specific. Soil and Tillage Research 120:99-111.

IPCC (2007). Assessment Report 4. In: Cooper JM, Butler G, Leifert C (2011). Life cycle analysis of greenhouse gas emissions from organic and conventional food production systems, with and without bio-energy options. NJAS - Wageningen Journal of Life Sciences 58:185-192.

Johnson JM-F, Franzluebbers AJ, Weyers SL, Reicosky DC (2007). Agricultural opportunities to mitigate greenhouse gas emissions. Environmental Pollution 150:107-124.

Karlen DL, Ditzler C, Andrews SS (2003). Soil quality: why and how? In: Fuentes M, Govaerts B, De Leónc F, Hidalgo C, Dendooven L, Sayre, KD, Etchevers J (2009). Fourteen years of applying zero and conventional tillage, crop rotation andresidue management systems and its effect on physical and chemicalsoil quality. European Journal of Agronomy 30:228-237.

Kludze H, Deen B, Weersink A, Van Acker R, Janovicek K, De Laporte A, McDonald I (2013). Estimating sustainable crop residue removal rates and costs based on soil organic matter dynamics and rotational complexity. Biomass and Bioenergy 56:607-618.

Kutcher HR, Malhi SS (2010). Residue burning and tillage effects on diseases and yield of barley (*Hordeum vulgare*) and canola (*Brassica napus*). Soil & Tillage Research 109:153-160.

Lal R (1995). The role of residues management in sustainable agricultural systems. In: Redy DD, Rao AS, Singh M (2001). Crop residue addition effects on myriad forms and sorbtion of phosphorus in a Vertisol. Bioresource Technology 80:92-99.

Lal R (2007). Carbon management in agricultural soils. Mitigation and Adaptation Strategies for Global Change. In: Johnson JM-F, Franzluebbers AJ, Weyers SL, Reicosky DC (2007). Agricultural opportunities to mitigate greenhouse gas emissions. Environmental Pollution 150:107-124.

Lal MM 2008. An overview to agricultural waste burning. Indian Journal of Air Pollution Control 8(1):48-50.

Li LJ, Han X-Z, You M-Y, Yuan Y-R, Ding X-L, Qiao Y-F (2013). Carbon and nitrogen mineralization patterns of two contrasting crop residues in a Mollisol: Effects of residue type and placement in soils. European Journal of Soil Biology 54:1-6.

Liu H, Jiang GM, Zhuang HY, Wang KJ (2008). Distribution, utilization structure and potential of biomass resources in rural China: With special references of crop residues. Renewable and Sustainable Energy Reviews 12:1402-1418.

Malhi SS, Lemke R, Wang ZH, Chhabra BS (2006). Tillage, nitrogen and crop residue effects on crop yield, nutrient uptake, soil quality and greenhouse gas emissions. Soil and Tillage Research 90:171-183.

Malhi SS, Lemke R (2007). Tillage, crop residue and Nitrogen effects on crop yield, nutrient uptake, soil quality and nitrous gas emissions in a second 4-yr rotation cycle. Soil and Tillage Research 96:269-283.

Ministry of Agriculture and Rural Development, Romanian Agency for Payements and Interventions in Agriculture (2012) – Guide for farmers regarding the compulsory cross-compliance, 2nd Version.

Monforti F, Bódis K, Scarlat N, Dallemand J-F (2013). The possible contribution of agricultural crop residues to renewable energy targets in Europe: A spatially explicit study. Renewable and Sustainable Energy Reviews 19:666-677.

Navarro-Noya YE, Gómez-Acata S, Montoya-Ciriaco N, Rojas-Valdez A, Suárez-Arriaga MC, Valenzuela-Encinas C, Jiménez-Bueno N, Verhulst N, Govaerts B, Dendooven L (2013). Relative impacts of tillage, residue management and crop-rotation on soil bacterial communities in a semi-arid agroecosystem. Soil Biology and Biochemistry 65:86-95.

Nguyen TLT, Hermansen JE, Mogensen L (2013). Environmental performance of crop residues as an energy source for electricity production: The case of wheat straw in Denmark. Applied Energy 104:633-641.

NIS, 2013. Annual yearbook of the Romanian National Institut of Statistics for the period 1991-2009.

Pintilie C, Romoşan Şt, Pop L, Timariu Gh, Sebök P, Guş M (1985). Agrotehnică. Ediţia a doua. Editura Didactică şi Pedagogică, Bucureşti.

Rayment GE (2003). Water quality in sugar catchment of Queensland. In: Thorburn PJ, Meier EA, Collins K, Robertson FA (2012). Changes in soil carbon sequestration, fractionation and soil fertility in response to sugarcane residue retention are site-specific. Soil and Tillage Research 120:99-111.

Redy DD, Rao AS, Singh M (2001). Crop residue addition effects on myriad forms and sorbtion of phosphorus in a Vertisol. Bioresource Technology 80:92-99.

Sahai S, Sharma C, Singh DP, Dixit CK, Singh N, Sharma P, Singh K, Bhatt S, Ghude S, Gupta V, Gupta RK, Tiwari MK, Garg SC, Mitra AP, Gupta PK (2007). A study for development of emission factors for trace gases and carbonaceous particulate species from in situ burning of wheat straw in agricultural fields in India. Atmospheric Environment 41:9173-9186.

Singh G, Jalota SK, Singh Y (2007). Manuring and residue management effects on physical properties of a soil under the rice-wheat system in Punjab, India. Soil and Tillage Research 94:229-238.

Soon YK, Lupwayi NZ (2012). Straw management in a cold semi-arid region: Impact on soil quality and cropproductivity. Field Crops Research 139:39-46.

Sparrow SD, Lewis CE, Knight CW (2006). Soil quality response to tillage and crop residue removal under subarctic conditions. Soil and Tillage Research 91:15-21.

Streets DG, Yarber KF, Woo J-H, Carmichael GR (2003). Biomass burning in Asia: annual and seasonal estimates and atmospheric emissions. In: Gadde B, Bonnet S, Menke C, Garivait S (2009). Air pollutant emissions from rice straw open field burning in India, Thailand and the Philippines. Environmental Pollution 157:1554-1558.

Szabó M, Jaeger-Waldau A, Monforti F, Scarlat N, Bloem J, QuicheronM, *Huld T,* Ossenbrink H (2011). Technical assessment of the Renewable Action Plans. Eur 24926 EN, ISBN 978-92-79-21049-5.

Thorburn PJ, Meier EA, Collins K, Robertson FA (2012). Changes in soil carbon sequestration, fractionation and soil fertility in response to sugarcane residue retention are site-specific. Soil and Tillage Research 120:99-111.

Vadrevu KP, Giglio L, Justice C (2013). Satellite based analysis of fire-carbon monoxide relationships from forest and agricultural residue burning (2003-2011). Atmospheric Environment 64:179-191.

Verhulst N, Govaert B, Nelissen V, Sayre KD, Crossa J, Raes D, Deckers J (2011). The effect of tillage, crop rotation and residue management on maize and wheat growth and development evaluated with an optical sensor. Field Crops Research 120:58-67.

Viana, M, López JM, Querola X, Alastueya A, García-Gaciob D, Blanco-Herasb G, López -Mahía P, Piñeiro-Iglesiasc M, Sanz MJ, Sanz F, Chie X, Maenhaute W (2008). Tracers and impact of open burning of rice straw residues on PM in Eastern Spain. Atmospheric Environment 42:1941-1957.

Viana M, Reche C, Amato F, Alastuey A, Querol X, Moreno T, Lucarelli F, Nava S, Calzolai G, Chiari M, Rico M (2013). Evidence of biomass burning aerosols in the Barcelona urban environment during winter time. Atmospheric Environment 72:81-88.

Zhang B, Peng X-H (2006). Organic matter Enrichment and Aggregate Stabilization in a Severely Degraded Ultisol After Reforestation. Pedosphere 16(6):699-706.

Zhang H, Ye X, Cheng T, Chen J, Yang X, Wang L, Zhang R (2008). A laboratory study of agricultural crop residue combustion in China: Emission factors and emission inventory. Atmospheric Environment 42:8432-8441.

Zhang Q, Huang J, Yu G (2008). Polychlorinated dibenzo-p-dioxins and dibenzofurans emissions from open burning of crop residues in China between 1997 and 2004 Environmental Pollution 151:39-46.

Yao S, Qin J, Peng X, Zhang B (2009). The effects of vegetation on restoration of physical stability of a severely degraded soil in China. Ecological Engineering 35(5):723-734.

7

Genetic Diversity and Relationships in Local Varieties of Eggplant from Different Cultivar Groups as Assessed by Genomic SSR Markers

Santiago VILANOVA, Maria HURTADO, Adriana CARDONA, Mariola PLAZAS, Pietro GRAMAZIO, Francisco J. HERRAIZ, Isabel ANDÚJAR, Jaime PROHENS*

*Instituto de Conservación y Mejora de la Agrodiversidad Valenciana, Universitat Politècnica de València, Camino de Vera 14, 46022 Valencia, Spain; jprohens@btc.upv.es (*corresponding author)*

Abstract

Spain is a secondary center of diversity for eggplant (*Solanum melongena*). Spanish landraces of eggplant are normally classified in four cultivar groups: Round, Listada de Gandía, Semi-Long, and Long. We have used 19 genomic SSRs for the molecular characterization of 30 eggplant accessions corresponding to the four cultivar groups. Sixteen SSRs of which 15 were polymorphic could be amplified and 65 polymorphic alleles, with a range of two to 11 alleles/locus, were detected. The polymorphism information content (*PIC*) of SSR markers ranged from 0.07 to 0.77, with an average value of *PIC*=0.50. The mean observed heterozygosity (H_o) presented a very low value H_o=0.01, while the mean expected heterozygosity (H_e) had a value of H_e=0.57. Multivariate cluster analyses revealed that a considerable diversity exists within each of the cultivar groups. Listada de Gandía and Long cultivar groups were clearly separated from each other in different branches of phenogram. The principal coordinates analysis (PCoA) confirmed that each of the cultivar groups is genetically diverse and, with the exception of the Round group, they plot in different areas of the PCoA graph. Overall, the results indicate that Spanish eggplant landraces present a high degree of homozygosis, considerable intra-cultivar group diversity, and a certain degree of genetic differentiation. This information is of interest for selection and breeding of eggplant as well as for germplasm conservation.

Keywords: forage, breeding, cultivar groups, fruit shape, landraces, multivariate analysis, *Solanum melongena*

Introduction

Eggplant (*Solanum melongena* L.) was domesticated in Southeast Asia (Meyer *et al.* 2012). From there it spread to other tropical and subtropical areas of the world, where selection and other microevolutive forces led to the development of a wide array of locally adapted landraces of eggplant and to the emergence of secondary centers of diversity (Cericola *et al.*, 2013; Hurtado *et al.*, 2012; Prohens *et al.*, 2005; Tümbilen *et al.*, 2011). Differentiation between eggplants of Southeast Asia on one side and those of the Middle East, Africa and Mediterranean region of Europe on the other have led to the recognition of two major groups of eggplant cultivar groups denominated, respectively, Oriental and Occidental eggplants (Cericola *et al.*, 2013; Vilanova *et al.*, 2012).

The Mediterranean region of Spain is considered as a secondary center of diversity (Hurtado *et al.*, 2012; Prohens *et al.*, 2005) in which four cultivar groups are normally recognized: Round, Listada de Gandía, Semi-Long, and Long (Hurtado *et al.*, 2013; Muñoz-Falcón *et al.*, 2011; Prohens *et al.*, 2005). These four cultivar groups are distinguished mostly by the fruit length/width ratio (i.e., the Round, Semi-Long, and Long cultivar groups) (Cericola *et al.*, 2013; Hurtado *et al.*, 2013; Prohens *et al.*, 2005; Tümbilen *et al.*, 2011), and also for the presence of purple stripes on a white background combined with a fruit shape intermediate between the Round and Semi-Long types in the case of Listada de Gandía (Hurtado *et al.*, 2013; Muñoz-Falcón *et al.*, 2011).

Development of new eggplant varieties addressing old and new breeding objectives (Barchi *et al.*, 2012; Lebeau *et al.*, 2013; Sunseri *et al.*, 2003) requires of genetic diversity. Normally development of new commercial varieties, is based on intra-varietal group crossings, except in the case of the Semi-Long type, which can also be obtained by crossings between Round and Long types. Increasing the genetic base of new eggplant cultivars can be achieved through the incorporation of local landraces in the commercial breeding programmes (Muñoz-Falcón *et al.*, 2009a). Therefore, the study of genetic diversity and relationships of collections of local varieties provides information of relevance for the breeding programmes.

Several molecular studies (Cericola *et al.*, 2013; Prohens *et al.*, 2005; Tümbilen *et al.*, 2011) have shown that eggplant cultivar groups are genetically diverse. In the case of

Spanish varieties, Prohens *et al.* (2005) studied the AFLP diversity in a collection of eggplants from the Round, Listada de Gandía, Semi-Long, and Long types and found considerable intra-group diversity and genetic differentiation among groups. However, SSRs have proved as more powerful than AFLPs to study the relationships amongst closely related eggplant materials (Muñoz-Falcón *et al.*, 2009b). Furthermore, unlike AFLPs, which are dominant, SSRs are co-dominant, which allows determining the levels of observed heterozygosity. In particular, in eggplant, genomic SSRs have proved as much more informative than EST-SSRs (Muñoz-Falcón *et al.*, 2011). Therefore, in order to obtain information of interest for eggplant breeding and germplasm conservation, we studied the diversity, heterozygosity and relationships of 30 eggplant varieties from the Region of Valencia, situated in the Mediterranean region of Spain using genomic SSR markers.

Materials and methods

Plant material and DNA extraction

A total of 30 accessions corresponding to four cultivar groups (Round, Listada de Gandía, Semi-Long, and Long) were used (Tab. 1). All the accessions used, except the commercial selection Listada Clemente (Semillas Clemente, Vitoria, Spain), and the breeding line LF3-24 (INRA, France) are local landraces from the region of Valencia (Spain). A number of these accessions have been recently characterized by fruit shape (Hurtado *et al.*, 2013), and phenolics content (Plazas *et al.*, 2013).

For each accession, genomic DNA was extracted from 75 mg of young leaf tissue using the CTAB method (Doyle and Doyle, 1987). DNA concentration was quantified, after electrophoresis on a 1.0% agarose gel, using a Nanodrop ND-1000 (Nanodrop Technologies, Wilminton, Delaware, USA) spectrophotometer. Samples were adjusted to a DNA concentration of 20 ng/μl. The quality of DNA was evaluated through the 260/280 nm and 260/230 nm absorbance ratios (Sambrook *et al.*, 1989).

Molecular characterization

Nineteen genomic highly polymorphic SSR markers developed by Vilanova *et al.* (2012) were used to screen the 30 eggplant accessions (Tab. 2). SSRs were tested following the M13-tail PCR method of Schuelke *et al.* (2000) to facilitate the incorporation of a dye label during the PCR. An M13-tailed forward primer was used in combination with a standard M13 primer dye-labeled with FAM, NED, PET, or VIC fluorophores at its 5'-end.

PCR amplifications were performed in a total volume of 12 μl with 20 ng of DNA, 1.5 mM $MgCl_2$, 0.05 μM of forward primer, 0.25 μM of reverse primer, 0.2 μM of fluorescent M-13 primer, 0.2 mM dNTPs, and 0.04 units of Taq DNA polymerase. Amplifications were carried out in an Eppendorf Mastercycler ep gradient S (Eppendorf AG, Hamburg, Germany) thermocycler. Two different protocols for amplification (T or TD) were used depending on the marker (Tab. 2). The T protocol consisted in an initial step at 94 °C for 5 min, 35 cycles of 94 °C for 30 s, 1 min at the appropriated annealing temperature (Tab. 2), 72 °C for 2 min, and final 10 min extension at 72 °C. The TD protocol consisted in an initial step at 94 °C for 2 min, 7 cycles of 94 °C for 15 s, 55-48 °C (beginning with 55 °C in the first cycle and reducing 1 °C in each of the subsequent cycles) for 30 s, 72°C for 45 s, and final 10 min extension at 72 °C; subsequently, other 28 cycles are performed with the following conditions: 94 °C for 15 s, 48 °C for 30 s, 72°C for 45 s, and final 10 min extension at 72 °C.

PCR products were separated on an ABI Prism 3100 Avant (Applied Biosystems, Foster City, California, USA) genetic analyzer using GeneScan 3.7 (Applied Biosystems)

Tab. 1. Accessions used in the present work, fruit weight and fruit length/width ratio (mean±SE), grouped according to their varietal group

Accession[a]	Code	Fruit weight	Fruit length / width ratio
		Round	
B-31	B31	225±17	1.09±0.05
B-32	B32	346±33	1.63±0.05
V-S-9	VS9	233±18	0.83±0.02
V-S-13	VS13	180±12	0.97±0.03
		Listada de Gandía	
07-A25-01	07A	330±26	2.31±0.08
IVIA-025	I025	330±16	1.70±0.06
IVIA-347	I347	300±26	1.35±0.06
Listada Clemente	LC	209±23	2.47±0.07
V-S-1	VS1	239±32	1.59±0.07
V-S-2	VS2	278±15	1.60±0.05
V-S-7	VS7	346±20	1.47±0.06
V-S-8	VS8	288±24	2.36±0.10
V-S-10	VS10	169±12	1.56±0.07
V-S-11	VS11	182±9	1.87±0.05
V-S-15	VS15	228±12	1.86±0.03
		Semi-long	
B-33	B33	171±8	2.93±0.07
B-36	B36	176±16	2.99±0.12
V-S-14	VS14	250±16	2.88±0.09
V-S-16	VS16	290±19	2.61±0.35
V-S-17	VS17	228±20	2.99±0.12
V-S-18	VS18	168±14	3.47±0.11
		Long	
B-35	B35	218±15	3.98±0.22
LF3-24	LF3	181±13	3.99±0.12
V-S-3	VS3	164±12	3.92±0.13
V-S-4	VS4	206±18	3.84±0.14
V-S-5	VS5	232±19	4.43±0.13
V-S-6	VS6	223±13	4.02±0.13
V-S-12	VS12	200±12	5.36±0.11
V-S-19	VS19	252±15	4.52±0.11
V-S-21	VS21	301±24	3.88±0.23

[a]All accessions are local landraces of the Region of Valencia (Spain), except Listada Clemente, which is a commercial selection (Semillas Clemente, Vitoria, Spain) of the Listada de Gandía type, and LF3-24, which is a breeding line from INRA (France).

Tab. 2. SSR markers used in the present study along with their repeat motif, amplification protocol, annealing temperature, and linkage group in which they map (Vilanova *et al.*, 2010, 2012)

SSR locus	Repeat motif	Protocol[a]	Annealing temperature	Linkage group
CSM4	$(GA)_{15}$	TD	55-48	8
CSM7	$(CT)_{10}$	TD	55-48	Unknown
CSM12	$(AG)_{12}$	T	51	6
CSM27	$(GA)_{23}$	T	51	3
CSM29	$(AG)_{17}$	TD	55-48	12
CSM30	$(CT)_{20}$	T	51	9
CSM31	$(AG)_{28}$	T	51	1
CSM32	$(AG)_{23}$	T	51	4
CSM36	$(GA)_{27}$	T	51	9
CSM40	$(CT)_{45}$	T	51	4
CSM43	$(AG)_{14}$	TD	55-48	1
CSM44	$(AG)_{14}$	T	51	3
CSM45	$(AG)_{16}$	TD	55-48	5
CSM52	$(TC)_{12}$	T	50	Unknown
CSM54	$(GA)_{19}$	T	51	9
CSM57	$(CT)_{8}$	T	51	Unknown
CSM62	$(GA)_{27}$	T	51	Unknown
CSM74	$(GA)_{26}$	TD	55-48	12
CSM78	$(CT)_{19}$	TD	55-48	10

[a]See text for technical details of the amplification protocols

software. SSR alleles were precisely sized using GeneScan 500 Liz (Applied Biosystems) molecular size standards with Genotyper 3.7 software (Applied Biosystems).

Data analysis

For each SSR locus, the number of polymorphic alleles (N_a), frequency of the predominant allele (f), and effective number of alleles (N_e) was determined using the PowerMaker software (Liu and Muse, 2005).

The polymorphism information content (PIC) was calculated as $PIC = 1 - \sum_{i=1}^{n} p_i^2 - \sum_{i=i+1}^{n} 2p_i^2 p_i^2$, where n is the total number of alleles detected, p_i is the frequency of the ith allele, and p_j is the frequency of the jth allele (Botstein *et al.*, 1980). Also, the observed heterozygosity (H_o), expected heterozygosity (H_e), calculated as $He = 1 - \sum_{i=1}^{n} p_i^2$, were determined. Nei and Li (1979) genetic similarities were calculated and a neighbor-joining phenogram was built using genetic distances with the PowerMaker software (Liu and Muse, 2005) and plotted using TreeView software (Page, 1996). Genetic distances were also used to graphically represent genetic relationships among accessions by principal coordinates analysis (PCoA) using GenAlEx 6.5 software (Peakall and Smouse, 2012).

Results and discussion

SSR characterization and diversity

The 19 SSR markers could be amplified, but for three of them (CSM36, CSM44, and CSM78) the PCR products could not be successfully resolved. The 16 remaining SSR markers were polymorphic with the exception of CSM32, for which only one allele was detected. A total of 65 alleles were detected for the 15 polymorphic SSR loci, with an average number of alleles/locus (N_a) of 4.33 and a range between two (CSM52) and 11 (CSM31) alleles (Tab. 3).

The frequency of the predominant allele (f_p) ranged between 0.36 (CSM31) and 0.97 (CSM74), although with the exception of the latter SSR locus (CSM74) in all cases the f_p values have been below 0.65 (Tab. 3). The effective number of alleles (N_e) ranged between 1.07 (CSM74) and 4.85 (CSM31), and with the exception of CSM74 and

Tab. 3. SSR polymorphic markers, number of alleles per locus (N_a), frequency of the predominant allele (f), number of effective alleles per locus (N_e), polymorphism information content (PIC), observed heterozygosity (H_o), and expected heterozygosity (H_e) in the studied collection of 30 eggplant accessions

SSR locus	N_a	f	N_e	PIC	H_o	H_e
CSM4	6	0.43	3.03	0.61	0.04	0.67
CSM7	3	0.55	2.21	0.46	0.00	0.55
CSM12	3	0.50	2.14	0.42	0.00	0.53
CSM27	4	0.57	2.47	0.54	0.00	0.59
CSM29	3	0.56	2.31	0.49	0.00	0.57
CSM30	3	0.54	2.12	0.42	0.00	0.53
CSM31	11	0.36	4.85	0.77	0.00	0.79
CSM40	6	0.39	3.32	0.65	0.00	0.70
CSM43	3	0.59	2.04	0.41	0.00	0.51
CSM45	3	0.52	2.14	0.42	0.00	0.53
CSM52	2	0.58	1.95	0.37	0.00	0.49
CSM54	5	0.37	3.74	0.69	0.00	0.73
CSM57	5	0.63	2.a25	0.52	0.04	0.55
CSM62	6	0.38	3.38	0.65	0.00	0.70
CSM74	2	0.97	1.07	0.06	0.00	0.06
Mean	4.33	0.53	2.60	0.50	0.01	0.57

CSM52, which are the only polymorphic markers with N_a=2 (and therefore N_e<2), all loci had N_e>2 (Tab. 3).

The average value for the *PIC* value of the SSR markers tested was of 0.50, but the *PIC* value of individual SSR markers ranged between 0.07 (CSM74) and 0.77 (CSM31) (Tab. 3). The mean value for the observed heterozygosity (H_o) was very low (H_o=0.01), corresponding to H_o=0.00 values for 13 out of the 15 polymorphic SSR loci and to low values (H_o=0.04) for the two remaining loci (CSM4 and CSM57). Conversely, the mean value for the expected heterozygosity (H_e) was much higher (H_e=0.57), with values for individual SSR loci ranging from 0.06 (CSM74) to 0.79 (CSM31).

Multivariate analyses

The cluster analysis performed showed that the four cultivar groups present a considerable diversity (Fig. 1). However, a clear separation of the Long and Listada de Gandía accessions in different basal branches of the phenogram was observed (Fig. 1). All the Listada de Gandía accessions, except accession VS11, are clustered together in one of the sub-branches of the phenogram. Also, all Long accessions, except B35, are clustered together in another sub-branch in which there are also a Round (B32) and a Semi-Long (B33) accessions. Semi-Long, and particularly, the Round accessions, do not present this pattern of clustering, and are scattered in different branches of the phenogram.

When considering the principal coordinates analysis (PCoA), the first and second principal coordinates account, respectively, for 47 % and 13 % of the total variation. The representation of accessions in the PCoA graph shows that the cultivar groups are diverse, although each of them plot in different parts of the graph (Fig. 2).

In this respect, all Listada de Gandía accessions present positive values of the first coordinate and either low positive or negative values of the second coordinate and plot together, not being intermingled with other cultivar groups. Only the odd VS11 accession plots closer to a Round accession (VS9) than to the other Listada de Gandía accessions. All Long accessions present negative values for the first coordinate and either low positive or negative values for the second coordinate (Fig. 2).

Also, all Semi-Long accessions plot in the same area of

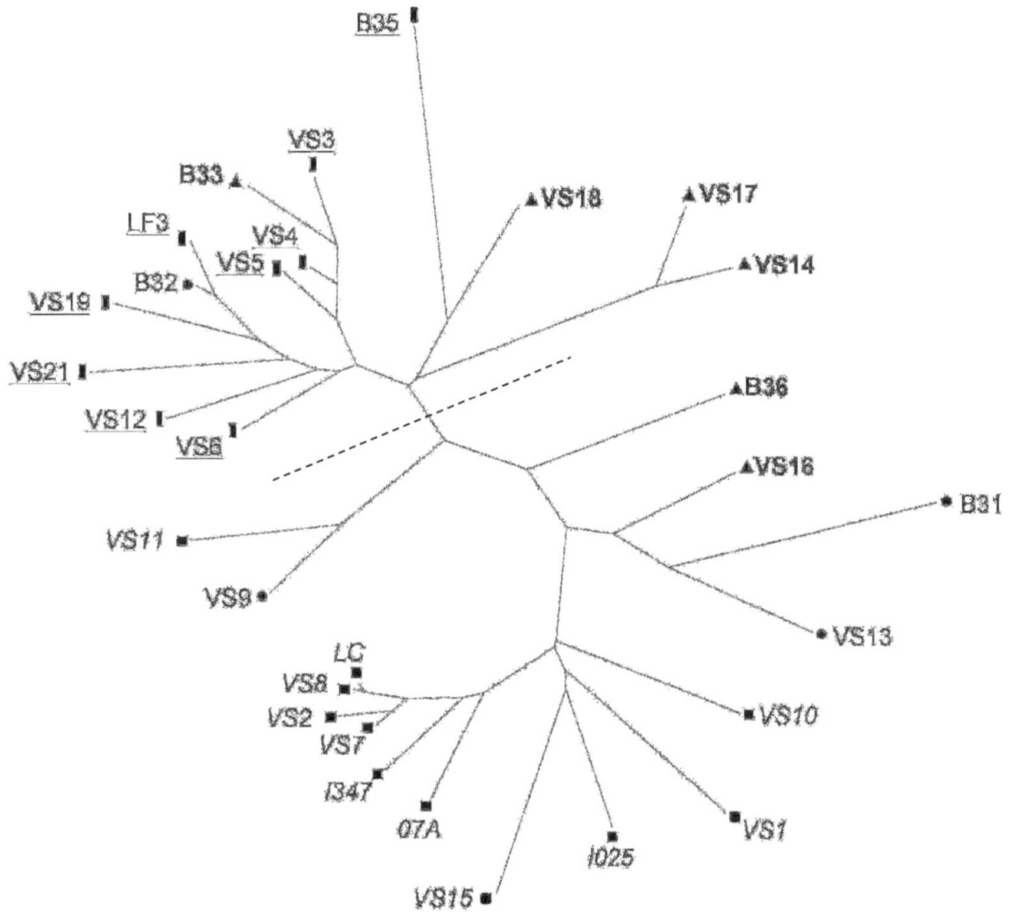

Fig. 1. Unrooted neighbor-joining phenogram of 30 eggplant accessions based on 15 polymorphic SSR markers. Phenetic relationships were derived from genetic distances (Nei and Li, 1979). The different groups of accessions are represented by different symbols and font types: Round (● and normal font); Listada de Gandía (▪ and italics font); Semi-long (▲ and bold font); Long (▌ and underlined font). The dashed line separates the phenogram branches that contain all Long accessions (above) and all the Listada de Gandía accessions

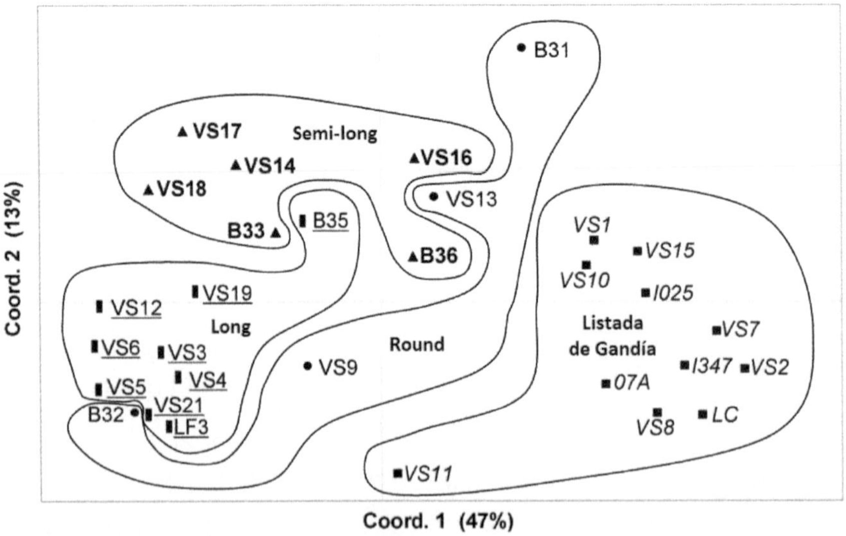

Fig. 2. Relationships between the 30 eggplant accessions based on 15 polymorphic SSRs according to the first and second principal coordinates obtained from a principal coordinates analysis (PCoA) (47% and 13% of the total variation accounted by the first and second coordinates, respectively). The different groups of accessions are represented by different symbols and font types: Round (● and normal font); Listada de Gandía (▪ and italics font); Semi-long (▲ and bold font); long (▌ and underlined font)

the PCoA graph, with either low positive or negative values for the first coordinate and positive values for the second coordinate. Conversely to the other cultivar groups, the four Round accessions are scattered in different parts of the PCoA graph, with two accessions presenting positive values for both the first and second coordinates and situated close to Semi-Long accessions, and two accessions with negative values for both coordinates, with one accession (B32) close to Long accessions and the other (VS9) close to the odd VS11 Listada de Gandía accession (Fig. 2).

Discussion

Genomic SSRs that previously proved to be highly polymorphic in eggplant (Vilanova *et al.*, 2012) have been found to be of great value for evaluating the genetic diversity and relationships in a collection of eggplants from different cultivar groups. The screening of this collection has revealed a high degree of polymorphism for the SSR markers tested, confirming that the Mediterranean region is a secondary center of diversity for eggplant (Cericola *et al.*, 2013; Hurtado *et al.*, 2012; Prohens *et al.*, 2005). Muñoz-Falcón *et al.* (2011) have shown that in eggplant genomic SSR markers are more polymorphic than EST-SSRs. Our results seem to confirm these results and that the genomic SSRs developed from an enriched genomic library by Vilanova *et al.* (2012) are particularly useful for the study of relationships in germplasm collections.

The same SSRs used by us were tested by Vilanova *et al.* (2012) in a collection of 22 *S. melongena* from different origins, including Occidental and Oriental types of eggplant. For the 15 polymorphic loci used in our study these authors found an average number of alleles per locus (6.47) higher than ours (4.33). Hurtado *et al.* (2012) in a study of 52 accessions from China, Spain, and Sri Lanka shared seven of the SSR markers that have been polymorphic in our study and also found a higher average number of alleles per locus (8.86) than we found for these seven SSR markers in our study (4.43). Also, Cericola *et al.* (2013) in a wide study of 238 eggplant materials from different origins shared four SSRs with our study. For these four SSRs, Cericola *et al.* (2013) found on average 7 alleles/locus, while these same alleles in our collection presented a mean of 5.5 alleles/locus.

The higher diversity found by Vilanova *et al.* (2012), Hurtado *et al.* (2012), and Cericola *et al.* (2013) probably is a consequence of the fact that our collection comes from a single geographic area (Region of Valencia, in the Mediterranean coast of Spain). In fact, when considering only the 14 Spanish accessions used in the Hurtado *et al.* (2012) study the number of alleles per locus (3.86) is similar to ours (4.43). Amazingly, for the CSM32 loci, which has been monomorphic in our collection, Hurtado *et al.* (2012) and Vilanova *et al.* (2012) found 12 and 8 alleles, respectively, indicating that selection or genetic drift in the materials of the collection we have evaluated have led to fixation of one specific allele.

SSR loci are highly informative when they present a great number of alleles (N_a), and have low values for the predominant allele frequency (f_p), i.e., with f_p values close to the theoretical minimum of $f_p=1/N_a$ (Botstein *et al.*, 1980; Powell *et al.*, 1996). In our case, some markers have presented a high number of alleles and presented relatively low values for the predominant allele (f_p), resulting in high values for the number of effective alleles (N_e) polymorphic information content (*PIC*), and expected heterozygosity (H_e) confirming that highly relevant information on the diversity and relationships of eggplant can be obtained with a relatively low number of genomic SSR markers (Hurtado *et al.*, 2012; Muñoz-Falcón *et al.*, 2009, 2011).

The low values of observed heterozigosity (H_o) were

expected as eggplant is fundamentally autogamous (Pessarakli and Dris, 2004), and the materials used are non-hybrid. In this respect, Muñoz-Falcón (2009) also found very low values for H_o in eggplant landraces (H_o<0.03), but substantially higher values (H_o=0.38) for commercial F1 hybrids. Cericola *et al.* (2013) also found that most eggplant landraces used in their study had low heterozygosity values, and only 38 out of 238 materials had H_o>0.10. The high level of homozygosity in eggplant landraces shows that pure lines can easily be derived by individual selection from these materials.

The multivariate analysis with SSR markers shows that, as already found by Prohens *et al.* (2005) and Tümbilen *et al.* (2011) with AFLP markers, and by Cericola *et al.* (2013) that considerable genetic diversity exists within each of the cultivar groups studied, which were mostly distinguished by the fruit shape. Also, Muñoz-Falcón *et al.* (2011) using SSR markers also found that one of the groups studied here (Listada de Gandía) was genetically diverse.

The existence of intra-varietal group genetic diversity has important implications for selection and breeding as it indicates that important genetic advances can be obtained with intra-cultivar group selection, and also that hybrids heterotic for yield may be obtained when crossing genetically different accessions of the same cultivar group (Rodríguez-Burruezo *et al.*, 2008).

Here we have found that a certain degree of genetic differentiation exists among the four cultivar groups. This is in agreement with previous reports (Cericola *et al.*, 2013; Prohens *et al.*, 2005; Tümbilen *et al.*, 2011), in which cultivar groups were found to present a moderate degree of differentiation. However, the Round group accessions were intermingled in the multivariate analyses with accessions of other groups indicating that this cultivar group is genetically highly variable. Wild relatives of eggplant have round, ovoid, or obovoid fruit shape (Knapp *et al.*, 2013), indicating that the fruit shape characteristic of the Round cultivar group (Hurtado *et al.*, 2013) is an ancestral trait probably present in the first domesticated eggplants. As occurred in tomato (Brewer *et al.*, 2007), artificial selection of mutations affecting fruit shape has led to other fruit shapes, like those characteristic of the Listada de Gandía, Semi-Long and Long types, which may have undergone genetic bottlenecks resulting in a lower diversity and higher degree of genetic differentiation. However, further studies should be undertaken to confirm this hypothesis.

In our study, the Long and Listada de Gandía groups are clearly differentiated at the genetic level. In this respect, Muñoz-Falcón *et al.* (2011) and Prohens *et al.* (2005) found that the Listada de Gandía cultivar group was genetically clearly differentiated from the rest of eggplant materials of other types. In our study, only one of the Listada de Gandía accessions used (VS11) seems to be genetically closer to one Round accession (VS9) that to other Listada de Gandía materials. This accession is characterized by an odd shape for the Listada de Gandía type, as it is ovoid instead of having the normal obovoid shape characteristic of the Listada de Gandía materials (Hurtado *et al.*, 2013) and may have been derived from introgression of the striped trait characteristic of Listada de Gandía into a different genetic background.

Conclusions

A reduced number of selected genomic SSR markers have allowed detecting considerable genetic variation in a collection of eggplants including different cultivar types. These markers have also been useful for studying relationships among four cultivar groups differentiated by fruit shape, showing that they are highly homozygous, have considerable intra-varietal group diversity, and present a certain degree of genetic differentiation. In particular the Listada de Gandía and Long varietal groups are clearly differentiated from each other. This information is of interest for the genetic improvement and conservation of genetic resources of eggplant.

Acknowledgements

This research has been partially funded by Ministerio de Economía y Competitividad and FEDER (grant AGL2012-34213) and by Universitat Politècnica de Valencia (grants SP20120681 and PAID-06-11 Nr. 2082).

References

Botstein D, White RI, Skolnich M, Davis RW (1980). Construction of a genetic linkage map in man using restriction fragment polymorphism. Amer J Human Genet 32:324-331.

Barchi L, Lanteri S, Portis E, Valè G; Volante A; Pulcini L, Ciriaci T, Acciarri N, Barbierato V, Toppino L, Rotino GL (2012). A RAD tag derived marker based eggplant linkage map and the location of QTLs determining anthocyanin pigmentation. PLoS ONE 7:43740.

Brewer MT, Moyseenko JB, Monforte AJ, van der Knaap E (2007). Morphological variation in tomato: A comprehensive study of quantitative trait loci controlling fruit shape and development. J Exp Bot 58:1339-1349.

Cericola F, Portis E, Toppino L, Barchi L, Acciarri N, Ciriaci T, Sala T, Rotino GL, Lanteri S (2013). The population structure and diversity of eggplant from Asia and the Mediterranean basin. PLoS ONE 8:73702.

Dice LR (1945). Measures of the amount of ecologic association between species. Ecology 26:297-302.

Doyle JJ, Doyle JL (1987). A rapid DNA isolation procedure for small quantities of fresh leaf tissue. Phytochem Bull 19:11-15.

Hurtado M, Vilanova S, Plazas M, Gramazio P, Fonseka HH, Fonseka R, Prohens J (2012). Diversity and relationships of eggplants from three geographically distant centers of diversity. PLoS ONE 7:41748.

Hurtado M, Vilanova S, Plazas M, Gramazio P, Herraiz FJ, Andújar I, Prohens J (2013). Phenomics of fruit shape in eggplant (*Solanum melongena* L.) using Tomato Analyzer software. Sci Hort 164:625-632.

Knapp S, Vorontsova MS, Prohens J (2013). Wild relatives of eggplant (*Solanum melongena* L.: Solanaceae): New understanding of species names in a complex group. PLoS ONE 8:57039.

Lebeau A, Gouy M, Daunay MC, Wicker E, Chiroleu F, Prior P,

Frary A, Dintinger J (2013). Genetic mapping of a major dominant gene for resistance to *Ralstonia solanacearum* in eggplant. Theor Appl Genet 126:143-158.

Liu K, Muse S (2005). PowerMarker: an integrated analysis environment for genetic marker analysis. Bioinformatics 21:2128-2129.

Meyer RS, Karol KG, Little DP, Nee MH, Litt A (2012). Phylogeographic relationships among Asian eggplants and new perspectives on eggplant domestication. Mol Phylogenet Evol 63:685-701.

Muñoz-Falcón JE, Prohens J, Vilanova S, Nuez F (2009a). Diversity in commercial varieties and landraces of black eggplants and implications for broadening the breeders' gene pool. Ann Appl Biol 154:453-465.

Muñoz-Falcón JE, Prohens J, Vilanova S, Ribas F, Castro A, Nuez F (2009b). Distinguishing a protected geographical indication vegetable (*Almagro* eggplant) from closely related varieties using morphological traits and molecular markers. J Sci Food Agric 89:320-328.

Muñoz-Falcón JE, Vilanova S, Plazas M, Prohens J (2011). Diversity, relationships, and genetic fingerprinting of the *Listada de Gandía* eggplant landrace using genomic SSRs and EST-SSRs. Sci Hort 129:238-246.

Nei M, Li W.H. (1979). Mathematical modelfor studying genetic variation in terms of restriction endonucleases. Proc Natl Acad Sci USA 76:5269-5273.

Page RDM (1996). TreeView: An application to display phylogenetic trees on personal computers. Comp Appl Biosci 12:357-358.

Peakall R, Smouse PE (2006). GenAlEx 6.5: Genetic analysis in Excel. Population genetic software for teaching and research - an update. Bioinformatics 28:2537-2539.

Pessarakli MM, Dris R (2004). Pollination and breeding of eggplant. J Food Agric Environ 2:218-219.

Plazas M, López-Gresa MP, Vilanova S, Torres C, Hurtado M, Gramazio P, Andújar I, Herráiz FJ, Bellés JM, Prohens J (2013). Diversity and relationships for key traits for functional and apparent quality in a collection of eggplant: Fruit phenolics content, antioxidant activity, polyphenol oxidase activity, and browning. J Agric Food Chem 61:8871-8879.

Powell W, Machray GC, Provan J (1996). Polymorphism revealed by simple sequence repeats. Trends Plant Sci 1:215-222.

Prohens J, Blanca JM, Nuez F (2005). Morphological and molecular variation in a collection of eggplants from a secondary center of diversity: Implications for conservation and breeding. J Amer Soc Hort Sci 130:54-63.

Rodríguez-Burruezo A, Prohens J, Nuez F (2008). Performance of hybrids between local varieties of eggplant (*Solanum melongena*) and its relation to the mean of parents and to morphological and genetic distances among parents. Eur J Hort Sci 73:76-83.

Sambrook J, Fritsch EF, Maniatis T (1989). Molecular cloning: A laboratory manual, 2nd ed. Cold Spring Harbor Laboratory Press, Cold Spring Harbor, NY, USA, p. 1626.

Schuelke M (2000). An economic method for the fluorescent labeling of PCR fragments. Nature Biotechnol 18:233-234.

Sunseri F, Sciancalepore A, Martelli G, Acciarri N, Rotino GL, Valentino D, Tamietti G (2003). Development of RAPD-AFLP map of eggplant and improvement of tolerance to *Verticillium* wilt. Acta Hort 625:107-115.

Tümbilen Y, Frary A, Mutlu S, Doganlar S (2011). Genetic diversity in Turkish eggplant (*Solanum melongena*) varieties as determined by morphological and molecular analyses. Intl Res J Biotechnol 2:16-25.

Vilanova S, Blasco M, Hurtado M, Muñoz-Falcón JE, Prohens J, Nuez F (2010). Development of a linkage map of eggplant based on a *S. incanum* x *S. melongena* backcross generation, p. 435-439. In: Prohens J, Rodríguez-Burruezo A (Eds.). Advances in genetics and breeding of *Capsicum* and eggplant. Ed. Universitat Politècnica de València, Valencia, Spain.

Vilanova S, Manzur JP, Prohens J (2012). Development and characterization of genomic simple sequence repeat markers in eggplant and their application to the study of diversity and relationships in a collection of different cultivar types and origins. Mol Breed 30:647-660.

8

Phenolic Profile of the Kernel of Selected Persian Walnut (*Juglans regia* L.) Cultivars

Géza BUJDOSÓ[1,2*], György VÉGVÁRI[3*], Veronika HAJNAL[1], Gitta FICZEK[1], Magdolna TÓTH[1]

[1]*Corvinus University of Budapest, Department of Pomology, 29-43, Villányi út, Budapest, Hungary; geza.bujdoso@uni-corvinus.hu, gitta.ficzek@uni-corvinus.hu, veronika.hajnal@uni-corvinus.hu, magdolna.toth@uni-corvinus.hu*
[2]*National Agricultural Research and Innovation Centre, Fruitculture Research Institute, 2, Park utca, Budapest, Hungary; resinfru@yahoo.com*
[3]*Corvinus University of Budapest, Department of Soil Science and Water Management, 29-43, Villányi út, Budapest, Hungary; gyorgy.vegvari@uni-corvinus.hu (*corresponding authors)*

Abstract

The Persian walnut (*Juglans regia* L.) also known as English, Carpathian or common walnut is an important fruit species for Central and Eastern Europe, as its productivity increases annually due to its good market situation. Phenolic compounds of fifteen Persian walnut cultivars and two selected genotypes were examined by HPLC in this research paper. If we rank the phenolic compounds of the examined cultivars based on their quantity into increasing relative order, the quantity of Vanillic acid was the highest compound followed by Cathechin, Pyrocathechin, Epicathechin, Rutin, Syringic acid, Gallic acid, Juglone and Cinnamic acid. As results it can be considered that the Hungarian bred walnut cultivars had higher phenolic content compared to foreign cultivars. Among foreign bred cultivars evaluated in this paper the Serbian cultivars 'Srem' and 'Sampion' had the best results.

Keywords: dried fruit samples, HPLC analysis, Hungarian bred cultivars, Serbian bred cultivars, ten phenolic compounds, US bred cultivars

Introduction

The Persian walnut (*Juglans regia* L.) plays an important role within the fruit producing industry of Central and Eastern European countries and the production is increasing yearly because of favourable markets. Ukraine has the largest Persian walnut dried in-shell production in the region with about 80,000 metric tons produced annually, followed by Romania with approximately 35,000 tons, Serbia with more or less 22,000 tons and Hungary with around 5,500 tons (FAO, 2013). The growers in Romania, Serbia and Hungary use state registered cultivars, while the Ukrainian growers mostly collect walnut from the seedling populations.

Walnut growing is a major tree crop in Central Europe with active breeding programs supporting this industry. In Hungary the Persian walnut breeding started at the Fruit Growing Research Institute in 1950, and from this program eight walnut cultivars are registered on the Hungarian National Variety List. In Romania new Persian walnut cultivars were either selected from local populations or by cross breeding (Botu *et al.*, 2010). In Serbia the "breeding" method was through selection by organizing a competition "Searching for quality walnut" (Korac *et al.*, 1986).

The walnut kernel has a high oil content (45-72% of the kernel weight), that consists mainly of unsaturated fatty acids. In addition to oil, the kernel contains proteins (13-25% of the kernel weight), available carbohydrate (4-15% of the kernel weight), fibre (4.6-7.5% of the kernel weight), and minerals (1.65-2.4% of the kernel weight) (Bujdosó *et al.*, 2010; Germain *et al.*, 1999). Studies of Bujdosó *et al.* (2010) and Solar *et al.* (2006) on walnut fruit quality concluded that the environment plays a major role to influence the ratio of these compounds of kernels yearly.

In recent studies reports on the phenolic profiles of walnut trees describe the total phenolic content in different parts of walnut, namely, shoots (Cosmulescu and Trandafir, 2011a; Solar *et al.*, 2006), leaves, and the green husk (Cosmulescu and Trandafir, 2011b; Jakopic *et al.*, 2007; Stampar *et al.*, 2006; Cosmulescu and Trandafir, 2012). Some phenolic compounds protect Persian walnut against bacterial and fungal diseases (Mikulic-Petkovsek *et al.*, 2011). Chevallier *et al.* (2009) stated that flavonoids and hydroxi-cinnamic acid are the main responsible compounds of different walnut cultivars and genotypes susceptibility or tolerance against *Xanthomonas arboricola* pv. *juglandis*. Furthermore, the extract of fresh walnut leaves inhibited selectively the growth of Gram positive bacteria (*Bacillus cereus, B. subtilis, Staphylococcus aureus*), but Gram negative bacteria (*Pseudomonas aeruginosa, Escherichia coli, Klebsiella pneumoniae*) and fungi (*Candida albicans, Cryptococcus neoformans*) were resistant to the extract at 100 mg/mL (Pereira *et al.*, 2007).

The five major phenolic compounds in the walnut kernel are gallic and cinnamic acid, catechin, juglone and rutin. Gallic acid is the most readily bio available compound for the human body, and is followed by cathechin

(Williamson and Manach, 2007). Cinnamic acid appears to improve glycaemic parameters in diabetic patients suffering from type two diabetes inconsistently (Rudkowska, 2009). A few studies show that juglone has anti-cancer activity against human tumour lines and has anti-cancer effect against chemo/radio resistant melanomas among *in vivo* conditions (Babich and Stern, 1993; Bonifazi *et al.*, 2010). Rutin has many physiological functions and is known for its anti-oxidant activity (Harborne and Williams, 2000). Other bioactive phenols include vanillic acid, epicathechin, cinnamic acid that are characteristic for higher plants.

The phenolic compounds in the seed coat are higher compared to the walnut flesh. For example, total phenolic content of a dehulled 'Chandler' kernel was 16,300 μg/g in the kernel and 403 000 μg/g in the seed coat (Labucas *et al.*, 2008). In addition, Colaric *et al.* (2005) measured a 38-fold higher phenolic content in the pellicle of 'Alsószentiváni 117' (727 mg/g) than in the walnut kernel (19.16 mg/g), thus the pellicle is much better source of phenolic compounds than the kernel.

This investigation is a follow-up from previous results to refine the analysis of the walnut kernel contents. The aim of the study was to identify and quantify the phenolic compounds in the kernel of a few selected Hungarian, Serbian, US, and French bred walnut cultivars using HPLC.

Materials and methods

Fruit samples

The fruit samples were collected in fall of 2012, and Tab. 1 shows the walnut cultivars with their most important characteristics included in this study. Fruit of the eight state registered Hungarian bred Persian walnut cultivars ('Alsószentiváni 117', 'Milotai 10', 'Tiszacsécsi 83', 'Milotai bőtermő', 'Milotai kései', 'Milotai intenzív', 'Bonifác', 'Alsószentiváni kései') and two selected genotypes (BD06, red kernel genotype) were examined in the trial. Fruit of state registered cultivars and selected genotypes were collected from experimental orchards of the NARIC Fruitculture Research Institute (average annual sunshine hours are 2,094 hours, average annual temperatures are 11.1 °C, average annual precipitation is 560 mm, type of the soil is Chernozem soil Ka= 40, pH=8, lime content 5%, humus content 2.3%).

Fruit samples of the American bred 'Chandler', and 'Hartley' and the French bred 'Fernor' cultivars were collected at Juglans Hungaria Ltd. in Lengyeltóti (Hungary) (average annual sunshine hours are 2,098 hours, average annual temperatures are 11.2 °C, average annual precipitation is 550 mm, type of the soil is Chernozem soil Ka= 38, pH=8, lime content 5%, humus content 1.9%). The Serbian bred cultivars were grown near Senta (Serbia) (average annual sunshine hours are 2,104 hours, average annual temperatures are 11.5 °C, average annual precipitation is 580 mm, type of the soil is Chernozem soil Ka= 42, pH=7.5, lime content 2.5%, humus content 2.0%).

In order to analyze the chemical content of the fruit, the husks were removed immediately after the harvest. The samples were washed and dried up to 10% of moisture content and were stored as a walnut in its shell at +8 °C until further processed in the laboratory. The harvest time was when 50% of husks had been opened.

Chemicals

Analytical HPLC grade standards of different phenolic compounds such as juglone [481-39-0], gallic acid [149-91-7], chlorogenic acid [327-97-9], pyrocatechin [120-80-9], catechin [154-23-4], syringic acid [530-57-4], rutin [153-18-4], cinnamic acid [140-10-3], epicathechin [154-234-], vanillic acid [121-34-6], and the solvents phosphoric acid and methanol (MeOH), were purchased from Sigma Aldrich Chemical Co. (St. Louis, MO, USA). The standards (0.5 g/ml) were dissolved in methanol and a 100× dilution was used as the working standard for HPLC.

Tab. 1. Walnut cultivars and genotypes* included in the research

Cultivar	Origin	Ripening time (in decade)	Fruit size (mm)	Kernel ratio (%)	Kernel colour
'Alsószentiváni 117'	Open-pollinated selection from Hungary	IX. 1st d	33-36	50	Light brown
'Milotai 10'	Selection from Hungarian population	IX. 2nd d	32-34	48	Yellowish brown
'Tiszacsécsi 83'	Selection from Hungarian population	IX. 3rd d	32-34	50	Yellowish brown
'Milotai bőtermő'	'Milotai 10' x 'Pedro', Hungary	X. 1st d	34-36	49	Yellowish brown
'Milotai intenzív'	'Milotai 10' x 'Pedro', Hungary	IX. 3rd d	32-34	52	Yellowish brown
'Milotai kései'	'Milotai 10' x 'Pedro', Hungary	X. 1st d	32-34	44	Light brown
'Bonifác'	'A 117' x 'Pedro', Hungary	X. 1st d	32-34	46	Light brown
'Alsószentiváni kései'	'A 117' x 'Pedro' Hungary	X. 1st d	32-34	48	Yellowish brown
'BD06'*	Selection from Hungarian population	IX. 1st d	32-34	46	Yellowish brown
Red kernel genotype*	Selection from Hungarian population	IX. 3rd d	32	44	Red
'Chandler'	'Pedro' x 'UC 56-224' (bred in USA)	IX 3rd d	28-30	49	Light brown
'Fernor'	'Franquette' x 'Lara' (bred in France)	X. 1st d	31-34	40-44	Yellowish brown
'Hartley'	Landscape selected from Napa Valley (USA)	X. 2nd d	28-30	42	Yellowish brown
'Sampion'	Selection from Serbian population	X. 2nd d	32-34	58	Light brown
'Srem	Selection from Serbian population	X. 2nd d	32-34	57	Light brown

Sources: Szentiványi, 2006; Cerovic *et al.*, 2010; Hendricks *et al.*, 1998.

Preparation of samples

Whole kernels with their seed coats (1000 g) were used as starter material and 1 g samples were extracted in 10 ml methanol for 12 hours in the dark at 4 °C, using an Edmund Bühler SM 30 control shaker (200 rpm/min). The supernatant was decanted and centrifuged in Eppendorf tubes in a Hettich Mikro 22R centrifuge (15000 rpm/min for 5 min). The supernatant was filtered on a 0.45 µm MILLEX®-HV Syringe Driven Filter Unit (SLHV 013 NL, PVDF Durapore), purchased from Millipore Co. (Bedford, MA, USA), and injected into the HPLC system. The quantities of the individual phenolic compounds are given in µg/g.

HPLC conditions

The WATERS High Performance Liquid Chromatograph (Waters Co., 34 Maple Street, Milford, MA, USA) was equipped with an absorbance detector (2487 Dual λ), a binary HPLC pump (1525), and in-line degasser, a column thermostat (set at 40 °C) and an 717plus auto sampler (set at 5 °C) and was controlled using EMPOWERTM2 software. A KINETEX C18 2.6µm 150×4.6 mm column (Phenomenex 411 Madrid Avenue Torrance, CA 90501-1430 USA) was installed, and the gradient mobile phase was A: H_2O: MeOH: H_3PO_4=940:50:1, B: MeOH (0-30 min: A 100%-10%, 30-30.1 min: 10%-100%, 30.1-31: A 100%) with a flow rate 1 cm^3/min, the pressure in the column was 4200±10 psi at a column temperature of 30°C. The running time was 22 minutes. Each injected volume was 20 µl.

The sampling rate was 10pt/sec, and the phenolic components were monitored at a wavelength of 280nm (gallic acid, pyrocatechin, catechin, syringic acid, rutin, cinnamic acid, vanillic acid, epicathechin) and 350 nm (juglone). The retention times of the standards were gallic acid (7.7 min), pyrocatcechin (7.5 min), cathechin 9.6 min, chlorogenic acid (11.4 min), vanillic acid (11.8 min), epicatchechin (12.7 min), syringiyc acid (12.9 min), rutin (18.5 min), juglone (20.6 min) and cinnamic acid (21.6 min) respectively.

Statistical analysis

Three replicates were made in all cases. The statistical analysis was performed using PASW STATISTICS 18.0. Univariate analysis was carried out to separate the homogeneous groups using Duncan's test. The RSD value of the determination was 5% (n=5). The results were expressed as µm of 1 g of dried weight kernel.

Results and discussions

The phenolic compound profiles of the fifteen tested walnut cultivars are shown in Figs. 1 to 9. Six of the nine tested phenolic compounds were present in all fifteen cultivars, and varied in their concentration levels. The phenolic compounds not present in all the cultivars are catechin (Fig. 3), vanillic acid (Fig. 4), epicatechin (Fig. 5), and chlorogenic acid (not shown). The gallic acid concentration was outstandingly high in 'Milotai bőtermő' (601 µm/g) followed by 'Alsószentiváni 117' (422 µm/g), red kernel genotype (399 µm/g) and 'Milotai kései' (380 µm/g). 'Hartley' (37 µm/g) and 'Milotai intenzív' (51 µm/g) had the lowest gallic acid content (Fig. 1).

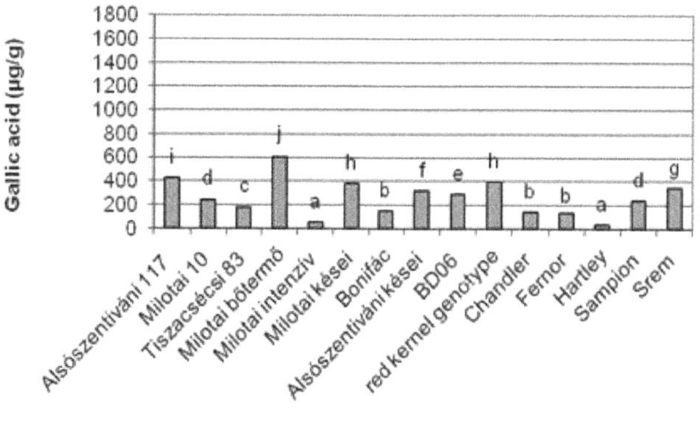

Fig. 1. Gallic acid content of kernels in analysed cultivars ($SD_{5\%}$=70)

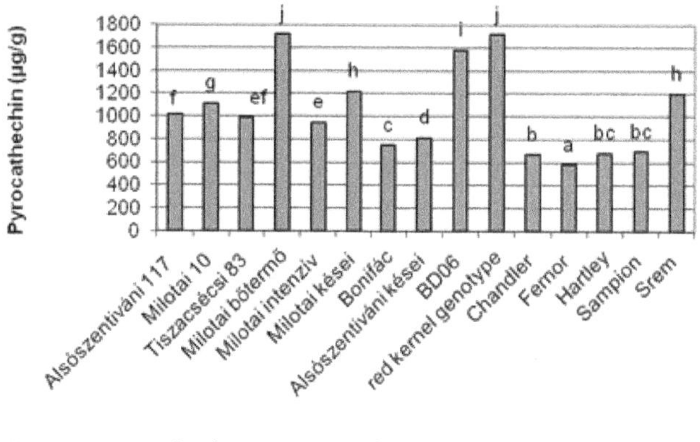

Fig. 2. Pyrocathechin content of kernels in analysed cultivars ($SD_{5\%}$=58)

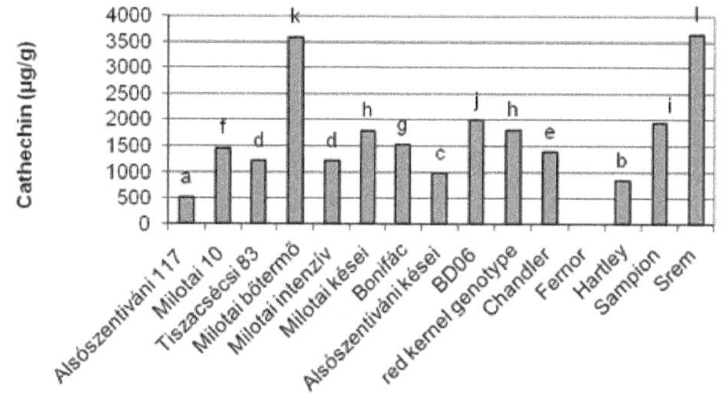

Fig. 3. Cathechin content of kernels in analysed cultivars ($SD_{5\%}$=212)

Red kernel genotype (1,717 µm/g), 'Milotai bőtermő' (1,715 µm/g) and 'BD06' (1,580 µm/g) genotype had the largest pyrocathecin concentration in our trial; the lowest ones were produced by foreign bred cultivars except 'Srem' (1 202 µm/g) (Fig. 2).

The Serbian bred 'Srem' (3 646 µm/g) and the Hungarian bred 'Milotai bőtermő' (3 587 µm/g) had 2 to 7-fold higher cathechin content compared to other Persian walnut cultivars examined in the trial. Cathechin content was not present in French bred 'Fernor' (Fig. 3).

Chlorogenic acid content was also examined in the samples but red kernel genotype contained this compound only, which quantity was equal to 428 µg/ of kernel.

Both Serbian bred cultivars ('Srem', 'Sampion'), 'Bonifác', 'Milotai kései' and 'Milotai intenzív' had Vanillic acid content among examined samples. 'Srem' had an

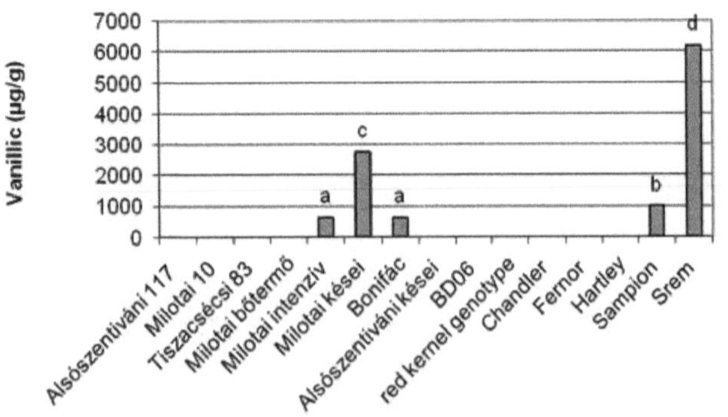

Fig. 4. Vanillic acid content of kernels in analysed cultivars ($SD_{5\%}$=256)

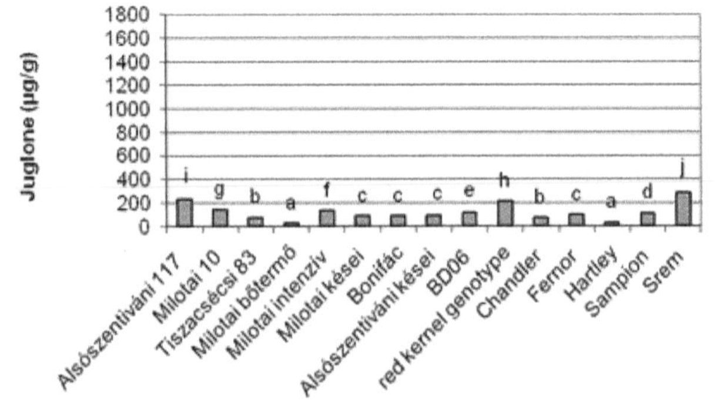

Fig. 7. Juglone content of kernels in analysed cultivars ($SD_{5\%}$=22)

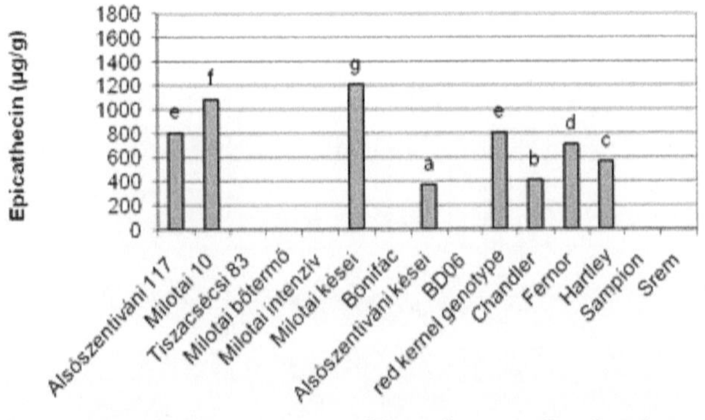

Fig. 5. Epicathechin content of kernels in analysed cultivars ($SD_{5\%}$=38)

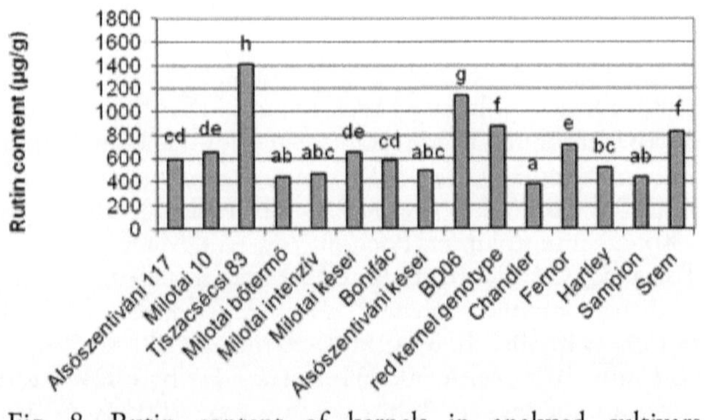

Fig. 8. Rutin content of kernels in analysed cultivars ($SD_{5\%}$=52)

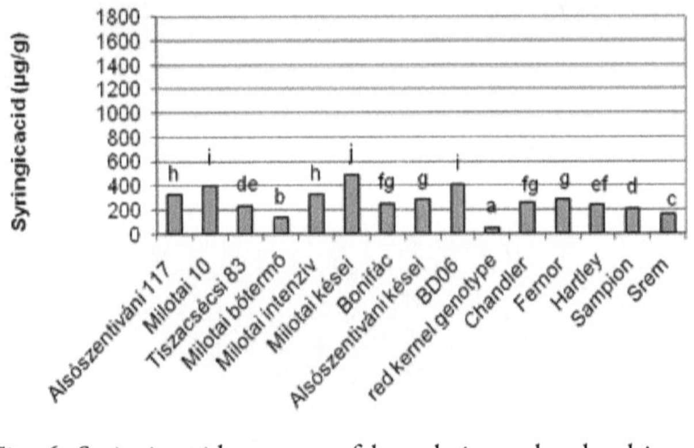

Fig. 6. Syringic acid content of kernels in analysed cultivars ($SD_{5\%}$=70)

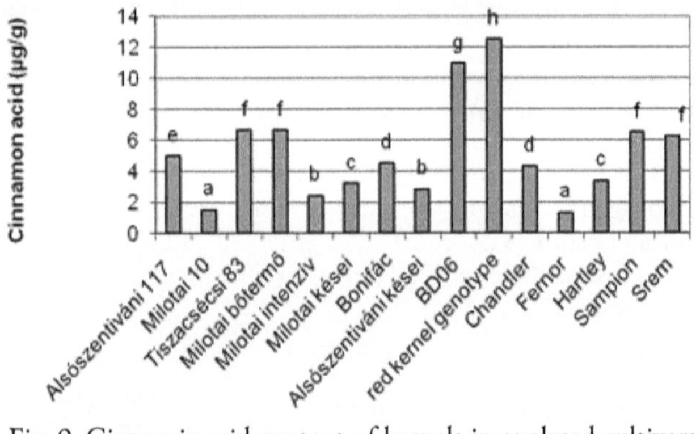

Fig. 9. Cinnamic acid content of kernels in analysed cultivars ($SD_{5\%}$=0.46)

outstanding Vanillic acid content, which was higher with 130% compared to 'Milotai kései' ranked on second place. The Vanillic acid content of other cultivars was equal or lower to 1 µg/g of kernel (Fig. 4).

Eight samples contained epicathechin among fifteen involved in the trial. 'Milotai kései' had the highest epicathechin content (1.20 µm/g) followed by 'Milotai 10' (1 088 µm/g) and 'Alsószentiváni 117' (807 µm/g). Lower than 1µg/g epicathechin content was measured in the kernel of the Serbian bred Persian walnut cultivars (Fig. 5).

The highest syringic acid content was in 'Milotai kései' (489 µm/g) followed by BD06 (405 µm/g) and 'Milotai 10' (399 µm/g). The red kernel genotype (48 µm/g) had the lowest syringic acid content (Fig. 6).

The highest juglone content was measured in 'Srem' (285 µm/g) followed by 'Alsószentiváni 117' (235 µm/g) and red kernel genotype (212 µm/g). The lowest juglone concentration was detected in 'Hartley' (31 µm/g) and 'Milotai bőtermő' (27 µm/g) (Fig. 7).

'Tiszacsécsi 83' (1 407 µm/g), BD06 (1 141 µm/g), red kernel genotype (874 µm/g) and 'Srem' (826 µm/g) had outstanding rutin content, these taxons had almost double rutin concentration than the other cultivars. Beside the named cultivars, 'Srem' (826 µm/g) and 'Fernor' (718 µm/g) had high rutin content. 'Chandler' (381 µm/g), 'Milotai intenzív' (471 µm/g) and 'Milotai bőtermő' (441 µm/g) had the lowest content (Fig. 8).

The highest cinnamic acid was produced by the red kernel genotype (12 µm/g) followed by BD06 (11 µm/g) and Serbian bred cultivars, 'Milotai bőtermő' (6 µm/g) and 'Tiszacsécsi 83' (6 µm/g) (Fig. 9).

There are few results available on the phenolic contents

of walnut kernel available in literature. The quantity in the phenolic compounds of Chinese genotypes, reported by Zhang *et al.* (2009), is equal or higher compared to the values of the Hungarian bred walnut cultivars. In retrospect, the Hungarian bred cultivars had higher levels of phenolic compounds compared to the Serbian, French and USA bred ones. The reason of this phenomenon may be the poor adaptation of Persian walnut to the different climate conditions, varying from their origin. From this study it was found that the Serbian bred 'Srem' was the only cultivar with high phenolic contents because its cathecin, vanillic acid, rutin and juglone values were almost the highest among the examined cultivars. This is better than for 'Sampion', the second Serbian cultivar reported by Solar *et al.* (2005) for its high phenolic contents in its kernel.

If the phenolic compounds are ranked according to their quantity, vanillic acid was the highest followed by cathechin, pyrocathechin, epicathechin, rutin, syringin acid, gallic acid, juglone and cinnamic acid in sequential decreasing order. Christopoulos *et al.* (2010) and Miltiadis *et al.* (2012) reported that vanillic acid was present in 'Franquette' but at lower levels than found in this study. In our research only five ('Milotai intenzív', 'Milotai kései', 'Bonifác', 'Srem' and 'Sampoin') of the fifteen cultivars contained vanillic acid. Christopolous *et al.* (2010) reported that the syringic acid content of 'Franquette' at lower levels than some of the Hungarian bred cultivars ('Milotai 10', BD06, 'Milotai kései'). Colaric *et al.* (2005) detected more juglone content in kernel of 'Alsószentiváni 117' than the other nine examined cultivars ('Adams', 'Cisco', 'Chandler', 'Elit', 'Fernette', 'Fernor', 'Franquette', 'Lara', 'Rasna'). In the cited research 'Alsószentiváni 117' had 727,000 μm/g of kernel juglone in the pellicle and 19,160 μm/g of kernel in the walnut flesh; both data were much more higher than measured in our study (235 μm/g of kernel). Labuckas *et al.* (2008) confirmed that 'Franquette' had a higher total phenolic content compared to 'Chandler'.

Conclusions

It is clear that this study shows that walnut kernels of Hungarian bred cultivars contain high phenolic content levels. These results are from one year of study. Therefore, more replications are required from another few years of study to provide better information of the Persian walnut cultivars. The phenolic compounds content is linked with susceptibility to different diseases and should be studied in more detail. According to our results genetic clarification of examined Persian walnut genotypes needs to be checked in the near future.

Acknowledgement

Our research was supported by TÁMOP 4.2.1./B-09/01/KMR/2010-0005 project.

References

Babich H, Stern A (1993). *In vitro* cytotoxicities of 1,4-naphthoquinone and hydroxylated 1,4-naphthoquinones to replicating cells. J Appl Toxicol 13:353-358.

Bonifazi EL, Ríos-Luci C, León LG, Burton G, Padrón JM, Misico RI (2010). Antiproliferative activity of synthetic naphthoquinones related to lapachol. First synthesis of 5-hydroxylapachol. Bioorg Med Chem 18:2621-2630.

Botu M, Tudor M, Botu I, Cosmulescu S, Papachatzis A (2010). Evaluation of walnut cultivars in the conditions of the Oltenia's hill area regarding functioning potential. Annals of the University of Craiova, the series Biology, Horticulture, Food Produce Processing Technology, Environment Engineering XV (XLXI):94-103.

Bujdosó G, Tóth-Márkus M, Daood HG, Adányi N, Szentiványi P (2010). Fruit quality and composition of Hungarian bred walnut cultivars. Acta Aliment 1(39):35-47.

Cerovic, S, Golosin, B, Ninic, Todorovic, J, Bijelic, S, Ognanos, V (2010). Walnut (*Juglans regia* L.) selection in Serbia. Hort Sci (Prague) 37(1):1-5.

Chevallier A, Bray O, Prunet JP, Giraud M (2009). Factors influencing walnut blight symptoms emergence and development. Acta Hort 861:473-478.

Christopoulos MV, Tsantili E, Papageorgiou V, Komaitis M, Rouskas D (2010). Effects of package atmosphere and temperature on phenolics, total antioxidant capacity and colour in kernels of 'Franquette' walnuts during 8-month storage. Acta Hort 858:75-81.

Colaric M, Veberic R, Solar A, Hudina M, Stampar F (2005). Phenolic acids, syringaldehide, and juglone in fruits of different cultivars of *Juglans regia* L. J Agric Food Chem 53:6390-6396.

Cosmulescu S, Trandafir I (2011a). Variation of phenols content in walnut (*Juglans regia* L.). South Western Journal of Horticulture, Biology and Environment 1(2):25-33.

Cosmulescu SN, Trandafir I (2011b). Seasonal variation of total phenols in leaves of walnut (*Juglans regia* L.). J Med Plants Res 5(19):4938-4942.

Cosmulescu S, Trandafir I (2012). Anti-oxidant activities and total phenolics content of leaf extracts from 14 cultivars of walnut (*Juglans regia*). J Hortic Sci Biotech 87(5):504-508.

Food and Agriculture Organization of the United Nations (web-page accessed: 7th February 2013).

Germain E, Prunet JP, Garcin A (1999). Le noyer. Ctifl, Paris, 26 p.

Harborne T, Williams A (2000). Advances in flavonoid research since 1992. Phytochemistry 55:481-504.

Hendricks E, Coates L, William W, Elkins B Rachel, McGranahan HG, Phillips, Herb A, Ramos E, Reil D, Wilbur O, Snyder RG (1998). Selection of varieties. In: Ramos, E. David (ed.). Walnut production manual University of California, Oakland 84-89.

Jakopic J, Colaric M, Verebic R, Hudina M, Solar A, Stampar F (2007). How much do the cultivar and preparing time influence the phenolics content in walnut liqueur? Food Chem 104:100-105.

Korac M, Slovic D, Rudic M, Cerovic S, Golosin B (1986).

Rezultati hibridizacije i selekcije oraha Na Poljoprivrednom fakultetu u Novom Sadu. Proceedings of the Yugoslav Symposium on the Selection and Breeding of Fruit, Cacak 100-107.

Labuckas DO, Maestri DM, Perelló M, Martinez LM, Lamarque AL (2008). Phenolics from walnut (*Juglans regia* L.) kernels: Antioxidant capacity and interactions with proteins. Food Chem 107(2):607-612.

Mikulic-Petkovsek M, Slatnar A, Verebic R, Stampar F, Solar A (2011). Phenolic response in green walnut husk after infection with bacteria *Xanthomonas arboricola PV. Juglandis*. Physiol Molecul Plant Pathol 3-4(76):159-165.

Miltiadis V, Christopoulos MV, Tsantili E (2012). Storage of fresh walnuts (*Juglans regia* L.)-Low temperature and phenolic compounds. Postharvest Biol Tec 73:80-88.

Pereira JA, Oliveira I, Sousa A, Valentao P, Andrade PB, Ferreira ICFR, Ferreres F, Bento A, Seabra R, Estevinho L (2007). Walnut (*Juglans regia* L.) leaves: Phenolic compounds, antibacterial activity and antioxidant potential of different cultivars. Food Chem Toxicol 11(45):2287-2295.

Rudkowska I (2009). Functional food for health: Focus on diabetes. Maturitas 3(62):263-269.

Solar A, Colaric M, Hudina M, Stampar F (2005). Phenolic content of walnut fruit as affected by cultivars and developmental stage. Acta Hort 705:231-240.

Solar A. Colaric M, Usenik V, Stampar F (2006). Seasonal variations of selected flavoniods, phenolic acids and quinines in annual shoots of common walnut (*Juglans regia* L.). Plant Sci 170(3): 453-461.

Stampar F, Solar A, Hudina M, Verebic R, Colaric M (2006). Traditional walnut liquere cocktail of phenolics. Food Chem 4(95): 627-631.

Szentiványi P (2006). Diónemesítés és fajtakutatás [Walnut breeding and variety research]. In. Szentiványi P.-Kállay T. -né (eds.): Dió [Walnut]. Mezőgazda Kiadó, Budapest 60-65.

Williamson G, Manach C (2007). Bioavailability and bioefficacy of polyphenols in human: II. Review of 93 intervention studies. Am J Clin Nut 81:243-255.

Zhang Z, Liao L, Moore J, Wu T, Wang Z (2009). Antioxidant phenolic compounds from walnut kernels (*Juglans regia* L.). Food Chem 113(1):160-165.

9

Regulation of Root Length and Lateral Root Number in Trifoliate Orange Applied by Peroxide Hydrogen and Arbuscular Mycorrhizal Fungi

Chun-Yan LIU, Yong-Ming HUANG, Ying-Ning ZOU*, Qiang-Sheng WU

*Yangtze University, College of Horticulture and Gardening, No 88 Jingmi Raod, Jingzhou, Hubei 434025, China; zouyingning@163.com (*corresponding author)*

Abstract

Root system morphology (RSM) in plants plays a key role in acquiring nutrients from the soil and is also altered by abiotic or biotic factors including soil microorganisms and signal molecules. The present study was made to evaluate the effects of an arbuscular mycorrhizal fungus (AMF, *Glomus versiforme*) and exogenous peroxide hydrogen (H_2O_2, 0, 1 and 100 μM) on root length, lateral root number and activities of polyamine-metabolized enzymes in trifoliate orange (*Poncirus trifoliata*) seedlings. After 5 months of inoculation with AMF, root mycorrhizal colonization was significantly increased by application of 1 μM H_2O_2, but markedly restrained by 100 μM H_2O_2. Inoculation with AMF significantly increased the taproot length and the number of second- and third-order lateral roots under 1 and 100 μM H_2O_2 application. The AMF infection significantly increased 0-1 cm classed root length and total root length, regardless of H_2O_2 concentration. In general, inoculation with AMF increased arginine decarboxylase (ADC) and ornithine decarboxylase (ODC) activity of roots under 0, 1 and 100 μM H_2O_2, increased diamine oxidase (DAO) activity of roots under 0 μM H_2O_2 and decreased DAO activity of roots under 1 and 100 μM H_2O_2. Root polyamine oxidase (PAO) activity was similar between AMF and non-AMF seedlings, irrespectively of H_2O_2 concentration. Results suggest that lower concentration of H_2O_2 (1 μM) might be regarded as a signal to stimulate mycorrhizal and lateral root development through increase of ADC and ODC and decrease of DAO, while high concentration of H_2O_2 (100 μM) as a toxic compound of reactive oxygen species restricted AMF colonization.

Keywords: arginine decarboxylase, *Glomus versiforme*, ornithine decarboxylase, root morphology

Introduction

Root is a vital plant organ which absorbs water and nutrients from the soil. Meanwhile, root system morphology (RSM), the configuration and structure of the root system, can determine the transport capacity of root in soil nutrient and water (de Dorlodot *et al.*, 2007). In general, rhizospheric nutritional status and microorganisms would highly impact RSM (Osmont *et al.*, 2007; Chapman *et al.*, 2011; Sato and Miura, 2011).

Arbuscular mycorrhizal fungi (AMF), a kind of soil inhabitant microorganism, can form a mutual symbiont with the roots of ~80% of land plants, thus enhancing the uptake of mineral nutrition and water from the soil to the host plant, as well as withstanding various adversities (Koide and Mosse, 2004). Studies showed that inoculation with AMF could alter the RSM traits and promote the formation of lateral roots of higher order for the host plant (Yao *et al.*, 2009). The AMF–mediated RSM improvement may be related to increased plant hormone biosynthesis proposed by Kaldorf and Ludwig-Müllerb (2000) in maize colonized by *Glomus intraradices*. Wu *et al.* (2011) found that the RSM alteration by mycorrhization may be due to the AMF-induced alteration of carbohydrates and peroxidase. However, other studies also revealed that AMF colonization did not alter RSM traits. Atkinson *et al.* (2003) found that root branching of *Trifolium repens* was not significantly affected by inoculation with *G. mosseae.* Herdlera *et al.* (2008) reported that the total root length and the root diameter of rice plants were notably decreased by the *G. intraradices* inoculation.

Hydrogen peroxide (H_2O_2) as one of reactive oxygen species (ROS) generally acts a key regulator in a broad range of physiological processes, such as senescence (Peng *et al.*, 2005), stomatal movement, cell cycle and growth and development (Deng *et al.*, 2012). The H_2O_2 excess accumulation in plants could be used as ROS to induce oxidative burst in cells, thereby triggering cell death. On the other hand, H_2O_2 is also regarded as an important signal compound concerning the interaction of plants with pathogenic microorganisms (Apel and Hirt, 2004; Laloi *et al.*, 2004), as well as many other processes referring to root biology, such as gravitropism, root elongation growth and root hair development (Zolla *et al.*, 2010; Jiang *et al.*, 2012). However, the interaction between AMF and H_2O_2 on RSM traits of the host plant is poorly known.

Diamine putrescine (Put), triamine spermidine (Spd) and tetraamine spermine (Spm), three types of polyamine (PAs), are formed either by the direct decarboxylation of L–ornithine through ornithine decarboxylase (ODC) or by decarboxylation from L–arginine to agmatine by arginine decarboxylase (ADC). The decomposition of PAs takes place under the two enzymes, diamine oxidase (DAO) and polyamine oxidase (PAO) (Kaur-Sawhney *et al.*, 2003). It is well documented that PAs play important roles in mycorrhizal formation, plant growth, organ development, fruit formation and leaf senescence (Kusano *et al.*, 2008; Wu *et al.*, 2012b). In root development, accumulation of PAs is involved in the formation of adventitious root (Couée *et al.*, 2004). Exogenous PAs could significantly promote the spore germination and hyphal growth of *G. mosseae* and *Gigaspora margarita, in vitro* culture conditions (Zhang *et al.*, 2003). Wu *et al.* (2012a) observed that AMF-mediated RSM improvement was related only to ADC and ODC, without any PAs concentrations. It seems that PA synthetase and degraded enzymes, except PA levels, were associated with AMF-mediated RSM. In addition, PA degradation may also act as a pathway to generate H_2O_2, which is considered to be an active secondary messenger (Vandenabeele *et al.*, 2003). However, so far, in plants, the potential relationship between RSM and PA-metabolized enzymes under H_2O_2 and mycorrhization conditions has not been examined in detail.

The aims of the present work were to (1) clarify the integrated effects of AMF and H_2O_2 on root length, lateral root number and PA-metabolized enzymes of citrus and to (2) evaluate the relationship between RSM and PA-metabolized enzymes.

Materials and methods

Plant culture

Seeds of trifoliate orange [*Poncirus trifoliata* (L.) Raf.] were surface-sterilized with 70% ethanol solution for 5 min, washed with distilled water and then sown into a plastic pot (18 cm upper diameter × 13 cm below diameter × 17 cm height) including 2.5 kg autoclaved (121 °C, 0.11 Mpa, 2 h) soil. The inoculums of *Glomus versiforme* (40 g) including spore and extraradical hyphae were placed below 5 cm of the soil surface at the time of sowing. The non-AMF pots were also supplied with 40 g sterilized inoculums as the control.

Experimental design

The experiment consisted in a completely randomized design with two factors: inoculation with or without an AMF (*G. versiforme*) and application of exogenous H_2O_2 with 0, 1 and 100 μM. Each treatment had three replicates, reaching a total of 18 pots. Three months after mycorrhizal inoculation, different concentrations (0, 1 and 100 μM) of H_2O_2 were weekly applied into the designed pots, nine times during the experiment.

Measurement of plant growth and root colonization

The AM and non-AM plants were harvested after 63 days of exogenous H_2O_2 treatments. Before harvest, plant height, stem diameter and leaf number were directly determined.

Root mycorrhizal colonization was measured by the method of Phillips and Hayman (1970) after clearing with 10% KOH and staining with 0.05% trypan blue.

Determinations of root system morphology and lateral root number

The root systems were carefully taken out from the plastic pots and the intact root systems were scanned immediately by the Epson Perfection V700 Photo Dual Lens System (Seiko Epson Corp, Japan). The images of root systems were analyzed by a WinRHIZO version from 2007 (Regent Instruments Incorporated, Canada) and the 0-1, 1-2, 2-3, 3-4 and >4 cm classed root lengths were automatically obtained. Taproot lengths were measured by a flexible rule and the number of lateral roots was artificially quantified.

Determinations of PA-metabolized enzyme activities

Activities of PA synthetases (ADC and ODC) and PA-degrading enzymes (PAO and DAO) were determined according to the methods previously described by Wu *et al.* (2012a).

Statistical analyses

Data were subjected to the analysis of variance (ANOVA) with SAS v8.1, and the significant differences among treatments were used for comparison to the Duncan test at $P<0.05$.

Results and discussion

Mycorrhizal colonization

After inoculation with *G. versiforme*, root colonization of the trifoliate orange seedlings ranged from 55.28 to 75.91% (Fig. 1). Compared to 0 μM H_2O_2 control, 1 μM H_2O_2 significantly increased AMF colonization by 16.6%, whereas 100 μM H_2O_2 level notably decreased mycorrhizal infection by 15.1%. There is increasing evidence indicating that during the early establishment of a symbiotic relationship, many plant signals and fungal perception are involved in the process (Requena *et al.*, 2007). It seems that the low concentration of H_2O_2 (1 μM) might be regarded as a signal molecule to stimulate mycorrhizal development, while high concentration of H_2O_2 (100 μM) as a toxic compound of ROS restricted AMF colonization.

Growth performance

The present study showed that exogenous H_2O_2 and AMF colonization did not significantly alter the stem diameter of the seedlings (Tab. 1). Leaf number per plant was significantly higher in AMF than in non-AMF seedlings under 0 and 100 μM H_2O_2, except treatments under 1 μM H_2O_2 (Tab. 1). Compared to non-AMF seedlings, AMF seedlings showed notably higher plant height under 0 μM H_2O_2 conditions, but lower under 100 μM H_2O_2 conditions. In addition, compared to 0 μM H_2O_2, 1 and 100 μM H_2O_2 generally increased plant height and leaf number, suggesting that the low or high concentration of H_2O_2 did not absolutely interfere with the plant growth performance. Previous studies had shown that AMF could facilitate the growth of the host plants, attributed to extraradical hyphal absorption of water and mineral nutrients (Sharif and Claassen, 2011). Although low concentrations of H_2O_2 promoted root AMF colonization, the H_2O_2-induced increase of root colonization, they did not show significant effects on plant height, stem diameter and leaf number (Tab. 1). There was a significantly interacted effect of AMF and H_2O_2 on leaf number.

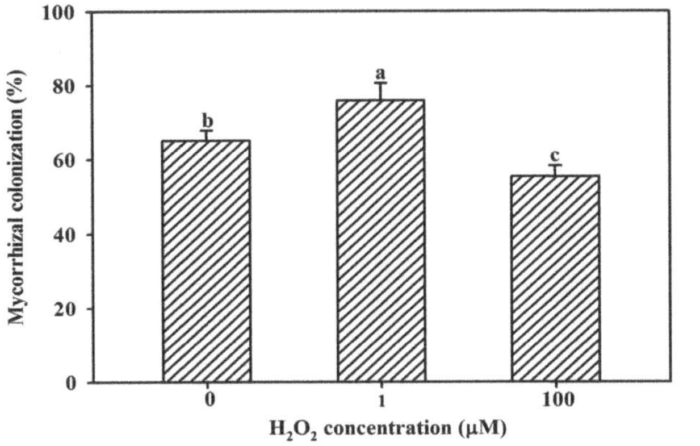

Fig. 1. Effect of exogenous H_2O_2 on AMF colonization of *Glomus versiforme*-inoculated trifoliate orange (*Poncirus trifoliata*) seedlings. Data (means±SE, $n=3$) followed by the different letter above the bars are significantly different at $P<0.05$.

Tab. 1. Effect of *Glomus versiforme* alone or in combination with exogenous H_2O_2 on growth performance of trifoliate orange (*Poncirus trifoliata*) seedlings

H_2O_2 concentration (μM)	Inoculation	Plant height (cm)	Stem diameter (cm)	Leaf number per plant
0	Non-AMF	17.0±1.7d	0.227±0.005a	15±0c
	AMF	20.9±0.3bc	0.237±0.018a	18±1b
1	Non-AMF	20.3±0.6c	0.237±0.011a	19±0ab
	AMF	21.5±1.1bc	0.236±0.016a	18±0b
100	Non-AMF	24.2±0.8a	0.249±0.027a	18±0b
	AMF	22.3±1.1b	0.250±0.013a	20±2a

Note: The data (means±SE, $n=3$) in a column followed by different letters are significantly different at 5% level.

Number of lateral roots

Our results showed that the root systems of the five months old trifoliate orange seedlings were primarily dominated by first-order and second-order lateral roots (Tab. 2). Mycorrhizal inoculation significantly increased taproot length compared to non-mycorrhizal seedlings, irrespectively of exogenous H_2O_2 levels.

A study showed that inoculation with *Paraglomus occultum* significantly increased number of the lateral roots in *Citrus tangerina* plants (Wu *et al.*, 2012b). The present study indicated that the AMF and exogenous H_2O_2 significantly influenced the number of the second-order and third-order lateral roots, but did not alter the number of the first-order lateral roots. This result is in compliance with the previous reports by Zai *et al.* (2007), who found that there was no significant influence on the number of first-order lateral roots between AM and non-AM trifoliate orange seedlings, but the number of the second-order lateral roots were markedly increased by mycorrhization.

Different classed root lengths

In the present study, the trifoliate orange seedlings in total root lengths depended on 0-1 cm classed root lengths (Tab. 3). Compared to non-AMF control, *G. versiforme* significantly increased 0-1 cm classed root lengths by 12.6, 10.6 and 7.3% and 3-4 cm classed root lengths by 100, 200 and 83.3% under 0, 1 and 100 μM H_2O_2, respectively. In addition, the AMF seedlings showed higher total root lengths than the non-AMF controls, regardless of exogenous H_2O_2 concentration. The results imply that AMF highly induced elongation of lateral roots, which is not dependent on exogenous H_2O_2 level. Yao *et al.* (2009) reported that the fine roots lengths in trifoliate orange seedlings colonized by *Gigaspora margarita*, *G. mosseae*, *G. versiforme* and *G. caledonium* were with 35.0, 35.1, 30.5 and 45.5% higher than in non-AMF seedlings. In addition, AMF and exogenous H_2O_2 generally did not alter 1-2, 2-3, and >4 cm classed root lengths, suggesting that the AMF and exogenous H_2O_2 mainly regulated the fine roots lengths (e.g., 0-1 cm classed root lengths).

Tab. 2. Effects of *Glomus versiforme* and exogenous H_2O_2 on taproot length and number of lateral roots in trifoliate orange (*Poncirus trifoliata*) seedlings

H_2O_2 concentration (μM)	Inoculation	Taproot length (cm)	Number of lateral roots		
			First-order	Second-order	Third-order
0	Non-AMF	18.2±0.6c	40±2a	74±3c	3±0cd
	AMF	26.5±1.5a	39±6a	60±3bc	7±0ab
1	Non-AMF	21.8±1.2b	38±11a	81±2b	4±2c
	AMF	26.2±0.9a	43±9a	89±5a	5±1bc
100	Non-AMF	22.7±0.9b	44±1a	66±3d	2±2d
	AMF	26.3±1.5a	44±1a	78±3bc	8±1a

Note: The data (means±SE, $n=3$) in a column followed by different letters are significantly different at 5% level.

Activities of PA synthetases ADC and ODC

Studies demonstrated that PAs are an important regulatory factor involved in the process of mycorrhizal formation (Wu *et al.*, 2012b) and that the formation of PAs derives either from L–ornithine by ODC or from L–arginine by ADC, being disintegrated by PAO or DAO (EI Meskaoui and Trembaly, 2009). In the present study, ADC activity of both leaves and roots was generally significantly higher under AMF than under non-AMF treatment under 0 and 100 μM H_2O_2 levels, while under 1 μM H_2O_2 concentration AMF seedlings recorded higher ADC activity of roots and lower ADC activity of leaves than non-AMF seedlings (Fig. 2a, 2b). ODC activity of leaves was similar between AM and non-AM seedlings under 1 and 100 μM and was significantly higher in AM than in non-AM seedlings grown in 0 μM H_2O_2 level (Fig. 2c). The *G. versiforme* increased the ODC activity of roots by 135.9, 3.8 and 7.8% under 0, 1 and 100 μM H_2O_2, respectively (Fig.

Tab. 3. Effect of *Glomus versiforme* and exogenous H_2O_2 on the different classed root lengths of trifoliate orange (*Poncirus trifoliata*) seedlings

H_2O_2 concentration (μM)	Inoculation	Lengths of different classed roots (cm)					
		0-1 cm	1-2 cm	2-3 cm	3-4 cm	>4 cm	Total
0	Non-AMF	183.5±4.4b	6.1±2.8a	3.9±1.4a	0.4±0.6b	0.0±0.0a	193.4±3.6c
	AMF	206.8±5.8a	6.7±1.6a	3.9±0.9a	0.8±0.5ab	0.1±0.1a	218.6±6.3a
1	Non-AMF	188.6±0.9b	7.2±2.9a	4.4±1.3a	0.5±0.3ab	0.1±0.2a	200.8±1.7bc
	AMF	208.6±8.7a	5.9±1.3a	4.4±0.7a	1.5±0.7a	0.2±0.2a	220.8±10.4a
100	Non-AMF	193.6±4.2b	7.0±0.2a	3.9±1.5a	0.6±0.1ab	0.1±0.1a	205.5±5.3b
	AMF	207.7±7.4a	6.4±2.4a	4.1±1.4a	1.1±0.7ab	0.1±0.1a	219.5±5.4a

Note: The data (means±SE, n=3) in a column followed by different letters are significantly different at 5% level.

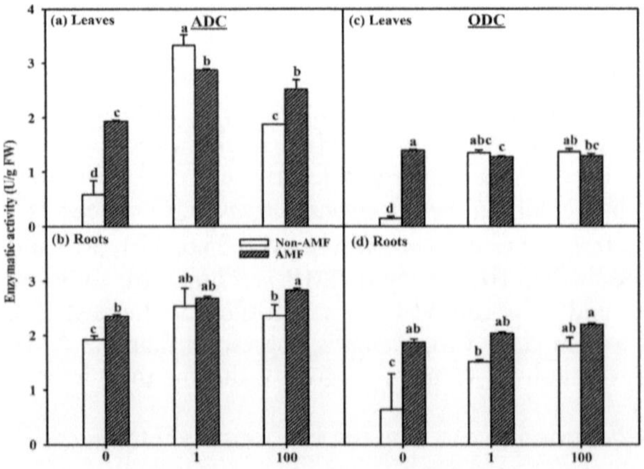

Fig. 2. Influence of *Glomus versiforme* and exogenous H_2O_2 on activity of PA synthetic enzymes (ADC and ODC) in leaves and roots of trifoliate orange (*Poncirus trifoliata*) seedlings. Data (means±SE, n=3) followed by the same letter above the bars are not significantly different at P<0.05

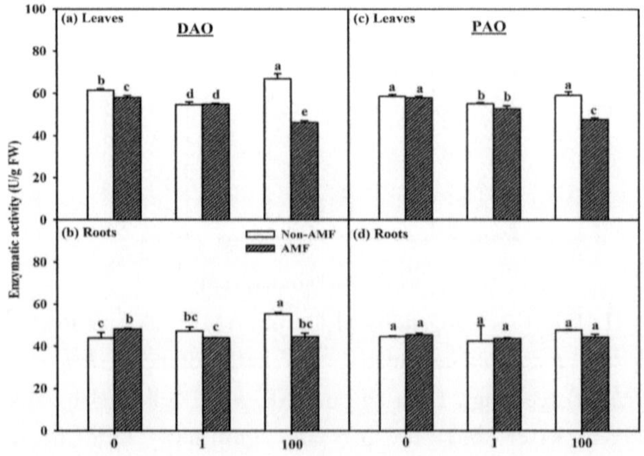

Fig. 3. Effects of *Glomus versiforme* and exogenous H_2O_2 on activity of PA-metabolic enzymes (DAO and PAO) in leaves and roots of trifoliate orange (*Poncirus trifoliata*) seedlings. Data (means±SE, n=3) followed by the same letter above the bars are not significantly different at P<0.05

2d). It seems that AMF plants presented higher ADC and ODC activity of roots, irrespectively of exogenous H_2O_2 concentrations. It is well known that ADC is generally associated with cell expansion and ODC with cell division (Paschalidis and Roubelakis-Angelakis, 2005). Wu *et al.* (2012a) revealed that AMF-mediated RSM was related to the change of ADC and ODC activity. Therefore, in AMF seedlings, greater activity of ADC and ODC would benefit the development of lateral roots.

Activities of PA-degrading enzyme DAO and PAO

As compared to that of the non-mycorrhizal seedlings, DAO and PAO activities in leaves of the mycorrhizal seedlings decreased by 31.0 and 19.4% (Fig. 3a, 3c) and DAO and PAO in roots by 19.6 and 7.0% under 100 μM H_2O_2 treatment, respectively (Fig. 3b, 3d). Under 1 μM H_2O_2 treatment, DAO and PAO activities in leaves and roots were similar between AMF and non-AMF seedlings. Under 0 μM H_2O_2 concentration conditions, PAO activity in leaves and roots of AMF and non-AMF seedlings was not significantly different, but lower leaf DAO and higher root DAO were found in AMF seedlings. DAO and PAO are not only involved in the degradation of PAs, but also in the production of H_2O_2, during root cell growth and differentiation (Wisniewski *et al.*, 2000). Since no significant differences in PAO of roots were observed between AMF and non-AMF trifoliate orange seedlings under 1 and 100 μM H_2O_2 application, higher total root length and number of second-order and third-order lateral roots in AMF seedlings were not related to PAO.

Conclusions

Low concentration (1 μM) of exogenous H_2O_2 may act as an important regulating factor of mycorrhizal development in trifoliate orange seedlings inoculated with *Glomus versiforme,* but high concentration (100 μM) of exogenous H_2O_2 may confer oxidative damage and thus restrict root mycorrhizal colonization. In addition, under application of a low H_2O_2 concentration (1 μM) conditions, AMF colonization significantly increased the number of second-order later root and total roots length (mainly 0-1 cm classed roots length), which were related to the increase of ADC and ODC and the decrease of DAO in roots.

Acknowledgements

This research was supported by the National Natural

Science Foundation of China (31101513).

References

Apel K, Hirt H (2004). Reactive oxygen species: metabolism, oxidative stress and signal transduction. Ann Rev Plant Biol 55:373-399.

Atkinson D, Black KE, Forbes PJ, Hooker JE, Baddeley JA, Watson CA (2003). The influence of arbuscular mycorrhizal colonisation and environment on root development in soil. Eur J Soil Sci 54:751-757.

Chapman N, Whalley WR, Lindsey K, Miller AJ (2011). Water supply and not nitrate concentration determines primary root growth in *Arabidopsis*. Plant Cell Environ 34:1630-1638.

Couée I, Hummel I, Sulmon C, Gouesbet G, Amrani AE (2004). Involvement of polyamines in root development. Plant Cell Tiss Org Cult 76:1-10.

de Dorlodot S, Forster B, Pages L, Price A, Tuberosa R, Draye X (2007). Root system architecture: opportunities and constraints for genetic improvement of crops. Trend Plant Sci 12:474-481.

Deng XP, Cheng YJ, Wu XB, Kwak SS, Chen W, Eneji AE (2012). Exogenous hydrogen peroxide positively influences root growth and metabolism in leaves of sweet potato seedlings. Aust J Crop Sci 6:1572-1578.

El Meskaoui A, Trembaly FM (2009). Effects of exogenous polyamines and inhibitors of polyamine biosynthesis on endogenous free polyamine contents and the maturation of white spruce somatic embryos. Afr J Biotech 8:6807-6816.

Herdlera S, Kreuzera K, Scheua S, Bonkowski M (2008). Interactions between arbuscular mycorrhizal fungi (*Glomus intraradices*, Glomeromycota) and amoebae (*Acanthamoeba* castellanii, Protozoa) in the rhizosphere of rice (*Oryza sativa*). Soil Biol Biochem 40:660-668.

Jiang J, Su M, Wang L, Jiao C, Sun Z, Cheng W, Li F, Wang C (2012). Exogenous hydrogen peroxide reversibly inhibits root gravitropism and induces horizontal curvature of primary root during grass pea germination. Plant Physiol Biochem 53:84-93.

Kaldorf M, Ludwig-Müllerb J (2000). AM fungi might affect the root morphology of maize by increasing indole–3–butyric acid biosynthesis. Physiol Plant 109:58-67.

Kaur-Sawhney R, Tiburcio AF, Altabella T, Galston AW (2003). Polyamines in plants: An overview. J Cell Mol Biol 2:1-12.

Koide RT, Mosse B (2004). A history of research on arbuscular mycorrhiza. Mycorrhiza 14:145-163.

Kusano T, Berberich T, Tateda C, Takahashi Y (2008). Polyamines: essential factors for growth and survival. Planta 228:367-381.

Laloi C, Apel K, Danon A (2004). Reactive oxygen signalling: the latest news. Curr Opin Plant Biol 7:323-328.

Osmont KS, Sibout R, Hardtke CS (2007). Hidden branches: developments in root system architecture. Ann Rev Plant Biol 58:93-113.

Peng LT, Jiang YM, Yang SZ, Pan SY (2005). Accelerated senescence of fresh-cut Chinese water chestnut tissues in relation to hydrogen peroxide accumulation. J Plant Physiol Mol Biol 31:527-532.

Phillips JM, Hayman DS (1970). Improved procedures for clearing roots and staining parasitic and vesicular–arbuscular mycorrhizal fungi for rapid assessment of infection. Trans Br Mycol Soc 55:158-161.

Requena N, Serrano E, Ocon A, Breuninger M (2007). Plant signals and fungal perception during arbuscular mycorrhiza establishment. Phytochem 68:33-40.

Sato A, Miura K (2011). Root architecture remodeling induced by phosphate starvation. Plant Signal Behav 6:1122-1126.

Sharif M, Claassen N (2011). Action mechanisms of arbuscular mycorrhizal fungi in phosphorus uptake by *Capsicum annuum* L. Pedosphere 21:502-511.

Vandenabeele S, Van Der Kelen K, Dat J, Gadjev I, Boonefaes T, Morsa S, Rottiers P, Slooten L, Van Montagu M, Zabeau M (2003). A comprehensive analysis of hydrogen peroxide-induced gene expression in tobacco. Proc Natl Acad Sci 100:16113-16118.

Wisniewski JP, Rathbun EA, Knox JP, Brewin NJ (2000). Involvement of diamine oxidase and peroxidase in insolubilization of the extracellular matrix: implications for pea nodule initiation by *Rhizobium leguminosarum*. Mol Plant Microbe Interact 13:413-420.

Wu QS, He XH, Zou YN, Liu CY, Xiao J, Li Y (2012a). Arbuscular mycorrhizas alter root system architecture of *Citrus tangerine* through regulating metabolism of endogenous polyamines. Plant Growth Regul 68:27-35.

Wu QS, Zou YN, He XH, Luo P (2011). Arbuscular mycorrhizal fungi can alter some root characters and physiological status in trifoliate orange (*Poncirus trifoliata* L. Raf.) seedlings. Plant Growth Regul 65:273-278.

Wu QS, Zou YN, Liu CY, Lu T (2012b). Interacted effect of arbuscular mycorrhizal fungi and polyamines on root system architecture of citrus seedlings. J Integr Agric 11:1675-1681.

Yao Q, Wang LR, Zhu HH, Chen JZ (2009). Effect of arbuscular mycorrhizal fungal inoculation on root system architecture of trifoliate orange (*Poncirus trifoliata* L. Raf.) seedlings. Sci Hortic 121:458-461.

Zhang Y, Xie LY, Xiong BQ, Zeng M (2003). Effect of polyamine on growth and development of arbuscular mycorrhizal fungi *in vitro* culture condition. Mycosyst 22:417-423.

Zai XM, Qin P, Wan SW, Zhao FG, Wang G, Yan DL, Zhou J (2007). Effects of arbuscular mycorrhizal fungi on the rooting and growth of beach plum (*Prunus maritima*) cuttings. J Hortic Sci Biotechnol 82:863-866.

Zolla G, Heimer YM, Barak S (2010). Mild salinity stimulates a stress-induced morphogenic response in *Arabidopsis thaliana* roots. J Exp Bot 61:211-224.

10

Kernel Quality of Some Sweet Corn Varieties in Relation to Processing

Ozlem ALAN[1]*, Gulcan KINACI[2], Engin KINACI[2], Zekiye BUDAK BASCIFTCI[2], Kenan SONMEZ[3], Yasemin EVRENOSOGLU[3], Imren KUTLU[4]

[1]*Ege University, Odemis Vocational Training School, 35760, Izmir, Turkey; ozlem.alan@ege.edu.tr (corresponding author)*
[2]*Eskisehir Osmangazi University, Faculty of Agriculture, Department of Field Crops, 26160, Eskisehir, Turkey; gkinaci@ogu.edu.tr; ekinaci@ogu.edu.tr; zbudak@ogu.edu.tr*
[3]*Eskisehir Osmangazi University, Faculty of Agriculture, Department of Horticulture, 26160, Eskisehir, Turkey; ksonmez@ogu.edu.tr; yevrenosoglu@ogu.edu.tr*
[4]*Eskisehir Osmangazi University, Faculty of Agriculture, 26160, Eskisehir, Turkey; ikutlu@ogu.edu.tr*

Abstract

The quality characteristics (dry matter content, soluble solid concentration, kernel colour, sugar, starch and protein content) of fresh, frozen and canned kernels of seven sweet corn varieties ('Lumina', 'Merit', 'Sunshine', 'Jubilee', 'Challenger', 'Yellow Baby' and '2201') were studied. The present research was conducted during 2009 and 2010 in Eskisehir, Midwestern Turkey. The trials were set up in randomised complete block design, with four replications. Ears were harvested and randomly selected for analysis as fresh, frozen and canned. Dry matter content ranged from 34.2% ('2201') to 39.5% ('Yellow Baby'), soluble solid concentration from 16.3% ('2201') to 27.4% ('Yellow Baby'). The sugar content of fresh kernels was higher than other treatments (frozen and canned) for all varieties. The starch content of the varieties was decreased after processing, except in 'Yellow Baby'. Fresh, frozen and canned sweet corn kernels had similar protein contents; the highest protein content was obtained, for all treatments, from 'Challenger', and it maintained its higher protein content after processing. '2201' had the highest sugar and the lowest starch content for fresh, frozen and canned varieties. When compared on a kernel basis, sweet corn marketed as fresh, frozen or canned, it may be feasible to select varieties for different sweet corn markets.

Keywords: canned, fresh, frozen, starch, sugar, *Zea mays* L. var. *saccharata*

Introduction

Sweet corn (*Zea mays* L. var. *saccharata*) is a cultivated plant grown fresh for human consumption and is a raw or processed material of the food industry throughout the world. It is an important source of fibre, minerals and certain vitamins (Lertrat and Pulam, 2007). Sweet corn differs from other corns (field, pop corn etc.). The primary difference is gene expression that determines endosperm carbohydrate content as well as many other genes that affect maize growth (Znidarcic, 2012). It is produced for three distinct markets; fresh, frozen and canned. Production within these markets is largely independent of each other (Kleinhenz, 2008). Both the total production and the value of processed sweet corn have increased 60% over the last 25 years in the world (Williams, 2006). The reasons for the boost in popularity are the domestic consumption, export development and import replacement (Najeeb *et al.*, 2011).

One of the goals of sweet corn producers is to produce sweet corn with a high quality. The quality of fresh, or processed whole kernels, canned or frozen, is determined by its unique combination of flavour, texture and aroma (Lertrat and Pulam, 2007). Sweetness is the major component of flavour and affected by the amounts of sugar and starch in the endosperm (Flora and Wiley, 1974; Azanza *et al.*, 1994). It has been reported that selection for more tender and crispy genotypes with higher sugar and lower starch concentration and an increased sweet corn aroma would increase the eating quality of the product (Azanza *et al.*, 1996).

Based on the nature of kernel sweetness, sweet corns can also be classified into three main genetic groups: standard or sugary (*su*), super sweet or shrunken (sh_2), sugary enhanced (*se*) (Hale *et al.*, 2005). Sugary varieties accumulate less sugar compared with other sweet corn types. They have a creamy texture and good corn flavour and are known for their good germination and seedling vigour. But sugary varieties can lose their kernel quality rapidly after harvest due to the convertion of sugars to starch and moisture loss (Carey *et al.*, 1982, 1984). Shrunken varieties are characterised by kernels that have two to three times more sugar content at typical harvest maturity and have less starch than sugary varieties. The fresh market shelf life is extended because of the slower conversion of sugars to starch after harvest (Azanza *et al.*, 1996). Sugary enhanced varieties have a sugar content twice as much as sugary varieties, extremely tender kernels, creamy texture and good corn flavor. The sweet taste of their kernels is retained longer after harvest. Despite the desirable attributes of shrunken and sugar enhanced varieties, their production has been hindered by reduced field emergence, seedling vigour and standard uniformity (Styer *et al.*, 1980; Andrew, 1982; Douglass *et al.*, 1993).

Sweetness is determined not only by genetics, but also by the way the respective varieties are managed and harvested. Each year, seed companies apply a great number of new sweet corn varieties to variety lists. Primary differences among the hybrid varieties include the duration of their vegetation period, content of sugars, and suitability for various uses (Szymanek, 2012).

In Turkey, no information is available regarding the acreage of sweet corn grown. Despite the above-mentioned knowledge, interest in sweet corn has grown in recent years. Frozen sweet corn kernels are in the first rank among the fresh vegetables for import and frozen sweet corn export value was 36 tons in Turkey in 2010 (Aydın, 2011). This is mainly due to the fact that today more and more farmers are looking for a new alternative to traditional vegetable growing, which is threatened by various cultivation and economic problems. Several studies have been reported in sweet corn kernel quality. Still, studies of quality traits on numerous sweet corn varieties are resource-demanding and rare, including in Turkey. Besides this, many sweet corn varieties are developed under environments which differ from those in target production areas. Therefore, assessing the consistency of variety performance across locations and/or time is important to farmers. Yet, such studies are needed to assist in selecting a variety, maintaining consumer satisfaction and improving our understanding of the relative contributions of variety and environment to variability in quality traits. The objective of this research was to evaluate available commercial sweet corn varieties for their kernels quality. In this way, a wide assortment of varieties (which differ in soluble solids and colour, sugar content and other kernel characteristics) were documented to assist in selecting varieties for different sweet corn markets and increasing the efficiency of future evaluations.

Materials and methods

The field experiments were carried out in the experimental fields of Faculty of Agriculture, Eskişehir Osmangazi University, Eskişehir (39°46'N; 30°33'E; 801 m above sea level), during 2009 and 2010. The soil was sandy clay (44% sand; 20% clay) with 1% organic matter, 0.05% total salt, 5% lime and pH 7.6. Seven sweet corn varieties ('Lumina', 'Merit', 'Sunshine', 'Jubilee', 'Challenger', 'Yellow Baby' and '2201') were used as materials. Each cultivar was sown in randomised complete block with four replications and each experimental plot was 29.4 m^2. Seeds were sown on April 29 in 2009 and on May 19 in 2010 in a spacing of 70 cm x 25 cm. Plants were fertilised with the equivalent of 280 kg N, 110 kg P_2O_5 and 110 kg K_2O per hectare (Turgut, 2000) during the growing season. The drip irrigation was applied as needed and weeds were controlled mechanically by hand. No fungicide or insecticide was applied during cultivation.

When the sweet corn reached harvest maturity (juice consistency of kernels) fifty ears from the centre of each replication were harvested randomly by hand in the morning. Then, the ears were taken to the processing lab and husked, sorted, inspected and divided into the following categories for analysis: twenty ears for fresh kernel measurements, ten ears for frozen kernel measurements and ten ears for canned kernel measurements. Within an hour after harvest, kernels were cut from the ear with machine and the following experiments were conducted.

Freezing kernels

After the water comes to a boil, the cut kernels were blanched 3-5 minutes and cooled. Then, the kernels were put into freezer bags to a level of 3 to 4 inches from the top and the air was squeezed out, leaving 1-inch of head space, labelled, and frozen at –20 °C. The frozen kernels were stored for 6 months. After storage time, the kernels were evaluated according to their sugar, starch and protein content.

Canning kernels

After the water comes to a boil, the cut kernels were blanched 3-5 minutes and then put into clean jars with hot corn and cooking liquid leaving 1-inch of head space. The sealing surfaces of the jars were wiped with a clean damp paper towel and the lids were added and processed in an autoclave at 115 °C for fifty minutes. The canned kernels were stored for 6 months. After storage time, the kernels were evaluated according to their sugar, starch and protein content.

Dry matter content (DMC %)

Kernels were dried in an oven at 65 °C and the weight loss between measurements was < 0.05 g. The percentage difference between the fresh and dry weights was used to calculate the dry matter content of the kernel. This procedure was done for fresh kernels.

Soluble solid concentration (SSC %)

Kernels were cut from the centre section of ten ears from each plot, after a 2-inch section was removed from each end of each ear. Fifteen grams from each ear were placed on a double layer square of cheesecloth. Extract was collected and placed on the ATAGO Mark Digital Refractometer for analysis. This procedure was done for fresh kernels.

Kernel colour

Kernel colour was measured with a colorimeter at the CIE (Commission Internationale De L'eclairage) L*a*b*. Kernels were measured for each replicate, and the colour was characterised by lightness (L*), hue angle ($H^\circ = \tan^{-1}(b^*/a^*)$) and chroma($\sqrt{a^{*2}+b^{*2}}$) (Tuncay *et al.*, 2005). This procedure was done for fresh kernels.

Sugar content (%)

Kernel sugar content was determined by the Lane-Eynon method (Horwitz, 2000). This procedure was done for fresh, frozen and canned kernels.

Starch content (%)

Kernel starch content was determinated by the polarimetric method (Earle and Milmer, 1944). This procedure was done for fresh, frozen and canned kernels.

Protein content (%)

Kernel protein content was determined by the Kjeldahl method (Kirk and Sawyer, 1991). Nitrogen content in the sample was converted to total protein content by multiplying the N percentage by the standard corn conversion factor (6.25). This procedure was done for fresh, frozen and canned kernels.

Data evaluation

Analysis of variance (SAS, 1996) was performed on DMC, SSC and color characteristics for each year. A factorial design was used to analyse for sugar, starch and protein content, with treatments (fresh, frozen and canned) as main factors. Mean differences were determined by an LSD test.

Results and discussion

Dry matter, soluble solid concentration and color characteristics of fresh sweet corn kernels

Dry matter content (DMC) and soluble solid concentration (SSC) of fresh kernels are presented in Tab. 1. DMC was affected significantly by variety in both years ($p<0.01$ and $p<0.05$, respectively). DMC was slightly higher in the second year of the experiment. The 'Yellow Baby' variety had a DMC of 39.1% (2009) and 39.9% (2010), which was significantly higher than other varieties. In the first year of the experiment, the lowest DMC were obtained from 'Merit' and '2201' with a percent of 33.9 while '2201' (34.7%) and 'Challenger' (34.9%) had the lowest DMC in 2010, which was in the range reported for sweet corn previously by Bozokalfa *et al.* (2004).

The effect of variety on SSC was found significant ($p<0.01$) in both years (Tab. 1). SSC was slightly higher in the second year of the experiment. SSC changed between 15.2% and 27.2% in 2009; and between 12.4% and 27.7% in 2010. In general, fresh kernels had higher SSC in 2010 compared to 2009. 'Yellow Baby' had the highest SSC while 'Challenger' had the lowest SSC in both years. SSC data reported here are similar to those found by Kleinhenz (2003) and Hale *et al.* (2005). Many authors stated that the refractometer, which measures SSC, has been utilised as a rapid, preharvest method to determine sweet corn sugar content (Kleinhenz, 2003; Randle *et al.*, 1984). Kumari *et al.* (2006) also reported total sugar, reducing sugar and non-reducing sugar content were positively and significantly correlated with SSC. On the other hand, Zhu *et al.* (1992) reported that an overall R^2 of -0.99 between SSC and total sugars; SSC cannot be used as a reliable indicator of sweet

Tab. 1. Dry matter content - DMC (%) and soluble solid concentration - SSC (%) of fresh sweet corn kernels

Varieties	DMC (%)			SSC (%)		
	2009	2010	Mean	2009	2010	Mean
'Lumina'	34.3^{cd}	35.4^{bc}	34.9^{cd}	23.7^{bc}	24.1^{b}	23.9^{b}
'Merit'	33.6^{d}	36.7^{abc}	35.2^{cd}	22.5^{cd}	24.8^{b}	23.7^{b}
'Sunshine'	35.7^{bc}	38.9^{ab}	37.3^{b}	21.3^{d}	26.2^{ab}	23.8^{b}
'Jubilee'	36.2^{bc}	36.0^{bc}	36.1^{bc}	24.3^{b}	25.4^{ab}	24.8^{b}
'Challenger'	35.9^{bc}	34.9^{c}	35.4^{bcd}	15.2^{f}	12.4^{d}	13.8^{d}
'Yellow Baby'	39.1^{a}	39.9^{a}	39.5^{a}	27.2^{a}	27.7^{a}	27.4^{a}
'2201'	33.6^{d}	34.7^{c}	34.2^{d}	17.6^{e}	15.0^{c}	16.3^{c}
Mean	35.5	36.7		21.7	22.2	
LSD (%5) Year			n.s.			n.s.
LSD (%5) Variety	1.72**	3.51*	1.89**	1.53**	2.33**	1.34**
LSD (%5) Year*Variety			n.s.			1.90**

*: $p<0.05$; **: $p<0.01$ n.s.: not significant; Within each column, different letters denote significant differences (LSD test)

Tab. 2. Colour characteristics of fresh sweet corn kernels

Varieties	Lightness			Hue			Chroma		
	2009	2010	Mean	2009	2010	Mean	2009	2010	Mean
'Lumina'	73.4	75.5	74.4	83.3	83.6^{bc}	83.4^{bc}	27.8^{a}	27.2^{cd}	27.5^{ab}
'Merit'	71.3	73.6	72.5	87.6	87.9^{a}	87.8^{a}	22.5^{c}	22.5^{e}	22.5^{c}
'Sunshine'	72.5	74.5	73.5	85.9	85.9^{ab}	85.9^{ab}	23.7^{bc}	23.7^{de}	23.7^{c}
'Jubilee'	71.0	68.9	69.9	82.6	81.3^{c}	81.9^{c}	27.6^{a}	29. 1^{bc}	28.4^{ab}
'Challenger'	75.5	77.9	76.7	85.0	86.2^{ab}	85.6^{ab}	23.5^{bc}	31.4^{ab}	27.5^{ab}
'Yellow Baby'	76.4	79.1	77.7	86.7	87.4^{a}	87.1^{a}	23.7^{bc}	29.4^{bc}	26.5^{b}
'2201'	69.9	71.3	70.6	83.2	84.7^{abc}	83.9^{bc}	24.9^{b}	33.6^{a}	29.3^{a}
Mean	72.9^{b}	74.4^{a}		84.9	85.3		24.8^{b}	28.1^{a}	
LSD (%5) Year			1.31*			n.s.			1.45**
LSD (%5) Variety	n.s.	n.s.	n.s.	n.s.	3.81*	3.03**	1.73**	3.85**	2.04**
LSD (%5) Year*Variety			n.s.			n.s.			2.88**

*: $p<0.05$; **: $p<0.01$, n.s.: not significant; Different letters, small in line and each column, denote significant differences (LSD test)

corn total sugar concentrations (Hale *et al.*, 2005).

Colour characteristics of fresh kernels are presented in Tab. 2. Kernel lightness values were not significantly affected by varieties in both years. Mean values ranged from 69.9 to 77.7. Lightness values were significantly higher in the second year of the experiment. This indicates that kernels grown in 2010 had lighter coloured kernels compared to those grown in 2009. In the first year of the experiment, non-significant differences in kernel hue values were found but in 2010, hue values were significantly ($p<0.05$) affected by variety. The lightest yellow kernels were obtained from 'Merit' (87.9) and 'Yellow Baby' (87.4), while 'Jubilee' (81.3) had the darkest kernels. Significant differences ($p<0.01$) were found in kernel chroma values between varieties in both years. 'Merit' had the lowest chroma value with 22.5 when compared to the other varieties. The highest chroma values were obtained from 'Lumina' (27.8) and 'Jubilee' (28.4) in 2009 and from '2201' (33.6) in 2010. Colour characteristics values are in close agreement with earlier report of Tuncay *et al.* (2005). Also, this trait was influenced by genotype, genotype x environment interactions and in-field management (Esiyok *et al.*, *2004*; Esiyok and Bozokalfa, 2005).

Sugar, starch and protein content of fresh, frozen and canned sweet corn kernels

The sugar, starch and protein content of fresh, frozen and canned sweet corn kernels are presented in Tab. 3. In both years, treatment, variety and the interaction between treatment and variety had a statistically significant ($p<0.01$) effect on sugar content.

In 2009, the highest sugar content was obtained from fresh kernels of '2201' (9.5%) and the lowest sugar content was found in canned kernels of 'Merit' (1.6%) and 'Lumina' (2.1%). In the second year of the experiment, fresh kernels of 'Yellow Baby' and frozen kernels of '2201' had similar values and higher sugar contents than all others (7.7% and 7.6%, respectively). The lowest sugar contents were obtained from canned kernels of 'Merit' (1.5%), 'Sunshine' (1.6%), 'Lumina' (1.7%) and 'Jubilee' (1.8%). According to the average results from the two years, the sugar content from fresh kernels was higher than other treatments for all varieties and the sugar content of all varieties decreased after processing, except '2201'. The highest sugar content was obtained, for all treatments, from '2201' (Fig. 1). The presented results revealed that the varieties differed significantly for sugar content and there were differences among the varieties compliance processes. Similar findings were reported by Lertrat and Pulam (2007).

Tab. 3. Sugar, starch and protein content (%) of fresh, frozen and canned sweet corn kernels

			Varieties							
		Treatments	'Lumina'	'Merit'	'Sunshine'	'Jubilee'	'Challenger'	'Yellow Baby'	'2201'	Mean
Sugar content (%)	2009	Fresh	4.1	4.6	3.9	3.5	6.1	6.2	9.5	5.4^{a}
		Frozen	2.8	2.0	2.5	3.3	5.3	3.9	8.4	4.0^{b}
		Canned	2.1	1.6	2.6	2.2	3.6	2.8	5.1	2.7^{c}
		Mean	3.0^{d}	2.7^{d}	3.0^{d}	3.0^{d}	5.0^{b}	4.3^{c}	7.7^{a}	
		LSD(%5)	T:0.22** V:0.33** T*V:0.58**							
	2010	Fresh	3.4	3.3	3.7	3.5	5.7	7.7	6.6	4.9^{a}
		Frozen	3.2	2.3	3.0	3.1	5.6	3.9	7.6	4.1^{b}
		Canned	1.7	1.5	1.6	1.8	3.0	2.0	5.2	2.4^{c}
		Mean	2.8^{d}	2.4^{e}	2.8^{d}	2.8^{d}	4.8^{d}	4.5^{c}	6.5^{a}	
		LSD(%5)	T: 0.17** V:0.25** T*V:0.44**							
Starch content (%)	2009	Fresh	18.0	19.3	18.3	19.3	13.6	17.7	16.3	17.^{a}
		Frozen	17.4	14.9	15.8	18.5	12.3	17.9	9.1	15.^{b}
		Canned	12.7	13.4	13.9	14.2	7.9	13.4	4.1	11.^{c}
		Mean	16.0^{b}	15.9^{b}	16.0^{b}	17.3^{a}	11.4^{c}	16.4^{b}	9.8^{d}	
		LSD(%5)	T:0.62** V:0.94** T*V:1.63**							
	2010	Fresh	17.8	16.9	17.0	18.2	12.2	14.5	14.4	15.^{a}
		Frozen	17.6	18.9	18.1	18.1	11.3	19.4	6.4	15.^{b}
		Canned	13.8	13.5	16.3	15.5	6.1	14.9	5.4	12^{c}
		Mean	16.4^{c}	16.5^{bc}	17.1^{ab}	17.3^{a}	8.3^{d}	16.2^{c}	8.7^{d}	
		LSD(%5)	T: 0.45** V:0.69** T*V:1.20**							
Protein content (%)	2009	Fresh	10.6	10.3	10.9	9.9	13.0	10.5	12.5	11.1
		Frozen	10.6	9.9	11.2	10.0	12.9	10.9	10.9	11.2
		Canned	10.4	10.1	10.5	10.0	12.8	10.0	12.9	11.0
		Mean	10.5^{c}	10.1^{d}	10.9^{b}	10.0^{d}	12.9^{a}	10.5^{c}	12.8^{a}	
		LSD(%5)	T:n.s. V:0.29** T*V:n.s.							
	2010	Fresh	10.7	10.7	11.2	9.7	13.3	10.4	11.6	11.1
		Frozen	10.2	9.5	10.1	9.8	14.5	9.5	12.7	10.9
		Canned	10.8	10.6	9.8	11.0	14.3	9.7	12.2	11.2
		Mean	10.6^{c}	10.3^{cd}	10.4^{c}	10.2^{cd}	14.0^{a}	9.8^{d}	12.1^{b}	
		LSD(%5)	T: n.s. V:0.47** T*V:0.82							

**: $p<0.01$; n.s.: not significant. T: Treatment; V: Variety; T*V: Treatment*Variety; Different letters, small in each line and each column, denote significant differences (LSD test)

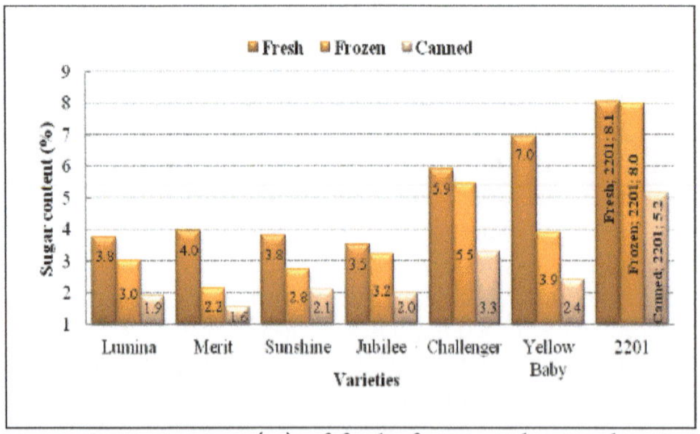

Fig. 1. Sugar content (%) of fresh, frozen and canned sweet corn kernels, as average values for two years

Both main effects and the interaction between the experimental factors on starch content were found to be statistically significant ($p<0.01$) in 2009 and 2010 (Tab. 3). In 2009, canned kernels of '2201' (4.1%) had the lowest starch content while fresh kernels of 'Jubilee' and 'Merit' with 19.3% gave the highest starch contents. In 2010, the lowest starch content was found in canned kernels of '2201' (5.3%), followed by canned kernels of 'Challenger' (6.1%) and frozen kernels of '2201' (6.4%). The highest starch content was obtained from frozen kernels of 'Yellow Baby' (19.4%), followed by frozen kernels of 'Merit' (18.9%) and fresh kernels of 'Jubilee' (18.2%). According to the average results from the two years, the starch content of the varieties was decreased after processing, except in 'Yellow Baby'. The lowest starch content, for all treatments was different according to varieties. The lowest starch content was obtained from 'Challenger' for fresh kernels whereas '2201' had the lowest starch content for frozen and canned kernels (Fig. 2). This indicates that the varieties differed significantly for starch content and there were differences among the varieties compliance processes. These results are in close agreement with earlier report of Lertrat and Pulam (2007). Also, the starch content values for fresh kernels reported here are similar to those found by Szymanek (2009).

In the first year of the experiment, the protein content was only affected significantly ($p<0.01$) by variety (Tab. 3). The highest protein content was attained, for all treatments, from 'Challenger' and '2201' with a mean content of 12.9% and 12.8%, respectively. 'Jubilee' (10.0%) and 'Merit' (10.1%) had the lowest protein content. In

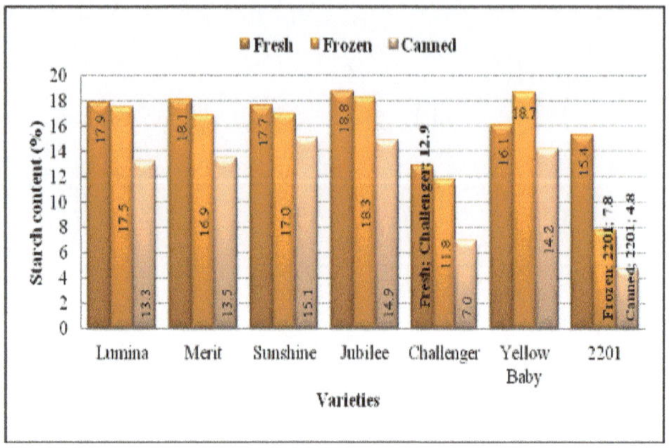

Fig. 2. Starch content (%) of fresh, frozen and canned sweet corn kernels, as average values for two years

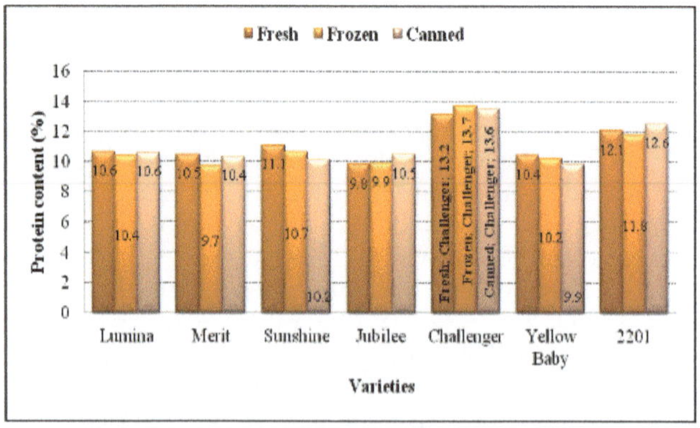

Fig. 3. Protein content (%) of fresh, frozen and canned sweet corn kernels, as average values for two years

2010, the effects of variety and 'treatment*variety' interaction on protein content were found significant ($p<0.01$). Protein contents changed between 9.8% ('Yellow Baby') and 14.0% ('Challenger'). According to the average results over the two years, fresh, frozen and canned sweet corn kernels had similar protein contents; the highest protein content was obtained, for all treatments, from 'Challenger', and it maintained its higher protein content after processing (Fig. 3). It has been reported that sweet corn is one of the most important sources of dietary protein among vegetables because of its relatively high protein concentration (3.5 g/100 g edible portion) (Goldman and Tracy, 1994). Our protein content results are similar to the results of Sanderson *et al.* (1979) and Goldman and Tracy (1994).

Conclusions

It can be concluded that kernel quality can be changed both among the varieties and treatments (fresh, frozen and canned). Sugar content of fresh kernels was higher than other treatments for all varieties. '2201' had the highest sugar and the lowest starch content for fresh, frozen and canned treatments. It could be suggested as suitable variety both for fresh consumption and processing. Fresh, frozen and canned sweet corn kernels had similar protein contents. Also, 'Yellow Baby' and 'Challenger' varieties can be advised for fresh consumption with moderate sugar and lower starch content. It has been reported that kernel quality is determined not only by varieties, but also by genetic groups (sugary, shrunken or sugary enhanced). Therefore, when compared on a kernel basis, sweet corn marketed as fresh, frozen or canned may be identified and it may be feasible to select varieties and genetic groups for different sweet corn markets.

Acknowledgement

This work was supported by The Eskişehir Osmangazi University, The Scientific Research Projects Commission (project number:200823026).

References

Andrew RH (1982). Factors influencing early seedling vigor of *shrunken*-2 maize. Crop Sci 22:263-266.

Aydın IS (2011). Frozen fruits and vegetables in Turkey. Available at http://tcp.gov.tr. Accessed in February 22, 2011.

Azanza F, Juvik JA, Klein BP (1994). Relationships between sensory quality attributes and kernel chemical composition. J Food Qual 17:150-172.

Azanza F, Bar-Zur A, Juvik JA (1996). Variation in sweet corn kernel characteristics associated with stand establishment and eating quality. J Food Sci 61:253-257.

Bozokalfa KM, Esiyok D, Ugur A (2004). Determination of yield quality and plant characteristic of some sweet corn (*Zea mays* L. var. *saccharata*) varieties as main and second crop in Ege region. The Journal of Agricultural Faculty 41(1):11-19.

Carey EE, Rhodes AM, Dickinson DB (1982). Post-harvest levels of sugars and sorbitol in *sugary enhancer (su1 se1)* and *sugary (su1 Se1)*maize. HortScience 17:241-242.

Carey EE, Dickinson DB, Rhodes AM (1984). Sugar characteristics of sweet corn populations from a *sugary enhancer* breeding program. Euphytica 33:609-622.

Douglass SK, Juvik JA, Splittstoesser WE (1993). Sweet corn seedling emergence and variation in kernel carbohydrate reverse. Seed Sci Technol 21:433-445.

Earle FR, Milmer RT (1944). Improvements in the determination of starch in corn and wheat. Cereal Chem 21:567-575.

Esiyok D, Bozokalfa KM, Ugur A (2004). Determination of yield quality and some plant characteristic of some sweet corn (*Zea mays* L. var. *saccharata*) varieties in different locations. The Journal of Agricultural Faculty 41(1):1-9.

Esiyok D, Bozokalfa KM (2005). The effects of sowing and planting dates on yield and some agronomic properties of sweet corn (*Zea mays* L. var. *saccharata*). The Journal of Agricultural Faculty 42(1):35-46.

Flora LF, Wiley RC (1974). Sweet corn aroma, chemical components, and relative importance in the overall flavor response. J Food Sci 39:770-773.

Goldman IL, Tracy WF (1994). Kernel protein concentration in *sugary*-1 and *shrunken*-2 sweet corn. HortScience 29(3):209-210.

Hale TA, Hassell RL, Phillips T (2005). Refractometer measurements of soluble solid concentration do not reliably predict sugar content in sweet corn. HortTechnology 15(3):668-672.

Horwitz W (2000). Official Method of Analysis of AOAC. International. 17th Edition. AOAC International, Maryland, USA.

Kirk RS, Sawyer R (1991). Pearson's composition and chemical analysis of foods. England, 9th Edition. Longman Scientific & Technical, Essex.

Kleinhenz MD (2003). Sweet corn variety trials in Ohio: recent top performers and suggestions for future evaluations. HortTechnology 13(4):711-718.

Kleinhenz MD (2008). Sweet corn quality-What is it?. Available at http://www.oardc.ohio-state.edu/kleinhenz. Accessed in May 25, 2008.

Kumari J, Gadag RN, Jha GK (2006). Heritability and correlation studies in sweet corn for quality traits, field emergence and grain yield. MNL 80:18-19.

Lertrat K, Pulam T (2007). Breeding for increased sweetness in sweet corn. Int J Plant Breed 1(1):27-30.

Najeeb S, Sheikh FA, Ahangar MA, Teli NA (2011). Popularization of sweet corn (*Zea mays* L. *saccharata*) under temperate conditions to boost the socieconomic conditions. Maize Genetic Cooperation Newsletter 85.

Randle WM, Davis DW, Groth JV (1984). The effects of corn leaf rust on maturity and quality of fresh market ears of sweet corn. J Amer Soc Hort Sci 109:654-648.

Sanderson JE, Paulis JW, Porcuna FN, Wall JS (1979). Sweet corn: varietal and developmental differences in amino acid content and composition of grain. J Food Sci 44:386-388.

SAS (1996). Statical analysis system. Institute Inc SAS Version. 6.12. Cary NC (Computer program).

Styer RC, Cantliffe DJ, Hannah LC (1980). Differential seed and seedling vigor in *shrunken-2* compared to three other genotypes of corn at various stages of development. J Am Soc Hort Sci 105:329-332.

Szymanek M (2009). Influence of sweet corn harvest date on kernel quality. Res Agr Eng 55(1):10-17.

Szymanek M (2012). Processing of sweet corn, trends in vital food and control engineering. Ayman Amer Eissa (Ed.). InTech. Available at http://www.intechopen.com/books/trends-in-vital-food-and-control engineering/processing-of-sweet-corn. Accessed in May 22, 2013.

Tuncay O, Bozokalfa KM, Esiyok D (2005). Determination of ear characteristic and technological properties of some sweet corn (*Zea mays* L. var. *saccharata*) varieties as main and second crop. The Journal of Agricultural Faculty 42(1):47-58.

Turgut I (2000). Effects of plant population and nitrogen doses on fresh ear yield component of sweet corn (*Zea mays* L. var. *saccharata Sturt.*) grown under Bursa conditions. Turk J Agric For 24:341-347.

Williams II MM (2006). Functional relationships between giant ragweed (*Ambrosia trifida*) interference and sweet corn yield and ear traits. Weed Sci 54:948-953.

Zhu S, Mount JR, Collins JL (1992). Sugar and soluble solids changes in refrigerated sweet corn (*Zea mays* L.). J Food Sci 57:454-457.

Znidarcic D (2012). Performance and characterization of five sweet corn cultivars as influenced by soil properties. J Food Agric Environ 10(1):495-500.

11

Exogenous Proline and Betaine-induced Upregulation of Glutathione Transferase and Glyoxalase I in Lentil (*Lens culinaris*) under Drought Stress

Md. Rezwan MOLLA[1], M. Rawshan ALI[2], Mirza HASANUZZAMAN[3], Mahamud Hossain AL-MAMUN[4], Asgar AHMED[2], M.A.N. NAZIM-UD-DOWLA[1], Md. Motiar ROHMAN[2*]

[1]*Plant Genetic Resources Centre, Bangladesh Agricultural Research Institute, Gazipur-1701, Bangladesh*
[2]*Molecular Breeding Laboratory, Plant breeding Division, Bangladesh Agricultural Research Institute, Gazipur-1701, Bangladesh*
[3]*Department of Agronomy, Faculty of Agriculture, Sher-e-Bangla Agricultural University, Dhaka-1207, Bangladesh; motiar_1@yahoo.com (*corresponding author)*
[4]*Olericulture Division, Horticultural Research Centre, Bangladesh Agricultural Research Institute, Gazipur-1701, Bangladesh*

Abstract

This study was conducted to investigate the proline (Pro) and betaine (Bet) driven modulation of glutathione *S*-transferase (GST) and glyoxalase I (Gly I) in drought stressed lentil seedlings. Among the seven lentil varieties tested, BARI mosur 2 was found have higher activities of GST and Gly 1 which was selected for further time-course investigation. Drought stress resulted in a significant increase in glutathione (GSH), glutathione disulfide (GSSG) and hydrogen peroxide (H_2O_2) contents and GST and Gly I activities at different period of the stress, while the GST activity decreased significantly at one and three days drought. However, exogenous application of 15 mM Bet or Pro with drought stress resulted in an increase in GSH content, maintenance of high activities of GST and Gly I as compared to the control and mostly also drought stressed plants, with a concomitant decrease in GSSG content and H_2O_2 level suggesting important role in protecting cell from the toxic effects of reactive oxygen species (ROS) and methyl glyoxal (MG) detoxification system. The application of Pro and Bet with the stress period revealed that they have important role in upregulating the homeostasis in lentil under stress condition. However, Pro exhibited better protection than Bet under drought stress. These findings suggest that both Pro and Bet provided a protective role in drought induced oxidative stress by reducing H_2O_2 levels and by increasing the antioxidant defense systems.

Keywords: pulse crops, antioxidant defense, water stress, reactive oxygen species

Introduction

Plants are constantly challenged by various abiotic stresses, such as drought, salinity, cold, high temperature, chemical toxicity, high light intensity may trigger in plants oxidative stress, generating the formation of reactive oxygen species (ROS). These species are partially reduced or activated derivatives of oxygen, comprising both free radical ($O_2^{\bullet-}$, $^1O_2$, $OH^{\cdot}$, $OH_2^{\bullet}$) and non-radical (H_2O_2) forms, leading to cellular damage, metabolic disorders and senescence processes, structural and functional loss of cell organelles, and eventually lead to death (Blokhina *et al.*, 2003). Osmotic stress caused by drought is one of the major abiotic factors limiting crop productivity because it affects almost all plant functions.

To counteract osmotic stress, many plants accumulate several kinds of compatible solutes such as Pro (Pro), betaine (Bet), sugars and polyols. Among them, Pro and Bet are the most common compatible solutes that contribute to osmotic adjustment (Greenway and Munns, 1980; Rhodes and Hanson, 1993; Hasegawa *et al.*, 2000; Ashraf and Foolad, 2007). In addition to their roles as osmoprotectants, Pro and Bet might perform a protective function by scavenging reactive oxygen species (ROS) (Hayat *et al.*, 2012). Both Pro and Bet improve salt tolerance in tobacco BY-2 cells by increasing the activity of enzymes involved in the antioxidant defense system (Hoque *et al.*, 2007a, b). Reactive oxygen species are highly reactive and toxic to plants and can lead to cell death by causing damage to proteins, lipids, DNA and carbohydrates (Noctor and Foyer, 1998; Apel and Hirt, 2004). Plants possess both enzymatic and non-enzymatic antioxidant defense systems to protect their cells against ROS (Noctor and Foyer, 1998;

Apel and Hirt, 2004). Thiol or thiol-containing compounds are chemically the most active groups found in cells and are known to act as antioxidants, and participate in the detoxification of xenobiotics. Reduced glutathione (GSH) is the most abundant low molecular weight thiol in plants and plays an important role in the detoxification of ROS (Noctor and Foyer, 1998; Noctor *et al.*, 2002; Smirnoff, 2005; Gill *et al.*, 2013). Glutathione dependent enzymes, glutathione *S*-transferases (GSTs; EC 2.5.1.18) and Glyoxalase I (Gly-1) play important role in stress combating through homeostatic and detoxification system. GSTs are the most studied multifunctional stress-inducible family enzyme which catalyze the conjugation of the tripeptide glutathione (γ-Glu-Cys-Gly; GSH) to a variety of electrophilic compounds to direct them to specific sites both intra- and extra-cellularly. Recently, these two enzyme systems have been reported as selection criteria for stress tolerant genotypes (Hefny and Abdul-Kader, 2007). On the other hand, pulse crop including lentil (*Lens culinaris*) is important source of sulphur which have important role in GSH metabolism. Since lentil is an important pulse crop and is sensitive to drought, for better understanding the mechanism of homeostatic role, this study was designed to examine the regulatory activities of the GST and Gly-I in lentil under draught condition with supplement of Pro and Bet.

Materials and Methods

Plant materials and stress treatments

Seeds of seven lentil (*Lens culinaris* L.) varieties viz. 'BARI' masur 1, 2, 3, 4, 5, 6 and 7 were collected from the Pulse Crop Research Division, Bangladesh Agricultural Research Institute. Seeds of uniform size were selected and surface-washed several times with sterile distilled water. The seeds were then sown in pot. Germinated seedlings were then allowed to grow under normal conditions (light, 100 µmol photon m^{-2} s^{-1}; temperature, 25±2 °C; RH, 65-70%) using Hogland solution as a source of nutrients. After 12 days, Lentil (BARI masur 2) seedlings were further grown without (control) and treated with 15 mM Pro and Bet solution at 1, 3, 5 and 7 day drought condition. The experiment was repeated three times under the same conditions.

Measurement of H_2O_2

H_2O_2 was assayed according to the method described by Yu *et al.*. (2003). H_2O_2 was extracted by homogenizing 0.5 g of leaf samples with 3 ml of 50 mM potassium-phosphate (K-P) buffer (pH 6.5) at 4 °C. The homogenate was centrifuged at 11500×*g* for 15 min. Three ml of supernatant was mixed with 1 ml of 0.1% $TiCl_4$ in 20% H_2SO_4 (v/v) and kept in room temperature for 10 min. After that the mixture was again centrifuged at 11500×*g* for 15 min. The optical absorption of the supernatant was measured spectrophotometrically (UV-1800, Simadzu, Japan) at 410 nm to determine the H_2O_2 content (Є=0.28 $\mu M^{-1}cm^{-1}$) and expressed as µmol g^{-1} fresh weight.

Extraction and measurement of glutathione

Fresh leaves with shoot were homogenized in ice-cold acidic extraction buffer (5% meta-phosphoric acid containing 1 mM EDTA) using a mortar and pestle. Homogenates were centrifuged at 11,500×*g* for 15 min at 4 °C and the supernatant was collected for analysis of glutathione.

The glutathione pool was assayed according to previously described methods (Yu *et al.*, 2003) with modifications (Paradiso *et al.*, 2008) utilizing 200 µl of aliquots of supernatant neutralized with 300 µl of 0.5 M K-P buffer (pH 7.0). Based on enzymatic recycling, GSH is oxidized by 5,5′-dithio-bis (2-nitrobenzoic acid) (DTNB) and reduced by NADPH in the presence of GR, and glutathione content is evaluated by the rate of absorption changes at 412 nm of 2-nitro-5-thiobenzoic acid (NTB) generated from the reduction of DTNB. GSSG was determined after removal of GSH by 2-vinylpyridine derivatization. Standard curves with known concentrations of GSH and GSSG were used. The content of GSH was calculated by subtracting GSSG from total glutathione.

Determination of protein

The protein concentration of each sample was determined following the method of Bradford (1976) using BSA as a protein standard.

Enzyme extraction and assays

Using a pre-cooled mortar and pestle, 0.5 g of leaf tissue was homogenized in 1 ml of 50 mM ice-cold K-P buffer (pH 7.0) containing 100 mM KCl, 1 mM ascorbate, 5 mM β-mercaptoethanol and 10% (w/v) glycerol. The homogenates were centrifuged at 11500×*g* for 10 min and the supernatants were used for determination of enzyme activity. All procedures were performed at 0-4 °C.

GST (EC:2.5.1.18) activity was determined spectrophotometrically by the method of Rohman *et al.*, (2009) with some modifications. The reaction mixture contained 100 mM Tris-HCl buffer (pH 6.5), 1.5 mM GSH, 1 mM 1-chloro-2,4-dinitrobenzene (CDNB), and enzyme solution in a final volume of 700 µl. The enzyme reaction was initiated by the addition of CDNB and the increase in absorbance was measured at 340 nm for 1 min. The activity was calculated using the extinction coefficient of 9.6 $mM^{-1}cm^{-1}$.

Glyoxalase I (EC:4.4.1.5) assay was carried out according to Hasanuzzaman and Fujita (2013). Briefly, the assay mixture contained 100 mM K-P buffer (pH 7.0), 15 mM magnesium sulphate, 1.7 mM GSH and 3.5 mM MG in a final volume of 700 µl. The reaction was started by the addition of MG and the increase in absorbance was recorded at 240 nm for 1 min. The activity was calculated using the extinction coefficient of 3.37 mM^{-1} cm^{-1}.

Statistical analysis

All data obtained were subjected to analysis of variance (ANOVA) and the mean differences were compared by a Tukey's Test using SAS software (Addinsoft 2010). Differences at $p<0.05$ were considered significant.

Results and Discussion

Before stress implementation, the activities of GST and Gly I were compared among the lentil varieties. Among the

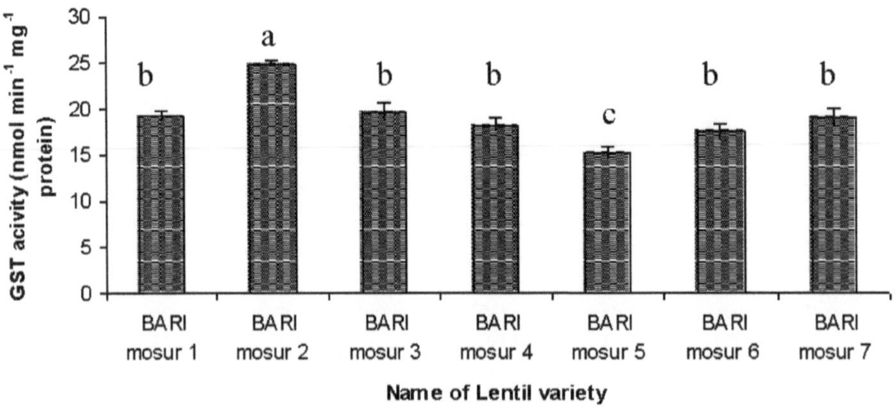

Fig. 1. Activities of glutathione *S*-transferase (GST) in lentil varieties under control condition

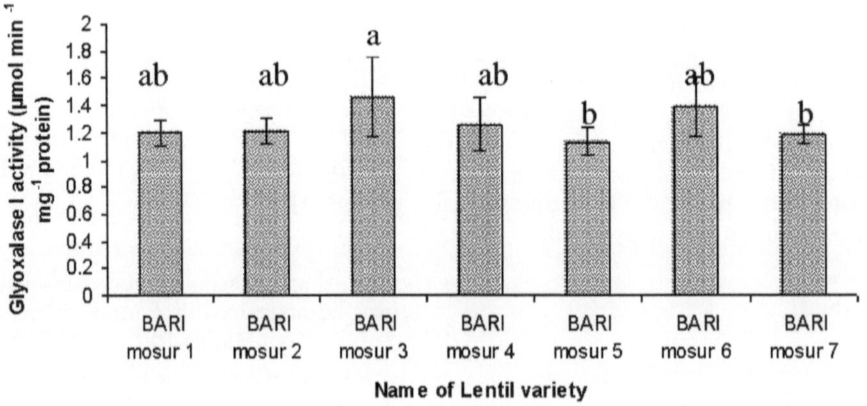

Fig. 2. Activities of glyoxalase I (Gly 1) in lentil varieties under control condition

lentil varieties, BARI mosur 2 exhibited the highest GST activity, while BARI masur 3 had the highest Gly I activity (Figs. 1 and 2). However, considering all, BARI masur 2 was selected for details study.

Activities of glutathione S-transferase

Incidences of drought stress in plants are might linked to osmoregulation problems and oxidative damage. Plant accumulates compatible solute such as Pro and Bet to mitigate the damaging effects of salt stress. Recent studies in plants have demonstrated that exogenous application of Bet or Pro at high concentration (1–20 mM) enhance tolerance to abiotic oxidative stress (Park *et al.*, 2006; Hoque *et al.*, 2007a, b; Huang *et al.*, 2009; Islam *et al.*, 2010; Hossain and Fujita, 2010). To check whether the upregulation of glutathione GST in lentil seedling plants in response to drought stress, its levels were measured in one, three, five and seven days seedlings under control as well as stressful conditions, It was found that GST activity decreased slightly (15 and 20%) in response to one and three days drought stress (Fig. 3). In contrary, GST activity increased sharply due to duration of stress followed by 24% at five days and 54% at seven days stress. Moreover, exogenous Pro and Bet exhibited further increase in GST activity by 61% and 41%, respectively, under five and seven days drought stress.

The plant GSTs are a large and diverse group of enzymes that catalyze the conjugation of electrophilic xenobiotic substrates with the GSH and it is associated with responses to various forms of abiotic stress and confers stress tolerance in plants (Hossain *et al.*, 2006; Dixon *et al.*, 2010). In our study, GST activity in lentil seedlings profoundly increased in response to drought stress (Fig. 3). More importantly, drought stress seedlings supplemented with Pro and Bet also increased the activities further. This enhanced activity of GST decreased the levels of H_2O_2 and lipid peroxidation in lentil seedlings under drought stress condition (Fig. 4). Similar increases in GST activity after Pro and Bet supplementation under stress were observed by other researchers (Hasanuzzaman and Fujita, 2011). The higher level of GSTs activities in lentil seedlings suggested that the enzymes are capable of detoxifying a maximum amount of toxins in the presence of sufficient GSH. Beside detoxification, they also play important role in other physiological functions such as GSH peroxidase activity (Bartling *et al.*, 1993; Mannervik and Danielson, 1988) and intracellular binding and transport of phytochemicals (Edwards *et al.*, 2000). However, the fact that exogenous Pro and Bet increased GST activities and maintained higher GSH level indicates that both Pro and Bet might play a significant role in GSH biosynthesis and metabolism. Transgenic tobacco plants overexpressing both GST and

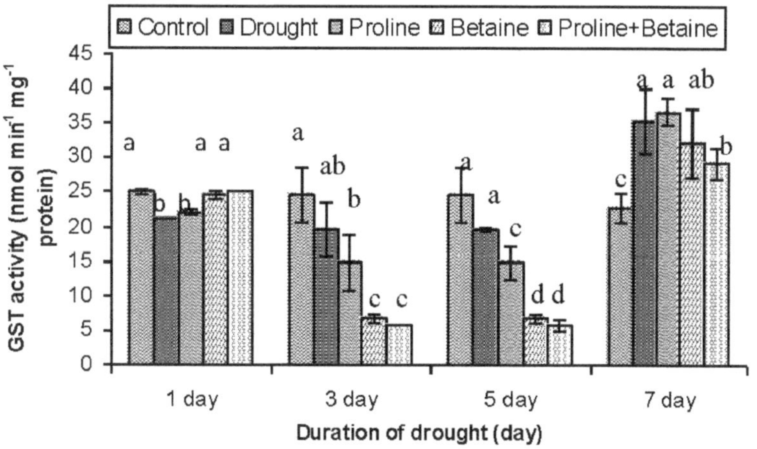

Fig. 3. Activities of glutathione *S*-transferase (GST) in lentil cv. BARI mosur 2 under control and stress condition. Fifteen mM of Proline (P) and betaine (B) was treated as foliar spray twice a day. Data were measured in soluble protein extracts of seedling after 1, 3, 5 and 7 days of stress implementation. Values represent the means ± SE from three independent experiments.

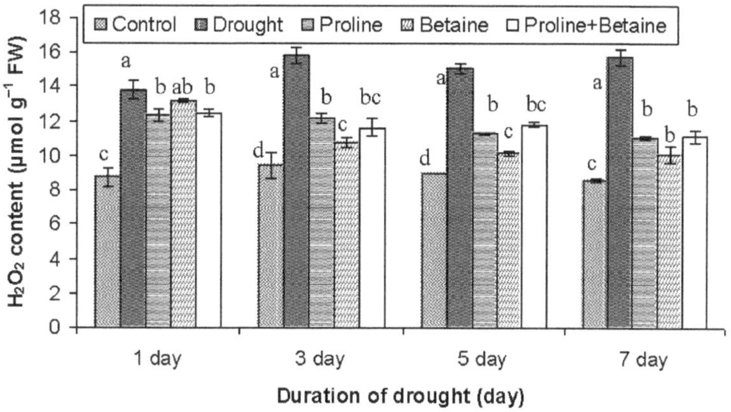

Fig. 4. Content of Hydrogen peroxide (H_2O_2) in lentil cv. BARI mosur 2 under control and stress condition. Fifteen mM of Proline (P) and betaine (B) was treated as foliar spray twice a day. Data were measured in soluble protein extracts of seedling after 1, 3, 5 and 7 days of stress implementation. Values represent the means ± SE from three independent experiments

GPX also displayed improved tolerance to salinity and chilling stress (Roxas *et al.*, 1997).

Level of hydrogen peroxide

A significant increase in the H_2O_2 level was observed in Lentil in response to drought stress and compared to control the levels was 69 and 84% higher at five and seven day drought treatment respectively (Fig. 4). Proline and Bet supplemented drought-stressed seedlings also showed continual increase in H_2O_2 content. However, the level of H_2O_2 content was significantly lower than the seedlings treated with drought alone after different days of treatment. After seven days of treatment Pro and Bet supplemented drought-stressed seedlings showed 30 % and 36 % decrease in H_2O_2 content as compared to the seedlings subjected to drought stress without Pro and Bet.

Glutathione contents

Reduced glutathione (GSH) is a key component of the antioxidant network that scavenges ROS either directly or indirectly by participating in the ascorbate glutathione cycle (Noctor and Foyer, 1998; Smirnoff, 2005).

The central role of GSH in the antioxidant defense system is due to its ability to regenerate ascorbate through reduction of dehydroascorbate via the ascorbate glutathione cycle (Noctor and Foyer, 1998). It is known that (GSH) is essential for effective scavenging of toxic compounds (such as H_2O_2 and organic H_2O_2) and for maintenance of other antioxidants.

In addition, GSH plays a vital role in the antioxidant defense system as well as the glyoxalase system by acting as a substrate or cofactor for certain enzymes like glutathione peroxidase (GPX), GST and Gly I and thus participate in removal of ROS, methylglyoxal (MG) and other endogenous toxic compounds (Noctor *et al.*, 2012).

To understand the GSH-dependent protective mechanisms offered by Pro and Bet in plant responses to oxidative stress, GSH level and its associated enzyme activities induced by Pro and Bet under drought stress are discussed. In this study, slightly decreases in GSH contents were observed in response to one day drought stress, as

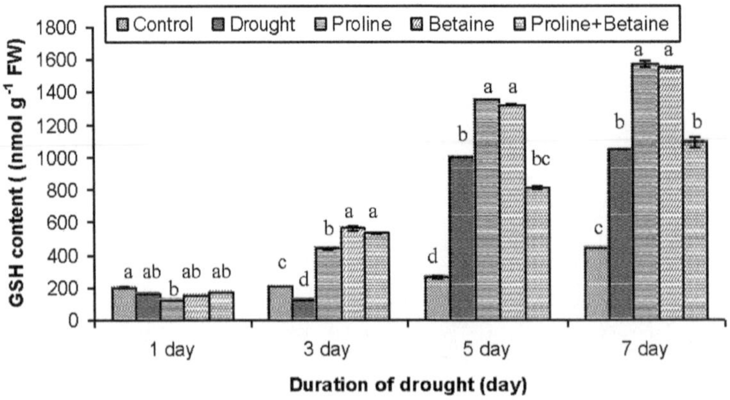

Fig. 5. Content of reduced glutathione (GSH) in lentil cv. BARI mosur 2 under control and stress condition. Fifteen mM of Proline (P) and betaine (B) was treated as foliar spray twice a day. Data were measured in soluble protein extracts of seedling after 1, 3, 5 and 7 days of stress implementation. Values represent the means ± SE from three independent experiments

compared to the control (Fig. 5). On the contrary, the amount of GSH increased dramatically after supplementation of Pro and Bet with extent of drought stress. The increased level of the GSH pool is generally regarded as a protective response against oxidative stress (May and Leaver, 1993; Xiang and Oliver, 1998), although defense against stress situations sometimes occurs irrespective of the GSH concentration (Potters *et al.*, 2004). The seedlings exposed to seven days of drought stress showed highest amount of GSH (1569.90 nmol/g FW) in presence of Pro compared to their respective control (Fig. 5). An increase in GSH content due to Pro pretreatment and lower lipid peroxidation has recently been reported (Xu *et al.*, 2009). Therefore, both Pro and Bet might play a significant role in maintaining higher GSH level either through efficient recycling or by modulating higher GSH synthesis (Kocsy *et al.*, 2005; Hossain and Fujita, 2010).

Glutathione disulfide content

Drought stress significantly increased the content of GSSG in lentil seedlings (Fig. 6). On the other hand, the seedlings in response to one day drought stress which were supplemented with Pro and Bet maintained the GSSG levels lower than the seedlings that were grown under drought condition without Pro and Bet supplementation. At seven days after supplementation, drought stress caused a significant decrease in the amount of GSSG, whereas, importantly, both Pro and Bet could not significantly contributed to the maintenance of the drought-induced redox state. Our results indicated that an increase GSH biosynthesis is an immediate response to drought stress and a drastic reduction of GSH concentration at seven days due to high drought stress and simultaneous increase in GSSG (Fig. 1a-C), a true indication of oxidative stress. The formation of GSSG in drought-treated seedlings might be due to the reaction of GSH with oxyradicals generated due to oxidative stress or due to enhancement of GPX activity that decompose H_2O_2 (Aravind and Prasad, 2005; Shalata *et al.*, 2001) as well as dehydroascorbate reductase (DHAR) activity (Hoque *et al.*, 2007b) which required GSH to regenerate ascorbate from dehydroascorbate. Proline and Bet supplemented drought stressed seedlings maintained higher GSH level and glutathione redox state with simultaneous decrease in the oxidized GSSG (Figs. 5 and 6).

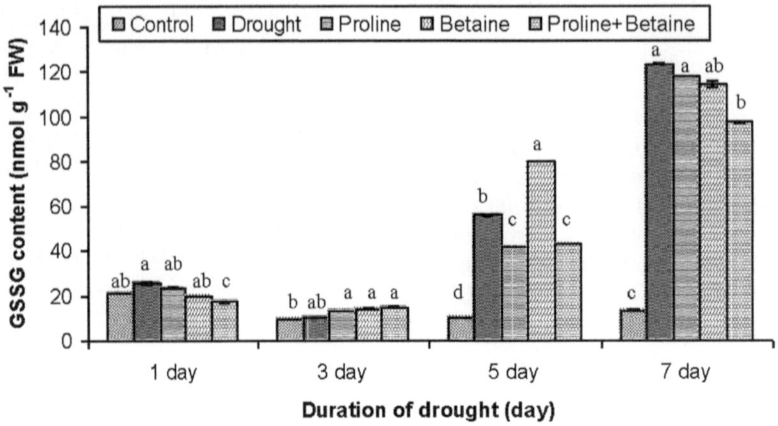

Fig. 6. Content of oxidized glutathione (GSSG) in lentil cv. BARI mosur 2 under control and stress condition. Fifteen mM of Proline (P) and betaine (B) was treated as foliar spray twice a day. Data were measured in soluble protein extracts of seedling after 1, 3, 5 and 7 days of stress implementation. Values represent the means ± SE from three independent experiments

Activity of glyoxalase I

Glyoxalase I could be expected to be a house-keeping protein present in all cells. Although up-regulation of glyoxalase I activity in response to salt, water deficit, ABA and heavy metal stress treatments has been reported earlier in plant (Espartero *et al.*, 1995; Veena *et al.*, 1999; Yadav *et al.*, 2005). It is conceivable that an elevated level of glyoxalase I activity is required to remove MG, a toxic and unavoidable by-product of triosephosphate metabolism produced in ample amounts under normal and various stressful conditions. In this study, dramatic increase in Gly I activity was observed in response to drought stress. As shown in Fig. 7, drought treatment for three days resulted in 455% increase in Gly I activities compared to control. However, exogenous Pro and Bet application in drought stressed seedlings caused more increase in the activities of Gly I compared to drought alone. Upon three days of drought treatment with Pro and Bet the activity increased by 773 and 573%, respectively compared to control (Fig. 7). Application of Pro with drought treatment at three days five and seven days enhanced the activity of Gly I compared to drought treatment with Bet or combination of Pro and Bet (Fig. 7).

Increase in Gly I activity during stress tolerance may also indicate active metabolic status of the cell, in which cell division and growth are compromised in order to conserve energy for mobilization of resources towards stress tolerance and defense strategies. The gene expression profile of Gly I also showed true reflection of possible changes in activity levels due to different abiotic stresses. Moreover, overexpression of glyoxalase genes involved in the regulation of glutathione homeostasis (namely, GR, GST / GPX) in transgenic plants has been shown to result in an increased tolerance against oxidative stress (Broadbent *et al.*, 1995; Noctor *et al.*, 1998; Roxas *et al.*, 1997; Yadav *et al.*, 2005).

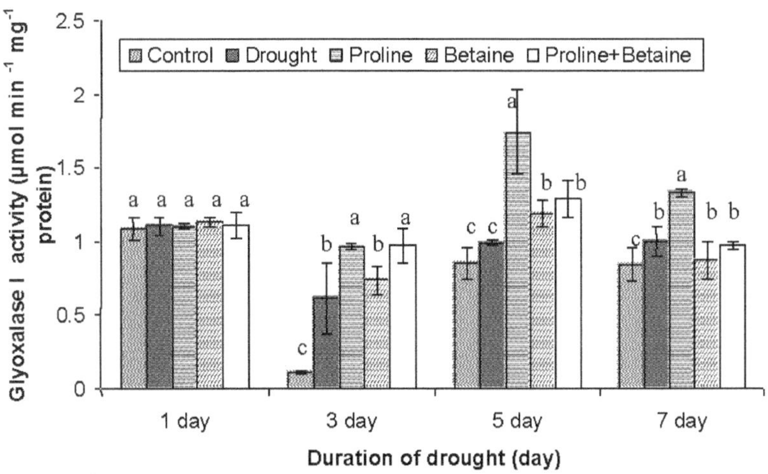

Fig. 7. Activities of glyoxalase I (Gly I) in lentil cv. BARI mosur 2 under control and stress condition. Fifteen mM of Proline (P) and betaine (B) was treated as foliar spray twice a day. Data were measured in soluble protein extracts of seedling after 1, 3, 5 and 7 days of stress implementation. Values represent the means ± SE from three independent experiments

Conclusion

In conclusion, drought stress induces a severe oxidative stress in lentil seedlings where the antioxidant defense and glyoxalase systems seemingly fail to combat with the stress-induced oxidative damage. Exogenous applications of Pro or Bet showed enhance tolerance to oxidative damage by enhancing ROS and MG detoxification systems. These findings suggest that both Pro and Bet provide protective effects against drought induced oxidative stress by reducing H_2O_2 and by increasing the antioxidant defense and glyoxalase systems. However, Pro provided better protective roles that Bet. Therefore, it was concluded that GST and glyoxalase system mediated detoxification and antioxidant activities played important roles with supplementation of Pro and Bet.

References

Apel K, Hirt H (2004). Reactive oxygen species: metabolism, oxidative stress, and signal transduction. Annu Rev Plant Biol 55:73-99.

Aravind PA, Prasad NV (2005). Modulation of cadmium induced oxidative stress in *Ceratophyllum demersum* by zinc involves ascorbate-glutathione cycle and glutathione metabolism. Plant Physiol Biochem 45:107-116.

Ashraf M, Foolad MR (2007). Roles of glycinebetaine and praline in improving plant abiotic stress resistance. Environ Exp Bot 59:206-16.

Bartling D, Radzio R, Steiner U, Weiler EW (1993). A glutathione *S*-transferase with glutathione-peroxidase activity from *Arabidopsis thaliana*: Molecular cloning and functional characterization. Eur J Biochem 216:579-586.

Blokhina O, Virolainen E, Gagerstedt KV (2003). Antioxidants, oxidative damage and oxygen deprivation stress: a review. Ann Bot 91:179-194.

Bradford MM (1976). A rapid and sensitive method for the quantitation of microgram quantities of protein utilizing the principle of protein-dye binding. Anal Biochem 72:248-254.

Broadbent P, Creissen GP, Kular B, Wellburn AR, Mullineaux P (1995). Oxidative stress responses in transgenic tobacco containing altered levels of glutathione reductase activity. Plant J 8: 247-255.

Dixon DP, Skipsey M, Edwards R (2010). Roles for glutathione transferases in plant secondary metabolism. Phytochem 71:338-350.

Edwards R, Dixon DP, Walbot V (2000). Plant glutathione *S*-transferases: Enzymes with multiple functions in sickness and health. Trends Plant Sci 5:193-198.

Espartero J, Aguayo IS, Pardo JM (1995). Molecular characterization of glyoxalase I from a higher plant; upregulation by stress. Plant Mol Biol 29:1223-1233.

Gill SS, Anjum NA, Hasanuzzaman M, Gill R, Trivedi DK, Ahmad I, Pereira, Tuteja N (2013) Glutathione reductase and glutathione: a boon in disguise for plant abiotic stress defense operations. Plant Physiol Biochem 70:20-212.

Greenway H, Munns R (1980). Mechanisms of salt tolerance in nonhalophytes. Annu Rev Plant Physiol 31:149-90.

Hasanuzzaman M, Fujita M (2011). Selenium pretreatment upregulates the antioxidant defense and methylglyoxal detoxification system and confers enhanced tolerance to drought stress in rapeseed seedlings. Biol Trace Elem Res 143:1758-1776.

Hasanuzzaman M, Fujita M (2013). Exogenous sodium nitroprusside alleviate arsenic-induced oxidative stress in wheat seedlings by enhancing antioxidant defense and glyoxalase system. Ecotoxicology 22:584-596.

Hasegawa PM, Bressan RA, Zhu JK, Bohnert HJ (2000). Plant cellular and molecular responses to high salinity. Annu Rev Plant Physiol Plant Mol Biol 51:463-99.

Hayat S, Hayat Q, Alyemeni MN, Wani AS, Pichtel J, Ahmad A (2012). Role of proline under changing environments: A review. Plant Signal Behav 7:1-11.

Hefny M, Abdel-Kader DZ (2007). Antioxidant-Enzyme System as Selection Criteria for Salt Tolerance in Forage Sorghum Genotypes (*Sorghum bicolor* L. Moench. Int J Plant Breed Genet 1:38-53.

Hoque MA, Banu MNA, Okuma E, Amako K, Nakamura Y, Shimoishi Y, Murata Y (2007b). Exogenous praline and glycinebetaine ingresses NaCl-induced ascorbateglutathione cycle enzyme activities and proline improves salt tolerance more than glycinebetaine in tobacco Bright yellow-2 suspension- cultured cells. J Plant Physiol 164:553-561.

Hoque MA, Okuma E, Banu MNA, Nakamura Y, Shimoishi Y, Murata Y (2007a). Exogenous proline mitigates the detrimental effects of salt stress more than exogenous betaine by increasing antioxidant enzyme activities. J Plant Physiol 164:553-561.

Hossain MA, Fujita M (2010). Evidence for a role of exogenous glycinebetaine and proline in antioxidant defense and methylglyoxal detoxification systems in mung bean seedlings under salt stress. Physiol Mol Biol Plant 16:19-29.

Hossain MZ, Hossain MD, Fujita M (2006). Induction of pumpkin glutathione *S*-transferase by different stresses and its possible mechanisms. Biol Plant 50:210-218.

Huang Y, Bie Z, Liu Z, Zhen A, Wang W (2009). Protective role of proline against salt stress is partially related to the improvement of water status and peroxidase enzyme activity in cucumber. Soil Sci Plant Nutr 55:698-704.

Islam MM, Hoque MA, Okuma E, Banu MNA, Shimoishi Y, Nakamura Y, Murata Y (2010). Exogenous proline and glycinebetaine increase antioxidant enzyme activities and confer tolerance to cadmium stress in cultured tobacco cells. J Plant Physiol 166: 1587-1597.

Kocsy G, Laurie R, Szalai G, Szilagyi V, Simon-Sarkadi L, Galiba G (2005). Genetic manipulation of proline levels affects antioxidants in soybean subjected to simultaneous drought and heat stresses. Physiol Plant 124:227-35.

Mannervik B, Danielson UH (1988). Glutathione transferases–structure and catalytic activity. CRC Crit Rev Biochem 23:283-337.

May MJ, Leaver CJ (1993). Oxidative stimulation of glutathione synthesis in *Arabidopsis thaliana* suspension cultures. Plant Physiol 103:621-627.

Noctor G, Foyer CH (1998). Ascorbate and glutathione: Keeping active oxygen under control. Annu Rev Plant Physiol Plant Mol Biol 49:249-279.

Noctor G, Gomez L, Vanacker H, Foyer CH (2002). Interactions between biosynthesis, compartmentation and transport in the control of glutathione homeostasis and signaling. J Exp Bot 53:1283-304.

Noctor G, Mhamdi A, Chaouch S, Han Y, Neukermans J, Marquez-Garcia B, Queval G, Foyer CH (2012). Glutathione in plants: an integrated overview. Plant Cell Environ 35:454-484.

Paradiso A, Berardino R, de Pinto M, di Toppi LS, Storelli FT, de Gara L (2008). Increase in ascorbate-glutathione metabolism as local and precocious systemic responses induced by cadmium in durum wheat plants. Plant Cell Physiol 49:362-374.

Park EJ, Jeknic Z, Chen THH (2006). Exogenous application of glycinebetaine increases chilling tolerance in tomato plants. Plant Cell Physiol 47:706-714.

Potters G, Horemans N, Bellone S, Caubergs RJ, Trost P, Guisez Y (2004). Dehydroascorbate influences the plant cell cycle through a glutathione-independent reduction mechanism. Plant Physiol 134:1479-1487.

Rhodes D, Hanson AD (1993). Quaternary ammonium and tertiary sulfonium compounds in higher plants. Annu Rev Plant Physiol Plant Mol Biol 44:357-84.

Rohman MM, Hossain MD, Suzuki T, Takada G, Fujita M (2009). Quercetin-4'-glucoside: a physiological inhibitor of the activities of dominant glutathione *S*-transferases in onion (*Allium cepa* L.) bulb. Acta Physiol Plant 31:301-309.

Roxas VP, Smith RK, Ellen ER, Allen RD (1997). Overexpression of glutathione *S*-transferase/ glutathione peroxidase enhances the growth of transgenic tobacco seedlings during stress. Plant Cell Physiol 41:1229-1234.

Shalata A, Mittova V, Volokita M, Guy M, Tal M (2001). Response of the cultivated tomato and its wild salt tolerant relative *Lycopersicon pennellii* to salt-dependent oxidative stress: the root antioxidative system. Physiol Plant 112:487-494.

Smirnoff N (2005). Antioxidants and reactive oxygen species in plants. Oxford: Blackwell Publishing.

Veena, Reddy VS, Sopory SK (1999). Glyoxalase I from *Brassica juncea*: molecular cloning, regulation and its over-expression confer tolerance in transgenic tobacco under stress. Plant J 17:385-395.

Xiang C, Oliver DJ (1998). Glutathione metabolic genes coordinately respond to heavy metals and jasmonic acid in *Arabidopsis*. Plant Cell 10:1539-1550.

Xu J, Yin HX, Li X (2009). Protective effects of proline against cadmium toxicity in micropropagated hyperaccumulator, *Solanum nigrum* L. Plant Cell Rep 28:325-353.

Yadav SK, Singla-Pareek SL, Ray M, Reddy MK, Sopory SK (2005). Transgenic tobacco plants overexpressing glyoxalase enzymes resist an increase in methylglyoxal and maintain higher reduced glutathione levels under salinity stress. FEBS Lett 579:6265-6271.

Yu CW, Murphy TM, Lin CH (2003). Hydrogen peroxide-induces chilling tolerance in mung beans mediated through ABA-independent glutathione accumulation. Funct Plant Biol 30:955-963.

Genetic Similarity Assessment among Selected Naked Oat Cultivars and Breeding Lines Using ISSR Markers

Edyta PACZOS-GRZEDA[1*], Piotr Tomasz BEDNAREK[2], Aneta KOROLUK[1], Zygmunt NITA[3], Zofia BANASZAK[4], Andrzej BICHONSKI[5], Marek CHMIEL[5], Agnieszka GRADZIELEWSKA[1], Katarzyna NOWACZYK[4], Aleksandra SZOLKOWSKA[4], Krystyna WERWINSKA[3], Patrycja WIECZOREK[5]

[1]*University of Life Sciences in Lublin, Institute of Plant Genetics, Breeding and Biotechnology, Akademicka Str 15, 20-950 Lublin, Poland; edyta.paczos@up.lublin.pl (*corresponding author); aneta.koroluk@up.lublin.pl; agnieszka.gradzielewska@up.lublin.pl*
[2]*Plant Breeding and Acclimatization Institute – NRI, Radzikow, 05-870 Blonie, Poland; p.bednarek@ihar.edu.pl*
[3]*Strzelce Plant Breeding Company, Glowna Str 20, 99-307 Strzelce, Poland; z_nita@hr-strzelce.pl; k_werwinska@hr-strzelce.pl*
[4]*DANKO Plant Breeding Company, Choryn 27, 64-000 Koscian, Poland; zofia.banaszak@danko.pl; katarzyna.nowaczyk@danko.pl, aleksandra.szolkowska@danko.pl*
[5]*Malopolska Plant Breeding Company, Polanowice, 32-090 Slomniki, Poland; andrzejb@hbp.pl; jare-polanowice@up.pl; patrycjawieczorek19@interia.pl*

Abstract

Naked oat refers to a variety of *Avena sativa* with lemma and palea separating from the grains. Its spikelets are multiflorous and morphologically different from the husked oat. Problems with preharvest sprouting, threshability, rancidity, a wide range of kernel sizes, as well as its relatively low tolerance to limited soil water content, are its main drawbacks. Nevertheless it could be an alternative to a conventional oat. Unfortunately, its genetic variation is still poorly recognized. In the given study a set of 26 naked oat cultivars and lines were analyzed with 25 inter-simple sequence repeats (ISSR) primers that amplified as many as 429 DNA fragments among which 204 were polymorphic. The average number of markers amplified per primer pair and polymorphism information content (PIC) value equaled to eight and 0.23, respectively. Forty four unique PCR products were identified for different genotypes. While Unweighted Pair-Group Method with Arithmetic Mean failed to distinguish the materials into main clusters it demonstrated that cultivars 'Akt', 'Polar', 'Cacko', 'Siwek', 'Nagus' and most of the DC lines were within a single group. Moreover, the cultivars that were closely related based on their breeding pedigree (related to 'Akt') were close to each other. Principal Coordinate Analysis explained 54.1% of variance and was in good agreement with the UPGMA. ISSR markers could be used for the evaluation of genetic similarity of cultivars and lines as well as the differentiation of individual genotypes. This study demonstrated that the available *A. sativa* naked type genetic pool is relatively wide and have the potential for further breeding progress.

Keywords: *Avena sativa*, genetic pool, molecular markers, genetic diversity

Introduction

Common oat is a cereal crop adapted to moderate climate, cultivated predominantly in temperate regions (Burrows, 1986). It is used for forage, feed and food production. Varieties of cultivated oat belong *to Avena sativa* ssp. *sativa* L. (spring oat, naked and husked type) (Valentine, 1995), and *A. sativa* ssp. *byzantina* L. (red oat) (Zeller, 1998).

High quality groat with excellent fat and amino acid composition and husk having an energy yield like that of straw determines the nutritional value of conventional, covered oat. Due to the husk, the yield of total energy is weaker in oat than in case of other cereals. Thus, the cultivation and use of oat is behind other plants rich in energy (Kirkkari, 2008). Moreover, husk decreases bulk density, reducing storage and transport efficiency (Burrows, 1986). Its percentage ranges from a minimum of 21% to a maximum of 41% (Ronald *et al.*, 1999). Husks comprise mainly of cell wall (> 83%) with nearly equal quantities of cellulose and hemicellulose (30-35%) (Welch *et al.*, 1983). Protein and oil contents are low (1.6-5% and 1-2.2%) (Salo and Kotilainen, 1970). Starch and water-soluble carbohydrates contents do not exceed 2% and 1%, respectively (Welch *et al.*, 1983). Husk composition within and among varieties may vary due to location and weather conditions (Ronald *et al.*, 1997; Sykut-Domańska, 2012). Oat husk is a poor quality feedstuff. The improvement in grain quality could be achieved by breeding for reduced husk content or breeding high yielding naked oat cultivars (Kirkkari, 2008).

Naked oat refers to a variety with lemma and palea separating from the grains during threshing. The spikelets of naked oat are multiflorous and morphologically different from the husked oat (Valentine, 1995). Whereas, husked oat spikelets may contain 1 to 3 (rarely 4) florets, naked oat consists of 4 to 7, but quite often may possess up to 12 florets (Burrows, 1986; Valentine, 1995). This trait, mostly controlled by a single dominant gene, is modified by several minor loci (Valentine, 1995; Doehlert *et al.*, 2006). The oat kernels are covered with fine hair termed trichomes. Trichomes are problematic, and efforts are being made in order to obtain cultivars with reduced pubescence (Kirkkari *et al.*, 2009). Existed differences of cultivars in pubescence resulted in attempts of breeding varieties with reduced trichome numbers.

Naked oat cultivars are characterized by lower mass of thousand grain and higher hectolitre weight than common oat (Sykut-Domańska, 2012). The yield of naked oats is ca. 25-30% lower than the yield of the common one. However, when the same characteristics are recalculated for seeds without husk, the results are comparable (Nita, 2003). Moreover, naked oat is enriched in fat and unsaturated fatty acids, water – soluble β – glucans forming dietary fibre and good quality proteins (Welch *et al.*, 1983). It should be mentioned that the chemical content of naked oat seeds may vary significantly on different continents. Canadian and the USA varieties have more proteins (> 14%) and fewer fats (< 6%) while European and Australian forms have fewer proteins (11-13%) but are richer in fats (7-10%) (Nita 2003). Unfortunately, problems with preharvest sprouting, threshability, rancidity, a wide range of kernel sizes as well as its relatively low tolerance to limited soil water content (Nita, 2003) are its main drawbacks (Forsberg and Reeves, 1992). Moreover, the husk plays a crucial role in protecting the groat from damage (Burrows, 1986) and pathogenic attack (Ronald *et al.*, 1999). Nevertheless, dehusked conventional oat, does not achieve comparable economic result as naked one, as dehusking costs and disposition of the husk waste is considerably large (Kirkkari, 2008). Naked oat could be a suitable alternative to conventional oat.

There is an interest in the evaluation of genetic diversity of the available oat genetic pool and use of this information in practical breeding (Tinker *et al.*, 2009). Genetic diversity of breeding materials depends largely on differentiation of parental lines used for crossing in conventional breeding programs during last decades. Moreover, selection directed to create improved locally adapted crop varieties, tends to erode genetic diversity. For pragmatic reasons, most oat breeders favour crosses among their own best varieties, which guarantees short-term success. Molecular studies using AFLP markers applied to a core set of cultivated oat germplasm (Fu *et al.*, 2005), as well as analyses performed with DArTs (Tinker *et al.*, 2009), indicated that most genetic relatedness was associated with geographical origin and development within the same breeding program. The evaluation of genetic diversity plays a prominent role in the characterization of breeding lines or varieties. It could be also applied for the selection of parents in the development of crossing schemes (Paczos-Grzeda, 2004). The higher genetic distance between parental forms, the greater probability to achieve new allele combinations (Cox and Murphy, 1990). Better knowledge of genetic relatedness of parents selected for crossing could help to maintain genetic diversity, evaluate the potential vulnerability to abiotic stresses and pests, sustain long-term selection gain and assure continued genetic improvement (Martin *et al.*, 1991; Chowdhury *et al.*, 2002). Therefore, understanding the genetic relationships among naked oat breeding lines, could maximize the rational use of the genetic resources and encourage breeders to use new germplasm. The primary focus of the study was to perform preliminary estimation of the available genetic pool of the naked oat cultivars registered in Poland and 18 breeding lines based on ISSR marker polymorphisms.

Materials and Methods

Plant material and DNA extractions

Twenty six naked oats (*A. sativa* L.) including five cultivars registered in Poland (cv. 'Akt' was represented by four sublines) and eighteen breeding lines from three breeding programs listed in Tab. 1 were analysed. Pedigree of oat cultivars was reproduced based on breeding records. Genomic DNA was extracted from 15-20 coleoptiles of several day old seedlings of each breeding form following the CTAB procedure (Doyle and Doyle, 1987).

Tab. 1. Arrangement of the analysed naked oat cultivars and lines with their descriptive data. D, M, S states for breeding programs run in commercial companies

Cultivar/ Line	Pedigree	Breeding program	Comments
'Akt'	'Adam'×'Adamo'	S	Registered in 1997
'Akt' 1	'Adam'×'Adamo'	S	Sister line of cv. 'Akt'
'Akt' 2	'Adam'×'Adamo'	S	Sister line of cv. 'Akt'
'Akt' bis	'Adam'×'Adamo'	S	Sister line of cv. 'Akt'
STH 721	'Akt'×AVE 2999	S	AVE 2999 – *A. sterilis*
'Polar'	'Ago'×'Ramiro'	S	Registered in 2002
'Cacko'	'Ago'×'Ramiro'	S	Registered in 2000
'Maczo'	STH 7520×STH 7376	S	Registered in 2010
STH 8307	STH 8070×STH 6102	S	Preliminary field trial
STH 6315	STH 14063×STH 13876	S	Preliminary field trial
'Siwek'	'Benquel' × 'Akt'	M	Registered in 2010
'Nagus'	'Auron'×POB-W 481	D	Registered in 2011
DC 2188	STH 3997×'Jakub'	D	Preliminary field trial
DC 2215	'Bullion'×'Mozart'	D	Preliminary field trial
DC 2612	'Samuel'×STH 6202	D	Preliminary field trial
DC 2711	('Izak'×'Bullion')× 'Akt'	D	Preliminary field trial
DC 2973	'Polar'×'Mozart'	D	Preliminary field trial
DC 3674	STH 4899×'Polar'	D	Preliminary field trial
DC 3080	P(508)19251×CHD 1743	D	Preliminary field trial
DC 3996	P(508)19251×CHD1743	D	Preliminary field trial
DC 3416	'Sallust'×'Akt'	D	Preliminary field trial
DC 3249	POB-W 481×'Jawor'	D	Preliminary field trial
DC 3170	P(508)19251×CHD 1743	D	Preliminary field trial
DC 2567	'Auron'×POB-W 481	D	Preliminary field trial
DC 2931	STH 4879 × 'Sallust'	D	Preliminary field trial
DC 2934	STH 4879×'Sallust'	D	Preliminary field trial

ISSR

Amplification was performed according to the ISSR method described by Zietkiewicz *et al.*, (1994) with modifications. Reaction mixture of 15 μl contained 1 × PCR Buffer (10 mM Tris pH 8.8; 50 mM KCl; 0.08% Nonidet P40), 160 μM of each dNTP, 0.35 pmol of primer, 1.3 mM of $MgCl_2$, 0.4 mM of spermidine, 0.5 U of Taq DNA Polymerase and 30 ng of template DNA. The twenty five ISSR random primers were used in the analysis. Sequences of the applied primers are presented in Tab. 2. Amplifications were carried out in a T1 Biometra thermal cycler with an initial denaturation step at 95 °C for 4 min, followed by 3 cycles of 30 sec at 95 °C, 45 sec at 54 °C, 2 min at 72 °C; 3 cycles of 30 sec at 95 °C, 45 sec at 53 °C, 2 min at 72 °C; 32 cycles of 30 sec at 95 °C, 45 sec at 52 °C, 2 min at 72 °C and a final extension step of 7 min at 72 °C. Amplification products were separated on 2.5% agarose gel containing 0.1% ethidium bromide in 1 × TBE Buffer (89 mM Tris – borate, 2.5 mM EDTA). DNA marker GeneRulerTM 100 bp Plus DNA Ladder was used.

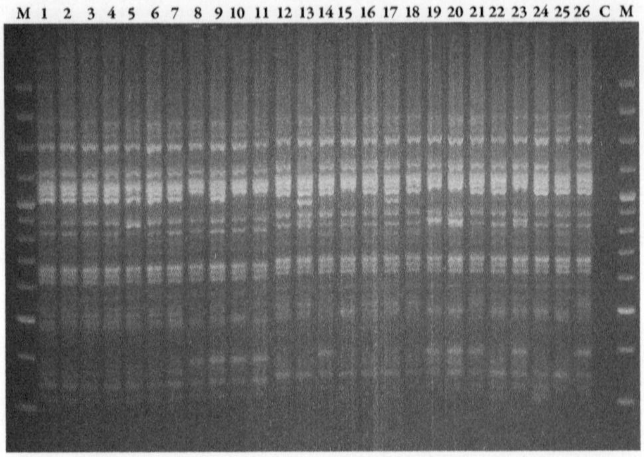

Fig. 1. An example of the DNA fragments profiling with primer SR - 55
M - 100bp Gene Ruler (Fermentas), 1 - 'Akt', 2 - 'Akt' 1, 3 - 'Akt' 2, 4 - 'Akt' bis, 5 - STH 721, 6 - 'Polar', 7 - 'Cacko', 8 - 'Maczo', 9 - STH 8307, 10 - STH 6315, 11 - 'Siwek', 12 - 'Nagus' 13 - DC 2188, 14 - DC 2215, 15 - DC 2612, 16 - DC 2711, 17 - DC 2973, 18 - DC 3674, 19 - DC 3080, 20 - DC 3996, 21 - DC 3416, 22 - DC 3249, 23 - DC 3170, 24 - DC 2567, 25 - DC 2931, 26 - DC 2934, C - negative control

Data analysis

The amplified ISSR fragments were scored as present (1) or absent (0) and converted into a form of a binary matrix. Genetic pairwise similarities were evaluated using Nei's coefficient (Nei and Li, 1979). Polymorphic information content (PIC) values were estimated for each marker (Powell *et al.*, 1996). UPGMA cluster analyses were performed using PAST software. The branches robustness was estimated using 1000 bootstrap replicates within PAST program. PCoA was performed in XlStat v.7.5.2.

Results and discussions

Although oat is a marginal cereal in Poland, by now, at least five naked domestic oats cultivars ('Polar', 'Cacko', 'Maczo', 'Siwek' and 'Nagus'), and many advanced lines have been tested by breeders. Naked forms of *A. sativa* seem to show promise for feeding and food production, but little is known on genetic variability of available cultivars and lines and their distinctiveness. This information is crucial for further breeding programs. Among many molecular marker systems useful for such studies, the ISSR approach has been used in wheat (Ammiraju *et al.*, 2001), barley (Hou *et al.*, 2005), maize (Carvalho *et al.*, 2002) and rice (Bao *et al.*, 2006; Qian *et al.*, 2001). In all cases, the ISSR markers could identify as few as 34% and as many as 98% of polymorphisms useful for plant material differentiation.

In the given study, five cultivars and 18 advanced breeding lines of oat were tested with the ISSR method. Clearly visible and repeatable profiles were obtained (Fig. 1). Out of fifty ISSR primers screened for repeatable polymorphisms, 25 produced 429 scorable fragments, where 204 (47%) were polymorphic (Tab. 2). The percentage of polymorphic bands is congruent with earlier data (Boczkowska and Tarczyk, 2013) - 59.3% obtained for Polish oat landraces.

The number of fragments obtained for 26 analysed genotypes ranged from five (amplified by SR-27) to twenty six (SR-37 and SR-55), with an average 17.2 fragments per primer. Detected polymorphism varied from twenty (SR-27) to a hundred percent (SR-22) among primers. The number of polymorphic bands amplified by a single ISSR primer ranged from 1 (SR- 27) to 15 (SR-23) with an average equal to 8.1.

PIC characterizes the given marker system demonstrating its informativeness. It is usually assumed that the higher its value the better it fits experimental requirements with a value around 0.3 being sufficient for most taxonomic purposes (Botstein *et al.*, 1980). PIC for the primers used in a given experiment ranged from 0.15 (SR-50) to 0.37 (SR-53) with an average equal to 0.23 per primer (Tab. 2). Previously, slightly lower values of PIC indices were obtained based on ISSRs (Boczkowska and Tarczyk, 2013) in case of Polish oat landraces. The average PIC for the particular marker ranged from 0.17 to 0.23 with the mean of 0.20. Higher indices (0.26-0.40) were evaluated in *A. fatua* L. (Paczos-Grzeda *et al.*, 2009). While PIC values evaluated by us were not too high, they were large enough to assume that ISSR marker system was capable of differentiating the analysed materials.

There were 44 markers unique to individual cultivars or lines (Tab. 2). Those markers constituted of 10.2% of the identified ISSR products. Out of all unique products recognized in this study, 21 were specific for the STH 721 line, that resulted from a cross between cv. 'Akt' and *A. sterilis* AVE 2999. The line specific markers originated from *A. sterilis* as they were not identified in cv. 'Akt'. Four line specific markers were amplified in cv. 'Maczo' and STH 6315, three in - DC 2188 and two in STH 8307, DC 2934 and DC 3416. Single unique amplicons were detected in 'Akt', DC 2612, DC 2711, DC 2931, DC 3996 and DC 3170. Twelve forms had no unique markers. Those cultivars and lines could be recognized by ISSRs fragment profiles amplified by one or more ISSR primers.

Tab. 2. Characteristic of primers and detected polymorphism

Primer	Sequence 5'→3'	No of DNA fragments				% of polymorphic fragments	PIC
		total	monomorphic	polymorphic	specific		
SR – 1	$(AG)_8G$	19	14	5	1	26.3	0.189
SR – 11	$(AC)_8G$	15	9	6	1	40.0	0.280
SR – 17	$(GA)_8C$	23	13	10	0	43.5	0.260
SR – 22	$(CA)_8G$	6	0	6	2	100.0	0.108
SR – 27	$(TC)_8G$	5	4	1	0	20.0	0.210
SR – 28	$(TG)_8G$	19	12	7	0	36.8	0.402
SR – 33	$(AG)_8T$	21	9	12	3	57.1	0.266
SR – 37	$(AC)_8C$	26	15	11	2	42.3	0.155
SR – 39	$(GA)_8GG$	19	10	9	4	47.4	0.180
SR – 40	$(AC)_8T$	20	15	5	1	25.0	0.243
SR – 42	$(AG)_8YA$	20	7	13	2	65.0	0.268
SR – 45	$(GA)_8T$	15	7	8	0	53.3	0.260
SR – 46	$(GA)_{10}A$	19	5	14	6	73.7	0.164
SR – 48	$(CA)_{10}T$	13	6	7	1	53.8	0.316
SR – 49	$(TC)_9A$	14	10	4	3	28.6	0.196
SR – 50	$(TC)_9C$	16	7	9	3	56.3	0.153
SR – 53	$(CT)_8A$	17	2	15	2	88.2	0.368
SR – 55	$(CAC)_6A$	26	12	14	2	53.8	0.236
SR – 56	$(CAC)_6G$	13	10	3	0	23.1	0.235
SR – 57	$(CTC)_6CG$	13	8	5	2	38.5	0.189
SR – 58	$(ACC)_6T$	20	11	9	2	45.0	0.228
SR – 61	$(CAC)_6G$	20	12	8	3	40.0	0.177
SR – 65	$(ATG)_6T$	13	6	7	0	53.8	0.278
SR – 69	$(AC)_7G$	19	11	8	3	42.1	0.157
SR – 70	$(AC)_8YG$	18	12	6	1	33.3	0.207
Total		429	227	204	44		
Average		17.2	9.1	8.1	1.8	47.5	0.229

Tab 3. ISSR markers number differentiating naked oat cultivars and lines

Cultivar/ Line	'Akt'	'Akt' 1	'Akt' 2	'Akt' bis	STH 721	'Polar'	'Cacko'	'Maczo'	STH 8307	STH 6315	'Siwek'	DC 2567	'Nagus'	DC 2188	DC 2215	DC 2612	DC 2711	DC 2931	DC 2934	DC 2973	DC 3674	DC 3080	DC 3996	DC 3416	DC 3249
'Akt' 1	3																								
'Akt' 2	4	3																							
'Akt' bis	1	2	3																						
STH 721	69	68	69	68																					
'Polar'	31	30	31	30	78																				
'Cacko'	35	34	35	34	76	8																			
'Maczo'	56	55	54	55	91	55	61																		
STH 8307	61	62	63	60	92	54	56	61																	
STH 6315	62	63	64	61	95	65	67	50	59																
'Siwek'	23	24	25	22	78	43	48	66	58	65															
DC 2567	38	39	40	37	83	41	45	58	63	54	47														
'Nagus'	31	32	33	30	74	40	38	63	60	67	42	25													
DC 2188	56	57	58	55	89	53	55	70	51	70	65	62	57												
DC 2215	46	47	48	45	87	45	43	62	57	64	41	46	49	66											
DC 2612	49	50	49	48	86	44	42	65	68	65	56	39	36	63	47										
DC 2711	24	23	24	23	67	27	27	56	55	64	37	42	35	54	44	47									
DC 2931	37	38	39	36	84	24	30	59	60	71	48	33	42	53	45	52	39								
DC 2934	37	38	37	36	82	28	36	53	56	65	46	39	42	57	45	52	35	20							
DC 2973	34	33	34	33	71	19	25	62	53	68	45	44	41	50	48	43	28	33	33						
DC 3674	29	28	29	28	74	10	14	59	58	67	44	43	36	55	49	44	25	28	28	17					
DC 3080	57	58	59	56	92	62	64	71	82	57	66	55	64	73	71	60	59	62	56	61	60				
DC 3996	56	57	58	55	93	63	65	72	81	58	65	56	65	72	72	59	60	63	57	62	61	3			
DC 3416	31	32	33	30	78	40	42	53	56	65	46	43	44	57	47	52	35	32	24	37	40	62	61		
DC 3249	49	50	51	48	86	46	44	67	66	63	54	19	30	63	45	28	45	40	42	43	44	56	57	46	
DC 3170	60	61	62	59	93	65	67	68	85	60	67	56	67	74	64	63	64	61	55	66	63	15	16	61	61

The lines and cultivars differed from each other by as few as one ('Akt' vs. 'Akt' bis) and as many as 95 markers (STH 6315 vs. STH 721) with the average number equal to fifty (Tab. 3). Lines representing cv. 'Akt' were distinguished from each other by four markers. Only three ISSRs discriminated the DC 3080 and 3996 lines while eight once 'Polar' and 'Cacko' cultivars. Ten markers distinguished the DC 3674 line and cultivar 'Polar' while 14 once from 'Cacko'. Some markers differentiated the DC 3170 and DC 3996 (15 markers), DC 3170 and DC 3416 (16), DC 3674 and DC 2973 (17), DC 3674/02 and 'Polar', DC 3249/02 and DC 2567 (19). The DC 2931 and DC 2934, DC 1368 and DC 2567 were distinguished by 20 and 25 markers, respectively. 'Siwek' that originated from a cross between 'Benquel' and 'Akt', had 25 markers discriminating it from 'Akt'. The highest number of markers (from 67 to 95) differentiated line STH 721 ('Akt' x *A. sterilis* AVE 229) from the other materials because it originated from an interspecific cross of *A. sativa* and *A. sterilis*. The latter is a wild relative of *A. sativa* and is being used as a source of valuable traits for oat improvement (Forsberg and Reeves, 1992).

Tab 4. Dice genetic similarity coefficients evaluated based on ISSR polymorphism

Cultivar/ Line	'Akt'	'Akt' 1	'Akt' 2	'Akt' bis	STH 721	'Polar'	'Cacko'	'Maczo'	STH 8307	STH 6315	'Siwek'	'Nagus'	DC 2188	DC 2215	DC 2612	DC 2711	DC 2973	DC 3674	DC 3080	DC 3996	DC 3416	DC 3249	DC3170	DC 2567	DC 2931
'Akt' 1	0.98																								
'Akt' 2	0.97	0.98																							
'Akt' bis	0.99	0.99	0.98																						
STH 721	0.56	0.56	0.56	0.56																					
'Polar'	0.77	0.78	0.77	0.78	0.49																				
'Cacko'	0.75	0.75	0.75	0.75	0.50	0.94																			
'Maczo'	0.62	0.63	0.63	0.62	0.44	0.62	0.57																		
STH 8307	0.58	0.58	0.57	0.58	0.43	0.62	0.61	0.60																	
STH 6315	0.58	0.58	0.57	0.59	0.42	0.55	0.54	0.68	0.61																
'Siwek'	0.84	0.83	0.82	0.84	0.49	0.68	0.65	0.55	0.60	0.56															
'Nagus'	0.77	0.77	0.76	0.78	0.51	0.70	0.71	0.55	0.57	0.53	0.69														
DC 2188	0.61	0.60	0.59	0.61	0.43	0.62	0.60	0.53	0.65	0.53	0.54	0.58													
DC 2215	0.67	0.66	0.65	0.67	0.43	0.67	0.68	0.57	0.60	0.56	0.70	0.63	0.53												
DC 2612	0.64	0.63	0.64	0.64	0.43	0.67	0.68	0.54	0.51	0.55	0.58	0.72	0.54	0.65											
DC 2711	0.83	0.84	0.83	0.84	0.58	0.81	0.81	0.63	0.63	0.58	0.74	0.75	0.63	0.69	0.66										
DC 2973	0.76	0.77	0.76	0.76	0.54	0.86	0.82	0.58	0.63	0.54	0.68	0.70	0.65	0.65	0.68	0.81									
DC 3674	0.79	0.80	0.79	0.80	0.53	0.93	0.90	0.60	0.60	0.55	0.69	0.74	0.62	0.65	0.68	0.83	0.88								
DC 3080	0.63	0.62	0.61	0.63	0.45	0.59	0.57	0.55	0.48	0.65	0.57	0.57	0.53	0.53	0.60	0.62	0.60	0.61							
DC 3996	0.64	0.63	0.62	0.64	0.45	0.58	0.57	0.55	0.49	0.64	0.58	0.56	0.54	0.53	0.60	0.62	0.60	0.61	0.98						
DC 3416	0.79	0.78	0.77	0.79	0.51	0.72	0.70	0.65	0.63	0.58	0.68	0.69	0.61	0.67	0.63	0.77	0.75	0.73	0.61	0.62					
DC 3249	0.65	0.64	0.63	0.65	0.44	0.66	0.68	0.54	0.54	0.57	0.61	0.78	0.55	0.67	0.79	0.69	0.69	0.69	0.63	0.63	0.68				
DC 3170	0.60	0.60	0.59	0.60	0.44	0.56	0.54	0.56	0.45	0.62	0.55	0.54	0.51	0.57	0.57	0.58	0.56	0.58	0.91	0.90	0.61	0.59			
DC 2567	0.73	0.72	0.71	0.73	0.46	0.70	0.67	0.60	0.56	0.63	0.66	0.81	0.56	0.66	0.71	0.70	0.68	0.69	0.64	0.63	0.70	0.86	0.62		
DC 2931	0.73	0.73	0.72	0.74	0.45	0.82	0.78	0.59	0.58	0.51	0.65	0.68	0.62	0.67	0.61	0.72	0.76	0.80	0.59	0.58	0.78	0.71	0.59	0.76	
DC 2934	0.74	0.74	0.74	0.75	0.48	0.80	0.74	0.64	0.62	0.57	0.68	0.70	0.61	0.68	0.62	0.76	0.77	0.81	0.64	0.64	0.84	0.70	0.64	0.72	0.86

The genetic similarity matrix was obtained based on ISSR data using the Dice's coefficient (Tab. 4). The lowest value of genetic similarity was observed for the STH 721 and STH 6315 lines (0.42), whereas the highest for 'Akt' and 'Akt' bis (0.99). The mean genetic similarity was 0.65. Very high genetic similarity coefficients were also found between cv. 'Akt' and its sister lines (0.98), cv. 'Polar' and 'Cacko' (0.94) as well as DC 3674 line and cv. 'Polar' (0.93). The observed high level of similarity among 'Akt' and its sister lines demonstrated that ISSR markers are capable of identifying even very few differences that may exhibit some minor variation within the above mentioned materials. Thus, cv. 'Akt' and its sublines are identical. In this context, it seems surprising that a high similarity of some cultivars do not have 'Akt' in its crossing formulas with 'Akt' may suggest mistakes during crossing or incorrect pedigree data.

While UPGMA failed to distinguish the materials into main clusters it demonstrated that cultivars 'Akt', 'Polar', 'Cacko', 'Siwek', 'Nagus' and most of the DC lines were within a single group. Moreover, the cultivars that were closely related based on their breeding pedigree (namely, those related to 'Akt'), were close to each other. Also three DC lines (DC 3080; DC 3996; DC 3170) that originated from the same cross formula (P(508)19251 × CHD 1743) were closely related. High level of similarity was observed between two lines DC 3080 and DC 3996. Although breeding pedigree between cultivar 'Maczo' and STH 6315 line, as well as STH 8307 and DC 2188 lines, is not known, the materials appeared to be relatively close to each other. In general, the DC lines have lower variation than the STH pool. Interestingly, that similar analyses performed on 23 old cultivars using AFLP, ISSR and RAPD markers failed in any significant grouping of the analysed materials (Boczkowska *et al.*, 2014). The data demonstrates that the analysed materials show enough variation useful for breeding purposes.

Interestingly, that STH 721 line located at the edge of the dendrogram (Fig. 2). It originated from a cross where *A.*

sterilis was a counterpart. This is the only such line used in the study. While the STH 721 is an advanced line, evidently, such grouping could be explained by the presence of *A. sterilis* DNA that was not eliminated despite years of selection.

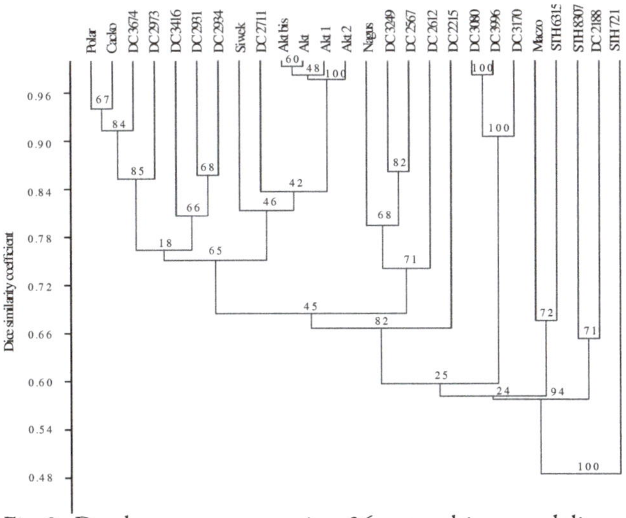

Fig 2. Dendrogram representing 26 oat cultivars and lines. UPGMA method and Dice similarity index evaluated on ISSR markers were used

In general, Principal Coordinate Analysis was in good agreement with the UPGMA (Fig. 3). The first two main components of PCoA explained 54.1 % of the variance and allowed partial differentiation of the materials. Evidently, DC 3170, DC 3080 and DC 3996 lines originated from the same cross formula (P(508)19251 x CHD 1743) established a separate group. Similarly DC 2567, DC 2612 and DC 3249 that have similar parental forms made the second one which was close to the group of cultivars related to 'Akt' as well as to the group of materials encompassing cv. 'Cacko', 'Polar' and some DC lines that have 'Polar' or 'Sallust' in pedigrees. As in the case of the UPGMA, the STH 721 line was separated from all the materials.

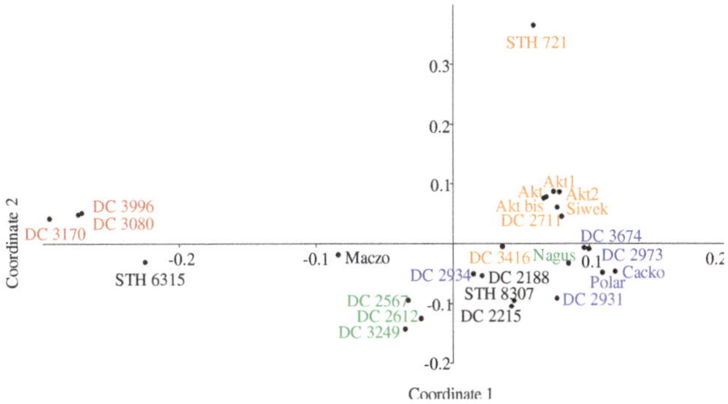

Fig 3. Principle component analyses of oat cultivars and lines originated from different breeding companies and evaluated based on ISSR markers. 'Polar', 'Cacko' and DC lines with 'Polar' or 'Sallust' in pedigrees are indicated in blue; the DC lines that have common origin are given in red; cv. 'Akt' and genotypes with 'Akt' in their pedigrees are depicted in orange; cv. 'Nagus' and lines with similar parental forms are indicated in green

Conclusions

Based on the current study it was demonstrated that ISSR markers could be used for the evaluation of available genetic pool of naked *A. sativa* lines and cultivars and differentiation of individual genotypes. Such approach seems to be a method of choice when limited funds are available. Moreover, the ISSR markers are cheap, cost effective and highly reproducible comparative to e.g. SNP markers. Evidently, the method could be successfully applied for expended analyses of oat accessions or genotype identification.

The presented results demonstrate that Polish breeding materials represent relatively rich genetic pool where the most variable, are lines that are currently under preliminary field experiments. According to our data, the naked oat genetic pool available for breeders, could be efficiently exploited in the future to create new cultivars.

References

Ammiraju JSS, Dholakia BB, Santra DK, Singh H, Lagu MD, Tamhankar SA, Dhaliwal HS, Rao VS, Gupta VS, Ranjekar PK (2001). Identification of inter simple sequence repeat (ISSR) markers associated with seed size in wheat. Theor Appl Genet 102:726-732.

Bao J, Corke H, Sun M (2006). Analysis of genetic diversity and relationships in waxy rice (*Oryza sativa* L.) using AFLP and ISSR markers. Genet Resour Crop Ev 53:323-330.

Boczkowska M, Nowosielski J, Nowosielska D, Podyma W (2014). Assessing genetic diversity in 23 early Polish oat cultivars base on molecular and morphological studies. Genet Resour Crop Ev DOI 101007/s10722-014-0087-4.

Boczkowska M, Tarczyk E (2013). Genetic diversity among Polish landraces of common oat (*Avena sativa* L.). Genet Resour Crop Ev 60:2157-2169.

Botstein D, White RL, Skolnick M, Davis RW (1980). Constructions of genetic linkage map in man using restriction fragment length polymorphisms. Am J Hum Genet 32:314-333.

Burrows VD (1986). Breeding oats for food and feed: conventional and new techniques and materials. 13-46 p. In: Webster FH (Ed.). Oats: chemistry and technology. St. Paul, Minnesota, USA.

Carvalho VP, Ruas PM, Ruas CF, Ferreira JM, Moreira RMP (2002). Assessment of genetic diversity in maize (*Zea mays* L.) landraces using inter simple sequence repeat (ISSR) markers. Crop Breed Appl Biot 2:557-568.

Chowdhury MA, Vandenberg B, Warkentin T (2000). Cultivar identification and genetic relationship among selected breeding lines and cultivars in chickpea (*Cicer arietinum* L.). Euphytica 127:317-325.

Cox TS, Murphy JP (1990). The effect of parental divergence on F_2 heterosis in winter wheat crosses. Theor Appl Genet 79:241-50.

Doehlert DC, Jannink JL, McMullen MS (2006). Kernel size variation in naked oat. Crop Sci. 46:1117-1123.

Doyle JJ, Doyle JL (1987). A rapid DNA isolation procedure for small quantities of fresh leaf tissue. Phytochemistry Bulletin 19:11-15.

Forsberg RA, Reeves DL (1992). Breeding oat cultivars for improved grain quality. 751-775 p. In: Sorrells ME, Marshall HG (Eds.). Oat Science and Technology. American Society of Agronomy. Agronomy Monograph No.33, Madison, Wis., USA.

Fu YB, Peterson GW, Williams D, Richards KW, Fetch JM (2005). Patterns of AFLP variation in a core subset of cultivated hexaploid oat germplasm. Theor Appl Genet 11:530-9.

Hou YC, Yan ZH, Wei YM, Zheng YL (2005). Genetic diversity in barley from west China based on RAPD and ISSR analysis. Barley Genetics Newsletter 35:9-22.

Kirkkari AM (2008). Comparative economic analysis for production of naked vs. conventional oat. Acta Agr Scand 58:305-313.

Kirkkari AM, Peltonen-Sainio P, Jauhiainen L (2009). Trichome numbers in naked-oat cultivars. J Agr Sci 147:63-69.

Martin JM, Blake TK, Hockett EA (1991). Diversity among North American spring barley cultivars based on coefficients of parentage. Crop Sci 31:1131-1137.

Nei M, Li WH (1979). Mathematical model for studying genetic variation in terms of restriction endonucleases. P Natl Acad Sci USA 76:5269–73.

Nita ZT (2003). Contemporary achievements and prospects of oat breeding in Poland. Biuletyn IHAR 229: 13-20 (in Polish with abstract in English).

Paczos–Grzeda E (2004). Pedigree, RAPD and simplified AFLP-based assessment of genetic relationship among *Avena sativa* L. cultivars. Euphytica 138:13-22.

Paczos-Grzeda E, Kruk K, Okon S (2009). Assessment of *Avena fatua* L. intraspecific genetic similarity based on DNA polymorphism identified by ISSR and REMAP Methods. Biuletyn IHAR 252:235-243 (in Polish with abstract in English).

Powell W, Morgante M, Andre C, Hanafey M (1996). Comparison of RFLP, RAPD, AFLP and SSR (Microsatellite) markers for germplasm analysis. Mol Breeding 2:225-238.

Qian W, Ge S, Hong DY (2001). Genetic variation within and among populations of a wild rice *Oryza granulata* from China detected by RAPD and ISSR markers. Theor Appl Genet 102:440-449.

Ronald PS, Brown PD, Penner GA, Brûlé-Babel A, Kibite S (1999). Heritability of hull percentage in oat. Crop Sci 39:52-57.

Ronald PS, Penner GA, Brown PD, Brule-Babel A (1997). Identification of RAPD markers for percent hull in oat. Genome 40:873-8.

Salo M-L, Kotilainen K (1970). On the carbohydrate composition and nutritive value of some cereals. J Agr Sci Finland 42:21-29.

Sykut-Domańska E (2012). Characterisation of selected physical properties of naked and common oat (*Avena sativa* L.) Acta Agrophysica 19:845-856.

Tinker NA, Kilian A, Wight CP, Heller-Uszynska K, Wenzl P, Rines HW, Bjornstad A, Howarth CJ, Jannink JL, Anderson JM, Rossnagel BG, Stuthman DD, Sorrells ME, Jackson EW, Tuvesson S, Kolb FL, Olsson O, Federizzi LC, Carson ML, Ohm HW, Molnar SJ, Scoles GJ, Eckstein PE, Bonman JM, Ceplitis A, Langdon T (2009). New DArT markers for oat provide enhanced map coverage and global germplasm characterization. BMC Genomics 10: 39.

Valentine J (1995). Naked oats. 504-532 p. In: Welch RW (Ed.). The Oat Crop: Production and Utilization. Chapman & Hall, England.

Welch R, Hayward MV, Jones DIH (1983). The composition of the oat husk and its variation due to genetic and other factors. Cereal Chem 34:417-426.

Zeller FJ (1998). Nutzung Des Genetischen Potentials Der *Avena* - Wildarten Zur Verbesserung Des Saathafers (*Avena sativa* L.). J Appl Bot 72: 180–185 (in German).

Zietkiewicz E, Rafalski A, Labuda D (1994). Genome fingerprinting by simple sequence repeat (SSR)-anchored polymerase chain reaction amplification. Genomics 20:176-83.

Phenotypic and Molecular Screening of Apple Genotypes to Woolly Apple Aphid Resistance

Saeid ABU-ROMMAN[1*], Mazen ATEYYAT[2]

[1]*Department of Biotechnology;* [2]*Department of Plant Production and Protection. Faculty of Agricultural Technology, Al-Balqa' Applied University, Al-Salt, 19117, Jordan; ssadroman@yahoo.com (*corresponding author)*

Abstract

Woolly apple aphid (WAA) is a major pest of apple orchards resulting in significant losses. In the present study, 22 apple (*Malus domestica* Borkh.) cultivars were tested in the filed for their relative resistance/susceptibility to WAA. These apple cultivars were found to respond differentially to WAA infestation. Based on tree infestation rating, cultivars were ranked into 6 relative resistance/susceptibility groups, as follows: immune - 'Golden Delicious', 'Delbarestivale', 'Golden Smoothie', 'Red Miracle' and 'Harmony'; resistant - 'Stark Gold', 'Early Gold' and 'Argi Gala'; moderately resistant - 'Evanee' and 'Black Stayman'; moderately susceptible - 'Vista Bella', 'Jonagold', 'Royal Gala', 'Jersey Mac', 'Granny Spur Type' and 'Summerred'; susceptible - 'Jonathan', 'Nagava 6', 'Florina', 'Red Chief' and 'Gold Iralis'; highly susceptible - 'Fuji 6'. Eight molecular markers linked to major WAA resistant genes (*Er1*, *Er2*, and *Er3*) were screened in apple cultivars using PCR. The markers NZms_EB145764, NZms_EB106753 and NZsc_E01 were ubiquitous in all cultivars under study, whereas, NZsn_O05 was absent. The results of other markers revealed distinct patterns of amplification among apple cultivars. No clear correlations can be made between the molecular data (marker presence and absence) and the phenotypic results (cultivar ranking). The differences among cultivars regarding WAA infestation can potentially be utilized by apple breeders and commercial growers to achieve effective, environmental-friendly, and low-cost pest control.

Keywords: cultivar, *Eriosoma lanigerum*, host plant resistance, infestation rating, *Malus domestica*, resistance, susceptibility

Introduction

Apple (*Malus domestica* Borkh.) is one of the most economically important fruit crops of the temperate zones of the world (Harris *et al.*, 2002). Apple trees are usually attacked by several types of arthropod pests (Shoonhoven *et al.*, 2005). Aphids (order Homoptera) are key pests in apple orchards worldwide (Bloomers, 1994; Prokopy and Croft, 1994; Beers *et al.*, 2003). Aphid infestation in apple causes deformation of foliage and fruits. In addition, some aphid species can act as vectors for virus transmission and the release of their honeydew promotes the development of mold infestation (Beers *et al.*, 2003; Blommers *et al.*, 2004; Arbab *et al.*, 2006).

The woolly apple aphid (WAA), *Eriosoma lanigerum* (Hausmann) (Hemiptera: Aphididae), is one of the most important aphid pests of apples in many of the world's apple growing regions (Hatton, 1937). WAA is becoming critical to the economics of apple industry in Jordan (Ateyyat and Antary, 2009; Ateyyat and Antary, 2010). WAA colonizes both the roots and the vegetative parts of apple trees (Alspach and Bus, 1999). This aphid pest overwinters in edaphic colonies on roots and produces its first instar nymphs (crawlers) in early spring (Thwaite and Bower, 1983). In spring, crawlers migrate upward and initiate new colonies in protected arboreal parts of the tree (Heunis and Pringle, 2006; Cockfield and Beers, 2007).

WAA infestation on roots resulted in the formation of hypertrophic galls which restricted water and nutrients uptake (Brown and Schmitt, 1990; Brown *et al.*, 1995). Strong WAA infestations resulted from population outbreaks in the arboreal part of the tree, and can lead to reduction in canopy vigor due to premature defoliation, reduction in yield due to reducing fruit buds and splitting of fruit bearing wood, and reduction in fruit quality as a result of contaminating fruits with honeydew and sticky wool (Bertus, 1986; Brown and Schmitt, 1990; Brown *et al.*, 1995; Pringle and Heunis 2001; Heunis and Pringle, 2006). WAA can be a significant phytosanitary issue for fruit exportation and an annoying pest during fruit harvesting (Cockfield and Beers, 2007).

WAA can be controlled chemically by organochloride and organophosphate insecticides (Cockfield and Beers, 2007; Shaw and Wallis, 2009). The international public trend toward the organic farming pushes against the use of chemical control of pests and prompting a need for integrated pest management (IPM). Therefore, adopting biological control through the introduction of the natural endoparasitoid *Aphelinus mali* is a good choice (Brown and Schmitt, 1994; Asante, 1997). However, its susceptibility to many of the insecticides and fungicides commonly used in apple orchards limits its use (Cohen *et al.*, 1996, Nicholas *et al.*, 2005; Rogers *et al.*, 2011). Cultural control still offers a valuable mean to control WAA in apple orchards through the use of host-plant-resistance (HPR) strategy (Webster, 2003). HPR is an important easy-to-apply tactic of IPM (Beers *et al.*, 1993) and can result from morphological factors and activation of defense and signaling responses (Pedigo, 2006). The presence of different apple cultivars and rootstocks provides the opportunity to select the most resistant genotypes available. Variation in susceptibility of apple cultivars to WAA infestations has been previously recorded (Asante *et al.*, 1993; Deng *et al.*, 1993; Sandanayaka *et al.*, 2003). The development of new resistant lines and understanding their genetic background of WAA resistance are significant aspects for foresighted breeding strategies (Brown and Maloney, 2003; Sandanayaka *et al.*, 2005). Molecular studies have been conducted for exploring the genetic basis of aphid resistance in apple. Some quantitative trait loci (QTLs) were identified associated with aphid resistance in apple. Molecular markers were discovered linked to QTLs for resistance to the rosy apple aphid, the leaf-curling aphid and the green apple aphid (Stoeckli *et al.*, 2008). Recently, progresses have been achieved regarding the development of molecular markers associated with WAA resistance genes (i.e. *Er1*, *Er2* and *Er3*) (Bus *et al.*, 2008). Current phenotyping methods for selection of plants resistant to WAA are cumbersome, environment-dependent, and cannot distinguish plants containing combinations of resistance genes from those carrying a single gene (Bus *et al.*, 2008). Therefore, screening for the presence of molecular markers linked to WAA resistance genes would be an alternative.

In the current research, different apple cultivars were screened phenotypically for their resistance to WAA infestation. Furthermore, a molecular study was performed for screening these cultivars for the presence of different molecular markers associated with *Er1*, *Er2* and *Er3* resistance genes.

Materials and methods

Monitoring of WAA infestation and data analysis

Monitoring part of the experiment was conducted at the Apple Genetic Complex in Ash-Shoubak Regional Center of Agriculture Research and Extension. The trees under investigations were trained under the central-leader system and were planted on Merton-Malling Series (MM) 106 rootstock. The experiment was conducted on 22 apple cultivars of different flowering time, fruit color, and time of fruit maturity.

WAA infestation was monitored in terms of aerial colonies formation. Monitoring was carried out every 2 weeks on five trees per cultivar for the duration of seven months from January to July. Each tree was rated on a 0-4 infestation rating scale, using a visual indexing technique and ratings modified from Bower (1987). Cultivars were ranked for their relative resistance/susceptibility to WAA based on the tree infestation rating according to Ateyyat and Al-Antary (2009) (Tab. 1).

Tab. 1. Rating scale for the assessment of level of resistance based on infestation level of WAA

Level of infestation (%)	Scale	Resistance rating
0	0	Immune
1-15%	0.5	Highly resistant
16-30%	1	Resistant
31-50%	1.5	Moderately resistant
51-65%	2	Moderately susceptible
66-80%	2.5	Susceptible
81-90%	3	Highly susceptible
91-100%	4	Very highly susceptible

Mean separation was performed by using Least Significant Differences test (LSD). All differences were compared at the 5 % level ($p \leq 0.05$) of significance and LSD test was used to separate means if significant differences found.

Molecular analysis

Healthy leaf tissues were collected from all cultivars under study and were stored at -20 °C for DNA extraction. Leaf tissues were ground into a fine powder in liquid nitrogen using a mortar with a pestle. Template DNA was extracted using a modified CTAB method (Saghi-Maroof *et al.*, 1984). For each cultivars, about 0.5 g of fine leaf powder were suspended in 2 ml of DNA extraction buffer (0.22 M EDTA pH 8, 0.8 M NaCl, 0.14 M sorbitol, 0.2 M Tris-HCl pH 8, 0.8% CTAB, 0.5% SDS, 0.5% β-mercaptoethanol). About 500 µl of chloroform were added to the suspension and mixed vigorously. The mixture was incubated at 65 °C for 40 min and then centrifuged. The supernatant was taken and the DNA was precipitated by the addition of 500 µl of isopropanol. The DNA was collected by centrifugation and the DNA pellet was washed twice with 70% ethanol. DNA pellets were dissolved in 100 µl TE buffer (10 mM Tris-HCl, 1 mM EDTA pH 8). The concentration and purity of DNA were assayed using UV spectrometry at 260 and 280 nm.

PCR reactions were performed in purpose of screening the presence of eight different molecular markers linked to major WAA resistance genes (*Er1*, *Er2*, and *Er3*). The molecular markers were amplified using specific primers as suggested by Bus *et al.* (2008) (Tab. 2). All PCR amplification reactions were performed in a total volume of 25 µl. The PCR reaction mixture contained about 300 ng of template DNA, 12.5 µl of 2x PCR master mix solution (iNtRON, Korea), 5 pmole of each primer and distilled H_2O was added to make up the final volume. Markers NZsc_GS327, NZsc_O05, NZsc_E01 and NZsc_A01 were amplified with an annealing temperature of 60 °C and an extension time of 2 min. Whereas, a touchdown PCR cycling regime was employed to amplify the rest of the markers with annealing temperature decreasing from 65 °C to 60 °C in decrements of 0.5 °C per cycle.

PCR amplified products were separated and analyzed by electrophoresis on 1-2% agarose gels stained with ethidium bromide. As molecular weight markers, 1-kp and 100-bp DNA ladders (New England Biolabs) were used. Results were scored for the presence and absence of the molecular markers.

Tab. 2. Primer sequences and product length in base pairs of the molecular markers used in the present study

Marker name	WAA resistance gene	Primer sequence (5'→3')		Product size (bp)
		Forward	Reverse	
NZsc_C20	*Er1*	TCTCTAACTCAATAACTCCCAAGAC	ACTTCGCCACCATTATCACTCCTGA	2,000
NZsc_GS327	*Er1*	GCCAAGCTTCAATGTCGGAGTAGAT	CAAGCTTCCCCTAAGGCTATTGCCA	1,600
NZsc_O05	*Er1*	CCCAGTCACTAACATAATTGGCACA	CCCAGTCACTGGCAAGAGAAATTAC	1,700
NZms_EB145764	*Er2*	TTCCAGCGATCCAAAACAAT	GCTCAGGAACACCTCGTTCT	198
NZsn_O05	*Er1/Er3*	AACGTCATGTCAATAT	CCCAGTCACTGGCAAGAGAAATTAC	880
NZms_EB106753	*Er1/Er3*	TCTGAGGCTCCCAAGTCC	TAGGAGCAGAAGAGGTGACG	175
NZsc_E01	*Er3*	CCCAAGGTCCGAACACAAATGAGAG	CCCAAGGTCCAAAACTATCCCGAAG	1,350
NZsc_A01	*Er3*	CAGGCCCTTCAGCAAAGAGGTGTCT	CAGGCCCTTCACTACTAATAAGAAC	1,250

Results and Discussion

WAA is a potentially detriment pest to apple production worldwide causing a range of diverse injuries to apple trees and fruits (Brown *et al.*, 1995). Management of WAA infestations using apple cultivars and rootstocks is considered an important strategy. In the present study, apple cultivars responded differentially to WAA infestation. Results represented in Fig. 1 showed that the highest number of WAA colonies were recorded for five cultivars, namely; 'Fuji 6', 'Jonathan', 'Red Chief', 'Gold Iralis', and 'Nagava 6'. Out of the 22 apple cultivars under investigation, cultivars 'Golden Delicious', 'Delbarestivale', 'Golden Smoothie', 'Red Miracle', and 'Harmony' were recorded with complete absence of WAA colonies.

Cultivars ranking based on the tree infestation rating were followed as suggested by Ateyyat and Al-Antary (2009). According to this ranking, 'Fuji 6' was recorded as highly susceptible cultivar, while 'Golden Delicious', 'Delbarestivale', 'Golden Smoothie', 'Red Miracle' and 'Harmony' were considered as immune (Fig. 2). Ateyyat and Al-Antary (2009) reported that WAA adults failed to colonize on 'Harmony' and no infestations were recorded either on its roots or shoots.

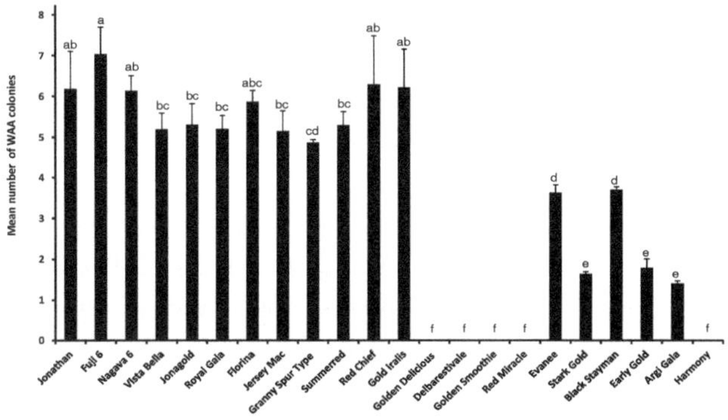

Fig. 1. Mean numbers of WAA colonies per 15 cm length limb of un-sprayed apple cultivars. Means with different letters are significantly different according to LSD

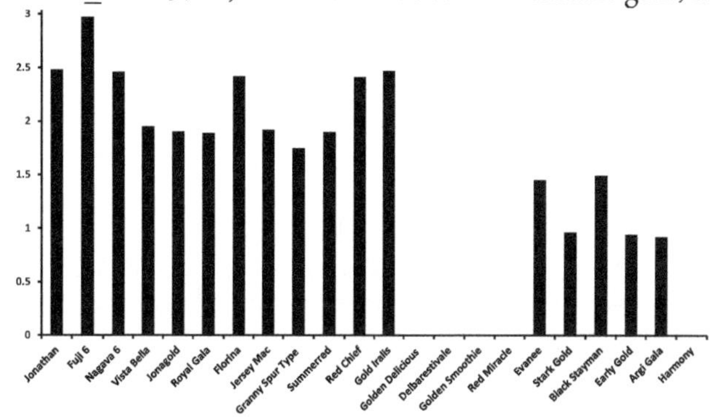

Fig. 2. Shoot infestation rating (0-4) of WAA on un-sprayed apple cultivars

Moreover, the present study yielded three resistant ('Stark Gold', 'Early Gold' and 'Argi Gala'), two moderately resistant ('Evanee', and 'Black Stayman'), six moderately susceptible ('Vista Bella', 'Jonagold', 'Royal Gala', 'Jersey Mac', 'Granny Spur Type' and 'Summerred'), and five susceptible ('Jonathan', 'Nagava 6', 'Florina', 'Red Chief' and 'Gold Iralis') cultivars. 'Royal Gala' is well-known cultivar for its susceptibility to WAA infestation. This cultivar was reported to display a range of WAA susceptible characteristics including high level of settlement, high survival rate, and long periods of phloem feeding (Sandanayaka *et al.*, 2005). None of the 22 apple cultivars tested were ranked as highly resistance or very highly susceptible (Fig. 2). Various WAA-resistant rootstocks were previously developed from the immune cultivars 'northern Spy', including the Malling-Merton rootstocks of the 100 series (Webster, 2003). These rootstocks may owe their resistance to high phenolic levels (Sen Gupta and Miles, 1975). However, some reports have indicated the breakdown of resistance of these rootstocks (Rock and Zeiger, 1974; Ateyyat and Al-Antary, 2009).

Three major WAA resistant genes (*Er1*, *Er2*, and *Er3*) have been characterized and used in apple resistance breeding (Bus *et al.*, 2008). Eight different molecular markers linked to these genes were screened by PCR in the apple cultivars under investigation (Tab. 3). The marker NZms_EB145764, which is linked to *Er2* resistant gene, is present in all cultivars. The same is applicable to NZms_EB106753 and NZsc_E01 which are linked to *Er1/Er3* and *Er3*, respectively. On the other hand, NZsn_O05 marker is absent from all cultivars under study. NZsc_A01 (linked to *Er3*) is present in all cultivars except for 'Fuji 6', 'Jonagold', and 'Granny Spur Type'. Sandanayaka *et al.* (2003) reported that *Er3* resistant gene was carried by the cultivar 'Aotea' and this type of mediated resistance was overcome by WAA, indicating that *Er3*-mediated resistance is distinct from *Er1* and *Er2*.

'Early Gold' is the only cultivar containing NZsc_C20 marker which is linked to *Er1*. The marker NZsc_GS327 is present in 11 cultivars, namely; 'Granny Spur Type',

Tab. 3. Detection of molecular markers linked to WAA resistance genes in different apple cultivars. Plus (+) indicates the presence of a marker

	Markers linked to WAA resistance genes							
Cultivars	NZsc_C20	NZsc_GS327	NZsc_O05	NZms_EB145764	NZsn_O05	NZms_EB106753	NZsc_E01	NZsc_A01
'Jonathan'	-	-	-	+	-	+	+	+
'Fuji 6'	-	-	-	+	-	+	+	+
'Nagava 6'	-	-	-	+	-	+	+	+
'Vista Bella'	-	-	-	+	-	+	+	+
'Jonagold'	-	-	-	+	-	+	+	+
'Royal Gala'	-	-	-	+	-	+	+	+
'Florina'	-	-	-	+	-	+	+	+
'Jersey Mac'	-	-	-	+	-	+	+	+
'Granny Spur Type'	-	+	-	+	-	+	+	-
'Summerred'	-	+	-	+	-	+	+	+
'Red Chief'	-	+	-	+	-	+	+	+
'Gold Iralis'	-	-	+	+	-	+	+	+
'Golden Delicious'	-	+	-	+	-	+	+	+
'Delbarestivale'	-	+	-	+	-	+	+	+
'Golden Smoothie'	-	+	-	+	-	+	+	+
'Red Miracle'	-	+	+	+	-	+	+	+
'Evanee'	-	+	+	+	-	+	+	+
'Stark Gold'	-	+	+	+	-	+	+	+
'Black Stayman'	-	+	+	+	-	+	+	+
'Early Gold'	+	-	+	+	-	+	+	+
'Argi Gala'	-	-	+	+	-	+	+	+
'Harmony'	-	+	+	+	-	+	+	+

'Summerred', 'Red Chief', 'Golden Delicious', 'Delbarestivale', 'Golden Smoothie', 'Red Miracle', 'Harmony', 'Stark Gold', 'Black Stayman', and 'Evanee'. NZsc_O05 marker is linked to *Er1* resistant gene and is present in the following eight cultivars: 'Harmony', 'Red Miracle', 'Evanee', 'Stark Gold', 'Black Stayman', 'Early Gold', 'Argi Gala', and 'Gold Iralis'. 'Northern Spy' and 'Robusta 5' are two WAA resistant genotypes and were reported to carry *Er1* and *Er2* resistant genes, respectively (King *et al.*, 1991). These genotypes possessed short duration of phloem ingestion suggesting the presence of resistant factors in their phloem tissues. A previous report indicated that the expression of resistance conditioned by *Er1* increased with tree maturity, while resistance conditioned by *Er2* was high at earlier growth stages (Sandanayaka *et al.*, 2003).

Although no clear correlations can be made between the molecular data (marker presence and absence) and the phenotypic results (cultivar ranking), the results of the current study indicated that resistance/susceptibility to WAA infestations varied among apple cultivars. These differences among cultivars can potentially be utilized by commercial apple growers to achieve effective pest control which in turn will reduce input costs and increase profitability.

Further studies are required to understand the mechanisms of responses/defense among different apple cultivars to WAA infestation. Such studies might employ the analysis of traits related to (i) pest-deterrent or repellent secondary metabolites, (ii) defense-related enzymes, and (iii) differential expression of defense genes. These traits may be utilized in the future as marker traits to select for resistance to WAA.

Acknowledgement

The authors greatly appreciate the technical assistance of Ahmad Sharab. This project was funded by Abdul Hameed Shoman Foundation.

References

Alspach PA, Bus VG, Bus M (1999). Spatial variation of woolly apple aphid (*Eriosoma lanigerum* Hausmann) in a genetically diverse apple planting. New Zeal J Ecol 23:39-44.

Arbab A, Kontodimas, DC, Sahragard A (2006). Estimating development of *Aphis pomi* (DeGeer) (Homoptera: Aphididae) using linear and nonlinear models. Environ Entomol 35:1208-1215.

Asante SK (1997). Natural enemies of the woolly apple aphid, *Eriosoma lanigerum* (Hausmann) (Hemiptera: Aphididae): a review of the world literature. Plant Prot Quart 12:166-170.

Asante SK, Danthanarayana W, Cairns SC (1993). Spatial and temporal distribution patterns of *Eriosoma lanigerum* (Homoptera: Aphididae) on apple. Environ Entomol 22:1060-1065.

Ateyyat MA, Al-Antary TM (2009). Susceptibility of nine apple cultivars to woolly apple aphid, *Eriosoma lanigerum* (Homoptera: Aphididae) in Jordan. Int J Pest Manage 55(1):79-84.

Ateyyat MA, Al-Antary TM (2010). Population Trends of Woolly Apple Aphid, *Eriosoma lanigerum* and its Parasitoid, *Aphelinus mali* on two Apple Cultivars in Jordan. Jordan J Agri Sci 6(3):333-341.

Beers EH, Brunner JF, Willett MJ, Warner GM (1993). Orchard Pest Management: A Resource Book for the Pacific Northwest. Good Fruit Grower, Yakima, WA.

Beers EH, Suckling DM, Prokopy RJ, Avilla J (2003). Ecology and management of apple arthropod pests, p. 489-514. In: Ferree DC, Warrington IJ (Eds.). Apples: Botany, Production and Uses. CABI Publishing, Wallingford, UK.

Bertus AL (1986). Sooty Mould. Agfact H1.AB.5, 2nd edn. Department of Agriculture NSW, Australia.

Blommers LHM (1994). Integrated pest-management in European apple orchards. Annu Rev Entomol 39:213-241.

Blommers LHM, Helsen HHM, Vaal FWNM (2004). Life history data of the rosy apple aphid (*Dysaphis plantaginea* Pass.) (Homoptera, Aphididae) on plantain and as migrant to apple. J Pest Sci 77:155-163.

Bower CC (1987). Control of San José scale (*Comstockaspis perniciosus* (Comstock) (Hemiptera: Diaspididae) and woolly aphid (*Eriosoma lanigerum* (Hausmann) (Hemiptera: Pemphigidae) in an integrated mite control program. Plant Prot Quart 2:55-58.

Brown SK, Maloney KE (2003). Genetic improvement of apple: breeding, markers, mapping and biotechnology, p. 31-61. In: Ferree DC, Warrington IJ (Eds.). Apples: Botany, Production and Uses. CABI Publishing, Wallingford, UK.

Brown MW, Schmitt JJ (1990). Growth reduction in nonbearing apple trees by woolly apple aphids (Homoptera, Aphididae) on roots. J Econ Entomol 83:1526-1530.

Brown MW, Schmitt JJ (1994). Population dynamics of woolly apple aphid (Homoptera: Aphididae) in West Virginia apple orchards. Environ Entomol 23:1182-1188.

Brown MW, Schmitt JJ, Ranger S, Hogmire HW (1995). Yield reduction in apple by edaphic woolly apple aphid (Homoptera, Aphididae) populations. J Econ Entomol 88:127-133.

Bus VGM, Chagné D, Bassett HCM, Bowatte D, Calenge F, Celton J-M, Durel C-E, Malone M. T, Patocchi A, Ranatunga AC, Rikkerink EHA, Tustin DS, Zhou J, Gardiner SE (2008). Genome mapping of three major resistance genes to woolly apple aphid (*Eriosoma lanigerum* Hausm.). Tree Genet Genom 4:223-236.

Cockfield SD, Beers EH (2007). Biology and management of woolly apple aphid, *Eriosoma lanigerum* (Hausmann), in Washington State, pp. 37-42. In: IOBC WPRS Bulletin 30(4):37.

Cohen H, Horowitz AR, Nestel D, Rosen D (1996). Susceptibility of the woolly apple aphid parasitoid, *Aphelinus mali* (Hym: Aphelinidae), to common pesticides used in apple orchards in Israel. Entomophaga 41:225-233.

Deng JQ, Rui GS, Guan YT., Yu YQ, Zhang D, Hong JY, Deng JQ, Rui-GS, Guan YT, Yu YQ, Zhang DM, Hong JY (1993). The selection of an apple stock line, Siberian crabapple Jin 67, immune to the woolly apple aphid. Acta Phytophyl Sin 20(3):217-222.

Harris SA, Robinson JP, Juniper BE (2002). Genetic clues to the origin of the apple. Trends Genet 18(8):426-430.

Hatton RG (1937). The problems raised by the woolly aphis (*Eriosoma lanigerum*) of the apple - a case for team research. Ann Appl Biol 24:169-210.

Heunis JM, Pringle KL (2006). Field biology of woolly apple aphid, *Eriosoma lanigerum* (Hausmann), and its natural enemy, *Aphelinus mali* (Halderman), in apple orchards in the Western Cape province. Afr Entomol 14:77-86.

King GJ, Alston FH, Battle I, Chevreau E, Gessler C, Janse J, Lindhout P, Manganaris AG, Sansavini S, Schmidt H, Tobutt KR (1991). The 'European Apple Genome Mapping Project' - developing a strategy for mapping genes coding for agronomic characters in tree species. Euphytica 56:89-94.

Nicholas AH, Spooner-Hart RN, Vickers RA (2005). Abundance and natural control of the woolly aphid *Eriosoma lanigerum* in an Australian apple orchard IPM program. BioControl 50:271-291.

Pedigo LP (2006). Entomology and Pest Management. Prentice Hall, Saddle River, NY.

Pringle KL, Heunis JM (2001). Woolly apple aphid in South Africa: biology, seasonal cycles, damage and control. Decid Fruit Grower 51:22-23.

Prokopy RJ, Croft BA (1994). Apple insect pest management, p. 543-585. In: Metcalf RL, Luckman WH (Eds). Introduction to Insect Pest Management. John Wiley & Sons Ltd., New York.

Rock G, Zeiger D (1974) Woolly apple aphid infests Malling and Malling-Merton rootstocks in propagation beds in North Carolina. J Econ Entomol 67:137-138.

Rogers DJ, Sharma N, Stretton DC, Walker JTS (2011). Toxicity of pesticides to *Aphelinus mali*, the parasitoid of woolly apple aphid. New Zeal Plant Prot 64:235-240.

Saghai-Maroof MA, Soliman KM, Jorgensen RA, Allard RW (1984). Ribosomal DNA spacer-length polymorphisms in barley: Mendelian inheritance, chromosomal location and population dynamics. Proc Natl Acad Sci USA 81:8014-8018.

Sandanayaka WRM, Bus VGM, Connolly P, Newcomb R (2003). Characteristics associated with woolly apple aphid *Eriosoma lanigerum*, resistance of three apple rootstocks. Entomol Exp Appl 109:63-72.

Sandanayaka WRM, Bus, VGM, Connolly P (2005). Mechanisms of woolly aphid [*Eriosoma lanigerum* (Hausm.)] resistance in apple. J Appl Entomol 129:534-541.

Schoonhoven LM, van Loon JJ, Dicke M (2005). Insect-Plant Biology. Oxford University Press, Oxford, UK.

Sen Gupta GC, Miles PW (1975) Studies on the susceptibility of varieties of apple to the feeding of two strains of woolly aphid (Homoptera) in relation to the chemical content of the tissues of the host. Aust J Agric Res 26:157-168.

Shaw PW, Wallis DR (2009). Early-season use of insecticides for management of woolly apple aphid (*Eriosoma lanigerum*) in Nelson apple orchards. New Zeal Plant Prot 62:291-295.

Stoeckli S, Mody K, Gessler C, Patocchi A, Jermini M, Dorn S (2008). QTL analysis for aphid resistance and growth traits in apple. Tree Genet Genom 4:833-847.

Thwaite WG, Bower CC (1983). Woolly Aphid. Agfact H4. AE.3. NSW Agriculture, Orange, Australia.

Webster AD (2003). Breeding and selection of apple and pear rootstocks. Acta Hort 622:499-512.

Grassland Degradation Decrease the Diversity of Arbuscular Mycorrhizal Fungi Species in Tibet Plateau

Xiao-Bu CAI[1], Yue-Lin PENG[2], Min-Na YANG[2], Tao ZHANG[3]*, Qiang ZHANG[4]*

[1]*Department of Resources and Environment, Tibet Agricultural and Animal Husbandry College, Tibet, 860000, China*
[2]*Department of Plant Science, Tibet Agricultural and Animal Husbandry College, Tibet, 860000, China*
[3]*Institute of Grassland Science, Northeast Normal University, Key Laboratory of Vegetation Ecology, Ministry of Education, Changchun, 130024, China; zhangt946@nenu.edu.cn (*corresponding author)*
[4]*Institute of Geographic Sciences and Natural Resources Research, Chinese Academy of Sciences, Beijing 100101, China; zhangqiang_zz@163.com (*corresponding author)*

Abstract

Arbuscular mycorrhizal (AM) fungi play a vital role in restoration of an ecosystem. Grassland degradation in alpine steppe is severe, but the influence of grassland degradation on AM fungi composition in Tibetan Plateau is still not well understood. This research studied the diversity of arbuscular mycorrhizal fungi in degraded alpine steppe, dominated by *Stipa purpurea*. The results showed that the species richness and abundance of AM fungi in degraded grasslands significantly decreased. In different typical grasslands species, diversity of AM fungi genus *Glomus* was much higher than other genera (*Acaulospora* and *Scutellospora*). Compared with normal grassland, the AM fungal species diversity in moderately degraded grassland decreased by 30%, but in lightly degraded and severely degraded grassland, it highly increased with 15.3% and 13.3%, respectively. The species diversity of genus *Glomus* in moderately degraded grassland and lightly degraded grassland were higher than for normal grassland. Significant differences were observed in a relative abundance of common dominant species among different degraded grasslands. The proportion of dominant species highly increased with the aggravation of grassland degradation. The results suggest that the grassland mild degradation increased the adaptability and stress resistance of species.

Keywords: alpine steppe, richness, species composition, *Stipa purpurea*

Introduction

Arbuscular mycorrhizal (AM) fungi are the most influential members of the soil microbiota (Smith and Read, 2008), which play a key role in improving the resistance of plants to drought (Wu *et al.*, 2008), salt (Evelin *et al.*, 2009) and nutrient deficiency (Smith and Read, 2008), restoration of degraded ecosystem (Bingham and Biondini, 2009; Zhang *et al.*, 2012) and sustaining plant diversity and the ecosystem' stability (van der Heijden *et al.*, 1998; O'Connor *et al.*, 2002; Wagg *et al.*, 2011). However, the composition and development of AM fungi are influenced by environmental changes and anthropogenic activity. A large number of studies have documented that overgrazing significantly reduced species diversity of AM fungi (Chaudhry *et al.*, 2005; Su and Guo, 2007; Wearn and Gange, 2007). Moreover, several studies found that some other factors, such as drought, wind erosion and poor soil factors affect species diversity of AM fungi in grassland ecosystems (Landis *et al.*, 2004; Uhlmann *et al.*, 2004; Liu *et al.*, 2009). Even so, the effects of grassland degradation on species composition of AM fungi in degraded ecosystem are not well understood, especially in extreme environment.

The Tibetan Plateau is called the third pole of the world, with an average altitude of 4,500 m above sea level (a.s.l.). The changes of plants and microorganisms can reflect the effects of climatic changes and human activities, on this distinctive ecosystem for the severe geographic and climatic conditions. Besides that, Tibetan Plateau is an important rangeland for the Tibetan herdsmen, occupying 82.07 million hectares of the Tibetan plateau (Gai *et al.*, 2009). In the last decade, several studies focused on AM fungal species diversity in Tibetan Plateau, which includes warm steppe, alpine meadow and alpine steppe (Gao *et al.*, 2006; Gai *et al.*, 2009; Cai *et al.*, 2010), and 111 species AM fungi belonging to 12 genera were observed. *Stipa* genus is one of the dominant species in arid and semiarid regions, which occupies the greatest area and plays a key role in the stability of the Tibetan Plateau ecosystem.

In recent years, the grassland degradation in Tibetan Plateau become more severe due to many exotic factors, which caused a significant decline of plant density and cover, and expanded desertification area. Although several previous projects have been studied the mycorrhizal structure of *Stipa breviflora* (Bao and Yan, 2004), the diversity of AM fungal species *Stipa capillata*, *S. purpurea*, *S. subsessiliflora* var. *basiplumosa*, *S. roborowskyi* and *S. glareosa*

(Shi *et al.*, 2007), also AM fungal species composition and diversity of *Stipa* in degraded Tibet Plateau, are still not clear. In order to verify the influence of grassland degradation on AM fungal species composition on alpine steppe, *Stipa purpurea* community was selected as research subject, while *S. purpurea* is a dominant species of Tibet Plateau.

Materials and methods

Study area

The average altitudes of plateau surface in the study area range from 4300 to 5300 m. The climate types in the study area vary from plateau frigid semi-arid climate type, to arid climate. There are two seasons, cold (October to May) and warm (June to September) every year, and the mean annual temperature is -1.2-3.0 °C. There is no absolute frost free period, and annual accumulated temperature (≥0 °C) is 800-1100 °C; annual precipitation is 100-200 mm, annual evaporation is about 2000 mm, and the days with instantaneous wind speed ≥17.0 m s^{-1} are 100.3-158.2 d yr^{-1}. Within the study area, *S. purpurea* is a major component of alpine steppe. Currently, alpine steppe has already degraded to varying degrees, and desertification and vegetation thinning gradually become severely. Previous studies indicated that under the fragile and extreme environment of north Tibet Plateau, once desertification happens, it will be extremely difficult for native vegetation to spontaneously recover. Moreover, the desertification may be irreversible, and even more, other plants are difficult to colonize.

Sample collection

Samples were collected in the south of northern Tibetan Plateau (30°57.290'-31°51.147'N, 86°56.173'-90°59.424'E). The most representative and typical plant species in alpine steppe, *S. purpurea*, was selected as research subject. Twelve zones (10 × 10 m) (3 for normal, 3 lightly degraded, 3 moderately degraded and 3 severely degraded grasslands) were randomly selected as sampling zones, according to the classification standard of grassland degradation (Zhang, 2003), having 30-50 km interval between each two sampling zones; 3 sampling plots (1 m × 1 m) were randomly selected in each sampling zone, leaving 100-150 m interval between each two sampling plots. In each sampling plot, all surface soil (2 cm) was removed, and separately, the 2-30 cm soil samples with roots were collected. Afterwards, the samples from the 3 sampling plots in each sampling zone were mixed well, to make 1 mixture sample (about 2.5 kg). The collected samples with roots were put into bags, with marks and records. All soil samples were calcareous soil. Soil parent material and soil type were lacustrine deposits and alpine steppe soil, respectively. Soil texture was low-sand soil.

AM fungi identification

The AM fungal spore numbers in soil samples were isolated and counted. AM fungal spores from 20 g mixed air-dried samples were extracted using wet screening-sucrose gradient centrifugation. The spore colour, size and connective mycelia were observed under stereo light microscope; spores were placed on glass slide (with 30% glycerol as flotation reagent cover slip) to conduct microscopy observation, and spore colour, size and connective mycelia were recorded; spores were then disrupted to observe inclusion, sporoderm stratification, colour and thickness. For identification, Melzer's reagents were applied to test spore specific reaction. The species were identified according to the manual for identification of Vesicular-Arbuscular mycorrhizal fungi (Schenck and Péréz, 1988) and INVNAM (*http://invam.caf.wvu.edu/Myc-lnfo/*), as well as identification materials and new species identified in recent years.

Data analysis

The following data were collected and analyzed in order to evaluate the composition and the diversity of AM fungi.

1) Spore density (SD): number of all AM fungal spores per 20 g air-dried soil.

2) Species richness (SR): the number of AM fungal species per 20 g air-dried soil.

3) Species diversity (*H*): it is calculated by Shannon-Weiner index. If one random sample includes *N* individuals, and the individual number of species *i* is N_i, then $P_i = N_i / N$. *H* can be calculated with this formula:

$$H = -\sum_{i=1}^{k} (P_i \ln P_i)$$

where *k* is the number of a certain sampling site; P_i is the percentage of spore density of AM fungal species *i* in total spore density, of one sampling site.

4) Species evenness (*J*): it is calculated by Pielou index, that is $J = H/\ln S$, where *H* is Shannon-Weiner index, and *S* is the number of AM fungal species at a certain sampling site.

5) Isolation frequency (*F*): occurrence frequency of certain AM fungal genus or species in total samples, that is *F* = occurrence number of certain AM fungal genus or species/sample number ×100%. According to *F* value, AM fungi are classified to 4 grades: dominant genus or species ($F > 50\%$), most common genus or species ($30\% < F < 50\%$), common genus or species ($10\% < F < 30\%$), rare genus or species ($F < 10\%$).

6) Relative abundance (*RA*): number of certain AM fungal genus or species spores at a certain sampling site / number of all AM fungal spores at a certain sampling site.

7) Importance value (I): the mean value of isolation frequency (F) and relative abundance (RA), that is I = (F + RA) / 2.

Correlation analysis and significance test were performed using DPS software and *LSD* method.

Results

Grass degradation had negative effects on AM fungi population composition in soil of *S. purpurea* community. AM fungi were composed of *Acaulospora*, *Glomus* and *Scutellospora* in normal, lightly, moderately and severely degraded grassland, and the species richness followed the order: *Glomus* > *Acaulospora* > *Scutellospora* (Tab. 1).

Compared to normal grassland, the composition of AM fungi in degraded grasslands changed significantly (Tab. 2); the proportion of *Acaulospora* in lightly and severely degraded grasslands increased by 92.3% and 179.0%

respectively, compared with normal grassland, while it decreased by 36.5% in moderately degraded grasslands. The change of *Glomus* was totally different from *Acaulospora.* The proportion of *Scutellospora* species significantly increased with the grassland degradation (Tab. 2). Grassland degradation had important effects on AM fungi composition in soil of *S. purpurea* community.

Meanwhile, in all these four types of grasslands, the relative abundance of common species of AM fungi *Glomus* were higher than other dominant species.

As dominant species, *A. scrobiculata* and *A. spinosa* only occurred in normal and lightly degraded grasslands (Tab.1).

Glomus was dominant genus with a frequency of occurrence of 100% over all spore samples examined for all types of grasslands, and the relative abundance (77.0-96.1%) and importance value (88.5-98.0%) were higher than for other genus. In different degraded grasslands, the changes of AM fungi spore density, relative abundance and importance value were all the same (Tab. 2) and they showed a specific tendency: *Glomus*>*Acaulospora*>*Scutellospora.* Significant differences among the genus were observed ($P<0.05$).

Tab. 1. Genera and species of AM fungi in soil from different degraded grasslands of *S. purpurea*

Genus	Species	Normal grassland	Lightly degraded grassland	Moderately degraded grassland	Severely degraded grassland
Acaulospora	*A. delicat*	+	+	-	-
	A. laevis	+	+	+	+
	A. scrobiculata	+	-	-	-
	A. spinosa	-	+	-	-
Glomus	*G. aggregatum*	+	+	+	+
	G. claroideum	+	+	+	+
	G. convolutum	+	+	+	-
	G. dominikii	+	-	+	+
	G. etunicatum	+	+	+	-
	G. geosporum	+	+	+	+
	G. mosseae	+	+	+	+
	G. versiforme	+	+	+	+
Scutellospora	*S. calospora*	+	+	+	+

Note: "+" indicates that the AM fungi occurred in the site sample; "-" indicates that no AM fungi appeared in the site.

Tab. 2. Spore density, composition and importance value of AM fungi in soil samples from *S. purpurea* community, under different degraded grasslands

State of grasslands	AM fungal genera	Spore density (20 g^{-1} dry soil)	Species richness	Diversity (*H*)	Species evenness (*J*)	Relative Abundance (%)	Importance value
Normal grassland	*Acaulospora*	2.25 ± 1.90 ^{b}	0.41 ± 0.26 ^{b}	0.1940 ^{b}	0.0781 ^{b}	5.87 ^{b}	52.94 ^{b}
	Glomus	34.00± 20.35 ^{a}	1.42 ± 0.38 ^{a}	1.2367 ^{a}	0.4977 ^{a}	88.70 ^{a}	94.35 ^{a}
	Scutellospora	2.08 ± 1.44 ^{b}	0.17 ± 0.17 ^{c}	0.1576 ^{b}	0.0634 ^{b}	5.43 ^{b}	36.05 ^{c}
Lightly degraded grassland	*Acaulospora*	3.00 ± 2.35 ^{b}	0.42 ± 0.26 ^{b}	0.3153 ^{b}	0.1315 ^{b}	11.29 ^{b}	55.65 ^{b}
	Glomus	23.00 ± 11.98 ^{a}	1.25 ± 0.37 ^{a}	1.3200 ^{a}	0.5505 ^{a}	86.52 ^{a}	93.26 ^{a}
	Scutellospora	0.58 ± 0.58 ^{c}	0.08 ± 0.08 ^{c}	0.0840 ^{c}	0.0350 ^{c}	2.19 ^{c}	17.76 ^{c}
Moderate degraded grassland	*Acaulospora*	3.08 ± 3.08 ^{b}	0.17 ± 0.17 ^{b}	0.1220 ^{b}	0.0530 ^{b}	3.73 ^{b}	35.20 ^{b}
	Glomus	79.42 ± 44.29 ^{a}	1.33 ± 0.36 ^{a}	1.3374 ^{a}	0.5808 ^{a}	96.07 ^{a}	98.04 ^{a}
	Scutellospora	0.17 ± 0.17 ^{c}	0.08 ± 0.08 ^{c}	0.0124 ^{c}	0.0054 ^{c}	0.20 ^{c}	16.77 ^{c}
Severely degraded grassland	*Acaulospora*	3.92 ± 3.92 ^{b}	0.25 ± 0.25 ^{b}	0.2965 ^{b}	0.1426 ^{b}	16.38 ^{b}	58.19 ^{b}
	Glomus	18.42 ± 9.03 ^{a}	0.92 ± 0.31 ^{a}	1.2289 ^{a}	0.5910 ^{a}	77.00 ^{a}	88.50 ^{a}
	Scutellospora	1.58 ± 1.58 ^{b}	0.08 ± 0.08 ^{c}	0.1794 ^{c}	0.0863 ^{c}	6.62 ^{c}	19.98 ^{c}

Note: these data were means ± SD; different lowercase letters on the same column means the significant difference at 5% level.

Among all grasslands, there was no consistent variation in spore density, relative abundance and importance value of AM fungi. The spore density of *Glomus* in moderately degraded grassland were 133%, 245.2% and 341.2% higher than that in normal grassland, lightly degraded grassland and severely degraded grassland, respectively. The spore density of *Acaulospora* in degraded grasslands was higher than that in normal grassland. The relative abundance of *Acaulospora* in severely degraded grassland was averagely higher by 187% than other grassland, and the importance value were similar, except in moderately degraded grassland (Tab. 2).

Grassland degradation had negative effects on AM fungi population composition. AM fungi species diversity in soil samples of *S. purpurea* community was different under the four grasslands degraded states. The AM fungi species richness of degraded grasslands was much lower than that of normal grassland, and the decreased tendency increased with the degradation of grassland. The species diversity in lightly and severely degraded grasslands was higher than that in moderately degraded and normal grassland (Fig. 1b).

The species richness of *Acaulospora, Glomus* and *Scutellospora* in severely degraded grasslands decreased by 39.0%, 33.1% and 52.9% compared with normal grassland

(Tab. 2). There was no significant difference for species diversity and evenness between lightly and moderately degraded grasslands. Species diversity and evenness of *Glomus*, *Acaulospora* and *Scutellospora* in severely degraded grassland increased compared to normal grassland (Tab. 2). Different degraded grassland had negative effects on AM fungi genus and species richness, and the grassland degradation facilitated the increase of species diversity and evenness of *Glomus*.

The spore density and relative abundance of dominant species *Glomus* in moderately degraded grassland were much higher than other grasslands (Fig. 2).

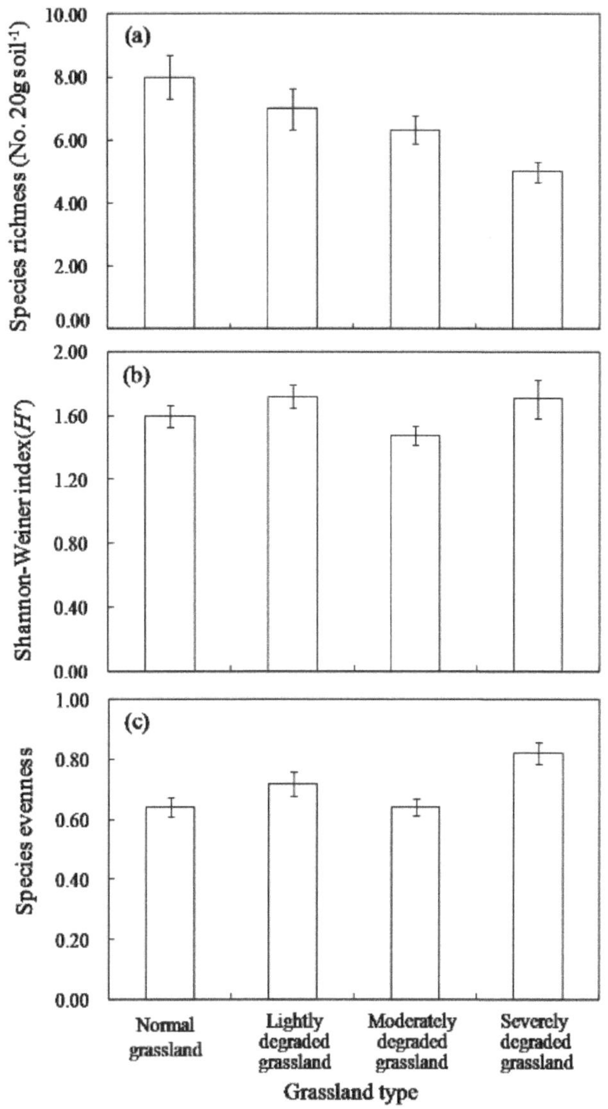

Fig. 1. Species richness (a), diversity (b) and evenness (c) of AM fungi in different grasslands

Compared with normal grassland, the dominant species of *Glomus* in moderately degraded grassland increased, while the genus *Acaulospora* decreased significantly (Tab. 2). Compared with normal grassland, the common dominant species of *Glomus* significantly increased in lightly and moderately degraded grassland, while it decreased in severely degraded grassland (Fig. 3).

In all these four grasslands, importance value of dominant species was greater than that of the most common species. Importance values of dominant species in degraded grasslands were much lower than the ones for normal grassland. The importance value of dominant species decreased lightly with the degradation aggravation, while the value of the most common species did not change significantly (Tab. 2). The importance values of *Acaulospora* in lightly and severely degraded grasslands were scant higher than that of *Glomus*, while the importance value of *Glomus* dominant species in moderately degraded grasslands was significantly higher than that of *Acaulospora*.

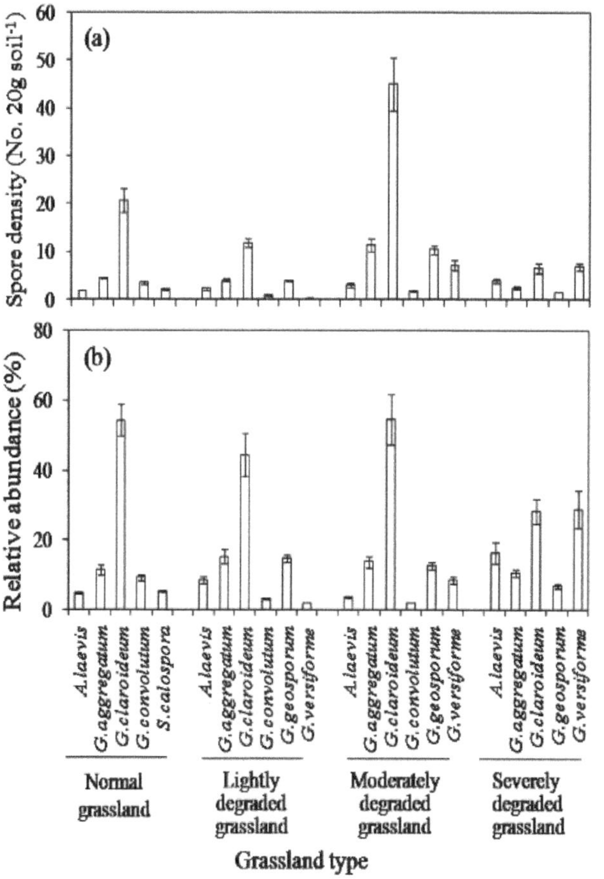

Fig. 2. Spore density (a) and relative abundance (b) of dominant AM fungi species in different grasslands

The relative abundance of dominant AM fungi in normal, lightly, moderately and severely degraded grasslands were 94.6%, 88.7%, 95.5% and 91.3%, respectively (Fig. 2b). The relative abundance of dominant species was significantly different among *Glomus,* as the proportion of *G. claroideum* species occupied about 30.9-57.5%.

Significantly positive correlations were observed between spore density and soil organic matter, spore density and altitude. There were significant negative correlations between diversity of AM fungi and soil available phosphorus, organic matter and altitude (Tab. 3).

Tab. 3. Correlation between environmental factors and AM fungal species density and the diversity of *S. purpurea*

	Soil water content	pH	Available phosphorus	Organic matters	Altitude
Spore density	0.1283	0.3051	0.7972**	0.9890**	0.9943**
Diversity	-0.2949	-0.3173	-0.9370**	-0.9223**	-0.9470**

Note: "**"denotes the significant difference at 1% level.

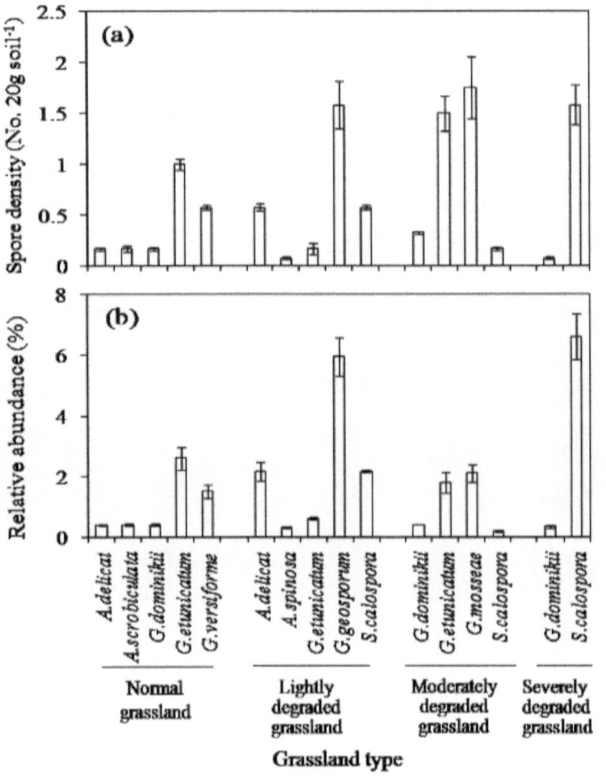

Fig. 3. Spore density (a) and relative abundance (b) of common dominant AM fungi species in different grasslands

Discussion

Several studies documented that degradation in Pakistan (Chaudhry *et al.*, 2005), Inner Mongolia steppe (Su and Guo, 2007) and southern England (Wearn and Gange, 2007) caused by overgrazing, declined species diversity of AM fungi. The current study found that grassland degradation had significant effects on AM fungal diversity of *S. purpurea* in Tibetan Plateau. Compared with normal grassland, AM fungal species richness in degraded grasslands significant declined with grassland degradation aggravation. However, AM fungi species diversity and evenness increased significantly in lightly and severely degraded grasslands, while it only decreased in moderately degraded grassland. The core of the grassland degradation is soil degradation, although the degradation of Tibet alpine grassland (which basically belongs to a natural process) is different from those caused by overgrazing (Gai *et al.*, 2006). Species diversity of AM fungi in Tibet alpine steppe showed the trend of lightly degraded grassland > severely degraded grassland > normal grassland > moderately degraded grassland, even though it was inconsistent with other previous results, that conclude that species diversity of AM fungi decline can be caused by overgrazing (Chaudhry *et al.*, 2005; Su and Guo, 2007; Wearn and Gange, 2007). This inconsistency might be related to the disruption of plant species caused by degradation in these areas, which grassland degradation may mainly decline with the proportion of non-mycorrhizal plant species, and little effects on mycorrhizal plant species. So that, in degraded grassland, the diversity of AM fungi till maintain at a higher level. Moreover, this also might be related to soil conditions and other environmental factors, caused by degradation, which can determine AM fungi community composition (Johnson *et al.*, 1992; Kernaghan, 2004). To obtain more information upon the impacts of degradation on alpine ecosystem, further studies on the influence of degradation on plant community composition, AM fungal colonization and soil conditions in Tibetan Plateau, are needed.

Previous results showed that AM fungi species *G. fasciculatum*, *G. geosporum*, *G. intraradices* and *G. mossea* were sensitive to overgrazing, while other *Glomus* fungi were not, obvious in Inner Mongolia (Su and Guo, 2007). The research on Namibia arid region showed that various grazing intensity affected the distribution of AM fungi, and *G. geosporum*, *G. mosseae* appeared in lightly grazing region, while they were not observed in moderately and severely grazing districts (Uhlmann *et al.*, 2006). These studies illustrate that AM fungi species have different reaction to environmental changes and grassland degradation in different ecosystems. In this study, *A. delicat* was found in normal grassland and lightly degraded grassland. *A. scrobiculata* was only observed in general grassland and it was not found in various degraded grasslands. However, 7 kinds of AM fungi, such as *A. laevis*, *G. aggregatum*, *G. claroideum*, *G. geosporum*, *G. mosseae*, *G. versiforme* and *S. calospora* had very strong adaptability, and were distributed also in normal grassland and all types of degraded grassland. Therefore, it might be concluded that different species of AM fungi have different response and sensitivity to grassland degradation.

Glomus species plays a dominant role, under different conditions of grasslands. In normal and degraded grasslands, the change trends of AM fungal species richness, spore density, relative abundance, importance value, diversity and evenness all followed the rule order of the three genera: *Glomus*>*Acaulospora*>*Scutellospora*. Unlike the relatively even distribution of AM fungal species in degraded steppe in Inner Mongolia (Wang and Liu, 2002), *Glomus* species had a dominant role (species richness, 63.6-80.0%; relative abundance, 77.0-96.1%) in AM fungal community structure of the normal and degraded grasslands in the ecological condition of alpine steppe of north Tibet Plateau. Moreover, *Glomus* species distributed widely (isolation frequency reached 100%). The result (Fig. 4) suggests that AM fungi in this region have a great tolerance to cold and arid environment, and also a high ability to adapt to degraded steppe, for long time co-evolution.

Although the composition of AM fungi dominant genera and species was simple, the percentage of dominant species increased with grassland degradation degree. The

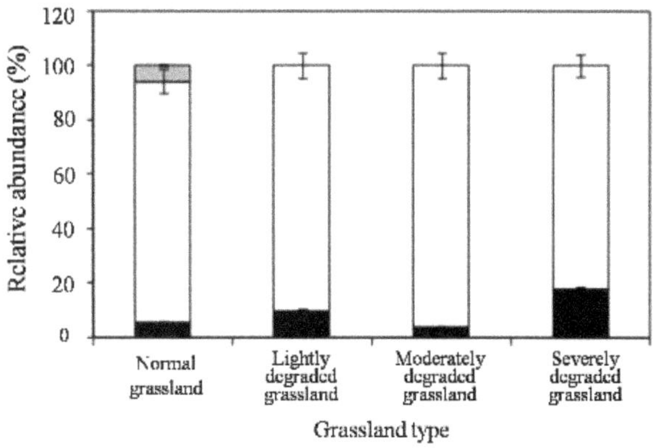

Fig. 4. Relative abundance of dominant AM fungi species in different grasslands. Black part in each column represents genus *Acaulospora*, white part in each column represents genus *Glomus* and grey part in each column represents genus *Scutellospora*

percentage of dominant species of *Glomus* genus (80.0-83.3%) in degraded grassland was significantly higher than for the normal one. This indicated that grassland degradation decreased AM fungal species richness. However, it promoted AM fungi adaptive ability to such an environment, and enhanced the survival of AM fungi of *Glomus* genera.

The role of AM fungal dominant species was very obvious in normal and degraded grasslands, with relative abundance of 88.7-95.5%. The relative abundances of the AM fungi and *Glomus* genera dominant species significantly decreased with degradation aggravation. It indicates that there is a significant relationship between plant species *S. purpurea* and AM fungal community composition. Generally, grassland degradation would thin and dwarf the vegetation, reduce the dominance of root system, which they all influence the survival and growth of AM fungi and inhibit the AM fungi variety and its infection on roots (Wang and Liu, 2002; Su and Guo, 2007; Wearn and Gange, 2007).

The study on alpine steppe of north Tibet Plateau found that the species abundance and richness of AM fungi in moderately degraded grassland reduced significantly, but species diversity and evenness were higher than the data for normal grassland. This might be related with the increase of soil organic matter. Under freezing low temperature in alpine steppe of north Tibet Plateau, the contact coverage of normal alpine grassland would lead to high soil water content and long term of soil freeze, as the soil organic matter is hard to decomposition and transform to sod layer, and therefore the formation and accumulation of organic matter would need a long time too. When the grass land were moderately degraded, the reduced soil water content might be suitable for microbe, and speed the decomposition of soil organic matter, and therefore increase the soil organic matter content, which would facilitate the propagation spore production of AM fungi, especially for *Glomus* genus.

It is worth to note that spore density, relative abundance, importance value, species diversity and evenness of *Acaulospora* and *Scutellospora* genera in moderately degraded grassland were lower compared to those in normal, lightly degraded and severely degraded grasslands, indicating that grassland degradation had adverse effects on AM fungal diversity.

Temperature (including soil temperature) is one of the most important ecological factors that determine AM fungal community composition, its development and infection in nature ecosystems (Heinemeyer and Fitter, 2004; Hawkes *et al.*, 2008). In the present study, with the increase of altitude, spore density increased significantly, while mycorrhizal infection decreased significantly ($r = -0.6502^*$). This result is inconsistent with previous results obtained from warm steppe of south Tibet (Cai *et al.*, 2010). The ecological phenomenon may be associated with a series of factors, such as soil humidity, soil pH value, organic matter, available phosphorus content and bacteria number in soil, in cold and arid environment. The results indicate that altitude has an important impact on soil factors, soil temperature, soil pH value, organic matter, available phosphorus content and bacteria number. Species diversity and evenness decreased with the increase of altitude, which may ascribe to the decrease of temperature. This was consistent with the results that AM fungal diversity increased with the increase of temperature within a certain range (Shi *et al.*, 2007), suggesting that an environment at a higher altitude had adverse effects on AM fungal diversity.

Conclusions

The major conclusion from this study is that grassland degradation decreased AM fungal species richness and abundance, but the diversity of AM fungi in degraded grassland increased. The results suggest that grassland degradation increased the adaptability and stress resistance of AM fungi alpine grassland in Tibet Plateau. In spite this, reduced human caused disturbance (overgrazing, picking traditional Chinese medicinal materials, etc.) in this area is imperative, in order to sustain the species richness of AM fungi. Furthermore, the effects of AM fungi on revegetation in Tibetan Plateau need more consistent study in future.

Acknowledgments

This research was funded by grants from the China's National Science Foundation (No. 41161043 and 31300097), the Fundamental Research Funds for the Central Universities (No. 12QNJJ017) and State Key Laboratory of Desert and Oasis Ecology, Xinjiang Institute of Ecology and Geography, Chinese Academy of Sciences.

References

BaoYY, Yan W (2004). Arbuscular mycorrhizae and their structural types in grasslands of mid-western Inner Mongolia. Biodiv Sci 12:501-508.

Bingham MA, Biondini M (2009). Mycorrhizal hyphal length as a function of plant community richness and composition in restored northern tallgrass prairies (USA). Rangeland Ecol Manag 62:60-67.

Cai XB, Peng YL, Gai JP (2010). Ecological distribution of arbuscular mycorrhizal fungi in alpine grasslands of Tibet Plateau. Chinese J Appl Ecol 21:2635-2644.

Evelin H, Kapoor R, Giri B (2009). Arbuscular mycorrhizal fungi in alleviation of salt stress: a review. Ann Bot 104:1263-1280.

Gai JP, Christie P, Cai XB, Fan JQ, Zhang JL, Feng G, Li XL (2009). Occurrence and distribution of arbuscular mycorrhizal fungal species in three types of grassland community of the Tibetan Plateau. Ecol Res 24:1345-1350.

Gai JP, Cai XB, Feng G, Christie P, Li XL (2006). Arbuscular mycorrhizal fungi associated with sedges on the Tibetan plateau. Mycorrhiza 16:151-157.

Gao QM, Zhang Y, Guo LD (2006). Arbuscular mycorrhizal fungi in the southeast region of Tibet. Mycosystema 25:234-243.

Chaudhry MS, Batool Z, Khan AG (2005). Preliminary assessment of plant community structure and arbuscular mycorrhizas in rangeland habitats of Cholistan desert, Pakistan. Mycorrhiza 15:606-611.

Hawkes CV, Hartley IP, Ineson P, Fitter AH (2008). Soil temperature affects carbon allocation within arbuscular mycorrhizal networks and carbon transport from plant to fungus. Global Change Biol 14:1181-1190.

Heinemeyer A, Fitter AH (2004). Impact of temperature on the arbuscular mycorrhizal (AM) symbiosis: growth responses of the host plant and its AM fungal partner. J Exp Bot 55:525-534.

Helgason T, Fitter AH (2009). Natural selection and the evolutionary ecology of the arbuscular mycorrhizal fungi (*Phylum glomeromycota*). J Exp Bot 60:2465-2480.

Johnson N, Tilman D, Wedin D (1992). Plant and soil controls on mycorrhizal fungal communities. Ecology 73:2034-2042.

Kernaghan G (2004). Mycorrhizal diversity: cause and effect?. Pedobiologia 49:511-520.

Landis FC, Gargas A, Givnish TJ (2004). Relationships among arbuscular mycorrhizal fungi, vascular plants and environmental condition in oak savannas. New Phytol 164:493-504.

Liu RJ, Jiao H, Li Y, Li M, Zhu XC (2009). Research advances in species diversity of arbuscular mycorrhizal fungi. Chinese J Appl Ecol 20:2301-2307.

Öpic M, Moora M, Liira J, Zobel M (2006). Composition of root-colonizing arbuscular mycorrhizal fungal communities in different ecosystems around the globe. J Ecol 94:778-90.

O'Connor PJ, Smith SE, Smith EA (2002). Arbuscular mycorrhizas influence plant diversity and community structure in a semiarid herbland. New Phytol 154:209-218.

Schenck NC, Péréz Y (1988). Manual for identification of vesicular-arbuscular mycorrhizal fungi. 2nd ed. Synergistic Publications, Gainseville, Florida, 1-137 p.

Shi ZY, Chen ZC, Zhang LY, Feng G, Christie P, Tian CY, Li XL (2007). Diversity and zonal distribution of arbuscular mycorrhizal fungi on the northern slopes of the Tianshan Mountains. Sci China Ser D 50:135-141.

Smith SE, Read DJ (2008). Mycorrhizal symbiosis. 3rd ed., Academic Press, San Diego, CA, USA. 27 p.

Su YY, Guo LD (2007). Arbuscular mycorrhizal fungi in nongrazed, restored and over-grazed grassland in the Innner Mongolia steppe. Mycorrhiza 17:689-693.

Uhlnann E, Görke C, Petersen A, Oberwinkler F (2004). Arbuscular mycorrhizae from semiarid region of Namibia. Can J Bot 82:645-653.

Uhlmann E, Görke C, Petersen A, Oberwinkler F (2006). Arbuscular mycorrhizae from arid parts of Namibia. J Arid Environ 64:221-237.

van der Heijden MGA, Klironomos JN, Ursic M, Moutoglis P, Streitwolf-Engel R, Boller T, Wiemken A, Sanders IR (1998). Mycorrhizal fungal diversity determines plant biodiversity, ecosystem variability and productivity. Nature 396:69-72.

Wagg C, Jansa J, Schmid B, van der Heijden MGA (2011). Belowground biodiversity effects of plant symbionts support aboveground productivity. Ecol Lett 14:1001-1009.

Wang FY, Liu RJ (2002). Effects of biological factors on the diversity of arbuscular mycorrhizal fungi. Acta Ecol Sin 22:403-408.

Wearn JA, Gange AC (2007). Above-ground herbivory causes rapid and sustained change in mycorrhizal colonization of grasses. Oecologia 153:959-971.

Wu QS, Xia RX, Zou YN (2008). Improved soil structure and citrus growth after inoculation with three arbuscular mycorrhizal fungi under drought stress. Eur J Soil Biol 44:122-128.

Zhang JT (2003). Applied ecology. Science Press, Beijing, 173-174 p.

Zhang T, Sun Y, Shi ZY, Feng G (2012). Arbuscular mycorrhizal fungi can accelerate the restoration of degraded spring grassland in central Asia. Rangeland Ecol Manag 65:426-432.

15

Nitrogen and Zinc Interaction Improves Yield and Quality of Submerged Basmati Rice (*Oryza sativa* L.)

Hakoomat ALI[1], Zuhair HASNAIN[1], Ahmad N. SHAHZAD[1*], Naeem SARWAR[1], Muhammad K. QURESHI[2], Shazia KHALIQ[3], Muhammad F. QAYYUM[4]

[1]*Bahauddin Zakariya University, Department of Agronomy, Faculty of Agricultural Sciences and Technology, Multan, 60800 Pakistan; hha94@bzu.edu.pk; z.hasnain@irri.org; anaeems@gmail.com (*corresponding author); bajwa834@gmail.com*

[2]*Bahauddin Zakariya University, Department of Plant Breeding and Genetics, Faculty of Agricultural Sciences and Technology, Multan, 60800 Pakistan; m.k.qureshi81@gmail.com*

[3]*Bahauddin Zakariya University, Department of Environmental Sciences, Multan, 60800 Pakistan; humza38@hotmail.com*

[4]*Bahauddin Zakariya University, Department of Soil Science, Multan, 60800 Pakistan; farooq.qayyum@bzu.edu.pk*

Abstract

Nitrogen (N) and zinc (Zn) are the two major yield-limiting factors of flooded rice cultivation systems. Both nutrients interact and affect availability of the other in alkaline calcareous soils. In order to evaluate the interactive effects of N and Zn on yield and quality of Basmati rice, a field study was conducted at Sheikhupura (Site I) and Sargodha (Site II), Pakistan. Nitrogen treatments (0, 40, 80, 120 and 160 kg/ha) were kept in main blocks while zinc levels (0, 8, 10, 12 and 14 kg/ha) were kept in sub blocks. The main effects of N and Zn levels were significant for grain yield and yield attributes. N and Zn interaction significantly improved the grain yield, yield components and all kernel quality parameters, except kernel amylose contents. Combined application of 120 kg N/ha and 14 kg Zn/ha produced the maximum grain yields at both sites (6.12 and 5.78 t/ha). This combination also yielded the maximum kernel lengths and widths, water absorption ratio and kernel protein contents. There was a significant positive correlation between grain yield and total dry matter, panicle-bearing tillers, spikelets panicle, grain weight, and harvest index. Application of 160 kg N/ha was detrimental to yield and quality attributes and reduced the agronomic efficiency of N use at both sites. Site comparison showed that soil pH and soil nutrient contents play a significant role in determination of the optimum nitrogen and zinc fertilizer doses for maximum yields.

Keywords: agronomic efficiency, aromatic rice, fertilizer, kernel quality, paddy yield

Introduction

Rice is staple food for more than half of the world's population (Fageria *et al.*, 2008). In Pakistan, rice with an average yield of 2398 kg ha^{-1} (Anonymous, 2013) is ranked as the second most essential food of the human diet. Basmati (aromatic) rice is known for its long fluffy grains with minimum kernel dimension, aroma intensity, texture of the cooked rice, high elongation during cooking, palatability and longer shelf life (Ahmad *et al.*, 2005). Most of the Basmati rice in Pakistan is produced in the Punjab province. Both yield potential and quality characteristics of new Basmati varieties such as 'Super Basmati', 'Basmati 2000' and 'Shaheen Basmati' has been improved by rigorous breeding efforts. There is a big difference in the yield potential of different rice varieties and the actual yields obtained by the farmers. The yield potential of Basmati varieties ranges from 4.5 to 6 t ha^{-1} while the farm yields vary between 2.0 to 2.8 t ha^{-1} (Ahmad *et al.*, 2005; Shivay *et al.*, 2010). Basmati rice is also an important export commodity of Pakistan. So, quality of Basmati rice is next to rice production. Among major constraints of quality and paddy yields in rice, imbalanced nitrogen fertilization (Wang *et al.*, 2008) and zinc deficiency (Shivay *et al.*, 2010) in soils are the most important variables.

Zinc (Zn) deficiency in staple foods is a major threat in combating malnutrition related health problems in many developing countries (Impa *et al.*, 2013). After nitrogen, zinc is the second most yield-limiting nutrient in rice (Quijano-Guerta *et al.*, *2002*). Most of the rice crop in Pakistan is cultivated in alkaline calcareous soils with high clay contents. Due to these soil characteristics, most of the zinc (Zn) applied as fertilizer is adsorbed and only little is recovered by rice crop during the growth period (Tahir *et al.*, 1992; Rahman *et al.*, 2002). High soil pH is considered to be the most crucial factor in causing Zn deficiency in Pakistani soils (Qadar, 2002). In addition, Zn deficiency is also wide spread in these soils (Rashid *et al.*, 1999). Zn fertilization has also been shown to affect kernel protein contents of rice (Mirzavand, 2007). Since Zn is a costly fertilizer, its economic use is important from the point of view of poor farmers.

Nitrogen (N) has been an important yield determinant of the flooded rice production systems. High prices, food shortages in developing world and the adverse effects of heavy N losses from flooded fields however necessitate optimal use of N. Nitrogen losses through ammonia volatilization, leaching and runoff have been recorded from 20-80% in the rice production (Singh *et al.*, 1998; Griggs *et al.*, 2007; Norman *et al.*, 2009). Economic use of nitrogen fertilizer is especially important for small growers as urea prices are continuously on rise. The response of rice crop to nitrogen fertilizers have been well documented, however the effects of N interaction with zinc application rates on yield and quality of rice are lacking. Non-judicious use of nitrogen fertilizer could further aggravate Zn deficiency in alkaline soils and suppress paddy growth and yield (Rashid, 1996). Nitrogen fertilization is found to enhance zinc contents of flooded rice, where urea performed better than NH_4NO_3 in increasing Zn contents (Chaudhry *et al.*, 1977; Singh and Singh, 1981). They observed that higher Zn contents in plant tissues by nitrogen application were due to better growth, increased Zn solubility and increased root efficiency for Zn uptake. Under submerged conditions, urea is more efficient as compared to nitrate containing fertilizers. Urea-N application lowers soil pH and in addition it forms NH_3-Zn complex and thus increases Zn availability under alkaline conditions (Gao *et al.*, 2012). Keeping in view the possible N × Zn interactions in alkaline soils, present study was aimed to identify optimum doses of N and Zn for higher grain yields and better quality of Basmati rice.

Materials and methods

The field experiments were conducted at Adaptive Research Farm Sheikhupura (31.6°N, 74.6°E, 217 m) and Sargodha (32.04°N, 72.67°E, 188 m) during 2009 and 2010. Weather from July to September in 2009 was dry with the rainfall of about 42.1% below average and in 2010 was wet with rainfall of about 10.5% above average (Fig. 1). Experiments were laid out in a randomized complete block design with split arrangements and were replicated thrice. Five nitrogen (N) levels (0, 40, 80, 120 and 160 kg/ha) were kept in main plots while zinc (Zn) levels (0, 8, 10, 12 and 14 kg/ha) were kept in sub plots. The net plot size was 1.8m x 12m. The soil texture at Sheikhupura (site I) was loam (12% Sand, 70% Silt and 18% Clay) with 8.38 pH, 10.45% total soluble salts, 0.82% organic matter, 0.06% total nitrogen, 10.6 ppm available phosphorus, 216 ppm available potassium and 0.51 mg AB-DPTA extractable Zn kg^{-1} soil. While at Sargodha (site II), soil was silt loam (22% Sand, 59% Silt and 18% Clay) with 7.59 pH, 10.2% total soluble salts, 0.94% organic matter, 0.06% total nitrogen, 15.9 ppm available phosphorus, 239 ppm available potassium and 0.32 mg AB-DPTA extractable Zn kg^{-1} soil.

Rice nursery (cv. 'Super Basmati') was sown during first week of June using hand drill at both sites. Nursery seedlings of 28-30 days old were manually transplanted in puddled fields by keeping a row to row and plant to plant distance of 22.5 cm on 5th and 7th July of both growing seasons at Sheikhupura and Sargodha, respectively. Before transplantation, half of the total nitrogen and half of the total Zn (doses as per treatments), 80 kg/ha phosphorus and 60 kg/ha potassium were applied in the puddle field in the form of urea, $ZnSO_4$, SSP and SOP, respectively. The remaining doses of nitrogen and zinc fertilizers were applied during active tillering stage. Butachlor (60% EC) @800 ml/ha was applied in standing water for weed control (Reddy, 2004). To control borers and leaf folders, Carbofuran (10% GR) @25 kg/ha was broadcasted at 55 day after transplantation. Irrigation water was maintained at a depth of 3-4 cm at transplantation and one week afterwards at a depth of 5-6 cm throughout the growing season till one week before harvesting. An area of 7.5 m^2 was harvested from each plot and 20 plants were randomly selected for determination of yield attributes. Final grain and dry matter yields were estimated from 7.5 m^2. Manual threshing was done separately for each plot. Agronomic efficiencies of N and Zn were calculated as:

AE_N = grain yield (kg/ha) with N application - grain yield (kg/ha) without N application / N application (kg/ha)

AE_{Zn} = grain yield (kg/ha) with Zn application - grain yield (kg/ha) without Zn application / Zn application (kg/ha)

Kernel dimensions (length and width) of 100 normal kernels out of each treatment were measured using a dial caliper and then averaged. Water absorption ratio of kernels was determined by dividing the weight of cooked rice with that of raw rice. Protein contents were calculated by micro-Kjeldahl digestion method to estimate nitrogen content, which was multiplied with factor 5.9 in order to convert into protein (Jacobs, 1958). Kernel amylose contents were determined according to the method of Juliano (1971) by using Spectrophotometer. The data collected were subjected to Fischer's ANOVA using MSTAT-C program. The treatment means were compared using Duncan Multiple Range Test at 5% probability level.

Results and discussion

Paddy yield components

Statistical analysis showed that year effect for different parameters studied in this work at both experimental sites was not significant. Therefore data presented in this work represent the averages of both years (2009 and 2010) for either site. Tillering of Basmati rice was significantly influenced by nitrogen (N) application. Comparison of means (Tab. 1) showed that N significantly increased the number of panicle-bearing tillers up to 120 kg N/ha; thereafter panicle bearing tillers were decreased at a higher dose of N (160 kg/ha). Similar trend was observed at both sites. Promotion of growth and tillering in rice by N application has been reported by many studies (Mae, 1997; Samonte *et al.* 2006). Reduction in number of productive tillers at higher N dose could be the result of excessive vegetative growth (Gebrekidan and Seyoum, 2006). On the other hand, significant effect of Zn on tiller production was only evident at site I (Sheikhupura) where 12 kg Zn/ha increased the number of panicle bearing tillers in comparison to control (Tab. 1; 0 kg Zn/ha). While at site II (Sargodha), increase in Zn rates did not significantly affect the number of panicle-bearing tillers. These varying responses to

Zn supply at two sites of the present experiment could be related to differences in soils characteristics. Zn nutrition has been shown to have positive influence on tiller production of rice (Impa *et al.*, 2013; Sarwar *et al.*, 2013). The interaction between N and Zn regarding panicle-bearing tillers was significant (Tab. 2). Highest number of panicle-bearing tillers hill^{-1} (13.8 and 13.1) at sites I and II were produced by 120 N + 14 Zn and 120 N + 12 Zn (kg/ha), respectively.

The number of spikelets panicle^{-1} varied from 88.1 to 110.7 across treatments, sites and years. At both sites, an increase in

Tab. 1. Main effect of nitrogen and zinc rates on yield and yield components of Basmati rice (*Oryza sativa* L.)

Treatments	PBT hill^{-1}		S panicle^{-1}		1000- GW (g)		TDM (t ha^{-1})		GY (t ha^{-1})		HI	
N kg ha^{-1}/ Site	I	II	I	II	I	II	I	II	I	II	I	II
0	9.79c	9.34c	101.72c	99.33d	19.87d	19.86e	9.19c	8.81c	2.72c	2.73c	29.41d	30.77e
40	10.47c	10.03c	107.65b	105.43c	20.44c	20.53d	9.79c	10.01c	3.00c	3.29c	30.64d	32.63d
80	11.37b	10.60bc	108.30b	106.90b	21.27b	21.32c	11.73b	11.58b	3.95b	4.05b	33.42c	34.68c
120	13.08a	12.06a	110.07a	108.06a	23.02a	22.67a	15.07a	14.03a	5.57a	5.40a	36.94a	38.48a
160	11.74b	11.34ab	108.65b	106.23bc	21.65b	22.04b	12.62b	13.71a	4.46b	4.99a	35.18b	36.44b
LSD (P ≤ 0.05)	0.88	0.95	1.03	0.87	0.42	0.37	1.21	1.43	0.77	0.69	1.18	1.47
Zn kg ha^{-1}/ Site												
0	10.53b	10.50	104.78c	102.75c	20.54d	20.73c	10.68c	10.82c	3.48d	3.72d	31.73c	33.52d
8	11.18ab	10.51	107.50b	105.25b	21.24c	21.24b	11.53b	11.41bc	3.88c	3.99c	33.25b	34.46c
10	11.48ab	10.73	107.77ab	106.16a	21.53a	21.46a	11.83ab	11.75ab	3.97bc	4.10bc	33.14b	34.38c
12	11.73a	11.07	108.21a	106.04a	21.40b	21.47a	12.12ab	12.03ab	4.10ab	4.27ab	33.23b	34.96b
14	11.52ab	10.86	108.13a	105.76ab	21.54a	21.53a	12.25a	12.13a	4.26a	4.40a	34.12a	35.69a
LSD (P ≤ 0.05)	1.09	NS	0.58	0.52	0.12	0.15	0.71	0.65	0.21	0.18	0.57	0.44

N = nitrogen; Zn = zinc; TDM = total dry matter; GY = grain yield; PBT = panicle-bearing tillers; S = spikelets; GW = grain weight; HI = harvest index; NS = Not Significant.
Treatment means were separated with Duncan Multiple Range Test at 5% probability. Different letters on treatment means within each column show statistical differences at P≤0.05.

Tab. 2. Interactive effect of nitrogen and zinc rates on yield components of Basmati rice (*Oryza sativa* L.)

Treatments		PBT hill^{-1}		S panicle^{-1}		1000-GW	
N (kg ha^{-1})	Zn (kg ha^{-1})	Site I	Site II	Site I	Site II	Site I	Site II
0	0	7.8	7.7	90.2	88.1	18.1	18.2
0	8	9.9	9.5	103.1	99.4	20.2	20.0
0	10	10.3	9.7	104.9	103.7	21.1	20.8
0	12	11.4	10.4	106.0	103.3	20.1	20.3
0	14	9.4	9.4	104.5	101.6	20.0	20.1
40	0	10.5	10.1	108.8	105.8	20.3	20.2
40	8	10.4	9.8	107.9	105.5	20.6	20.5
40	10	10.6	10.1	106.5	105.3	20.3	20.3
40	12	10.3	9.8	107.1	105.3	20.5	20.8
40	14	10.6	10.3	108.0	105.3	20.4	20.9
80	0	10.4	10.1	107.3	105.6	20.5	21.0
80	8	11.3	10.4	108.2	107.4	21.2	21.3
80	10	11.4	10.7	108.4	107.1	21.3	21.2
80	12	11.8	10.8	108.7	107.1	21.5	21.5
80	14	12.1	11.0	108.8	107.3	21.9	21.7
120	0	12.5	11.4	109.4	107.9	22.5	22.3
120	8	12.7	11.7	109.8	107.6	22.7	22.4
120	10	13.1	11.8	110.1	108.2	23.1	22.8
120	12	13.3	13.1	110.3	108.0	23.1	22.9
120	14	13.8	12.4	110.7	108.6	23.6	23.0
160	0	11.4	11.7	108.3	105.8	21.5	21.9
160	8	11.7	11.2	108.5	106.3	21.5	22.0
160	10	12.0	11.3	108.9	106.5	21.9	22.2
160	12	11.8	11.3	108.9	106.4	21.8	22.1
160	14	11.8	11.3	108.6	106.1	21.8	22.0
LSD (P ≤ 0.05)		1.21	1.45	1.14	1.04	0.54	0.49
F-Significant		**	**	**	**	*	*

N = nitrogen; Zn = zinc; PBT = panicle-bearing tillers; S = spikelets; GW = grain weight; NS = not significant; *, **: Significant at 0.05 and 0.01, respectively

the rate of N application up to 120 kg ha^{-1} progressively increased the number of spikelets $panicle^{-1}$ over control (Tab. 1). Gebrekidan and Seyoum (2006) revealed that increase in spikelet number by N is the result of increase in panicle length. In a recent study Ding *et al.* (2014) showed that the increase in the number of spikelets $panicle^{-1}$ by nitrogen fertilization is associated with an increase in cytokinen contents. Whereas at 160 kg N/ha, the decrease in spikelets $panicle^{-1}$ as compared to that of 120 kg N/ha (Tab. 1) may be because of increased competition among tillers for carbohydrate supply (Wu *et al.*, 1998). Different Zn application rates also significantly affected the number of spikelets $panicle^{-1}$ at both sites. Though spikelets $panicle^{-1}$ at Zn rates of 12-14 kg/ha were significantly higher over control, the gain at higher Zn doses was relatively small at both sites (Tab. 1). Zinc deficiency affects floral development of plants by lowering enzyme activity (Pandey *et al.*, 2001). Improvement in different grain yield components with an increase in Zn levels could be attributed to adequate Zn supply that might have affected the metabolic and enzymatic activity (Hafeez *et al.*, 2010). Combined effect of N and Zn on spikelets $panicle^{-1}$ was also significant. Maximum number of spikelets $panicle^{-1}$ at both sites were recorded where 120 N + 14 Zn (kg/ha) were applied (Tab. 2). While control plots (where N and Zn were not applied) showed the lowest number of spikelets $panicle^{-1}$.

Grain weight of rice kernels is an important yield attribute, which is mostly determined by genetic potential rather than environmental conditions (Ashraf *et al.*, 1999). Data showed that thousand grain weight significantly increased with an increase in N fertilizer up till 120 kg/ha, then declined at 160 kg N/ha (Tab. 1). Similar response was evident at both sites. Plots not supplied with N showed the lowest grain weights. Limited availability of nitrogen during grain filling stages affects carbohydrates supply to developing seeds and therefore decreases grain weight (Mae, 1997). Mean comparisons showed that Zinc application significantly increased the grain weight. At site I (Sheikhupura) increment in Zn fertilizer progressively increased the thousand grain weight, whereas at site II (Sargodha) grain weight was significantly increased till 10 kg Zn/ha and then remained static onwards (Tab. 1). Combined application of N and Zn also had a significant positive effect on thousand grain weight at both sites. Highest thousand grain weight was found where 120 N + 14 Zn (kg/ha) was applied (Tab. 2).

Total dry matter, paddy yield and harvest index

The nitrogen fertilizer significantly increased the total dry matter (TDM) yield with increasing rate of N application at both sites up to 120 kg ha^{-1}, thereafter TDM was slightly decreased at 160 kg ha^{-1} and difference between the two treatments were significant only at Sheikhupura site (Tab. 1). Increase in dry matter yield by N application is apparently the result of growth promotion by nitrogen. Similarly, increasing rates of Zn application significantly increased TDM up to 10 kg ha^{-1} at both sites (Tab. 1). Increment in Zn application from 10 kg Zn/ha onwards did not significantly affect the TDM yield. Wissuwa *et al.* (2006) observed that Zn deficiency caused stunted growth and reduced the TDM yield of many rice cultivars. They further exhibited that lower TDM yields under Zn deficiency are well correlated with lower Zn contents in rice plants. Interaction between N and Zn application rate affecting TDM yield was significant at both sites (Tab. 3). The maximum TDM yields 16.34 t ha^{-1} and 14.79 t ha^{-1} at Sheikhupura and Sargodha sites, respectively, were recorded where 120 N + 14 Zn (kg/ha) was applied.

Grain/paddy yields varied between 1.93-6.12 t/ha across sites, years and fertilizer treatments. Nitrogen application significantly increased the rice grain yield with each increment up to 120 kg N/ha at both sites (Tab. 1). The increase in grain yield with increasing N rates could be ascribed to more productive tillers and a higher grain weight. Grain yield had a significant positive correlation with total dry matter, panicle-bearing tillers, spikelets $panicle^{-1}$, grain weight, and harvest index at both sites (Tab. 4). The variances for six traits (total dry matter, grain yield, panicle-bearing tillers, spikelets $panicle^{-1}$, grain weight, and harvest index) were homogenous across 2009 and 2010; thus data for both years at each site were combined for correlation analysis. Comparable paddy yields (5.34 t ha^{-1}) for rice cultivar Basmati 2000 have been achieved at 175 kg N ha^{-1} (Manzoor *et al.*, 2006). They showed that increases in paddy yields at higher N supply were well correlated with an increase in number of grains $panicle^{-1}$ and 1000-grain weight. Highest N dose (160 kg N/ha) in the experiment significantly reduced the grain yield when compared with 120 kg N/ha at site I (Tab. 1). However, the differences in grain yield between 120 and 160 kg N/ha at site II were statistically at par. Reduction in grain yield at relatively higher N application might be the result of excessive vegetative growth and poor carbohydrate supply to developing grains. Nitrogen supply in excess is also known to increase the incidence of lodging and poor grain filling in rice (Saito, 1991).

Mean comparisons for Zn application showed that grain yield was significantly increased with each increment in Zn fertilizer. The highest grain yields 4.26 and 4.40 t/ha were recorded at site I and II, respectively, where 14 kg Zn/ha was applied (Tab. 1). The gain in grain yields at higher Zn levels could probably be the result of enhanced Zn concentration in the plant parts especially in rice grains (Rashid *et al.*, 2002). Interactive effects of N and Zn application on grain yield at both sites were also significant. Application of 120 N + 14 Zn (kg/ha) produced the maximum grain yields of 6.12 and 5.78 t/ha at site I (Sheikhupura) and Site II (Sargodha), respectively (Tab. 3). The lowest grain yields were obtained where nitrogen and zinc fertilizers were not applied. Higher paddy grain yields at treatments 120 N + 14 Zn (kg/ha) could be associated to higher tillering, more number of spikelets $panicle^{-1}$, higher grain weight as compared to unfertilized plots or plots where less amount of N and Zn were applied (Tab. 4).

Increasing rates of N application significantly increased harvest index (HI) up to 120 kg N/ha, thereafter HI significantly decreased at higher rate of N application 160 kg N/ha at both sites (Tab. 1). Reduction in HI at higher N dose is possibly the result of more carbohydrates supply towards dry matter accumulation than grain filling (Gebrekidan and Seyoum, 2006). Data also showed that Zn application at the rate of 14 kg/ha significantly increased HI compared to other Zn treatments at both the sites (Tab. 1). Statistical differences in HI between 8, 10 and 12 kg Zn/ha were however not significant at Sheikhupura site. A significant interaction between N x Zn showed that 120 N + 12 Zn (kg/ha) combination gave the maximum HI (37.5, 39.24) at Sheikhupura and Sargodha, respectively, whereas the minimum HI values 25.8 and 27.2 were observed at either site when N and Zn were not applied (Tab. 3). Site comparison showed that grain yields and HI at all Zn levels were slightly higher at Sargodha (Site II) than that of Sheikhupura (Site I) despite relatively lower soil Zn contents at the former site. Lower relative soil pH at Site II might have increased Zn solubility and thus produced better grain yields. Khan *et al.* (2007) compared the response of paddy yields to Zn fertilizer applied under different soil conditions. They noticed that the soils having low pH and lower $CaCO_3$ contents gave the highest yield response to Zn.

Tab. 3. Interactive effect of nitrogen and zinc rates on total dry matter, grain yield and harvest index of Basmati rice (*Oryza sativa* L.)

Treatments		TDM (t ha^{-1})		GY (t ha^{-1})		HI	
N (kg ha^{-1})	Zn (kg ha^{-1})	Site I	Site II	Site I	Site II	Site I	Site II
0	0	7.5	7.3	1.93	2.00	25.8	27.2
0	8	9.3	8.6	2.88	2.75	30.9	31.7
0	10	9.8	9.1	3.08	2.89	31.5	31.6
0	12	10.2	9.6	2.93	3.00	28.7	31.1
0	14	9.3	9.3	2.80	3.03	30.2	32.3
40	0	10.1	10.0	3.21	3.37	31.8	33.6
40	8	9.7	10.0	2.92	3.24	30.3	32.4
40	10	9.8	10.1	2.89	3.24	29.6	32.0
40	12	9.6	9.9	2.92	3.25	30.3	32.5
40	14	9.9	10.1	3.04	3.32	30.6	32.7
80	0	9.8	10.4	2.90	3.38	29.6	32.4
80	8	11.7	11.1	3.88	3.70	33.3	33.3
80	10	11.7	11.5	3.90	3.86	33.4	33.6
80	12	12.4	12.2	4.29	4.41	34.5	36.1
80	14	13.1	12.8	4.76	4.88	36.3	38.1
120	0	13.9	13.1	5.04	4.95	36.3	37.9
120	8	14.4	13.8	5.33	5.25	37.0	38.1
120	10	15.1	14.1	5.50	5.39	36.4	38.1
120	12	15.6	14.4	5.86	5.64	37.5	39.2
120	14	16.3	14.8	6.12	5.78	37.5	39.1
160	0	12.2	13.3	4.32	4.87	35.1	36.6
160	8	12.6	13.6	4.41	5.00	34.8	36.8
160	10	12.9	14.0	4.48	5.11	34.8	36.6
160	12	12.8	14.0	4.50	5.03	35.2	35.9
160	14	12.6	13.7	4.58	4.98	36.1	36.4
LSD ($P \leq 0.05$)		1.30	1.27	0.91	0.85	3.13	2.88
F-Significant		**	**	**	**	**	**

N = nitrogen; Zn = zinc; TDM = total dry matter; GY = grain yield; HI = harvest index; NS = not significant; *, **: Significant at 0.05 and 0.01, respectively.

Agronomic efficiency

Agronomic efficiency of nitrogen use (AE_N) or zinc use (AE_{Zn}) describes the capability of yield increase per kg nitrogen or zinc applied, respectively. Agronomic nutrient use efficiency depends on various factors such as nutrient availability, soil conditions, fertilizer source and dose, and most importantly crop acquisition efficiency (Mengel and Kirkby, 2001). Nitrogen application increased AE_N up to 120 kg N/ha, thereafter further increase in nitrogen rate (160 kg N/ha) reduced AE_N at both sites (Fig. 2). The minimum AE_N 7.0 and 14.1 kg grain/ kg N were obtained with the application of 40 kg N/ha whereas the maximum AE_N 23.8 and 22.3 kg grain/ kg N were recorded at application of 120 kg N/ha at site I and site II, respectively (Fig. 2). Agronomic efficiencies for N use (AEN) for irrigated rice varied between 11.5 kg grain/kg N in on-farm studies (Dobermann *et al.*, 2002) to 21.6 kg grain/kg

Tab. 4. Correlation between yield and yield components of Basmati rice (*Oryza sativa* L.)

Correlated Traits	Site I (n=12)	Site II (n=12)	Pooled (n=24)
GY vs. TDM	0.999**	0.996**	0.989**
GY vs. PBT hill-1	0.981**	0.728**	0.747**
GY vs. S Panicle-1	0.788**	0.802**	0.663**
GY vs. 1000-GW	0.986**	0.944**	0.961**
GY vs. HI	0.705*	0.990**	0.840**

TDM = total dry matter; GY = grain yield; PBT = panicle-bearing tillers; S = spikelets; GW = grain weight; HI = harvest index; *, **: Significant at 0.05 and 0.01, respectively.

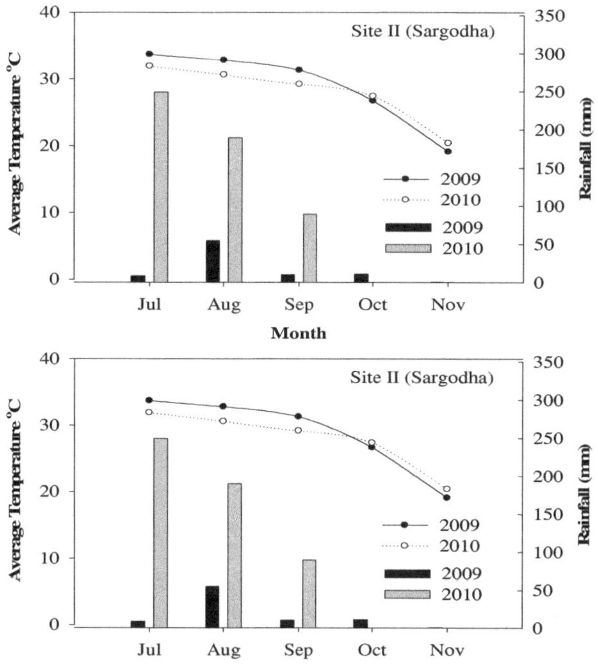

Fig. 1. Meteorological data for growing seasons during 2009-2010 at two cultivation sites

N in research trials (Ladha *et al.*, 2005) due to greater variability in soil and climatic conditions and incidence of insect/pest attack at farmer's field. Conversely, increase in Zn rates reduced the agronomic efficiency of Zn use (AE_{Zn}) at both sites.

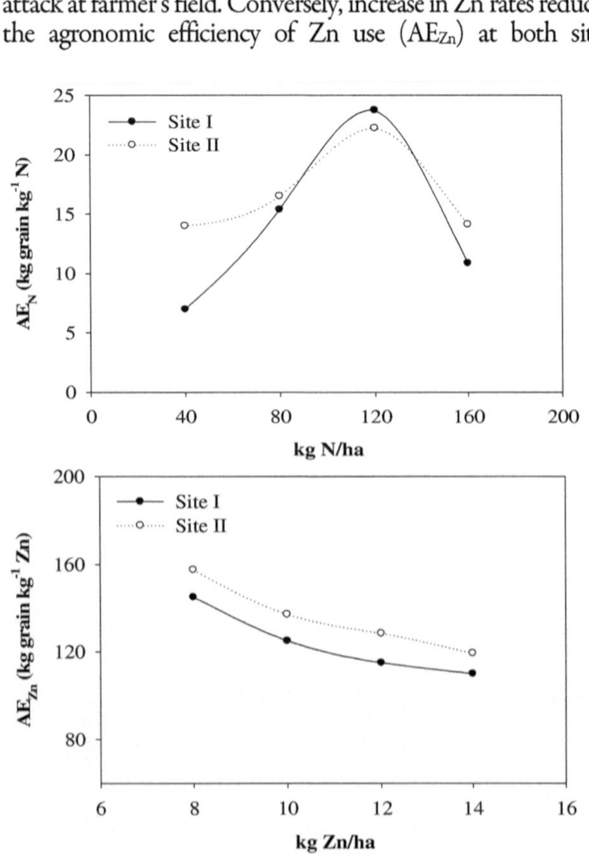

Fig. 2. Agronomic efficiency of Basmati rice (*Oryza sativa* L.) for nitrogen (AE_N) and zinc (AE_{Zn}) utilization at experimental site I (Sheikhupura) and II (Sargodha)

Highest AE_{Zn} were recorded at lower Zn doses at both sites (Fig. 2). High Zn use efficiency and high agronomic Zn efficiency at lower Zn doses have also been reported for barley (Genc *et al.*, 2002) and rice (Muthukumararaj and Sriramachandrasekharan, 2012), respectively. Comparison of both sites showed that AE_{Zn} values were higher at site II (Sargodha) than site I (Sheikhupura) under all Zn treatments. This varied response could probably be due to differences in inherent soil zinc contents at both sites. It seems that lower Zn contents (0.32 mg Zn/ kg soil) at site II as compared to site I (0.51 mg Zn/ kg soil) caused the relatively better yield response to applied Zn at site II.

Kernel quality

Combined application of N and Zn significantly improved the kernel dimensions (kernel length and width) at both sites (Tab. 5). Nitrogen application significantly increased the kernel length at each increment up to 120 kg N/ha where significant Zn response on kernel length was only evident at the highest dose of the experiment (14 kg Zn/ha). Application of 120 N + 14 Zn (kg/ha) produced the highest kernel lengths 7.04 mm and 6.31 mm at site I and II, respectively (Tab. 5). On the other hand, increase in both N and Zn progressively increased the kernel widths at both sites. Kernel width varied between 1.07 to 2.99 mm across years, N and Zn treatments and sites with maximum values obtained at 120 N + 14 Zn (kg/ha) and minimum values at 0 N + 0 Zn (kg/ha). An improvement in kernel dimensions (kernel length and kernel width) by increased N fertilization has also been shown by Maqsood *et al* (2013). Both Khan *et al.* (2009) and Shivay *et al.* (2005) reported that Zn application @ 10-15 kg/ha improved the yield and quality of Basmati rice in Pakistan and India. Water absorption ratio and protein contents of kernels were also significantly improved by combined application of N and Zn at both sites.

Tab. 5. Interactive effect of nitrogen and zinc rates on kernel quality of Basmati rice (*Oryza sativa* L.)

Treatments		KL (mm)		KW (mm)		WAR		PC (%)		AC (%)	
N (kg ha⁻¹)	Zn (kg ha⁻¹)	Site I	Site II	Site I	Site II	Site I	Site II	Site I	Site II	Site I	Site II
0	0	4.95	4.60	1.07	1.36	3.85	3.66	5.54	8.36	21.0	19.9
0	8	5.05	4.67	1.17	1.41	4.09	3.76	5.97	8.49	21.4	20.0
0	10	5.22	4.76	1.21	1.34	4.18	3.79	6.54	8.66	22.1	20.0
0	12	5.18	4.86	1.21	1.35	4.14	3.79	6.22	8.83	21.7	20.1
0	14	5.14	4.91	1.21	1.44	4.06	3.85	6.03	8.93	21.6	20.3
40	0	5.22	5.04	1.21	1.36	4.18	4.07	6.28	9.16	22.1	20.4
40	8	5.26	5.14	1.23	1.35	4.18	3.94	6.31	9.35	22.1	20.6
40	10	5.19	5.20	1.19	1.41	4.12	4.00	6.17	9.45	21.9	20.6
40	12	5.17	5.14	1.21	1.43	4.12	4.04	6.19	9.35	22.0	20.8
40	14	5.21	5.20	1.23	1.44	4.14	4.08	6.23	9.45	22.0	21.0
80	0	5.22	5.34	1.21	1.50	4.15	4.71	6.24	9.71	21.9	21.1
80	8	5.65	5.41	1.71	1.62	5.12	4.97	7.19	9.84	22.7	21.2
80	10	5.73	5.54	1.73	1.63	5.11	4.86	7.20	10.07	22.9	21.5
80	12	5.99	5.61	1.90	1.75	5.42	4.87	7.50	10.19	23.0	21.6
80	14	6.13	5.75	2.09	1.92	5.70	5.28	7.78	10.46	22.9	21.7
120	0	6.37	5.70	2.30	1.95	6.17	5.20	8.25	10.37	23.1	22.0
120	8	6.49	5.81	2.47	2.07	6.38	5.36	8.46	10.57	23.1	22.0
120	10	6.63	5.98	2.68	2.26	6.73	5.68	8.81	10.87	22.9	22.1
120	12	6.79	6.09	2.54	2.26	7.12	5.61	9.20	11.08	23.2	22.1
120	14	7.04	6.31	2.99	2.27	7.40	5.65	9.14	11.47	23.3	22.2
160	0	5.69	5.94	1.79	2.21	5.02	5.11	7.10	10.80	22.8	21.9
160	8	5.95	6.26	1.93	2.06	5.31	4.85	7.40	11.39	22.9	21.9
160	10	6.07	6.12	2.10	2.11	5.65	5.00	7.73	11.13	23.1	21.8
160	12	6.01	6.23	2.05	1.99	5.36	4.87	7.44	11.32	23.0	21.9
160	14	5.90	6.18	1.96	2.12	5.42	4.99	7.51	11.24	22.9	21.9
LSD ($P \leq 0.05$)		0.56	0.49	0.31	0.26	0.69	0.57	1.19	1.16	2.41	2.38
F-Significant		*	*	**	*	*	*	*	*	NS	NS

N = nitrogen; Zn = zinc; KL = kernel length; KW = kernel width; WAR = water absorption ratio; PC = protein content; AC = amylose content; NS = not significant; *, **: Significant at 0.05 and 0.01, respectively.

The maximum kernel protein contents were recorded with the application of 120 N + 12 Zn (kg/ha) at site I while at site II with the application of 120 N + 14 Zn (kg/ha; Tab. 5). Poor N availability during flowering can severely affect rice protein contents (Perez *et al.*, 1996). To attain high paddy yields with high protein contents, sufficient N supply is necessary (Perez *et al.*, 1996). High protein contents at higher N doses (120 kg N/ha) indicate that plants were well supplied with N and sufficient N was mobilized from vegetative parts to grains (Samonte *et al.*, 2006). This study is also in agreement with the findings of others (Schinir *et al.*, 1990; Ali *et al.* 1992; Ahmad *et al.*, 2009) who reported a significant improvement in kernel dimensions, kernel water absorption ratio, protein contents and milling quality of rice by optimum N fertilization. It was observed that both nitrogen and zinc treatments did not affect kernel amylose contents at both sites. It is interesting to note that kernel protein contents were higher at site II (Sheikhupura) as compared to site I (Sargodha) in almost all treatments (Tab. 5). While kernel amylose contents were slightly lower at site II than that of site I in most of the treatments. It seems that differences in soil factors such as pH, organic matter and available phosphorus and potassium other than treatments at both experimental sites produced the contrasting differences in these two quality parameters.

Acknowledgments

Financial support from Higher Education Commission of Pakistan under Indigenous Scholarship Program is highly acknowledged.

References

Ahmad S, Hussain A, Ali H, Ahmad A (2005). Grain yield of transplanted rice (*Oryza sativa* L.) as influenced by plant density and nitrogen fertilization. Journal of Agriculture and Social Sciences 1:212-215.

Ahmad S, Zia-ul-Haq M, Ali H, Ahmad A, Khan MA, Khaliq M, Husnain Z, Hussain A, Hoogenboom G (2009). Morphological and quality parameters of *Oryza sativa* L. as affected by population dynamics, nitrogen fertilization and irrigation regimes. Pak J Bot 41(3):1259-1269.

Ali A, Karim MA, Hussain G, Ali L, Ali SS, Majid A (1992). Rice grain quality as influenced by split application of nitrogen fertilizer. Int Rice Res Newslett 17(3):7-11.

Anonymous (2013). Economic Survey, Economic Affairs Division, Govt. Pakistan, Islamabad.

Ashraf M, Khalid A, Ali K (1999). Effect of seedling age and density on growth and yield of rice in saline soil. Pakistan Journal of Biological Sciences 2:860-862.

Chaudhry FM, Kausar MA, Rashid A (1977). Mechanism of nitrogen effect on zinc nutrition of flooded rice. Plant Soil 46(3):649-654.

Ding C, You J, Chen L, Wang S, Ding Y (2014). Nitrogen fertilizer increases spikelet number per panicle by enhancing cytokinin synthesis in rice. Plant Cell Rep 33(2):363-71.

Dobermann A, Cassman KG (2002). Plant nutrient management for enhanced productivity in intensive grain production systems of the United States and Asia. Plant Soil 247:153-175.

Fageria NK, Santos AB, Cutrim VA (2008). Dry matter and yield of lowland rice genotypes as influence by nitrogen fertilization. J Plant Nut 31(4):788-795.

Gao X, Hoffland E, Jan T, Cynthia S, Grant A, Zou C Zhang F (2012). Improving zinc bioavailability in transition from flooded to aerobic rice. A review. Agron Sustain Dev 32:465-478.

Gebrekidan H, Seyoum M (2006). Effects of mineral N and P fertilizers on yield and yield components of flooded lowland rice on vertisols of Fogera Plain, Ethiopia. J Agr Rural Dev Trop 107 (2):161-176.

Genc Y, McDonald GK, Graham RD (2002). Critical deficiency concentration of zinc in barley genotypes differing in zinc efficiency and its relations to growth responses. J Plant Nutr 25:545-560.

Griggs BR, Norman RJ, Wilson CE, Slaton NA (2007). Ammonia volatilization and nitrogen uptake for conventional and conservation tilled dry-seeded, delayed-flood rice. Soil Sci Soc Am J 71:745-751.

Hafeez B, Khanif YM, Samsuri AW, Radziah O, Zakaria W, Saleem M (2010). Evaluation of Rice Genotypes for Zinc Efficiency under Acidic Flooded Condition. In 19th World Congress of Soil Science, Soil Solutions for a Changing World 1-6 August 2010, Brisbane, Australia.

Impa SM, Morete MJ, Ismail AM, Schulin R, Johnson-Beebout ES (2013). Zn uptake, translocation and grain Zn loading in rice (*Oryza sativa* L.) genotypes selected for Zn deficiency tolerance and high grain Zn. J Exp Bot 64(10):2739-2751.

Jacobs MB (1958). The chemical analysis of food and food products. Roberts AV Pub Co NY 32-35.

Juliano BO (1971). A simplized assay for milled rice amylase. Cereal Science Today 16(10):334-340.

Khan MU, Qasim M, Khan I (2007). Effect of Zn fertilizer on rice grown in different soils of Dera Ismail Khan. Sarhad Journal of Agriculture 23(4):1033-1040.

Khan R, Gurmani AR, Khan MS, Gurmani AH (2009). Residual, direct and cumulative effect of zinc application on wheat and rice yield under rice-wheat system. Soil and Environment 28(1):24-28.

Ladha JK, Pathak H, Krupnik TJ, Six J, van Kessel C (2005). Efficiency of fertilizer nitrogen in cereal production: retrospect's and prospects. Adv Agron 87:85-156.

Mae T (1997). Physiological nitrogen efficiency in rice: Nitrogen utilization, photosynthesis, and yield potential. Plant Soil 196:201-210.

Manzoor Z, Awan TH, Safdar ME, Ali RI, Ashraf MM, Ahmad M (2006). Effect of nitrogen levels on yield and yield components of Basmati 2000. J Agri Res 44(2):115-120.

Maqsood M, Shehzad MA, Ali SNA, Iqbal M (2013). Rice cultures and nitrogen rate effects on yield and quality of rice (*Oryza sativa* L.) Turk J Agric For 37:665-673.

Mengel K and Kirkby EA (2001). Principles of Plant Nutrition. 5th ed., Kluwer Academic Publishers, Dordrecht.

Mirzavand J (2007). Effects of soil application of zinc sulfate and immersing of seedling roots in zinc oxide suspension on yield and chemical composition of rice (Cv. Ghasrodashti). Iranian Journal of Soil and Waters Sciences 21(1):23-31.

Muthukumararaja TM, Sriramachandrasekharan MV (2012). Effect of zinc on yield, zinc nutrition and zinc use efficiency of lowland rice. International Journal of Agriculture Techechnology 8(2):551-561.

Norman RJ, Wilson CE, Slaton NA, Griggs BR, Bushong JT, Gbur EE (2009). Nitrogen fertilizer sources and timing before flooding dry-seeded, delayed-flood rice. Soil Sci Soc Am J 73:2184-2190.

Pandey N, Gupta M, Sharma CP (2001). Zinc efficiency affects pollen structure and viability in green gram (*Vigna radiate* L. cv T-44). Geophytology 28:31-34.

Perez CM, Juliano OB, Liboon SP, Alcantara JM, Cassman KG (1996). Effects of late nitrogen fertilizer application on head rice yield, protein content, and grain quality of rice. Cereal Chem 73(5):556-560.

Qadar A (2002). Selecting rice genotypes tolerant to zinc deficiency and sodicity stresses. Differences in zinc, iron, manganese, copper, phosphorus concentrations, and phosphorus/zinc ratio in their leaves. J Plant Nutr 25:457-473.

Quijano-Guerta C, Kirk GJD, Portugal AM, Bartolome VI, McLaren GC (2002). Tolerance of rice germplasm to zinc deficiency. Field Crops Res 76:123-130.

Rahman A, Yasin M, Akram M, Awan ZI (2002). Response of rice to zinc-application and different N-sources in calcareous soil. Soil Conservation, Government of Punjab, Rawalpindi. National Agricultural Research Centre, Islamabad Quarterly. Science Vision 8(1):100-104.

Rashid A (1996). Secondary and micronutrient. In: A. Rashid and K. S. Memon, (eds). Soil Science. National Book Foundation, Islamabad.

Rashid A, Hussain F, Tahir M (1999). Alleviating zinc deficiency in transplanted flooded rice grown in alkaline soils of Pakistan. International Rice Research Notes (IRRI, Philippines) 24:32-33.

Rashid A, Yasin M (2002). Zinc enrichment of the mat-type rice nursery. International Rice Research Notes (IRRI, Philippines) 27(2):32-33.

Reddy S (2004). Agronomy of field crops. Kalyani Publishers, New Delhi, India.

Saito M (1991). Soil management for the conservation of soil nitrogen. Extension Bulletins 341. Food and Fertilizer Technology Center, Taiwan.

Samonte SO, Wilson LT, Medley JC, Pinson SRM, McClung AM, Lales JS (2006). Nitrogen Utilization Efficiency: Relationships with Grain Yield, Grain Protein, and Yield-Related Traits in Rice. Agron J 98:168-176.

Sarwar N, Ali H, Ahmad A, Ullah E, Ahmad S, Mubeen K, Hill JE (2013). Water wise rice cultivation on calcareous soil with the addition of essential micronutrients. J Anim Plant Sci 23(1):244-250.

Schinir HF, Dingkuhn M, DeDatta SK, Mengal K, Faronilo JE (1990). Nitrogen fertilization of direct seeded vs transplanted rice. Nitrogen uptake photosynthesis, growth and yield. Crop Sci 30:1276-84.

Shivay YS, Kumar D, Prasad R (2005). Effects of zinc fertilization on physical grain quality of Basmati rice. International Rice Research Notes (IRRI, Philippines) 32(1):41-42.

Shivay YS, Prasad R, Rahal A (2010). Genotypic variation for productivity, zinc utilization efficiencies and kernel quality in aromatic rice under low available zinc conditions. J Plant Nut 33(12):1835-1848.

Singh M, Singh SP (1981). Effect of nitrogen and zinc on the yield of submerged rice and uptake of N and Zn on unlimed and limed soils. Plant Soil 62 (2):183-192.

Singh U, Ladha JK, Castillo EG, Punzalan G, Tirol-Padre A, Duqueza M (1998). Genotypic variation in nitrogen use efficiency in medium- and long-duration rice. Field Crop Res 58:35-53.

Tahir M, Kausar MA, Bhatti AS (1992). Micronutrient status and effectively ameliorating zinc deficiency in rice. Pakistan Journal of Scientific and Industrial Research 35:133-137.

Wang YY, Zhu B, Shi Y, Hu CS (2008). Effect of nitrogen fertilization on upland rice based on pot experiments. Commun Soil Sci Plan 39(11-12):1733-1749.

Wissuwa M, Ismail AM, Yanagihara S (2006). Effects of Zinc Deficiency on Rice Growth and Genetic Factors Contributing to Tolerance. Plant Physiol 142(2):731-741.

Wu G, Wilson LT, McClung AM (1998). Contribution of rice tillers to dry matter accumulation and yield. Agron J 90:317-323.

16

Microsporogenesis of Apricot Cultivars in Hungary

Veronika HAJNAL[1*], Zarif OMID[1], Márta LADÁNYI[2], Magdolna TÓTH[1], László SZALAY[1]

[1]*Corvinus University of Budapest, Department of Pomology, 29-43 Villányi út, Budapest, Hungary; veronika.hajnal@uni-corvinus.hu (*corresponding author), eng.zarifomid@gmail.com, magdolna.toth@uni-corvinus.hu, laszlo.szalay@uni-corvinus.hu*

[2]*Corvinus University of Budapest, Department of Mathematics and Informatics, 29-43 Villányi út, Budapest, Hungary; marta.ladanyi@uni-corvinus.hu*

Abstract

Like all phenological processes, microsporogenesis is mainly determined genetically, but its phenotypical expression is greatly influenced by environmental factors, particularly the temperature. During the 3-year experimental period, the process of microsporogenesis was examined in eight apricot cultivars originated in North America and in Romania. The Hungarian cultivar 'Gönci magyar kajszi' was used as the control. Based on the results it was possible to rank the cultivars in terms of their microsporogenesis schedule. The same order was found in all three years. This order also indicates the yield reliability of the cultivars. Endodormancy ended between January 10th and 25th in the buds of the earliest cultivar 'Pinkcot', but not until February 5-10th in the latest cultivar 'Harlayne'. Three of the cultivars had faster flower bud development than the control, in the order 'Pinkcot', 'Orange Red' and 'Harcot'. Growing these cultivars thus involves greater risk than for 'Gönci magyar kajszi' in Hungary, due to their rapid winter flower bud development. Some cultivars found to have slower flower bud development than 'Gönci magyar kajszi'; 'Litoral', 'Harogem', 'Comandor', 'Sirena' and 'Harlayne' – thus these cultivars can be grown more reliably.

Keywords: anthers, ecodormancy, endodormancy, flower bud development, *Prunus armeniaca* L.

Introduction

The flower bud development of apricot cultivars during the winter dormancy period can be monitored by examining microsporogenesis. Like all phenological processes, microsporogenesis is mainly determined genetically, but its phenotypical expression is influenced to a great extent by environmental factors, particularly by temperature.

Detailed analyses were made in orchards in the Mediterranean region on the microsporogenesis of the most widely grown apricot cultivars (Andreini *et al.*, 2012; Bartolini and Viti, 1999; Bartolini *et al.*, 2006a, 2006b; Julian *et al.*, 2009, 2011; Scalabrelli *et al.*, 1991; Viti and Monteleone, 1991). In Hungary, Banainé (1981) and Nyujtó and Banainé (1975) were the first to publish similar research on Hungarian-bred cultivars. Later, in addition to Hungarian cultivars, foreign cultivars were also examined from this point of view in Hungarian orchards (Németh *et al.*, 2009; Sebők, 1993; Szalay, 2001; Szalay and Németh, 2010; Szalay *et al.*, 1999; Szalay *et al.*, 2004; Szalay *et al.*, 2006a, 2006b).

All in all, the results achieved till now show that the rate of microsporogenesis differs greatly from one apricot genotype to the other, while there are also considerable differences among locations and years.

Hungary is close to the northern limit for the cultivation of apricot. To ensure reliable yields, it is thus important to choose cultivars with slow winter flower bud development and a long endodormancy period (Szalay, 2001, 2008).

Efforts are constantly made to discover marketable cultivars with satisfactory yield reliability to extend the list of cultivars suitable to Hungarian conditions.

The microsporogenesis of three Romanian and five North American apricot cultivars was investigated in three consecutive years in the gene bank collection. The main aim was to determine whether the flower bud development traits of these cultivars made them suitable for reliable production under the ecological conditions in Hungary.

Materials and methods

The experiments were carried out in the Gene Bank Collection of Apricot Cultivars in the Experimental Orchard of the Department of Pomology, Corvinus University of Budapest, which is located in Soroksár, on the outskirts of Budapest.

Experiments were performed in the winters of three consecutive years, 2010/11, 2011/12 and 2012/13, on three cultivars from Romania (Ro) and five from North America (Am). These were as follows: 'Comandor' (Ro), 'Harcot' (Am), 'Harlayne' (Am), 'Harogem' (Am), 'Litoral' (Ro), 'Orange Red' (Am), 'Pinkcot' (Am) and 'Sirena' (Ro). The cultivar 'Gönci magyar kajszi' was included as a control.

Experimental methods

Fruiting branches were collected from designated trees every week (or if necessary, several times a week) throughout the winter dormancy period, and the flower buds located on the spurs were examined. The anthers were removed from the flower buds, placed on slides, stained with carmine acetic acid and covered with a cover slide. After gentle pressure, the slides were put under a microscope to examine the tissue structure of the anthers. On each occasion anthers from 8-10 flower buds were examined for each genotype. Phenological stages of microsporogenesis and their ratio were registered on each slide. Six developmental stages were distinguished:

(1)**Archesporium stage** – the tissues in the anthers were still undifferentiated.

(2)**String stage (string)** – the differentiation of the archespores in the anthers had begun, but the developing pollen mother cells were still stuck together in strings.

(3)**Pollen mother cell stage (p. m. cells)** – the pollen mother cells in the anthers were fully developed and could be seen separately on the slides.

(4)**Tetrad stage (tetrad)** – reduction division into four parts had taken place in the pollen mother cells, leading to the presence of tetrads in the anthers.

(5)**Microspore stage (msp)** – four microspores had formed from each pollen mother cell and could be seen separately in the anthers.

(6)**Pollen stage (pollen grains)** – pollen grain formation was complete; pollen grains of the shape and pattern characteristic of the genotype could be seen in the anthers.

The process of microsporogenesis can be modelled by sigmoid type functions of form

$$Y(X) = 100/\left[1 + EXP\left(-s * (X - m)\right)\right] + \varepsilon$$

where:

X - denotes the date expressed in Julian day;

m - is the inflexion point of the curve which corresponds to the median (the 50% quantile);

s - is a velocity factor i.e. the slope of the curve at point $X = m \quad s/4$;

ε is a normally distributed error term with expected value of zero.

The regression models fitted to the observations were evaluated by their F values, the t values of the parameter estimations together with the explained variances (R^2) and the significance of all F, t and R^2 values was proved ($p<0.05$). Normality of the error term was proved by Kolmogorov-Smirnov's test ($p > 0.05$).

The 50 % quantile of the sigmoid curve of a given stage was considered as the starting date of that stage in case of the stages beginning from string one up to the pollen one, while the start of blooming was expressed as the 5 % quantile since the day on which 5% of the flowers had opened was recorded in the orchards as the start of blooming.

First we compared the three years considering the starting dates of the different phenological stages using repeated measures ANOVA model separately for all phenological stages. Multivariate tests were used to test the significance of the unexplained variances, Wilk's lambda. Since the sphericity assumption was violated, we used Greenhouse-Geisser's correction. Tests of within subjects effects were executed which was followed by Bonferroni's pairwise comparisons. Normality of the error term was proved then by Kolmogorov-Smirnov's test ($p > 0.05$).

To test the similarity of the orders of the cultivars considering the occurrence of the phenological phases, we used Friedman's rank test.

To detect the year dependency, coefficients of variations (CV) were calculated as the standard errors divided by the means of starting dates of phenological stages for each cultivar and phenological stage. The CV data were then compared with 2-way ANOVA model where the two fixed factors were cultivar and phenological stage. Normality of the error term was proved then by Kolmogorov-Smirnov's test ($p > 0.05$).

We also tested the significance of the differences between the velocity factors of the phenological stages and cultivars with a 2-way analysis of variance model where the fixed factors were again the cultivar and phenological stage and the random factor was the year. Since the normality assumption was violated in this case, we applied an inverse transformation. Normality of the error term was then proved by Kolmogorov-Smirnov's test again ($p > 0.05$).

In both ANOVA tests, homogeneity of variances was checked by Levene's test ($p = 0.05$). Post hoc tests were performed by Tukey's or Games-Howell' test, according to the result of test Levene's test.

Statistical analysis was performed using IBM SPSS 20.

Results and discussions

Substantial differences were observed among cultivars regarding the schedule and the speed of the microsporogenesis process, and large differences were also noted between the years. The starting dates of each phenological stage are taken as the inflexion points of the sigmoid curves from the beginning of ecodormancy up to pollen stage. The start of blooming is taken as the value of 5% quantile (Tab. 1.). The dates are expressed as Julian days, i.e. the number of days elapsed from January 1st.

First we compared the years considering the starting dates of the different phenological stages using repeated measures ANOVA model separately for all phenological stages. Multivariate tests were all significant except for pollen mother cell stage (Wilk's lambda were all under 0.3 with $p < 0.01$ except for pollen mother cell stage when it was as high as 0.95 with $p = 0.83$). Since the sphericity assumption was violated, we used Greenhouse-Geisser's correction (ε were all above 0.53). Tests of within subjects effects revealed highly significant year effect ($F(2^* \varepsilon;16^* \varepsilon)$ were all above 20 with $p < 0.01$ except for pollen mother cell stage for which $F(1.1;8.7) = 0.17$ with $p = 0.71$). Bonferroni's pairwise comparisons yielded that every year was significantly different from all the others for blooming and pollen grains ($p < 0.01$), year 2012 differed significantly ($p < 0.01$) from the more similar ones 2011 and 2013 ($p > 0.05$) in cases of string, tetrads and microspore stage, while, of course, all the three years were similar for pollen mother cell stage ($p > 0.99$).

Tab. 1. Starting dates of the phenological stages in microsporogenesis in apricot cultivars in three consecutive years (number of days from January 1st) with the velocity factors (s) of the sigmoid curves. The starting dates of each phenological stage are taken as the inflexion points of the sigmoid curves from the beginning of ecodormancy up to pollen stage (m). The start of blooming is taken as the value of 5% quantile

Cultivar/year	string		p.m.cells		tetrads		msp		pollen grains		blooming	
2011	m	s	m	s	m	s	m	s	m	s	m*	s
Pinkcot	25	0.25	34	0.37	38	0.45	47	0.66	76	0.54	84	1.30
Orange Red	27	0.29	35	0.35	39	0.50	47	0.60	78	0.61	85	1.58
Harcot	30	0.29	37	0.38	41	0.47	49	0.50	78	0.62	84	1.58
Gönci magyar kajszi	32	0.28	40	0.37	43	0.45	52	0.56	79	0.57	86	1.55
Litoral	34	0.29	40	0.35	45	0.48	53	0.47	80	0.53	87	1.45
Harogem	36	0.30	43	0.38	48	0.50	56	0.51	81	0.52	91	1.45
Comandor	36	0.30	43	0.38	48	0.50	56	0.51	81	0.52	91	1.45
Sirena	38	0.30	44	0.32	49	0.46	57	0.50	82	0.64	92	1.45
Harlayne	40	0.29	46	0.32	51	0.47	59	0.74	84	0.56	93	1.45
2012	m	s	m	s	m	s	m	s	m	s	m*	s
Pinkcot	11	0.37	24	0.37	38	0.45	52	0.57	73	0.69	80	1.19
Orange Red	15	0.31	27	0.36	39	0.46	53	0.51	74	0.52	82	1.09
Harcot	18	0.35	32	0.39	46	0.45	58	0.33	76	0.75	83	0.92
Gönci magyar kajszi	20	0.31	37	0.36	51	0.47	64	0.60	77	0.77	83	1.95
Litoral	24	0.31	42	0.35	54	0.39	67	0.67	79	0.76	85	1.11
Harogem	28	0.31	45	0.35	57	0.40	70	0.57	80	0.75	86	1.73
Comandor	28	0.31	45	0.35	57	0.40	70	0.57	80	0.75	86	1.73
Sirena	31	0.33	48	0.34	61	0.39	72	0.63	81	0.70	88	2.19
Harlayne	36	0.33	54	0.39	66	0.54	75	0.57	83	0.89	89	2.19
2013	m	s	m	s	m	s	m	s	m	s	m*	s
Pinkcot	25	0.25	32	0.35	37	0.45	49	0.57	79	0.55	100	0.98
Orange Red	29	0.27	34	0.37	39	0.35	50	0.64	79	0.55	101	0.85
Harcot	31	0.26	38	0.44	42	0.59	52	0.54	80	0.59	102	0.74
Gönci magyar kajszi	33	0.24	39	0.34	43	0.61	53	0.54	81	0.57	103	0.88
Litoral	35	0.29	39	0.34	45	0.41	55	0.68	82	0.56	103	0.76
Harogem	36	0.28	42	0.37	46	0.41	55	0.45	82	0.56	104	0.77
Comandor	36	0.28	42	0.37	46	0.41	55	0.45	82	0.56	104	0.77
Sirena	38	0.30	44	0.38	48	0.45	58	0.54	83	0.59	104	0.64
Harlayne	41	0.31	49	0.39	53	0.45	61	0.56	84	0.54	104	0.65

*5 % quantile

Tab. 2. Coefficients of variations of the starting dates of the stages (m) in microsporogenesis in apricot cultivars with respect to the three years i.e. the standard errors divided by the means which can express the variability of the starting dates of phenological stages, depending on the effect of year. The different letters A to C represent phenological stages with significantly different CV values according to Games-Howell's post hoc test ($p < 0.05$)

Cultivar	string^{A}	p.m.cells^{BC}	tetrads^{AB}	msp^{AB}	pollen grains^{C}	blooming^{C}
Pinkcot	0.40	0.18	0.02	0.05	0.04	0.03
Orange Red	0.31	0.13	0.10	0.10	0.03	0.03
Harcot	0.27	0.09	0.06	0.09	0.03	0.04
Gönci magyar kajszi	0.26	0.04	0.10	0.12	0.03	0.02
Litoral	0.20	0.04	0.11	0.13	0.02	0.07
Harogem	0.14	0.04	0.12	0.14	0.01	0.03
Comandor	0.14	0.04	0.12	0.14	0.01	0.03
Sirena	0.11	0.05	0.14	0.13	0.01	0.04
Harlayne	0.07	0.08	0.14	0.13	0.01	0.02

To detect the year dependency, coefficients of variations (CV) were calculated as the standard errors divided by the means of starting dates of phenological stages for each cultivar and phenological stage. The 2-way ANOVA test revealed that the effect of cultivar was not significant ($F(8;40) = 0.53$; $p = 0.83$) while the one of phenological stage was significant ($F(5;40) = 13.93$; $p < 0.001$). The result of Games-Howell's post hoc test is presented in Tab. 2. We can conclude generally that the order of the phenological stages according to their year dependency is: string, microspore stage (msp), tetrads, mother cell stage (p. m. cells), blooming and pollen grains. Moreover, this can be stated quite similarly for the cultivars.

Three stages in microsporogenesis were highlighted to introduce the differences between cultivars and years: the string stage, which indicates the end of endodormancy in the flower buds; the microspore stage, when reduction division is completed and the formation of pollen grains has begun; and the start of blooming, which indicates the end of the microsporogenesis process.

In all three years the flower bud development of 'Pinkcot' was the earliest of all the cultivars, while that of 'Harlayne' was the latest. The microsporogenesis rates of the other cultivars tested, and also of the control cultivar 'Gönci magyar kajszi' (GMK), fell between these values. The results obtained for these three cultivars in 2011 are illustrated in Fig. 1 for the three stages focused (the string stage, the microspore stage and the start of flowering).

The order of the cultivars considering the occurrence of the phenological stages was very similar in each year for each phenological stage (Friedman's test: $\mathcal{X}^2(2)<0.4$; $p>0.8$). In a later phenological stage, however, the variance of the starting days measured for different cultivars was mostly lower. The negative linear trend was significant (Pearson's $R = -0.47$; $p < 0.01$). We also proved for string and blooming phases that the later the stage starts in a year the more quickly it passes (Pearson's $R = -0.38$; $p < 0.05$; Pearson's $R = -0.67$; $p < 0.001$).

At the beginning of microsporogenesis each phenological stage takes a longer time to be completed. The development of the string stage from the archespore tissue took 20-24 days in all the cultivars tested, while the transition from the tetrad stage to the microspore stage required 10-14 days and blooming was completed within 3-5 days (Fig. 1).

We can prove this using a two-way ANOVA test of significance for the differences of the velocity factors. We detected that the in heterogeneity of velocity factors was highly significant for the phenological stages ($F(5;106) = 262.94$; $p < 0.001$) and insignificant for the cultivars ($F(8;106) = 0.43$; $p = 0.90$). The interaction phenological stage*cultivars was insignificant ($F(40;106) = 0.71$; $p = 0.89$). Tukey's post hoc test revealed that every single velocity factor of a phenological stage differs significantly from all of others but msp and pollen grains ($p < 0.001$; $p = 0.15$, respectively). We can conclude that the later the start of phenological stage is, the faster its process is, similarly for all cultivars.

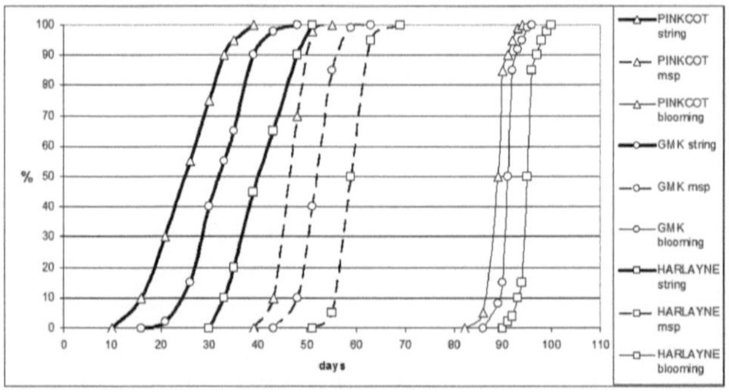

Fig. 1. Start and end of microsporogenesis and the microspore stages for three apricot cultivars ('Pinkot', 'Gönci Magyar kajszi', 'Harlayne') in 2011

The differences between the years are illustrated using the 3-year results recorded for the 'Litoral' cultivar, which has a medium rate of flower bud development (Fig. 2). The end of endodormancy (string stage) in the flower buds of this cultivar occurred on the same day, February 3rd, in 2011 and 2013, while in 2012 it was observed 10 days earlier. The time required for each stage of microsporogenesis to develop differed over the years, due to the diverse weather conditions.

In 2011, the microspores developed very quickly after the end of endodormancy in the flower buds of the 'Litoral' cultivar, the start of the microspore stage being recorded on February 22nd. Blooming, however, was not the earliest in this year, beginning on March 30th, which was intermediate between the other two years.

In 2012, the string stage was reached early, but the microspores developed latest in this year. Nevertheless, at the end of the microsporogenesis process, blooming started earlier than in the other two years, on March 25th.

In 2013, the process of microsporogenesis started similarly to that observed in 2011. The string stage was reached on the same date in 2013 as in 2011 and microspore development started two days later than in 2011. The blooming of 'Litoral', however, was latest in this year, not starting before than April 14th.

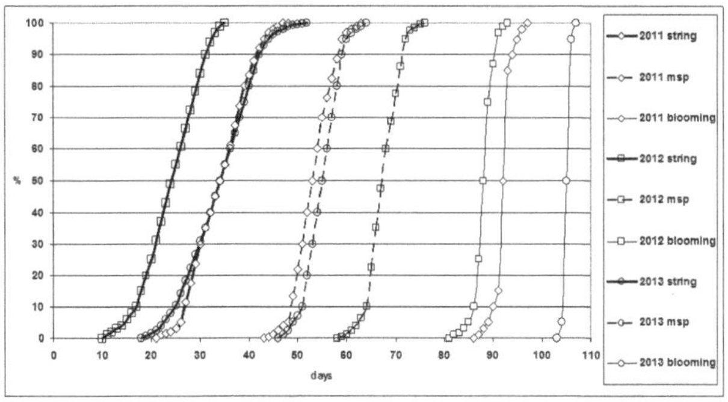

Fig. 2. Occurrence and characteristic of the phenological stages of microsporogenesis in the 'Litoral' cultivar in three consecutive years (2011-2013)

The rate of microsporogenesis is greatly influenced by environmental factors, especially the temperature (Szalay, 2008), resulting in substantial differences among years. This was confirmed by the present studies on nine cultivars over a 3-year period. However, the order of the cultivars was almost the same in each year, suggesting that, like all phenological processes, microsporogenesis is fundamentally genetically determined. This means that even analyses carried out over a few years are sufficient to allow the cultivars to be reliably ranked in terms of the end of endodormancy and the schedule of microsporogenesis. Before new cultivars are introduced into cultivation, it is essential to test their suitability for a given location. The data provided by phenological analysis are useful for this purpose if compared with those of control cultivars having known their frost tolerance and winter hardiness.

There is no close correlation between the schedule of microsporogenesis and the frost tolerance of the over-wintering organs in the flower buds, as they are subject to separate genetic regulation (Bassi *et al.*, 2006). However, the length of endodormancy has a considerable influence on the frost tolerance and winter hardiness of the cultivar, as the frost tolerance of the over-wintering organs decrease just during the ecodormancy period (Szalay, 2001, 2008).

As Hungary is one of the most northern countries where apricots can be grown economically, the best cultivars are those in which the over-wintering organs, particularly the flower buds, have slow winter development, while the end of endodormancy is late and frost tolerance is maintained for a long period (Szalay, 2001, 2008). Apricot cultivars with short endodormancy and rapid flower bud development can be successfully cultivated in more southern countries in the Mediterranean region (Guerriero *et al.*, 1988). From the point of view of yield reliability, cultivars with long endodormancy, bred under ecological conditions similar to those in Hungary, could be promising for widening the choice of cultivars. One aim of the apricot breeding programme in Romania was to develop late-flowering cultivars with good frost tolerance and winter hardiness for cultivation in the temperate zone (Cociu, 1982, 1991; Cociu and Hough, 1985). The cultivars examined in the present work originated from this breeding programme. It is planned to continue these tests on the latest cultivars developed by Romanian breeders (Topor *et al.*, 2010). In Canada, too, an improvement in frost tolerance and winter hardiness is an important aim in apricot breeding (Layne, 1978, 1979, 1981; Layne and Gadsby, 1995). Three cultivars ('Harcot', Harogem', 'Harlayne') originating from the Canadian breeding programme were also included in the present tests.

Conclusions

Over a three-year period the phenological processes of five cultivars from North America and three from Romania were analysed, using the cultivar 'Gönci magyar kajszi', which has moderate frost tolerance and winter hardiness and is grown widely in commercial orchards in Hungary, as a control. On the basis of the results, the cultivars were ranked in terms of the schedule of microsporogenesis. Almost the same order was observed in all three years, and this order can be regarded as indicative of the yield safety of cultivars. The end of endodormancy in the flower buds was recorded between January 10th and 25th for the first-ranking cultivar 'Pinkcot', but not before February 5-10th for the last cultivar, 'Harlayne'. Three of the cultivars tested had faster flower bud development than the control ('Pinkcot', 'Orange Red' and 'Harcot', in this order), indicating that their cultivation would be riskier than that of 'Gönci magyar kajszi' in Hungary, due to their rapid flower bud development in winter. Cultivars found to have slower flower bud development than 'Gönci magyar kajszi' (in the order 'Litoral', 'Harogem', 'Comandor', 'Sirena' and 'Harlayne') could be grown with less risk in Hungary.

Acknowledgement

The researches were supported by the TÁMOP 4.2.1./B-09/KMR/2010-0005, VM State Genereservation Tasks Project (Állami Génmegőrzési Feladatok) É-45343 and GOP-1.1.1-09/1-2009-0042.

References

Andreini L, Viti R, Bartolini S, Ruiz D, Egea J, Campoy JA (2012). The relationship between xilem differentiation and dormancy evolution in apricot flower buds (*Prunus armeniaca* L.): the influence of environmental conditions in two Mediterranean areas. Trees 26:919-928.

Banainé B (1981). Kajszi F1 hibridek populációgenetikai értékelése (in Hungarian). Population genetysc evaluation of apricot F1 hibrides. Horticultural University, Budapest, Doctoral thesis.

Bartolini S, Viti R (1999). Histological studies on flower buds of cultivar 'Stark Early Orange'. Acta Hort 488:335-339.

Bartolini S, Viti R, Guerriero R (2006a). Xilem differentiation and microsporogenesis during dormancy of apricot flower buds (*Prunus armeniaca* L.) European J Hort Sci 71:84-90.

Bartolini S, Viti R, Laghezali M, Olmez HA (2006b). Xilem vessel differntiation and microsporogenesis evolution in 'Canino' cultivar growing in tree different climatic areas. Italy, Marocco and Turkey. Acta Hort 701:135-140.

Bassi D, Bartolini S, Viti R (2006). Recent advences on environmental and physiological challenges in growing apricot. Acta Hort 717:23-31.

Cociu V (1982). New apricot varieties breeded in Romania. Acta Hort 121:211-216.

Cociu V, Hough LF (1985). Germplasm of apricot and some suggestions for its use. Acta Hort 192:291-297.

Cociu V (1991). Forty years of apricot breeding of Romania. Acta Hort 293:197-201.

Guerriero R, Audergon JM, Martinez CA (1988). Apricot cultivars for Mediterranean climate. Acta Hort. 209:39-47.

Julian C, Herrero M, Rodrigo J (2009). Pollen development and chilling requirements in apricot cultivars. Acta Hort 814:417-419.

Julian C, Rodrigo J, Herrero M (2011). Stamen development and winter dormancy in apricot (*Prunus armeniaca*). Annals of Botany 108:617-625.

Layne REC, Gadsby MF (1995). Determination of cold hardiness and estimation of potential breeding value of apricot germplasm. Fruit Varieties Journal 49(4):242-248.

Layne REC (1978). 'Harcot' apricot. HortScience 13:64-65.

Layne REC (1979). 'Harogem' apricot. HortScience 14:758-759.

Layne REC (1981). 'Harlayne' apricot. HortScience 16:97-98.

Németh Sz, Reményi ML, Szalay L (2009). Development of apricot flower buds on different types of fruiting branches. Acta Horticulturae et Regiotecturae 12:89-91.

Nyujtó F, Banainé B (1975). Előzetes közlemény a kajszibarack fajták termőrügyei téli morfogenézisének vizsgálatáról (in Hungarian). Preliminary study of winter flower bud morphogenesis of apricot cultivars. Gyümölcstermesztés. 2:15-20.

Scalabrelli G, Viti R, Cinelli F (1991). Change in catalase activity and dormancy of apricot buds in response to chilling. Acta Hort 293:267-274.

Sebők Sz (1993). A mikrosporogenézis jelentősége a kajszibarack nemesítési anyagainak megítélésében (in Hungarian). Importance of microsporogenesis in the apricot breeding. Horticultural University, Budapest, MsC thesis.

Szalay L (2001). Kajszi- és őszibarackfajták fagy- és téltűrése (in Hungarian). Cold hardiness and winter tolerance of apricot and peach cultivars. Faculty of Horticultural Sciences, Szent István University, Budapest, PhD thesis.

Szalay L (2008). Development and cold hardiness of flower buds of stone fruits, p. 63-82. In: Nyéki J, Soltész M, Szabó Z (Eds.). Morphology, biology and fertility of flowers in temperate zone fruits. Academic Press, Budapest. ISBN 978-963-05-8591-0

Szalay L, Pedryc A, Szabó Z (1999). Dormancy and cold hardiness of flower buds of some Hungarian apricot varieties. Acta Hort 488:315-319.

Szalay L, Balázs G, Nagy P (2004). Evaluation of foreign apricot cultivars in Hungary. Intern J Hort Sci 10(3):51-55.

Szalay L, Papp J, Pedryc A, Szabó Z (2006a). Diversity of apricot varieties based on traits determining winter hardiness and early spring frost tolerance of floral buds. Acta Hort 701:131-134.

Szalay L, Pedryc A, Szabó Z, Papp J (2006b). Influence of the changing climate on flower bud development of apricot varieties. Acta Hort 717:75-78.

Szalay L, Németh S, (2010). Phenological process of dormancy in apricot genotypes in the central part of the Carpatian Basin. Acta Hort 862:251-255.

Topor E, Vasilescu R, Balan V, Tudor V (2010). Apricot breeding programme for late and very late ripening period in Romania. Acta Hort 862:137-142.

Viti R, Monteleone P (1991). Observations on flower bud growth in some low yield varieties of apricot. Acta Hort 293:319-326.

17

Growth and Photosynthetic Response of Two Persimmon Rootstocks (*Diospyros kaki* and *D. virginiana*) under Different Salinity Levels

Meral INCESU, Berken CIMEN*, Turgut YESILOGLU, Bilge YILMAZ

*Cukurova University, Faculty of Agriculture, Department of Horticulture, 01330, Adana, Turkey; bcimen@cu.edu.tr (*corresponding author)*

Abstract

Salinity continues to be a major factor in reduced crop productivity and profit in many arid and semiarid regions. Seedlings of *Diospyros kaki* Thunb. and *D. virginiana* L. are commonly used as rootstock in persimmon cultivation. In this study we have evaluated the effects of different salinity levels on photosynthetic capacity and plant development of *D. kaki* and *D. virginiana*. Salinity was provided by adding 50 mM, 75 mM and 100 mM NaCl to nutrient solution. In order to determine the effects of different salinity levels on plant growth, leaf number, plant height, shoot and root dry mass were recorded. Besides leaf Na, Cl, K and Ca concentrations were determined. Also leaf chlorophyll concentration, chlorophyll fluorescence (*Fv'/Fm'*) and leaf gas exchange parameters including leaf net photosynthetic rate (P_N), stomatal conductance (g_S), leaf transpiration rate (E), and CO_2 substomatal concentration (Ci) were investigated. Significant decrease of leaf number, shoot length and plant dry mass by increasing salinity levels was observed in both rootstocks. *D. virginiana* was less affected in terms of plant growth under salinity stress. Leaf chlorophyll concentration reduction was higher in the leaves of *D. kaki* in comparison to *D. virginiana* in 100 mM NaCl treatment. By increasing salinity levels P_N, g_S and E markedly decreased in both rootstocks and *D. kaki* was more affected from salinity in terms of leaf gas exchange parameters. In addition there was no significant difference but slight decreases were recorded in leaf chlorophyll fluorescences of both rootstocks.

Keywords: abiotic stress, gas exchange parameters, growth, NaCl, persimmon, rootstock

Introduction

Persimmon (*Diospyros kaki*) is originally from China, and from there it spread to Korea and Japan. Its cultivation in western countries dates only since the second half of the 19th century. In recent years, the cultivation of persimmon has found renewed interest in various countries of the Mediterranean basin, as well as in Turkey. According to 2011 data, Turkey's persimmon production was 28295 tonnes on 2090 ha planted orchards (TUIK, 2014). Most of these orchards are located in the Mediterranean region where salt accumulation in soils is a natural process favoured by the ecological conditions and foremost by the water balance of the area (Zalidis *et al.,* 2002). Persimmons can be grown in a wide range of soils, but prefer well-drained loam. The tree does not tolerate high salinity. Persimmons grow better in heavier soils than the most deciduous tree fruits and are sensitive to boron and salts in the soil (Farrar, 1999).

Salinity is a major environmental factor affecting the performance of many crop plants (Munns, 2002). The effects of salinity on plants are evidenced by a severe reduction in plant growth and yield and, if the saline conditions persist, plant death can occur (Storey and Walker, 1999). Salinity causes a deficiency of water in plant tissue, and low water potential reduces growth by inhibiting cell division and cell expansion (Hasegawa *et al.,* 2000). The reduction in growth is mainly due to an osmotic effect of the accumulation of salts near the root zone, whereas the build up of toxic saline ions in plant tissues is responsible for the progressive impairment of several physiological processes (Munns, 2002). Salinity affects the crop during both the vegetative and the reproductive stage and, therefore, it causes reduction in plant growth and development with low water potential in the root medium (osmotic effect), too high internal ion concentration (ion excess/toxicity) and nutritional imbalance by depressing uptake and/or shoot transport (ion deficiency) (Levitt, 1980). Most of the salt stress in nature is due to sodium salts, particularly NaCl (Levitt, 1980; Muns and Termaat, 1986). High concentrations of Na^+ and Cl^- in the root medium saturation depress nutrient-ion activities and produce extreme ratios of Na^+/Ca^{2+}, Na^+/K^+, Ca^{2+}/Mg^{2+} and Cl^-/NO^{3-} (Grattan and Grieve, 1999). Osmotic effects resulting from salinity may cause disturbances in the water balance of the plant, including a reduction of turgor and an inhibition of growth, as well as stomatal closure and reduction of photosynthesis (Navarro *et al.,* 2000; Romero-Aranda *et al.,* 2001; Li and Stanghellini, 2001; Heuvelink *et al.,* 2003). The primary effect of high salt concentration in plants is stomatal closure. This causes a low transpiration rate and reduces the CO_2 availability for photosynthesis (Flexas *et al.,* 2007). Hussain *et al.* (2012) indicated that salinity reduced photosynthetic availability of some citrus species and genera. As a result, plants become susceptible to osmotic and specific-ion injury as well as to nutritional disorders that may result in reduced yield and quality.

The impact of using rootstocks to fruit crop production includes not only a stronger resistance against pathogens but also a higher tolerance to abiotic stress conditions such as salinity, heavy metal, nutrient stress, water stress and alkalinity (Rouphael *et al.,* 2012). The rootstocks used for persimmon cultivation belong essentially to four species of *Diospyros*: *lotus*, *virginiana*, *kaki* and *rhombifolia*. The most commonly used rootstock for persimmon production in Turkey is *D. kaki* (the oldest existing rootstock, of Japanese origin). However, *D. virginiana* (American origin) is an easily propagated rootstock, which has greater resistance to heavy and moist soils more than *D. kaki*. Persimmon is very sensitive to salinity so it is not suited to saline soil types or saline irrigation water (Mowat *et al.,* 1995).

Relatively few studies have been focused on the consequences induced by salinity stress on photosynthesis in persimmons. Besides there are only few studies carried out to determine the response of persimmon rootstocks to salinity stress by assessing gas exchange, to our knowledge. Fujita *et al.* (2003) measured the effects of salinity stress on photosynthetic rate, pre-dawn water potential, stomatal conductance, transpiration and Na and K contents of *D. kaki* leaves. The author indicated that salt stress reduced water potential, photosynthesis, transpiration and stomatal conductance. Specific salinity stress studies based on salinity levels are very limited for both *D. kaki* and *D. virginiana.* The aim of the present study is to determine the effects of different salinity levels on seedling growth rate and photosynthetic performances of both *D. kaki* and *D. virginiana* and to characterize their salt stress tolerances.

Materials and methods

Plant material, salinity treatment and salt stress symptom score

Seeds of *D. kaki* and *D. virginiana* were obtained from Cukurova University, Faculty of Agriculture, Department of Horticulture, Persimmon Germplasm orchard. Seeds were germinated in the dark at 22 °C in plastic trays with peat at the beginning of March. After germination, the seedlings were transferred and grown in moist peat for three months. Uniform sized seedlings were selected and transferred to quartz sand for an inert growth medium. Eight plants were used per treatment from each rootstock, distributed in a complete randomized design. The experiment was conducted in greenhouse conditions. The seedlings were fertilized for 15 days for adaptation to growth medium with the following solutions: 1.25 mM K_2SO_4, 0.625 mM KH_2PO_4, 2mM $MgSO_4$, 2mM $Ca(NO_3)_2$, 25 μM H_3BO_3, 2μM $MnSO_4$, 2μM $ZnSO_4$, 0.5 2μM $CuSO_4$, 0.065 μM $(NH_4)6Mo_7O_{24}$ and 50 μM Fe EDDHA. The pH of the nutrient solution was adjusted 6.0-6.5 with nitric acid. The plants were subjected to four levels of salinity with application of 0, 50, 75 and 100 mM NaCl. Seedlings were progressively adapted to salt stress in order to avoid osmotic shock. Stress conditions were maintained during 70 days. At the end of the experiment, severity of leaf symptoms were ranked as follows: 1 - normal healthy green plants without any injury; 2 - slightly wilted, damages on leaf tip; 3 - moderate to severe damages on leaves; 4 - most leaves with drying damages.

Photosynthetic measurements and chlorophyll concentration

At the end of the experiment, ten measurements were taken from each replicates of *D. virginiana* and *D. kaki* for each salinity treatment in the greenhouse. Transpiration rate (*E*) (mmol $m^{-2}s^{-1}$), stomatal conductance (g_S) [(mmol H_2O $m^{-2}s^{-1}$)], net photosynthetic rate (*Pn*) [μmol(CO_2) $m^{-2}s^{-1}$] and substomatal CO_2 concentration (ppm) were measured at the end of the experiment using a portable photosynthesis system (model LCA-4, ADC Bioscientific Ltd., Hoddesdon, UK). All gas exchange measurements were taken on attached fully expanded young leaves without visible injury symptoms and with an apparent homogeneous green colour. Measurements were recorded in the morning from 08:00 to 10:00 h to avoid high temperatures and low humidity in the afternoon. Average saturation irradiance was 1000 μmol $m^{-2}s^{-1}$ and leaf temperature ranged between 28 and 30 °C.

Leaf chlorophyll concentration was estimated using a portable SPAD meter. Chlorophyll fluorescence parameter (*Fv'/Fm'*) was measured with a portable fluorimeter in light adapted stage of the fully developed leaf of each plant. Readings were taken on the same leaves which the gas exchange measurements were recorded from.

Growth parameters and leaf tissue mineral analysis

After scale scoring, leaf chlorophyll concentration, PSII efficiency and leaf gas exchange measurements, the plants were harvested and leaf number and plant height (cm) were measured. Furthermore the plants were separated into shoots (all leaves and stem) and roots for dry matter assimilation and dried at 72 °C for 48 hours using a thermo ventilated oven.

At the end of the experiment, plants were removed from the pots and separated to root and shoots then washed with deionized water. Roots and shoots were oven dried for two days at 72 °C. The dry material was grounded. Chloride concentration was determined using a scientific chloride analyzer. K, Ca and Na were determined by atomic absorption spectrophotometry.

Experimentation and data analysis

The experiment was arranged as 4 x 2 x 8, four treatments, two rootstocks, eight replicates, respectively, in a 'Complete Randomized Design'. Data were subjected to two-way analysis of variance (ANOVA) in order to determine salinity effects on each rootstock. The means and calculated standard deviations are reported. Significant differences between means were evaluated by using LSD multiple range test at $P \leq 0.05$. In addition, two rootstocks were compared for each NaCl treatments by using t test at $P \leq 0.05$ in order to examine if there are significant differences between two rootstocks for each salinity levels in itself. All statistical analyses were performed by using SAS statistical software (v9.00, SAS Institute Inc., NC 27513-2414, USA) and SigmaPlot® (version 11.00, Systat Software, San Jose, CA, USA) was used for data presentation.

Results and discussions

At the end of the experiment, *D. kaki* and *D. virginiana* showed significant differences in their response to salt treatments. A two-way ANOVA indicated a significant main effect of rootstock and salinity treatments and also their interaction ($p \leq 0.01$) on the leaf salinity symptom (Tab. 1). Both genotypes showed salinity symptoms with increasing salinity levels. *D. virginiana* was less affected than *D. kaki*, based on severity of leaf symptoms. Furthermore, significant differences were obtained according to a t test between *D. kaki* and *D. virginiana* in the 50, 75 and 100 mM NaCl treatments (Tab. 2). Persimmon is recorded as very sensitive to salinity so it is not suited to saline soil types or saline irrigation water (Mowat *et al.,* 1995). In the present study, salinity symptoms were determined in different salinity levels.

In terms of leaf number and plant height, only the main effect of salinity treatments was significant ($p \leq 0.01$) according to a two-way ANOVA (Tab. 1). Significant decreases were recorded in plant leaf number and plant height for both genotypes by the increasing salinity levels (Tab. 3). Although no significant differences were determined regarding t-test between genotypes, the reduction in plant leaf number and height were higher in *D. kaki* then *D. virginiana* when the difference between 0 mM NaCl and 100 mM treatments were considered. Several authors claimed that NaCl reduced plant growth (Hassan and Galal, 1989; Abbas *et al.,* 1991; Franco *et al.,* 1993; Meneguzzo *et al.,* 2000; Zapata *et al.,* 2004) and similarly in this study, plant leaf number and plant height were significantly reduced by increasing salinity levels.

In 50, 75 and 100 mM saline conditions shoot and root dry matter productions of *D. kaki* and *D. virginiana* were decreased in comparison to 0 mM NaCl treatment (Tab. 4). On the other hand, the shoot and root dry weights in 50 and 75 mM NaCl treatments had similar values in both genotypes. A significant difference between genotypes was only obtained in 100 mM NaCl treatment, regarding t tests. Based on shoot and root dry matter productions, of *D. virginiana* was less affected than *D. kaki* regarding 100 mM NaCl treatments. It is well known that one of the first plant responses to salinity stress is a reduction in plant growth rate with associated reductions in leaf area available for photosynthesis. Subsequently, excessive accumulation of salts can lead to death of tissues, organs, and whole plants (Munns and Termaat, 1986). Similarly in the present study, the decrease in plant leaf number and leaf chlorophyll concentration of both genotypes under saline conditions resulted as a reduction in plant dry matter production. Cruz and Cuartero (1990) reported that both dry weights of leaf and stem are diminished in saline conditions, but the reduction of leaf dry weight is greater than that of dry shoot weight, which is in agreement with the present study. Furthermore, the relationship between shoot and root dry matter production and symptom scores of NaCl treatments for both *D. kaki* (Fig. 1A, B) and *D. virginiana* (Fig. 1C, D) were investigated in the present study. Although coefficients of determination were low, the relationship between plant dry matter productions and salinity symptoms were significant.

Tab. 1. Results of two-way analysis of variance (ANOVA) of rootstock (R) and salt treatment (S) effects and their interaction (R × S) for the dependent variables considered

Dependent variable	Independent variable		
	R	S	R x S
Salinity symptom	28.90**	28.01**	5.01**
Leaf number	3.65	18.83**	0.08
Plant height	0.42	7.24**	0.31
Shoot dry mass	0.91	22.83**	0.17
Root dry mass	2.15	10.26**	0.71
Na	413.54**	196.86**	65.87**
Cl	0.07	76.26**	1.06
K	37.90**	18.20**	15.10**
Ca	2.07	3.82*	1.22
SPAD	1.90	4.26*	3.29*
Fv'/Fm'	0.18	0.10	0.31
P_N	33.64**	102.98**	0.83
E	0.37	31.51**	2.08
g_s	1.23	490.86**	4.79*
C_i	11.61**	30.45**	4.87*

Note: Numbers represent *F* values. * $p \leq 0.05$; ** $p \leq 0.01$.

Tab. 2. Score scale of *D. kaki* and *D. virginiana* leaves under 0, 50, 75 and 100 mM NaCl treatments

NaCl treatment	Salinity symptom		t-test
	D. kaki	*D. virginiana*	
0 mM	1.00[y] b[z]	1.00 c	-
50 mM	2.60 a	1.64 b	*
75 mM	2.83 a	1.86 ab	*
100 mM	3.00 a	2.17 a	*
Prob>F	*<.0001*	*<.0001*	
LSD (0.05)	*0.638*	*0.393*	

[z] Means followed by different letters in the same column are significantly different (LSD, $P \leq 0.05$).
[y] Symptom score: 1 - normal healthy green plants without any injury; 2 - slightly wilted, damages on leaf tip; 3 - moderate to severe damages on leaves; 4 - most leaves with drying damages

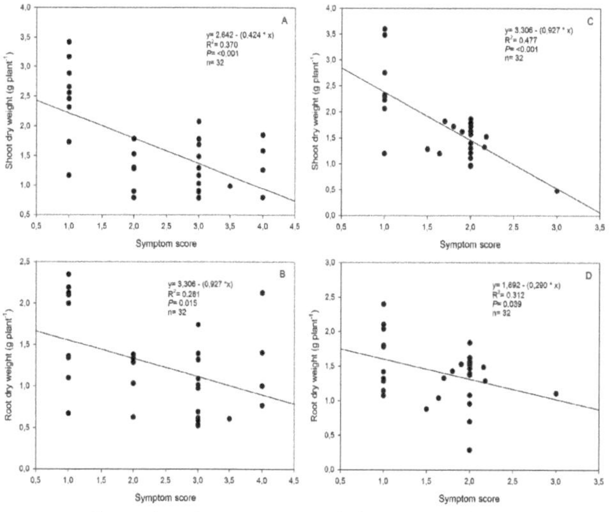

Fig. 1. Relationship between shoot (A) and root (B) dry matter production and symptom scores of *D. kaki*. Relationship between shoot (C) and root (D) dry matter production and symptom scores of *D. virginiana*. Symptom score: 1-normal healthy green plants without injury; 2-slightly wilted, damages on leaf tip; 3-moderate to severe damages on leaves; 4-most leaves with drying damages

Tab. 3. Leaf number and plant height (cm) of *D. kaki* and *D. virginiana* under 0, 50, 75 and 100 mM NaCl treatments

NaCl treatment	Leaf number *D. kaki*	Leaf number *D. virginiana*	t-test	Plant height (cm) *D. kaki*	Plant height (cm) *D. virginiana*	t-test
0 mM	13.00 a[z]	13.71 a	NS[y]	25.60 a	26.86 a	NS
50 mM	11.00 a	12.57 ab	NS	24.00 a	23.86 ab	NS
75 mM	8.80 b	10.00 bc	NS	23.80 a	22.95 ab	NS
100 mM	6.60 c	7.86 c	NS	18.83 b	21.00 b	*
Prob>F	*<.0001*	*0.0019*		*0.0101*	*0.0385*	
LSD (0.05)	*2.092*	*2.953*		*3.944*	*3.933*	

[z]Means followed by different letters in the same column are significantly different (LSD, $P \leq 0.05$).
[y]*: significant at $P \leq 0.05$, NS: not significant according to t-test between *D. kaki* and *D. virginiana*

Tab. 4. Dry matter production (g plant^{-1}) of *D. kaki* and *D. virginiana* under 0, 50, 75 and 100 mM NaCl treatments

NaCl treatment	Shoot *D. kaki*	Shoot *D. virginiana*	t-test	Root *D. kaki*	Root *D. virginiana*	t-test
0 mM	2.66 a[z]	2.75 a	NS[y]	1.81 a	1.79 a	NS
50 mM	1.69 b	1.65 b	NS	1.39 ab	1.43 ab	NS
75 mM	1.29 bc	1.45 b	NS	1.08 bc	1.39 ab	NS
100 mM	0.89 c	1.20 b	*	0.62 c	1.04 b	*
Prob>F	*0.0004*	*0.0002*		*0.0036*	*0.0397*	
LSD (0.05)	*0.616*	*0.535*		*0.538*	*0.454*	

[z]Means followed by different letters in the same column are significantly different (LSD, $P \leq 0.05$).
[y]*: significant at $P \leq 0.05$, NS: not significant according to t-test between *D. kaki* and *D. virginiana*

Reduction in leaf chlorophyll concentration based on SPAD readings was observed in both *D. kaki* and *D. virginiana* under saline conditions compared to the control (Fig. 2A). At the end of the experiment *D. virginiana* leaves had higher chlorophyll concentration than *D. kaki* in 100 mM NaCl treatment. In addition, SPAD readings were higher in *D. virginiana* at 0, 50 and 75 mM salinity levels. Chlorophyll content can be considered as one of the few physiological parameters that show a good correlation with salinity tolerance (Ashrafuzzaman *et al.,* 2000; Hakam *et al.,* 2000; Ali *et al.,* 2004). Similar findings were reported for other plants such as in maize (Ashrafuzzaman *et al.,* 2000) and rice (Ali *et al.,* 2004). Furthermore NaCl stress decreases total chlorophyll content of the plant by increasing the activity of the chlorophyll degrading enzyme: chlorophyllase (Rao and Rao, 1981), inducing the destruction of the chloroplast structure and the instability of pigment protein complexes (Sing and Dubey, 1995).

Chlorophyll fluorescence analysis has proven to be a sensitive method for the detection and quantification of stress-induced changes in PSII (Mehta *et al.,* 2010), and both light- and dark-adapted measurements can be used to determine whether photodamage has occurred in leaves or not (Naumann *et al.,* 2008). In the present study, slight decreases were determined in terms PSII in both genotypes with increasing salinity levels (Fig. 2B). Although there were no significant differences in both genotypes, *Fv'/Fm'* of *D. virginiana* was less affected than *D. kaki* in 100 mM NaCl treatment. It has been concluded that salt stress affects reaction centers of PSII either directly or indirectly (Masojidek and Hall, 1992; Moradi and Ismail, 2007).

Different NaCl treatments had significant reductions on photosynthetic performances of *D. kaki* and *D. virginiana* (Fig. 3). Of different gas exchange attributes, net CO_2 assimilation rate (P_N) (Fig. 3A) and transpiration rate (E) (Fig. 3B) of both genotypes were reduced significantly due to increasing salinity levels, respectively. Salt stress caused a marked reduction in stomatal conductance (g_S) too (Fig. 3C). Because of the stomatal closure under saline conditions, the availability of leaf sub-stomatal CO_2 (Ci) concentration was also decreased (Fig. 3D). Similar observations were also determined by Fujita *et al.* (2003) who reported that leaf gas exchange parameters of *D. kaki* were significantly reduced under salinity stress. Salinity affects plant growth due to changes in many physiological processes including photosynthesis (Kalaji and Pietkiewicz, 1993; Kalaji and Guo, 2008). A reduction in chlorophyll content, stomatal conductance, ribulose-1,5-bisphosphate carboxylase/oxygenase (Rubisco) activity and an increase in the chlorophyll a/b ratio had been observed earlier (Kalaji and Nalborczyk, 1991; Delfine *et al.,* 1999). Netondo *et al.* (2004) found that the photosynthesis rate of sorghum grown under salinity stress was affected primarily by stomatal closure. Banuls *et al.* (1997) associated reduction in photosynthetic capacity and stomatal conductance of citrus leaves treated with NaCl with high concentration of Cl. Garcia-Sanchez *et al.* (2002) studied effects of high salinity on different citrus rootstocks and reported that the higher reduction in leaf gas exchange parameters occurred in trees on 'Carrizo' where leaves accumulated more Cl and Na than in leaves on 'Cleopatra'. It has been proposed that the reduction of leaf gas exchange in response to salinity is due to the increase in leaf Na concentration (Garcia-Legaz *et al.,* 1993; Walker *et al.,* 1993).

A two-way ANOVA indicated a significant main effect of rootstock and salinity treatments and also their interaction ($p \leq 0.01$) on the leaf Na and K concentrations (Tab. 1). NaCl treatments resulted in accumulations of Na and Cl in leaves for both *D. kaki* and *D. virginiana.* The data shown in Tab. 5 suggested that, Cl content of leaves increased with increasing salinity in both genotypes. Cl accumulations in leaves were grouped in the same subset for 50 and 75 mM saline treatments in both genotypes whereas 100 mM NaCl treatment remarkably increased. The highest Na and Cl concentrations were determined in the leaves of *D. kaki.* Similarly, Fujita *el al.* (2003) reported high accumulation of Na in leaves of *D. kaki* under 100 mM NaCl treatment. Also at 200 mM NaCl, a great accumulation of Cl in both leaves and roots of sugar beet was reported (Ghoulam *et al.,* 2002). The K concentrations of the leaves of both genotypes were gradually decreased in response to NaCl and significantly affected by the increasing NaCl treatments (Tab. 6).

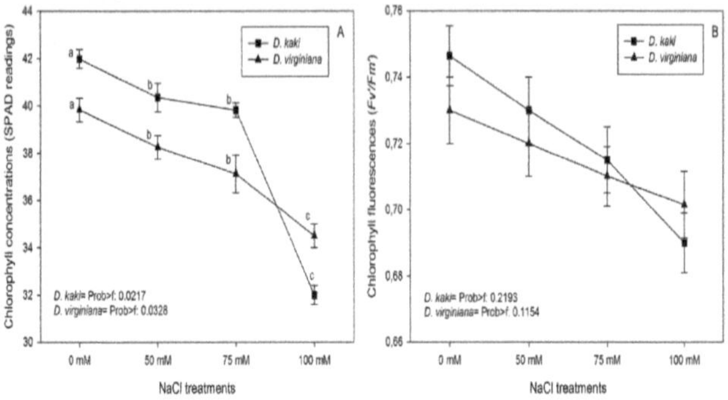

Fig. 2. Chlorophyll concentrations (SPAD readings) (A) and chlorophyll fluorescence (*Fv'/Fm'*, in light adapted stage) efficiency (B) of *D. kaki* and *D. virginiana* under 0, 50, 75 and 100 mM NaCl treatments; bars indicate SD

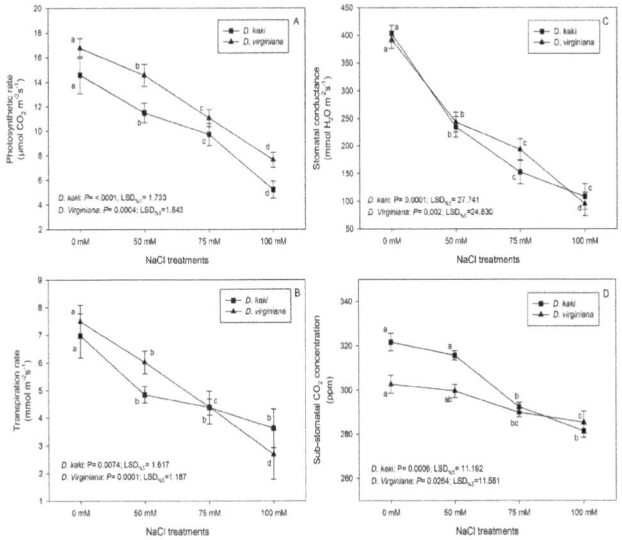

Fig. 3. Net photosynthetic rate (A), transpiration rate (B), stomatal conductance (C) and sub-stomatal CO_2 concentration (D) of *D. kaki* and *D. virginiana* under 0, 50, 75 and 100 mM NaCl treatments; bars indicate SD

Tab. 5. Leaf Na (%) and Cl (mg l^{-1}) concentrations of *D. kaki* and *D. virginiana* under 0, 50, 75 and 100 mM NaCl treatments

NaCl	Na		*t*	Cl		*t*
treatment	*D. kaki*	*D.virginiana*	test	*D. kaki*	*D.virginiana*	test
0 mM	0.20 a[z]	0.18 c	NS[y]	23.00 a	30.33 c	NS
50 mM	0.87 b	0.28 bc	*	115.67 b	125.00 b	*
75 mM	1.03 c	0.40 b	*	134.33 b	134.00 b	NS
100 mM	1.80 d	0.60 a	*	200.00 c	175.00 a	*
Prob>F	*<.0001*	*0.0029*		*<.0001*	*0.0001*	
LSD (0.05)	*0.0851*	*0.1744*		*34.367*	*36.295*	

[z]Means followed by different letters in the same column are significantly different (LSD, *P*≤0.05).
[y]*: significant at *P*≤0.05, NS: not significant according to t-test between *D. kaki* and *D. virginiana*

Tab. 6. Leaf K (%) and Ca (%) concentrations of *D. kaki* and *D. virginiana* under 0, 50, 75 and 100 mM NaCl treatments

NaCl	K		*t*	Ca		*t*
treatment	*D. kaki*	*D.virginiana*	test	*D. kaki*	*D.virginiana*	test
0 mM	2.35 a[z]	3.58 a	*[y]	1.10	1.35 b	NS
50 mM	2.34 a	2.69 b	NS	1.42	2.08 a	NS
75 mM	2.25 ab	2.30 bc	NS	1.88	1.85 a	NS
100 mM	1.92 b	2.25 c	*	1.77	1.74 a	NS
Prob>F	*<.0001*	*0.0001*		*0.2402*	*0.0119*	
LSD (0.05)	*0.3485*	*0.3839*		-	*0.3722*	

[z]Means followed by different letters in the same column are significantly different (LSD, *P*≤0.05).
[y]*: significant at *P*≤0.05, NS: not significant according to t-test between *D. kaki* and *D. virginiana*

Ghoulam *et al.* (2002) reported that K concentration in sugar beet was decreased under saline conditions. K is an essential macronutrient and the most abundant cation in higher plants, whereas the closely related ion Na is toxic to most plants at high concentrations (Mäser *et al.,* 2002). There were no statistically significant differences in Ca concentration of *D. kaki* in all NaCl treatments. On the other hand a slight increase in Ca concentration of the leaves of *D. virginiana* under saline conditions was recorded (Tab. 6).

Conclusions

In this article, we described how different salinity levels affected the growth and the photosynthetic performance of two persimmon rootstocks. According to the data obtained from plant growth, mineral nutrients uptake and leaf gas exchange parameters, we conclude that *D. kaki* and *D. virginiana* did not respond in the same way. In this study, it can be suggested that *D. virginiana* has more tolerance to salinity than *D. kaki.* However, the plant materials utilized in this experiment were from un-grafted rootstocks. Therefore, when a commercial cultivar of persimmon is grafted, the results could differ. Also, further studies are needed to evaluate these rootstocks under field conditions.

Acknowledgements

The authors are grateful to Dr. Ben Faber for critical reading and language correction of the manuscript.

References

Abbas MA, Younis ME, Shukry WM (1991). Plant growth, metabolism and adaptation in relation to stress conditions: XIV. Effect of salinity on the internal solute concentrations in *Phaseolus vulgaris*. J Plant Physiol 138:722-727.

Ali Y, Aslam Z, Ashraf MY, Tahir GR (2004). Effect of salinity on chlorophyll concentration, leaf area, yield and yield components of rice genotypes grown under saline environment. International Journal of Environmental Science and Technology 1:221-225.

Ashrafuzzaman M, Khan MAH, Shohidullah SM, Rahman MS (2000). Effect of salinity on the chlorophyll content, yield and yield components of QPM Cv. Nutricta. Pakistan Journal of Biological Sciences 3:43-46.

Banuls J, Serna MD, Legaz F, Primo-Millo E (1997). Growth and gas exchange parameters of citrus plants stressed with different salts. J Plant Physiol 150:194-199.

Cruz V, Cuartero J (1990). Effects of salinity at several developmental stages of six genotypes of tomato (*Lycopersicon* spp). In: Cuarteroi J, Gomez-Guillamon ML, Fernandez-Munoz R (Eds.). Eucarpia Tomato. XIth Eucarpia Meeting on Tomato Genetics and Breeding Malaga, Spain pp. 81-86.

Farrar K (1999). Crop profile for persimmons in California. http://www.ipmcenters.org/cropprofiles/docs/capersimmons.pdf.

Flexas J, Diaz-Espejo A, Galmes J, Kaldenhoff R, Medrano HOL, Ribas-Carbo M (2007). Rapid variations of mesophyll conductance in response to changes in CO_2 concentration around leaves. Plant Cell and Environment 30:1284-1298.

Franco JA, Esteban C, Rodriguez C (1993). Effects of salinity on various growth stages of muskmelon cv. Revigal. J Hort Sci 68(6):899-904.

Fujita K, Ito J, Mohapatra PK, Saneoka H, Lee K, Kurban H, Kawai K, Ohkura K (2003). Circadian rhytm of stem and fruit diameter dynamics of Japanese persimmon (*Diospyros kaki* Thunb.) is affected by deficiency of water in saline environments. Functional Plant Biology 30:747-754.

Garcia-Legaz MF, Ortiz JM, Garcia-Lidon AG, Cerda A (1993). Effect of salinity on growth, ion content and CO_2 assimilation rate in lemon varieties on different rootstock. Physiol Plant 89:427-432.

Garcia-Sanchez F, Jifon JL, Carvajal M, Syvertsen JP (2002). Gas exchange, chlorophyll and nutrient contents in relation to Na^+ and Cl accumulation in 'Sunburst' mandarin grafted on different rootstocks. Plant Sci 162:705-712.

Ghoulam C, Foursy A, Fares K (2002). Effects of salt stress on growth, inorganic ions and proline accumulation in relation to osmotic adjustment in five sugar beet cultivars. Environ Exp Bot 47:39-50.

Grattan SR, Grieve CM (1999). Salinity-mineral nutrient relations in horticultural crops. Sci Hort 68:127-157.

Navarro MJ, Martinez V, Carvajal M (2000). Ammonium, bicarbonate and calcium effects on tomato plants grown under saline conditions. Plant Sci 157:89-96.

Hakam N, Khanizadeh S, DeEll JR, Richer C (2000). Assessing chilling tolerance in roses using chlorophyll flouresence. HortScience 35:184-186.

Hasegawa PM, Bressan RA, Zhu JK, Bohnert HJ (2000). Plant cellular and molecular responses to high salinity. Annu Rev Plant Physiol Plant Mol Biol 51:463-499.

Hassan MM, Galal MA (1989). Salt tolerance among some citrus rootstocks. J King Saud Univ Agric Sci 1:87-93.

Heuvelink E, Bakker M, Sthangellini C (2003). Salinity effects on fruit yield in vegetable crops: a Simulation Study. Acta Hort 609:133-140.

Hussain S, Luroa F, Costantinoa G, Ollitrault P, Morillon R (2012). Physiological analysis of salt stress behaviour of citrus species and genera: Low chloride accumulation as an indicator of salt tolerance. South African Journal of Botany 81:103-112.

Kalaji MH, Nalborczyk E (1991). Gas exchange of barley seedlings growing under salinity stress. Photosynthetica 25:197-202.

Levitt J (1980). Responses of plants to environmental stresses. Vol. I Chilling, freezing, and high temperature stress. London, New York, Toronto: Academic Press.

Li LY, Stanghellini C (2001). Analysis of the effect of EC and potential transpiration on vegetative growth of tomato. Sci Hortic 89:9-21.

Marschner H (1995). Saline soils. In: Mineral nutrition of higher plants. Academic Press, New York 1:657-680.

Mäser P, Gierth M, Schroeder JI (2002). Molecular mechanisms of potassium and sodium uptake in plants. Plant and Soil 247:43-54.

Masojidek J, Hall DO (1992). Salinity and drought stress are amplified by high irradiance in sorghum. Photosynthetica 27:159-171.

Mehta P, Jajoo A, Mathur S, Bharti S (2010). Chlorophyll a fluorescence study revealing effects of high salt stress on photosystem II in wheat leaves. Plant Physiol Biochem 48:16-20.

Meneguzzo S, Navari-Izzo F, Izzo R (2000). NaCl effects on water relations and accumulation of mineral nutrients in shoots, roots and cell sap of wheat seedlings. J Plant Physiol 156:711-716.

Moradi F, Ismail AM (2007). Responses of photosynthesis, chlorophyll fluorescence and ROS-scavenging systems to salt stress during seedling and reproductive stages in rice. Ann Bot 99:161-1173.

Mowat AD, George AP, Collins RJ (1995). Cultivation of persimmon *Diospyros kaki* under tropical conditions. Acta Hort 409:141-149.

Munns R (2002). Comparative physiology of salt and water stress. Plant Cell Environ 25:239-250.

Munns R, Termaat A (1986). Whole-plant responses to salinity. Aust J Plant Physiol 13:143-160.

Naumann JC, Young DR, Anderson JE (2008). Leaf chlorophyll fluorescence, reflectance, and physiological response to freshwater and saltwater flooding in the evergreen shrub, *Myrica cerifera*. Environ Exp Bot 63:402-409.

Netondo GW, Onyango JC, Beck E (2004). Sorghum and salinity: II. Gas exchange and chlorophyll fluorescence of sorghum under salt stress. Crop Sci 44:806-811.

Rao GG, Rao GR (1981). Pigment composition and chlorophyllase activity in pigeon pea (*Cajanus indicus*) and Gingelley (*Sesamum indicum*) under NaCl salinity. Indian J Experim Biol 19:768-770.

Romero-Aranda R, Soria T, Cuartero J (2001). Tomato water uptake and plant water relationships under saline growth conditions. Plant Sci 160:265-272.

Rouphael Y, Cardarelli M, Rea E, Colla G (2012). Improving melon and cucumber photosynthetic activity, mineral composition, and growth performance under salinity stress by grafting on to *Cucurbita* hybrid rootstocks. Photosynthetica 50:180-188.

Singh AK, Dubey RS (1995). Changes in chlorophyll *a* and *b* contents and activities of photosystems 1 and 2 in rice seedlings induced by NaCl. Photosynthetica 31:489-499.

Storey R, Walker RR (1999). Citrus and salinity. Sci Hort 78:39-81.

TUIK (2014). Turkish Statistical Institute. http: //www.turkstat.gov.tr/Start.do/.

Walker RR, Blackmore DH, Sung Q (1993). Carbon dioxide assimilation and foliar ion concentration in leaves of lemon (*Citrus limon* L.) trees irrigated with NaCl or Na_2SO_4. Aust J Plant Physiol 20:173-185.

Zalidis G, Stamatiadis S, Takavakoglou V, Eskridge K, Misopolinos N (2002). Impacts of agricultural 19 practices on soil and water quality in the Mediterranean region and proposed assessment 20 methodology. Agr Ecosyst Environ 88:137-146.

Zapata PJ, Serrano M, Pretel MT, Amorós A, Botella MA (2004). Polyamines and ethylene changes during germination of different plant species under salinity. Plant Sci 167:781-788.

18

Nutrition Intensity in Ternary Diagrams Interpretation for Some Ornamental Species Cultivated on Organic Substrate with Increased Biological Activity

Roxana Maria MADJAR[1], Gina VASILE SCĂEȚEANU[1], Mircea MIHALACHE[1], Cătălina CĂLIN[2], Valentin Sebastian DAN[3], Adrian PETICILĂ[4*]

[1]*University of Agronomic Sciences and Veterinary Medicine, Faculty of Agriculture, 59 Mărăşti Blvd., District 1, 011464, Bucharest, Romania; rmadjar@yahoo.com; ginavasile2000@yahoo.com; mihalachemircea@yahoo.com*

[2]*Petroleum-Gas University of Ploiesti, Faculty of Petroleum Refining and Petrochemistry, 39 Blvd. Bucureşti, 100680, Ploiesti, Romania; catalina.calin20@yahoo.com*

[3]*University of Agricultural Sciences and Veterinary Medicine, 3-5 Manastur St., Cluj-Napoca, Romania; dan.valentin@gmail.com*

[4]*University of Agronomic Sciences and Veterinary Medicine, Faculty of Horticulture, 59 Mărăşti Blvd., District 1, 011464, Bucharest, Romania; apeticila@yahoo.com (*corresponding author)*

Abstract

Nowadays, many biodegradable organic wastes no longer need to represent an environmental hazard and as a consequence, they could be recycled to obtain horticultural substrates. An experiment was conducted on two deciduous (*Tamarix tetrandra*, *Ligustrum ovalifolium* 'Aureum') and two coniferous species (*Chamaecyparis pisifera* 'Boulevard', *Chamaecyparis lawsoniana* 'Stardust') grown on a soil mixture of forestry compost, leaves compost, peat and grape marc compost. The aim of the research was to investigate the response to fertilization and to obtain valuable information regarding absorption rate of nutritive elements during vegetation. Nitrogen data show a lowering of its nutritive equilibrium point towards autumn in the leaves of deciduous shrubs species. Resorption of nutrients before leaves fall occurs due to conservation process in woody plants with deciduous leaves. In the case of coniferous species the lowering of nitrogen content is less intense. The monthly fertilization with Coïc solution indicates no influence on nitrogen metabolism of this species. The exception was on *Ligustrum* with differences between nitrogen content in fertilized and unfertilized plants. The phosphorus nutritive equilibrium point reveals a decrease during the summer months (July - August), the species presenting the lowest values in this period and the cause of this behaviour was the plants adaptation to high temperature and low humidity. Potassium nutritive equilibrium data indicates small differences in the unfertilized plants in comparison with those fertilized in all species. The novelty of the research is represented by the ternary diagrams N-P-K that were constructed, interpreted and reported for dendrologic species.

Keywords: dendrologic species, N-P-K diagrams, substrate, grape marc compost

Introduction

In European Union countries, where recycling and reintroduction of numerous waste vegetal circuits is a widespread activity, compost marketing is regulated by several laws. Composts for horticultural market are labelled mandatory wearing explanation of the origin, the components of the substrate, as any natural or synthetic materials, organic or mineral elements that may alter the mechanical, physical, chemical or biological character of the substrate. Feasibility of mixtures of substrates is virtually numerous, but substrates characterized by phytotoxicity, instability, high cost or limited availability are constantly rejected.

For intensive horticulture is mandatory to assure maximum quality and to use environmental friendly methods. A literature survey concerning different substrates used in horticulture revealed that until at a certain moment, the cultures on growing media have been carried out with a limited number of substrates (peat, rockwool, glasswool, perlite, vermiculite, charcoal chips) (Olle *et al.*, 2012; Restrepo *et al.*, 2013; Santiago *et al.*, 1989).

Lately, people are keen on the research and development of ecologically friendly substrates, which recently have been proven as a viable alternative to other traditional methods.

Nowadays, there is a growing interest in the exploitation of the residues generated by wine industry. Grape marc (waste skins, seeds and stalks), a valuable waste product in wine producing regions, is generated from the crushing, draining and pressing stages of the wine production, and contains approximately 8% seeds, 10% stems, 25% skins, 57% pulp (Laurenson *et al.*, 2012). There are researches that present the effects of composted grape marc as substrate for ornamental plants (Chen *et al.*, 1988), vegetables (Baran *et al.*, 2001; Carmona *et al.*, 2012; Leoni *et al.*, 1992). Therefore, identifying new materials suitable to obtain substrates for horticultural purposes is an important requirement. Grape marc composts possess chemical particularities that are responsible for fertility improvement (high contents of potassium and organic matter and significant contents of nitrogen and phosphorus) (Bustamante *et al.*, 2005, 2007, 2010), but there are some environmental problems related with its acidic character, high content of phenolic compounds and with phytotoxic and antimicrobial effects (Carmona *et al.*, 2012). Sometimes, substrates used for different ornamental species are insufficiently rich in nutrients. Providing ornamental plant nutrients can be handled by applying the fertilizer in the form of liquid or solid substrate.

Having in view that many biodegradable organic wastes can be recycled to ensure sustainability, this being an inexpensive alternative to obtain horticulture substrates, an experiment was designed and focused on four dendrologic species grown on a substrate composed of forestry compost, leaves compost, peat and grape marc compost, with the aim to investigate the plants' response to fertilization and to obtain ternary diagrams (N-P-K) that will provide information suitable to interpret species particularities, the absorption rate of nutritive elements during vegetation period.

Materials and methods

Substrate composition and plant material

The experiments were set up at the greenhouse from University of Agronomic Sciences and Veterinary Medicine Bucharest. Four dendrologic species were planted in vegetation pots (8 L capacity) that contained substrate formulated according to the recipe described below.

The organic materials used for the substrate composition were forestry compost (FC), leaves compost (LC), peat (P) and grape marc compost (GM) according to volumetric ratio FC: LC: P: GM = 1:1:1:0.5. The disinfection of the substrate components for pathogen destruction was made at 55 °C.

There were analyzed two deciduous species (*Tamarix tetrandra*, *Ligustrum ovalifolium* 'Aureum') and two coniferous species (*Chamaecyparis pisifera* 'Boulevard', *Chamaecyparis lawsoniana* 'Stardust'). During vegetation period it were performed the cutting of the branches to uniform the plants (in the middle of May), watering, mechanic weed control and application of phytosanitary treatments. In the experiments, it was used Coïc acidophile solution (Davidescu *et al.*, 2001) as source of macro- and microelements.

In June, the substrate was characterized from agrochemical point of view and it has been applied as the first fertilization treatment with Coïc acidophile solution, for a part from potted plants. At 30 days (July) the substrate was analyzed after fertilizer has been applied, the operation being repeated another 2 times (August, September).

Substrate agrochemical characterization and plant analysis

The substrate analyses were performed at the beginning of the experiment. The agrochemical characterization of the substrate was made by assessment of pH, EC (electrical conductivity), soluble forms (1:10, v/v ratio) as nitrate, ammonium, phosphorus, potassium and sodium (Tab. 1). Also, mobile phosphorus and exchangeable forms of main cations (K^+, Na^+, Ca^{+2}, Mg^{+2}) were determined in CH_3COONH_4 0.5M, pH=4.65, 1:3 v/v ratio (Tab. 2).

During the experiment at four moments (M1 – June, M2 – July, M3 – August, M4 – September) in order to establish the absorption rate of nutritive ions, for each plant species it were achieved total forms of nitrogen, phosphorus and potassium, using acknowledged methods (see Analytical methods and techniques section) (Tab. 3).

The nutritive equilibrium is given by the ratio of each element (N, P, K) expressed as percent against the sum of their concentrations, as it is presented below:

$$En_N = \frac{\%N}{\sum N,P,K} \cdot 100;\quad En_P = \frac{\%P}{\sum N,P,K} \cdot 100;$$

$$En_K = \frac{\%K}{\sum N,P,K} \cdot 100.$$

Biotic and enzymatic characterization of substrates

The biotic tests carried out on substrates have investigated the respiration and cellulolytic activity (Dashtaban *et al.*, 2010). Enzymatic tests were based on sugar hydrolysis (saccharase activity) (Balezentiene *et al.*, 2009; Shi *et al.*, 2008), urea hydrolysis (urease activity) (Shi *et al.*, 2008) and phosphorus hydrolysis from organic compounds (phosphatase activity) (Balezentiene *et al.*, 2009; Shi *et al.*, 2008). In addition, catalase activity (Shi *et al.*, 2008) was determined in order to obtain information concerning oxidation processes from substrates.

Analytical methods and techniques

Analytical grade chemicals were used throughout the study without any further purification. The pH values were carried out through potentiometric method, meanwhile EC (electrical conductivity) was determined using conductivity method. Soluble and total phosphorus contents were evaluated by spectrophotometric molybdenum blue method. The levels of nitrate and ammonium species were assessed by spectrophotometric means, also using phenoldisulfonic acid and Nessler reagent, respectively. Total form of nitrogen was determined using Kjeldahl method, meanwhile potassium and sodium contents were assessed using atomic emission spectrometry.

Calcium and magnesium concentrations were determined through flame atomic absorption spectrometry (FAAS). Measurements were carried out using a Thermo Scientific AA Spectrometer. Selected wavelengths were 422.7 nm for calcium and 285.2 nm for magnesium. The calibration curves for calcium and magnesium are linear for the studied ranges and were plotted by running different concentrations of standard solutions prepared from stock solutions of 1000 mg/L calcium and magnesium, respectively.

Results and discussions

Substrate characterization

The main agrochemical characteristics of the substrates are presented in Tabs. 1 and 2. Also, biotic and enzymatic tests were carried out for substrates and the results are presented in Tab. 3.

Respiration activity indicates the global biological activity of a substrate and the obtained value for this biotic test was 93.36 mg CO_2/100 g substrate (d.w.). Cellulolytic potential of the substrate is correlated with organic matter, the latest one representing the substrate for cellulolytic microflora. Catalase and saccharase activities are related with high organic matter content of the substrates. The urease potential of the analyzed substrate is high (662 mg NH_4^+/100 g substrate), this indicating the possibility of ammonia loses through volatilisation.

Phosphatase activity of the substrate is quantified through phosphorus that is released enzymatic and the obtained value is 5.72 mg P/100 g substrate (d.w.). The increased biological activity is due to the high C/N ratio (20.8-26.85) in comparison with humus level (10-12%), this indicating that organic matter is still decomposing.

Nutritive equilibrium

The establishment of nutritive equilibrium was done using the method based on the global nutrition calculation, according to experimental analysis performed during the vegetation period. The global nutrition is the sum of N, P, K concentrations (as total forms) expressed as percents from dry matter. The values obtained after plant analysis during vegetation period at certain moments are summarized in Tab. 4.

The interpretation was expressed through ternary diagram N-P-K using a triangular system. For each of the dendrologic specie that were subjects of the experiment, it were constructed ternary diagrams (N-P-K) that allowed to interpret the species particularities, the absorption rate of nutritive elements for fertilized and unfertilized variant.

The ternary diagrams with highlighted theoretic nutritive equilibrium obtained for *Tamarix tetrandra*, *Ligustrum ovalifolium* 'Aureum', *Chamaecyparis pisifera* 'Boulevard', *Chamaecyparis lawsoniana* 'Stardust' represent a new edge concerning nutritive equilibrium of N, P, K for ornamental species. Also, the values for nutritive equilibrium of the same elements after plant analysis were evidenced.

Nutritive equilibrium for Tamarix tetrandra

The points for nutritive equilibrium resulted for unfertilized variant *Tamarix tetrandra* describe the equilibrium area with trilinear coordinates for N between 48.36% and 61.67%, for phosphorus between 3.91% and 14.02% and for potassium between 24.32% and 40.46%.

The analysis revealed that at the beginning of the vegetation period the nitrogen absorption was higher and decreased as dry matter accumulation increased. The phosphorus absorption is intense at the beginning, and decrease to a third, during vegetation period, so that at the end to increase at values comparable with those recorded at vegetation start stage. Potassium absorption increase slightly during vegetation period as it is shown by the equilibrium points resulted from En_K coordinates (Fig. 1, a).

Equilibrium points resulted from the variant fertilized with acidophile Coïc solution marks the nutrient equilibrium for species *Tamarix tetrandra* by a shift of equilibrium in behalf of N absorption and less intense for P and K compared to the unfertilized (control) variant (Fig. 1, a).

Nutritive equilibrium for Ligustrum ovalifolium

Ligustrum ovalifolium equilibrium points obtained from the unfertilized variant describe the nutritive equilibrium area with trilinear coordinates for En_N between 41.43% and 57.13%, for En_P between 3.62% and 14.38% and for En_K between 32.87% and 50.50%. Nitrogen uptake is more intense at the early vegetation stage, then has an equilibrium plateau and finally decreases slightly.

Phosphorus uptake is intense in June and then decreases in July and August about three times; in early September it was observed the increase of the absorption intensity of phosphorus, which exceeds the values recorded at the beginning of the vegetation period.

During the vegetation, potassium absorption is intense and constantly increases, with highest values in July, until it records a slight decline marked by a stable balance at the end of the vegetation stage (Fig. 1, b).

Equilibrium area delimited by equilibrium points in fertilized variants exhibits a constant absorption of phosphorus, significant variation of nitrogen absorption intensity that is intense at the beginning of vegetation stage, but decreases as dry matter accumulate and grows again in the following stages. Potassium absorption intensity shows large increases to a maximum and decreases at the end of the vegetation period.

Nutritive equilibrium for Chamaecyparis pisifera 'Boulevard'

The equilibrium areas for *Chamaecyparis pisifera* 'Boulevard' bounded by the equilibrium points resulted after data calculation, presents differences between unfertilized and fertilized variants. It was observed that the uptake of N was more intense at the early vegetation period for unfertilized variant in comparison with the fertilized one. The absorption intensity values for P are undifferentiated by fertilization and constant for K at fertilized variant, meanwhile the unfertilized variant it registered an increasing trend. The coordinates for ternary diagram take values between 46.86% and 57.75% for En_N, 4.40% and 12.49% for En_P and 37.38% and 46.06% for En_K (Fig. 1, c).

Nutritive equilibrium for Chamaecyparis lawsoniana 'Stardust'

The equilibrium areas delimited by the equilibrium points shows that between unfertilized and fertilized variants there are no significant differences for *Chamaecyparis lawsoniana* 'Stardust'; the areas are almost overlapped, this behavior indicating that fertilization does not influence the absorption of nutritive elements N-P-K. The coordinates for nutritive equilibrium for unfertilized variant (as control), vary between 56.42% and 72.85% for En_N, 5.10% and 15.32% for En_P and between 13.05% and 38.20% for En_K (Fig. 1, d).

Tab. 1. The agrochemical characterization (soluble forms) of the substrates with potted plants

Substrate with potted plants	pH	EC $dS \cdot m^{-1}$	PO_4^{3-} ppm	NO_3^- ppm	NH_4^+ ppm	Ca^{2+} ppm	Mg^{2+} ppm	K^+ ppm	Na^+ ppm
Tamarix tetrandra	7.83	0.1359	22.00	0.50	25.00	112.23	13.693	140	20
Ligustrum ovalifolium 'Aureum'	8.14	0.2687	38.40	trace	23.00	163.24	24.604	290	17
Chamaecyparis pisifera 'Boulevard'	8.03	0.1468	42.40	4.00	23.00	193.85	24.192	145	11
Chamaecyparis lawsoniana 'Stardust'	7.80	0.1037	29.80	3.00	9.00	102.03	14.434	62	20

After chemical plant analysis at moments M1-M4 (June - September), there were obtained results that were compared with limits from literature data (Davidescu *et al.*, 2001). An overview based on the obtained results is detailed in Tab. 5. The studies of nutrition intensity of dendrologic species cultivated in substrate made of leaves compost, forestry compost, peat and grape marc compost (ratio 1: 1: 1: 0.5) reveals significant differences between deciduous (*Tamarix tetrandra, Ligustrum ovalifolium* 'Aureum') and coniferous species (*Chamaecyparis pisifera* 'Boulevard', *Chamaecyparis lawsoniana* 'Stardust'), but not significant differences in applied fertilizers as Coïc solution during vegetation period (June-September) in comparison with unfertilized plants for all studied species.

Discussion

Ternary diagrams are used in various areas (chemistry, geology, industry, biology, agriculture) and represent graphic tools that permit to evaluate the interaction between systems and the interactions between systems and environment (Almeida *et al.*, 2007; Giannetti *et al.*, 2006, 2013), allow the estimation of nutrients balance in soil (Parent *et al.*, 2012), visualization of relation among the three humus fractions (Watanabe *et al.*, 1991) or variation of sugar composition in nectars (Herrera *et al.*, 2006). In literature, there is lack of information concerning ternary diagrams correlated with nutritive equilibrium in the case of dendrology species. Notwithstanding, some similar researches have been presented for potato (Thomas, 1938).

The data obtained in this experiment conducted with the aim to evaluate the absorption rate of nutritive elements during vegetation period for different dendrologic species, allowed us to construct ternary diagrams N-P-K. The four studied species belonging to the two different groups of deciduous and coniferous trees, with distinct metabolism and diverse biological particularities, have responded different to the application of fertilizers. Their behaviour after nutritive

Tab. 2. Mobile phosphorus and exchangeable forms of main cations in substrate with potted plants

Substrate with potted plants	P (ppm)	Ca (mE/100g)	Mg (mE/100g)	K (mE/100g)	Na (mE/100g)
Tamarix tetrandra	262.00	59.05	2.566	0.910	0.1413
Ligustrum ovalifolium 'Aureum'	278.00	52.94	2.553	1.028	0.1217
Chamaecyparis pisifera 'Boulevard'	162.75	42.76	2.027	0.910	0.1000
Chamaecyparis lawsoniana 'Stardust'	290.00	54.47	2.537	0.769	0.1300

Tab. 3. Biotic and enzymatic tests developed on substrate

Biotic tests		Enzymatic tests			
Respiration activity ($mg\ CO_2/100\ g$)	Cellulolytic activity (g cellulose/ 100 g cotton)	Catalase activity ($cm^3\ O_2$/ 100 g)	Saccharase activity (mg saccharose/ 100 g)	Urease activity ($mg\ NH_4^+$/ 100 g)	Phosphatase activity (mg P/ 100 g)
93.36	76.20	953	2104	662	5.72

Tab. 4. Total forms of nutritive elements after plant analysis

Var	June			July			August			September		
	Total forms (%)											
	N	P	K	N	P	K	N	P	K	N	P	K
V1	4.1850	0.952	1.65	3.2517	0.218	2.10	2.5542	0.174	1.65	2.8686	0.663	2.40
V2	4.2734	0.872	1.50	3.2223	0.244	1.95	2.6132	0.168	1.50	2.7900	0.625	2.00
V3	3.3893	0.593	1.95	3.1339	0.247	3.45	2.9472	0.214	2.60	2.4265	0.844	2.60
V4	3.1339	0.614	1.70	2.2497	0.236	3.40	2.7212	0.235	2.65	2.6819	0.718	2.40
V5	2.2399	0.189	1.45	2.1514	0.191	2.00	2.3086	0.189	1.50	2.1907	0.582	1.90
V6	2.1908	0.171	1.60	2.2006	0.190	1.90	2.2890	0.156	1.90	2.3784	0.560	1.95
V7	2.7900	0.540	0.50	2.1416	0.204	1.45	2.4953	0.204	1.30	2.3086	0.653	1.30
V8	3.0651	0.558	0.65	2.1907	0.200	1.50	2.5542	0.191	1.25	2.3577	0.561	1.35

V1 - *Tamarix tetrandra* (unfertilized); V2 - *Tamarix tetrandra* (fertilized); V3 - *Ligustrum ovalifolium* (unfertilized); V4 - *Ligustrum ovalifolium* (fertilized); V5 - *Chamaecyparis pisifera* 'Boulevard' (unfertilized); V6 - *Chamaecyparis pisifera* 'Boulevard' (fertilized); V7 - *Chamaecyparis lawsoniana* 'Stardust' (unfertilized); V8 - *Chamaecyparis lawsoniana* 'Stardust' (fertilized)

Tab. 5. General overview on nutritive equilibrium for ornamental species

Var.	Specie	June			July			August			September		
		En (%)			En (%)			En (%)			En (%)		
		N	P	K	N	P	K	N	P	K	N	P	K
V1	*Tamarix tetrandra*												
V2	*Tamarix tetrandra* (fertilized)												
V3	*Ligustrum ovalifolium*												
V4	*Ligustrum ovalifolium* (fertilized)												
V5	*Chamaecyparis* 'Boulevard'												
V6	*Chamaecyparis* 'Boulevard' (fertilized)												
V7	*Chamaecyparis* 'Stardust'												
V8	*Chamaecyparis* 'Stardust' (fertilized)												

Legend: very low ; low ; normal ; high ; very high

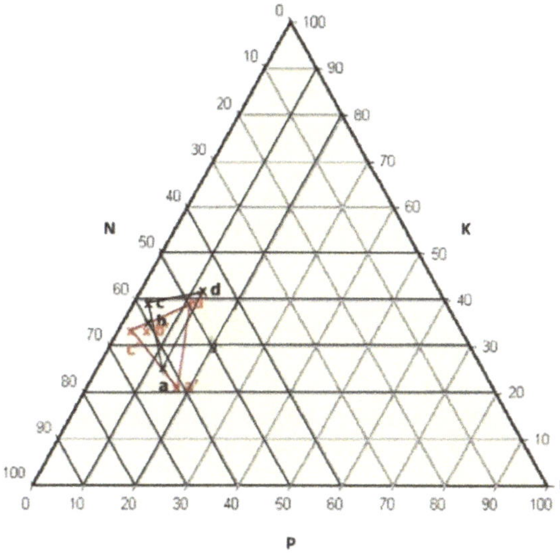

a) Ternary diagram N-P-K for *Tamarix tetrandra*

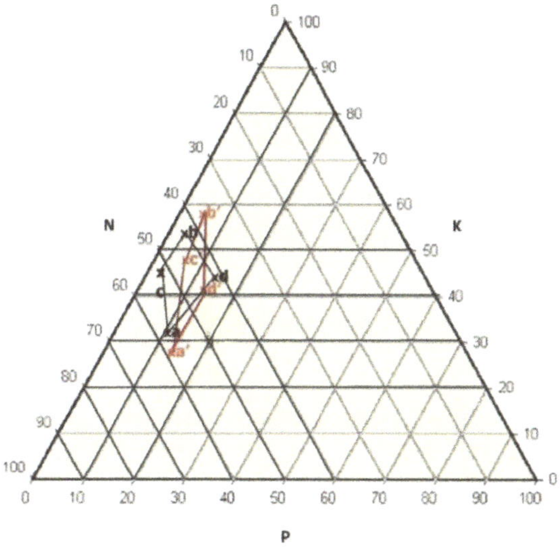

b) Ternary diagram N-P-K for *Ligustrum ovalifolium* 'Aureum'

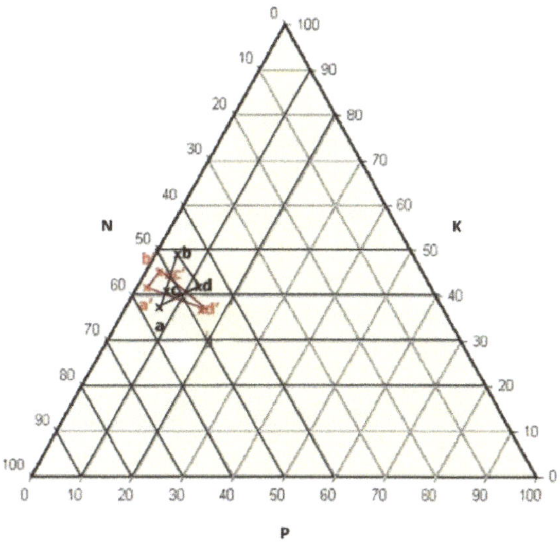

c) Ternary diagram N-P-K for *Chamaecyparis pisifera* 'Boulevard'

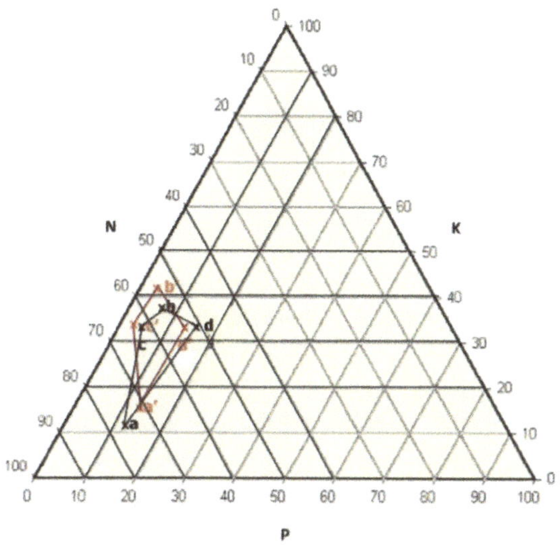

d) Ternary diagram N-P-K for *Chamaecyparis lawsoniana* 'Stardust'

Fig. 1. Ternary diagram for studied dendrologic species (black line - unfertilized; red line - fertilized)

solution application on the same substrate is different concerning nutrient absorption capacity, nutrient balance during the vegetation period and the relationship between species, enzyme substrate and biotic indicators of used substrate.

Equilibrium points represented in the ternary diagrams N-P-K resulted from fertilized variant with acidophile Coïc solution, marks the nutrient equilibrium area for specie *Tamarix tetrandra* by a shift of nutrient balance for the absorption of N and less intense for P and K compared to the control variant.

The equilibrium area delimited by the points of equilibrium for *Ligustrum ovalifolium* 'Aureum' in the case of fertilized variants present constant absorption of P, large variations of the intensity of N absorption that is intense at the beginning of the vegetation period and decrease as the dry matter accumulate, then increase again during the rest of the following period. The intensity of K absorption shows large increases up to a maxima and then decreases at the end of vegetation period.

The equilibrium areas for *Chamaecyparis pisifera* 'Boulevard' bounded by the equilibrium points resulted after calculation shows differences between unfertilized and fertilized variants with more intense consumption at the beginning of the vegetation period for N at unfertilized variant in comparison with the fertilized one. The values that describe the intensity of the P absorption are undifferentiated from fertilization and are constant for K at fertilized variant, with increasing values for unfertilized variant.

For *Chamaecyparis lawsoniana* 'Stardust' there is no differences between fertilized and unfertilized plants related with the nutrients absorption, the fertilization leading to an undetectable influence. On the other hand, there is a higher difference of nutritive elements absorption intensity in comparison with *Chamaecyparis pisifera* 'Boulevard'.

References

Almeida CMVB, Barrella FA, Giannetti BF (2007). Emergetic ternary diagrams: five examples for application in environmental accounting for decision-making. J Clean Prod 15:63-74.

Balezentiene L, Klimas E (2009). Effect of organic and mineral fertilizers and land management on soil enzyme activities. Agronomy Research 7(1):191-197.

Baran A, Çayci G, Kütük C, Hartmann R (2001). Composted grape marc as growing medium for hypostases (*Hypostases phyllostagya*). Bioresource Technol 78:103-106.

Bustamante MA, Paredes C, Moral R, Moreno-Caselles J, Perez-Espinosa A, Perez-Murcia MD (2005). Use of winery and distillery effluents in agriculture: characterisation of nutrient and hazardous components. Water Sci Technol 51:145-151.

Bustamante MA, Paredes C, Moral R, Moreno-Caselles J, Perez-Murcia MD, Perez-Espinosa A, Beral MP (2007). Co-composting of distillery and winery wastes with sewage sludge. Water Sci Technol 56(2):187-192.

Bustamante MA, Said-Pullicino D, Paredes C, Cecilia JA, Moral R (2010). Influences of winery-distillery waste compost stability and soil type on soil carbon dynamics in amended soils. Waste Manage 30:1966-1975.

Carmona E, Moreno MT, Aviles M, Ordovas J (2012). Use of grape marc compost as a substrate for vegetable seedlings. Sci Hortic-Amsterdam 137:69-74.

Carmona E, Moreno MT, Aviles M, Ordovas J (2012). Composting of wine industry wastes and their use as a substrate for growing soilless ornamental plants. Span J Agric Res 10(2):482-491.

Chen Y, Inbar Y, Hadar Y (1988). Composted agricultural wastes as potting media for ornamental plants. Soil Sci 145(4):298-303.

Dashtaban M, Maki M, Leung KT, Mao C, Qin W (2010). Cellulase activities in biomass conversion: measurement methods and comparison. Crc Cr Rev Biotechn 30(4):302-309.

Davidescu V, Costea G, Madjar R, Stanica F, Caretu G (2001). Substrates of culture (in Romanian). Ceres, Bucharest.

Giannetti BF, Barrella FA, Almeida CMVB (2006). A combined tool for environmental scientists and decision makers: ternary diagrams and emergy accounting. J Clean Prod 14:201-210.

Giannetti BF, Almeida CMVB, Agostinho F, Bonilla SH, Ulgiati S (2013). Primary evidences on robustness of environmental accounting from emergy. J Environ Acc Manag 1(2):203-212.

Herrera CM, Perez R, Alonso C (2006). Extreme intraplant variation in nectar sugar composition in an insect-pollinated perennial herb. Am J Bot 93(4):575-581.

Laurenson S, Houlbrooke D (2012). Review of guidelines for the management of winery wastewater and grape marc. Report prepared for Marlborough District Council.

Leoni S, Madeddu B (1992). The use of marc as a substrate for tomato cultivation in greenhouses. Colture Prot 21(6):67-71.

Olle M, Ngouajio M, Siomos A (2012). Vegetable quality and productivity as influenced by growing medium: a review. Agriculture 99(4):399-408.

Parent SE, Parent LE, Rozanne DE, Hernandez A, Natale W (2013). Nutrient balance as paradigm of soil and plant chemometrics. InTech 83-114. http://dx.doi.org/10.5772/53343.

Restrepo AP, Medina E, Perez-Espinosa A, Agullo E, Bustamante MA, Mininni C, Bernal MP, Moral R (2013). Substitution of peat in horticultural seedlings: suitability of digestate-derived compost from cattle manure and maize silage codigestion. Commun Soil Sci Plan 44:668-677.

Santiago A, Santiago L (1989). Charcoal chips as a practical substrate for container horticulture in the humid tropics. Acta Hort 238:141-148.

Shi ZJ, Lu Y, Xu ZG, Fu SL (2008). Enzyme activities of urban soils under different land use in the Shenzhen city, China. Plant Soil Environ 54(8):341-346.

Thomas W (1938). Foliar diagnosis: its relation to the optimum nutrition of the potato. Plant Physiol 13(4):677-694.

Watanabe A, Kuwatsuka S (1991). Triangular diagram for humus composition in various types of soils. Soil Sci Plant Nutr 37(1):167-170.

19

Quality Changes of Different Sweet Cherry Cultivars at Various Stages of the Supply Chain

Fatih SEN*, Rustu Efe OKSAR, Mina GOLKARIAN, Sevde YALDIZ

*Ege University, Faculty of Agriculture Department of Horticulture 35100 Bornova, İzmir, Turkey; fsenmacar@gmail.com (*corresponding author))*

Abstract

Transportation of sweet cherry fruits to distant markets and further marketing processes often takes approximately 2-3 weeks. The present study investigates the quality changes during this time period at three stages for three sweet cherry cultivars: 'Early Burlat', 'Napoleon', and '0900 Ziraat'. Following pre-cooling, the sweet cherries were placed in modified atmosphere packages and exposed to the following stages for the indicated durations: transportation (T) [8 days at 2 °C and 85% relative humidity (RH)]; distribution center (DC) (4 days at 6.5 °C and 80% RH), and shelf-life (SL) (2 days at 19 °C and 70% RH). Weight losses at the end of the SL stage were 3.11, 3.18, and 2.74%, respectively in 'Early Burlat', 'Napoleon' and '0900 Ziraat'. Fruit firmness decreased after SL as compared to that at other stages and was more remarkable in '0900 Ziraat'. Decreased Chroma values which indicates, the intensity or color saturation were observed in all cultivars, whereas decreased hue angle values colours expressed in degrees were observed in the 'Early Burlat' and '0900 Ziraat'. In addition, a decrease was noted in the titratable acidity of all cultivars at the end of SL. The total soluble solids, total phenolic content, and antioxidant activities were similar for all cultivars at all stages. The visual appearance scores of 'Early Burlat' cherries decreased at the end of SL, because of development of pitting on the fruit surface. The fruit quality changes were limited at T and DC stages; however, these changes became more distinctive during the SL period. It was thus concluded that the SL duration and conditions were of the highest significance with regard to maintenance of the sweet cherry fruit quality.

Keywords: antioxidants, firmness, marketing, pitting, *Prunus avium* L., transportation

Introduction

Sweet cherry is a member of the genus *Prunus* and generally grows in mild climates. It is a fleshy and non-climacteric fruit. The properties of sweet aroma, pleasant color, shape, high antioxidant activity, and nutritional content together enhance the consumer demand for sweet cherries. Accordingly, there is a rapid increase in sweet cherry production and global production sites parallel to the consumer demands. Turkey ranks first in the world's sweet cherry production (Anonymous, 2014). Higher cherry consumption plays a significant role in the prevention of diseases and maintenance of a healthy life. The health benefits of sweet cherry are linked to strong antioxidant activities (Yoo *et al.*, 2010). The antioxidant capacities of sweet cherry fruits depend on the presence of phenolic compounds such as anthocyanins and melatonin (Burkhardt *et al., 2001*; Vinson *et al., 2001*). Phenolic antioxidants have several positive effects on human health such as the anticarcinogenic and anti-inflammatory effects (Usenik *et al.*, 2008). The cultivar type and post-harvest conditions have direct impacts on antioxidant activities and phenolic contents of sweet cherries. While the total phenolic content of 'Van' and 'Tragana' cultivars increases, that of 'Burlat' does not significantly change during storage (Esti *et al.*, 2002).

Due to their high respiration rate, minimal reserve carbohydrates, and high susceptibility to mechanical damage, sweet cherries are highly perishable (Kupferman and Sanderson, 2001). Adverse conditions during transportation and marketing of sweet cherries may hinder the supply of high-quality fruits to consumers. Losses in weight, firmness, color, aroma, and acidity as well as desiccation and browning or discoloration of the green stem of the sweet cherry are some of the major quality losses (Alique *et al., 2005*; Bernalte *et al., 2003*). Susceptibility to fungal rotting, a high transpiration rate, and vulnerability to physiological disorders such as bruising and pitting aggravate the deterioration of sweet cherry fruits (Alique *et al., 2005*). Therefore, optimum conditions specific to the cultivar should be provided throughout the entire marketing chain of this fruit to protect its quality. Optimization of harvesting, handling, processing, packaging, storing, transporting, distribution, and marketing conditions play highly significant roles in quality preservation. Modified atmosphere packaging (MAP) technology has been successfully used to maintain the postharvest quality and to prolong the storage period of several fresh fruits and vegetables. MAP technology is more effective in combination with cooling and enables preservation of the fruit color and brightness and greenness of the stalks as well as reduces weight loss, fruit damage, and spoilage during transport and marketing activities (Kader and Watkins, 2000; Singh *et al.*, 2012; Wani *et al.*, 2014).

As sweet cherries are perishable fruits, relevant attention and great care should be taken while transporting them to distant markets so as to supply quality fruits to the consumers. In some cases, retail chains are not equipped to provide proper conditions to preserve the fruits (Wani *et al.*, 2014). Temperature and relative humidity (RH) are critical factors that influence the quality of sweet cherries during postharvest storage (Yaman and Bayindirli, 2002). Although these two factors are controlled during the storage and transportation processes, it is difficult to adjust them, especially to regulate the RH at the distribution centers and during sales at retailers (Wani *et al.*, 2014). Optimum storage conditions of sweet cherries were reported as 0°C temperature and 90-95% RH (Bernalte *et al.*, 1999). Despite several studies on the storage of sweet cherries, only a few articles simulated the transportation and marketing stages of the supply chain, and additionally the results were controversial perhaps due to the varietal and agro-ecological differences (Wang and Long, 2014). The present study is the first on major cherry cultivars that are exported to distant markets. The cultivar '0900 Ziraat' constitutes a significant proportion cherries traded of the worldwide (Demirtas and Sarisu, 2011).

The present study was conducted to investigate the changes in the physical, chemical, and sensory characteristics of 'Early Burlat', 'Napoleon', and '0900 Ziraat' cherry cultivars during their transportation (T), storage at the distribution centers (DC; wholesalers), and shelf-life (SL; retailers) stages.

Materials and methods

Fruit material and supply chain trial

The experiment was conducted in 2013 on fruits harvested from 13-year-old 'Early Burlat', 'Napoleon', and '0900 Ziraat' sweet cherry trees (*Prunus avium* L.) grafted onto '*Prunus mahaleb*' rootstock in Kemalpasa, Izmir Province, Turkey. 'Early Burlat', 'Napoleon', and '0900 Ziraat' sweet cherry fruits were harvested at full-ripe stage (08, 22, and 29 May, respectively) and hydro-cooled (fruit pulp temperature at 1-2 °C) the same day. Fruits of uniform size (26-28 mm diameter) that were disease-free and without other defects were selected, put into modified atmosphere packages, and the packages were sealed tightly. MAP were placed into cardboard boxes and subjected to the following stages: a) transportation (T; 8 days at 2 ± 0.5 °C and 85 ± 5% RH) that included the pre-stage at the packing house; b) distribution center (DC; 4 days at 6.5 ± 0.5 °C and 80 ± 5% RH), and c) shelf life (SL; 2 days at 19 ± 0.5 °C and 70 ± 5% RH) to simulate the real conditions faced during marketing.

Considering the actual transportation to distant markets (e.g., from Turkey to the United Kingdom) and distribution centers and the prevailing marketing conditions, the duration of exposure and ambient conditions were defined. Fruit samples were collected at the end of each stage and subjected to physical, chemical, and sensory analyses. The research was designed as a randomized block design with 5 replications and each MAP package (3 kg of cherry fruits) was considered as a single replication.

Quality attributes

The sweet cherry samples were weighted at the initial phase and at the end of the T, DC and SL stages on an electronic scale (XB 12100, 0.05 g accuracy), and weight loss was determined and expressed as the percent lost from the initial weight.

Fruit firmness was determined on 25 cherry fruit per replicate by using a penetrometer (FT 011) using a 4.0-mm diameter head and conical-shaped spear; the results were expressed in Newtons (N).

The external skin color was measured at the equatorial level on both sides of the 25 sweet cherry fruits by using a colorimeter (CR-300), and the average scores were recorded in terms of CIE-$L^*a^*b^*$ values. These values were then used to calculate

Chroma ($C^* = [a^{*2} + b^{*2}]^{1/2}$),

which indicates the intensity or color saturation and hue angle ($h° = \tan^{-1}[b^*/a^*]$) that is expressed in degrees and represents: 0° (red-purple), 90° (yellow), 180° (bluish-green), and 270° (blue) (McGuire, 1992).

Juice was extracted from 25 cherry fruits per replicate for further analysis. The total soluble solid (TSS) content of the fruit juice was determined by using a digital refractometer (PR-1) and expressed in percent. The titratable acidity (TA) was determined by titrating 10 mL of juice with 0.1 N NaOH up to pH 8.1. The results were expressed as grams malic acid per 100 mL of fruit juice, in accordance with the AOAC standards (1984).

Total phenolic content and antioxidant activity

Fruit extracts were prepared from 10 fruits according to Thaipong *et al.* (2006), with some modifications for total phenolic content and antioxidant activity (in methanol extract) analysis. Total phenolic content was determined by the Folin–Ciocalteu method according to the method of Swain and Hillis (1959), with an incubation time of 120 minutes for color development. The absorbance was measured at 725 nm using a spectrophotometer and the results were expressed as milligram gallic acid equivalent (GAE) per 100 gram of fresh weight (FW) with reference to a gallic acid (0-0.1 mg/mL) standard curve.

The ferric reducing ability of plasma (FRAP) assay was performed as previously described by Benzie and Strain (1996), where reductants ("antioxidants") in the sample reduce Fe (III)/tripyridyltriazine complex to a blue ferrous form, with an increase in the absorbance at 593 nm. The final results are expressed in μmol trolox equivalents (TE)/g FW, with reference to a trolox (25–500 μmol) standard curve.

Decay development

Decay development was examined on 3 kg fruit sample and their rates were expressed in percent.

Sensory evaluation

The sensory quality evaluation of appearance was conducted by an experienced six-member panel on a nine-point hedonic scale: appearance (9: displays the complete characteristics of the sweet cherry appearance at harvest; 7: good appearance; 5: acceptable appearance, slight pitting; 3: poor appearance, moderate pitting; 1: high pitting, decay and/or stem desiccation). The procedures for sensory evaluation of horticultural crops described by Heintz and Kader (1983) were utilized by the panelists.

Statistical analysis

All data were subjected to analyses of variance (ANOVA) by using the IBM® SPSS® Statistics 19 statistical software. Significant differences between the means for each sweet cherry cultivars were determined by Duncan's multiple range tests at $p<0.05$. Standard deviation of the mean (SD) was also calculated from the replicates.

Results and discussion

Weight loss and fruit firmness values of cherry cultivars during the transport and marketing stages are presented in Fig. 1. The increases observed in the weight loss of cherry cultivars during T (8 days at 2 °C), DC (4 days at 6.5 °C), and SL (2 days at 19 °C) stages were found to be significant ($p<0.01$). Such increases became especially distinctive after the SL stage. At the end of all stages, the weight loss in 'Early Burlat', 'Napoleon', and '0900 Ziraat' cherry cultivars were observed as 3.11, 3.18, and 2.74%, respectively (Fig. 1a). Lower weight loss in the T and DC stages were mainly because of the use of MAP packages (Kucukbasmaci *et al.*, 2008; Singh *et al.*, 2012; Wani *et al.*, 2014) and the limited moisture loss due to ambient conditions (temperature and RH). Weight loss during the SL stage was higher than that at the T+DC stages when the MAP packages were open, ambient temperature was high (19 °C), and RH was low (70%) during SL for 2 days. Similarly, the weight loss observed in the '0900 Ziraat' cherry cultivar during 6 weeks of storage was lower than that in 2 days of SL at 20 °C (Onursal *et al.*, 2013). Weight loss of the '0900 Ziraat' cherry cultivar at the end of SL stage was 13% lower than that of the other cultivars. Variations in weight loss of the different cherry cultivars in this study comply with those in previous reports (Diaz-Mula *et al.*, 2010), irrespective of the difference in the skin characteristics and composition of the fruits.

Fruit firmness is directly related to storage potential, mechanical resistance, and rot development (Esti *et al.*, 2002). The decreases observed in firmness of all three cherry cultivars during the transport and marketing stages were significant ($p<0.01$). Fruit firmness of 'Early Burlat', 'Napoleon' and '0900 Ziraat' cultivars were observed to be 9.9, 10.2, and 11.9 N, respectively, at harvest (Fig. 1b). While fruit firmness of 'Early Burlat' and '0900 Ziraat' after DC were similar to that after T, there was a decrease in fruit firmness of 'Napoleon'. Fruit firmness of the cherry cultivars exhibited a decrease after SL as compared to that in the other stages. At the end of SL and compared to the initial firmness values, there were 19.8, 21.2, and 16.5% decreases, respectively, in 'Early Burlat', 'Napoleon', and '0900 Ziraat' cherry fruit. Fruits began to soften following rapid loss of moisture from the surface (Wani *et al.*, 2014). The decreases in firmness during the transportation and marketing processes also resulted from moisture loss. Meheriuk *et al.* (1995) reported significant losses in firmness of 'Lapins' cherry cultivar during storage. Similarly, decreasing firmness values were also reported for other cherry cultivars (Kappel *et al.*, 2002; Clayton *et al.*, 2003; Onursal *et al.*, 2013; Wang and Long, 2014). At all stages, fruit firmness of '0900 Ziraat' was about 16.5% higher than that of the other cultivars. Differences in fruit firmness among cultivars indicate the differences in their cell structures, skin characteristics, compositions and/or respiration rates (Karacali, 2012).

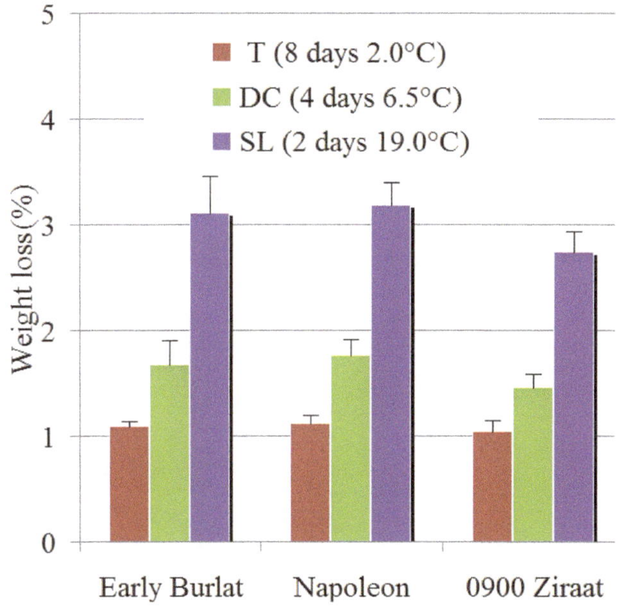

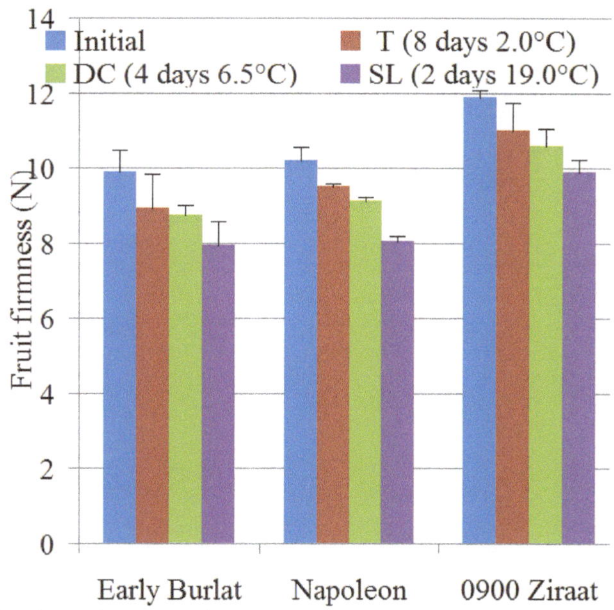

Fig. 1. Postharvest weight loss (a-left) and flesh firmness (b-right) values of 'Early Burlat', 'Napoleon', and '0900 Ziraat' cherry cultivars at three different stages of the supply chain

Tab. 1. Postharvest TSS and TA content of 'Early Burlat', 'Napoleon', and '0900 Ziraat' cherry cultivars at three different stages of the supply chain

Sampling stage	TSS content (%)			TA content (g malic acid/100 mL)		
	'Early Burlat'	'Napoleon'	'0900 Ziraat'	'Early Burlat'	'Napoleon'	'0900 Ziraat'
Initial	14.97^{NS}	14.97^{NS}	16.50^{NS}	0.91 a^{z**}	0.88 a^{**}	0.99 a^{**}
T (8 days, 2.0°C)	15.17	15.06	16.69	0.86 ab	0.86 a	0.92 b
DC (4 days, 6.5° C)	15.23	15.23	16.97	0.79 bc	0.86 a	0.93 b
SL (2 days, 19.0°C)	15.17	15.27	17.13	0.72 c	0.79 b	0.88 c

NS,*,**, Nonsignificance or significance at $p < 0.05$ or 0.01, respectively
[z]Means for each experiment were separated by different letters within columns by Duncan's multiple-range test, p<0.05

The changes in C^* and h values of different cherry cultivars during the transportation and marketing phases are presented in Fig. 2. The decreases in C^* values of the 'Early Burlat' cultivar at T and SL stages and decrease in C^* values of '0900 Ziraat' at the end of SL were statistically significant (p<0.01). The decrease in C^* values of 'Early Burlat', 'Napoleon', and '0900 Ziraat' cherry fruit during the SL stage were calculated as 15.7, 13.6, and 7.8%, respectively. The changes in $h°$ values of 'Early Burlat' and '0900 Ziraat' cultivars were similar to the changes observed in C^* values and the decreases at the end of SL were found to be significant (p<0.05). The changes in $h°$ values of 'Napoleon' cherries were highly limited and varied between 14.53 and 15.75%. The changes in fruit color parameters during the transportation and marketing phases are proportional to the ripening of the fruits. Decreased brightness values were also found in 'Lapins' cherry cultivar during the storage period (Meheriuk *et al.*, 1995). Distinctive changes were not observed in color values of cherries since the fruits were harvested at a fully ripe stage. Color darkening and increased anthocyanin contents were reported for cherries harvested at different ripening stages during storage and shelf life, based on the cultivar and ripening stage (Serrano *et al.*, 2009).

Changes in TSS and TA contents of the cherries during the transportation and marketing stages are given in Tab. 1. It has been reported that TSS and TA contents are related to the intensity of sweet cherry flavor and that consumer acceptability increases with high TSS and TA levels in the fruit (Crisosto *et al., 2003*; Kalyoncu *et al.*, 2009). TSS contents of 'Early Burlat', 'Napoleon', and '0900 Ziraat' cherry cultivars during transportation and marketing stages exhibited similar tendencies and they were 15.0-15.2, 15.0-15.3, and 16.5-17.1%, respectively (Tab. 1). Similar to the findings of the present study, significant changes were not observed in the TSS contents of cherries during the MAP storage of fruits (Meheriuk *et al.*, 1995; Meheriuk *et al.*, 1997; Remon *et al.*, 2000; Onursal *et al.*, 2012). The higher TSS contents of '0900 Ziraat' cultivar was mainly due to the cultivar type. Girard and Kopp (1998) reported that the TSS contents of 12 different cherry cultivars ranged between 13.5 and 24.5%. Several other studies also reported significantly different TSS contents for different sweet cherry cultivars (Kappel *et al.*, 2002; Diaz-Mula *et al.*, 2010).

The decreases in TA values of cherries before the SL stage were significant (p<0.01). Such decreases were especially remarkable in the 'Early Burlat' variety (21%). The post-harvest decreases in TA values are proportional to the ripening of the fruits. Similar decreases in TA values were also reported during the storage of '0900 Ziraat' (Kucukbasmaci *et al.*, 2008; Onursal *et al.*, 2013), 'Napoleon' (Esturk *et al.*, 2014), 'Burlat' (Remon *et al.*, 2000), 'Sciazza', 'Ferrovia' (Esti *et al.*, 2002), and 'Lapins' (Meheriuk *et al.*, 1995) cultivars. These decreases are closely related to the ripening level at harvest, storage conditions, and durations (Wani *et al.*, 2014). As sweet cherry fruit is a

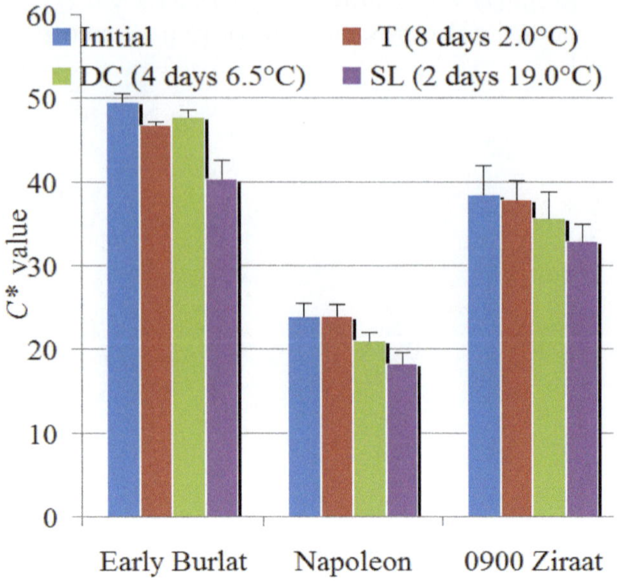

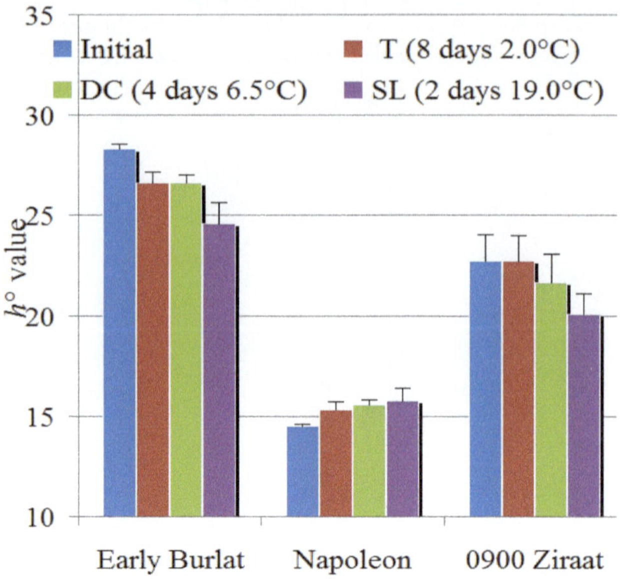

Fig. 2. Postharvest C^* (a-left) and $h°$ values (b-right) of 'Early Burlat', 'Napoleon', and '0900 Ziraat' cherry cultivars at three different stages of the supply chain

Tab. 2. Post-harvest total phenolic content and antioxidant activity of 'Early Burlat', 'Napoleon', and '0900 Ziraat' cherry cultivars at three different stages of the supply chain

Sampling stage	Total phenolic content (mg GAE/100 g FW)			Antioxidant activity (μmol TE/g FW)		
	'Early Burlat'	'Napoleon'	'0900 Ziraat'	'Early Burlat'	'Napoleon'	'0900 Ziraat'
Initial	101.3^{NS}	103.5^{NS}	94.6^{NS}	9.22^{NS}	13.32^{NS}	7.93^{NS}
T (8 days, 2.0°C)	95.7	106.5	95.0	8.38	13.27	7.49
DC (4 days, 6.5°C)	99.5	106.8	94.5	9.42	14.10	7.18
SL (2 days, 19.0°C)	102.2	102.6	88.0	9.21	12.25	6.75

NS, Nonsignificant

non-climacteric fruit, loss in TA tends to occur simultaneously with metabolic activities.

Total phenolic content and antioxidant activity of cherry fruit at harvest and at T, DC, and SL stages are presented in Tab. 2. Significant differences were observed in the total phenolic content and antioxidant activity of the tested cherry cultivars during the transportation and marketing stages. The average total phenolic contents of 'Early Burlat', 'Napoleon', and '0900 Ziraat' cultivars during the transport and marketing processes were 99.7, 104.9, and 93.0 mg GAE/100 g FW, and the antioxidant activities were 9.06, 13.24, and 7.34 μmol TE/g FW, respectively. Harvest at fully ripe stage resulted in limited changes in the total phenolic content and antioxidant activity. Serrano *et al.* (2009) reported increased post-harvest total phenolic and antioxidant activity values for cherries based on the ripening levels. Functional parameters reached the maximum levels with 2-day delay in harvest (Diaz-Mula *et al.*, 2010). Total phenolic content and especially the antioxidant activity of 'Napoleon' were significantly higher than those of '0900 Ziraat' cultivar. Previous studies also reported significantly different results for different sweet cherry cultivars (Usenik *et al.*, 2008; Serrano *et al.*, 2009; Diaz-Mula *et al.*, 2010; Kelebek and Selli, 2011).

Decay development was not observed in 'Early Burlat', 'Napoleon' or '0900 Ziraat' cherry fruit during the transportation and marketing stages. However, storage in MAP, climatic conditions during the growth, cultural practices, and disease and pest management practices had significant impacts on rot development.

Visual appearance scores of cherry fruits and stems at the end of T and DC stages were similar to the initial scores. The difference in the decrease in the visual appearance of 'Napoleon' and '0900 Ziraat' cherry fruit at the end of SL stage were not significant. On the other hand, such decreases were found to be significant ($p<0.01$) in 'Early Burlat' (Tab. 3) mainly because of development of pitting over the fruit skin at the end of SL stage. Stems of all cultivars preserved their green color and brightness during the T and DC stages because limited water loss occurred from the fruits and stems during these stages. Storage in MAP was effective in limiting water loss from the stems (Kucukbasmaci *et al.*, 2008; Singh *et al.*, 2012; Wani *et al.*, 2014). Very slight browning or discoloration of the green stem was observed at the end of SL stage.

Tab. 3. Postharvest visual appearance of 'Early Burlat', 'Napoleon', and '0900 Ziraat' cherry cultivars at three different stages of the supply chain

Sampling stage	'Early Burlat'	'Napoleon'	'0900 Ziraat'
Initial	9.0 a^{z**}	9.0^{NS}	9.0^{NS}
T (8 days,2.0°C)	9.0 a	9.0	9.0
DC (4 days, 6.5°C)	8.2 a	8.6	9.0
SL (2 days, 19.0°C)	5.8 b	7.8	8.2

NS,**, Nonsignificance or significance at $p<0.01$.
[z]Means for each experiment were separated within columns by Duncan's multiple-range test, $p<0.05$

Conclusion

Changes in the quality parameters of cherry cultivars were more remarkable during the 2-day SL period as compared to the 8-day T period or the 4-day DC period. While '0900 Ziraat' and 'Napoleon' cultivars showed preserved quality at the end of SL stages, some 'Early Burlat' fruits showed visible spoilage. With regard to the investigated quality parameters, '0900 Ziraat' cherry cultivar was found to be superior over the others. The present results clearly indicate the significance of the SL conditions and/or durations for preservation of the sweet cherry quality.

Acknowledgement

The authors wish to extend their thanks to Dr. Belit Balci from Alara, Turkey for his valuable inputs and comments.

References

Alique R, Zamorano JP, Martinez MA, Alonso J (2005). Effect of heat and cold treatments on respiratory metabolism and shelf-life of sweet cherry, type picota cv. 'Ambrunes'. Postharvest Biol Tec 35:153-165.

Anonymous (2014). Food and Agriculture Organization of the United Nations. Available at: http://faostat.fao.org (Accessed March 16, 2014).

Kucukbasmaci F, Ozkaya O, Agar T, Saks Y (2008). Effect of retail-size modified atmosphere packaging bags on postharvest storage and shelf-life quality of '0900 Ziraat' sweet cherry. Acta Hortic 795:775-780.

Benzie IEF, Strain JJ (1996). The ferric reducing ability of plasma (FRAP) as a measure of "antioxidant power": the FRAP assay. Anal Biochem 239:70-76.

Bernalte MJ, Hernandez MT, Vidal-Aragon MC, Sabio E (1999). Physical, chemical, flavor and sensory characteristics of two sweet cherry varieties grown in 'Valle del Jerte' (Spain). J Food Quality 22:403-416.

Bernalte MJ, Sabio E, Hernandez MT, Gervasini C (2003).

Influence of storage delay of Van sweet cherry. Postharvest Biol Tec 28:303-312.

Burkhardt S, Tan DX, Manchester LC, Hardeland R, Reiter RJ (2001). Detection and quantification of the antioxidant melatonin in Montmorency and Balaton tart cherries (*Prunus cerasus*). J Agr Food Chem 49:4898-4902.

Crisosto CH, Crisosto GM, Metheney P (2003). Consumer acceptance of Brooks and Bing cherries is mainly dependent on fruit SSC and visual skin colour. Postharvest Biol Tec 28:159-167.

Clayton M, Biasi WV, Agar IT, Southwick SM, Mitcham EJ (2003). Postharvest quality of 'Bing' cherries following preharvest treatment with hydrogen cyanamide, calcium ammonium nitrate, or gibberellic acid. J Hortic Sci 38:407-411.

Demirtas I, Sarisu HC (2011). Cherry Cultivation. Ministry of Food, Agriculture and Livestock Fruit Research Station, Isparta, Turkey 11:1-12.

Diaz-Mula HM, Castillo S, Martinez-Romero D, Valero D, Zapata PJ, Guillen F, Serrano M (2010). Sensory, nutritive and functional properties of sweet cherry as affected by cultivar and ripening stage. Food Sci Technol Int doi:10.1177/1082013209351868.

Esti M, Cinquanta L, Sinesio F, Moneta E, Mateo MD (2002). Physicochemical and sensory fruit characteristics of two sweet cherry cultivars after cool storage. Food Chem 76:399-405.

Esturk O, Ayhan Z and Ustunel MA (2014). Modified atmosphere packaging of 'Napoleon' cherry: effect of packaging material and storage time on physical, chemical, and sensory quality. Food Bioprocess Tech 5:1295-1304.

Girard B, Kopp TG (1998). Physicochemical characteristics of selected sweet cherry cultivars. J Agr Food Chem 46:471-476.

Heintz CM, Kader AA (1983). Procedures for the sensory evaluation of horticultural crops. J Hortic Sci 18:18-22.

Kader AA, Watkins CB (2000). Modified atmosphere packaging-toward 2000 and beyond. HortTechnology 10(3):483-486.

Kalyoncu IH, Ersoy N, Yilmaz M (2009). Some physicochemical properties and mineral contents of sweet cherry (*Prunus avium* L.) type grown in Konya. Afr J Biotechnol 8:2744-2749.

Kappel F, Toivonen P, McKenzie DL, Stan S (2002). Storage characteristics of new sweet cherry cultivars. J Hortic Sci 37:139-143.

Karacali I (2012). Storage and marketing of horticultural products. Ege University Agricultural Faculty Publication No: 494 İzmir, Turkey.

Kelebek H, Selli S (2011). Evaluation of chemical constituents and antioxidant activity of sweet cherry (*Prunus avium* L.) cultivars. Int J Food Sci Tech 46:2530-2537.

Kupferman G, Sanderson P (2001). Temperature management and modified atmosphere packing to preserve sweet cherry fruit quality. Acta Hort 667:523-528.

McGuire RG (1992). Reporting of objective color measurements. J Hortic Sci 27:1254-1255.

Meheriuk M, Girard B, Moyls L, Beveridge HJT, McKenzie DL, Harrison J, Weintraub S, Hocking R (1995). Modified atmosphere packaging of 'Lapins' sweet cherry. Food Res Int 28(3):239-244.

Meheriuk M, McKenzie B, Girard B, Molys AL, Weintraub S, Hocking R, Kopp T (1997). Storage of 'Sweetheart' cherries in sealed plastic film. J Food Quality 20:189-198.

Onursal CE, Calhan O, Eren I, Cetinbas M, Butar S, Demirtas I (2013). Effects of preharvest aminoethoxyvinylglycine (AVG) treatments on cold storage and shelf life quality of '0900 Ziraat' sweet cherry cultivar. J Agr Sci 6(1):91-96.

Ferrer A, Marquina P, Burgos J, Oria R (2000). Use of modified atmospheres to prolong the postharvest life of Burlat cherries at two different degrees of ripeness. J Sci Food Agr 80:1545-1552.

Serrano M, Diaz-Mula HM, Zapata PJ, Castillo S, Guillein F, Martinez-Romero D (2009). Maturity stage at harvest determines the fruit quality and antioxidant potential after storage of sweet cherry cultivars. J Agr Food Chem 57:3240-3246.

Singh P, Wani AA, Goyal GK (2012). Shelf life extension of fresh ready-to-bake pizza by the application of modified atmosphere packaging. Food Bioprocess Tech 5(3):1028-1037.

Swain T, Hillis WE (1959). The phenolic constituents of *Prunus domestica* I. The quantitative analysis of phenolic constituents. J Sci Food Agr 10:63-68.

Thaipong K, Boonprakob U, Crosby K, Cisneros-Zevallos L, Byrne DH (2006). Comparison of ABTS, DPPH, FRAP, and ORAC assays for estimating antioxidant activity from guava fruit extracts. J Food Compos Anal 19:669-675.

Usenik V, Fabcic J, Stampar F (2008). Sugars, organic acids, phenolic composition and antioxidant activity of sweet cherry (*Prunus avium* L.). Food Chem 107:185-192.

Vinson JA, Su X, Zubik L, Bose P (2001). Phenol antioxidants quantity and quality in foods: Fruits. J Agr Food Chem 49:5315-5321.

Wang Y, Long LE (2014). Respiration and quality responses of sweet cherry to different atmospheres during cold storage and shipping. Postharvest Biol Tec 92:62-69.

Wani AA, Singh P, Guld K, Wani MH, Langowski HC (2014). Sweet cherry (*Prunus avium*): Critical factors affecting the composition and shelf life. Food Packaging Shelf Life 1(1):86-99.

Yaman O, Bayindirli L (2002). Effects of an edible coating and cold storage on shelf-life and quality of cherries. LWT-Food Sci Technol 35:146-150.

Yoo KM, Al-Farsi M, Lee H, Yoon H, Lee CY (2010). Antiproliferative effects of cherry juice and wine in Chinese hamster lung fibroblast cells and their phenolic constituents and antioxidants activities. Food Chem 123:734-740.

20

Nutritional and Nutraceutical Components of Commercial Eggplant Types Grown in Sinaloa, Mexico

Guillermo NIÑO-MEDINA[1,2], Dolores MUY-RANGEL[1*], Alfonso GARDEA-BÉJAR[3], Gustavo GONZÁLEZ-AGUILAR[3], Basilio HEREDIA[1], Manuel BÁEZ-SAÑUDO[1], Jorge SILLER-CEPEDA[1], Rosabel VÉLEZ DE LA ROCHA[1]

[1]*Centro de Investigación en Alimentación y Desarrollo (CIAD) A.C., Unidad Culiacán, Carretera Culiacán a El Dorado, km 5.5, C.P. 80110, Culiacán, Sinaloa, Mexico; mdmuy@ciad.mx (*corresponding author); jbheredia@ciad.mx; manuel.baez@ciad.edu.mx; jhsiller@ciad.mx; rosabel@ciad.edu.mx*

[2]*Universidad Autónoma de Nuevo León, Facultad de Agronomía, Francisco Villa S/N, C.P. 66050, Col. Ex-Hacienda El Canadá, Escobedo, Nuevo León, Mexico (present address); guillermo.ninomd@uanl.edu.mx*

[3]*Centro de Investigación en Alimentación y Desarrollo, (CIAD) A.C., Unidad Hermosillo, Carretera a La Victoria km 0.6, C.P. 83304, Hermosillo, Sonora, Mexico; gardea@ciad.mx; gustavo@ciad.mx*

Abstract

The nutrimental composition and main nutraceutical components were determined in Chinese, Philippine, American, Hindu and Thai eggplant (*Solanum melongena* L.) types grown in Sinaloa, Mexico. Thai type showed the highest amount of protein (0.90%), crude (1.54%) and dietary (3.93%) fibre and the highest concentrations of total soluble phenolics (2049 mgCAE $100g^{-1}$) and chlorogenic acid (1700 mg $100g^{-1}$). Hindu type obtained the highest content of the minerals potassium (191.19 mg $100g^{-1}$), calcium (59.63 mg $100g^{-1}$), phosphorus (33.52 mg $100g^{-1}$), magnesium (28.96 mg $100g^{-1}$), manganese (0.44 mg $100g^{-1}$), zinc (0.78 mg $100g^{-1}$), in addition this eggplant type showed the highest levels of ascorbic acid (22 mg $100g^{-1}$). Philippine type showed the highest concentration in anthocyanins (161.10 mgC_3GE $100g^{-1}$) and the highest levels of radical scavenging activity in DPPH (92.50% of inhibition) and ORAC (538.90 µmolTE g^{-1}). Nutrient components with functional properties like biologically essential minerals, dietary fibre, ascorbic acid (vitamin C) and soluble phenols were found in higher amounts in the five eggplant types analyzed, as compared with reports from other parts in the world available in the literature.

Keywords: antioxidant capacity, ascorbic acid (Vitamin C), chlorogenic acid, mineral content, proximate composition, *Solanum melongena* L., total soluble phenols

Introduction

The eggplant (*Solanum melongena* L.) also known as aubergine, guinea squash or brinjal, is an economically important vegetable crop in the tropics and subtropics. Eggplant cultivars produce fruits with a wide diversity of shapes, sizes and colours (Kashyap *et al.*, 2003; Kantharajah and Golegaonkar, 2004). Sinaloa, in northwest Mexico, is the major eggplant producer in the country, growing an average of eight eggplant types, out of which American type is the most widely grown. Eggplants have an important nutritional value due to its composition, which includes minerals like potassium, calcium, sodium and iron (Mohamed *et al.*, 2003; Raigón *et al.*, 2008) as well as dietary fibre (USDA, 2014; Sanchez-Castillo *et al.*, 1999). Fruits and vegetables are the main dietary sources of phenolic compounds for humans, with phenolic acids and flavonoids being the most abundant (Scalbert and Williamson, 2000; Manach *et al.*, 2004). Eggplant fruits have shown high hydrophilic oxygen radical absorbance capacity (ORAC) (Cao *et al.*, 1996), which has been correlated to phenolic compounds presence, including delphinidin as a major component in peel (Wu *et al.*, 2006; Koponen *et al.*, 2007) and chlorogenic acid in flesh (Winter and Hermann, 1986; Whitaker and Stommel, 2003). Other studies of the main phenolics reported in eggplants, showed that they are well metabolized and absorbed (Olthof *et al.*, 2001; Olthof *et al.*, 2003), also they show a good antioxidant capacity (Noda *et al.*, 2000; Hanson *et al.*, 2006). The aim of this work was to characterise and compare the proximal composition, mineral content and antioxidant bioactive compounds of eggplant types grown in Sinaloa, Mexico.

Materials and methods

Sample preparation

Eggplant fruits from the representative types of Chinese, Philippine, American Hindu, and Thai were used in this work (Fig. 1). Fruit smples of each eggplant type were harvested at commercial ripeness stage from a local farm (24°40'50" N and 107°30'18" W), washed and sorted by colour and size and divided into two groups. The first group was used fresh for proximate and mineral analysis, while the second group was freeze dried and stored at -80 °C, for further evaluations of ascorbic acid, chlorogenic acid and total phenolics.

Fig. 1. Phenotypic differences among eggplant types (from left to right: Chinese, Philippine, American, Hindu, and Thai)

Proximate composition

Moisture, ash, protein (N x 6.25), crude fibre, fat and carbohydrates (by difference) contents were determined according to AOAC methods (AOAC, 1998). Total dietary fibre was evaluated by the Megazyme total dietary fibre assay procedure, which is based on the AOAC 991.43 method (AOAC, 1998).

Mineral content

Mineral analysis was done based on AOAC methods (AOAC, 1998). Potassium (K), sodium (Na), calcium (Ca), magnesium (Mg), iron (Fe), zinc (Zn), copper (Cu) and manganese (Mn) were determined by atomic absorption spectroscopy using a Varian SpectrAA 220. Phosphorus (P) was determined colorimetrically at 650 nm in a Varian Cary E1 UV-Vis spectrophotometer.

Ascorbic acid (Vitamin C)

Ascorbic acid extraction was done according to Özgür and Sungur (1995) with the following modifications; 100 mg of freeze dried sample were mixed with 5 mL of extraction solution at 5 °C (3% metaphosphoric acid, 8% acetic acid, 1mM EDTA). The extract was filtered through a Whatman No. 1 filter paper and passed through 45 µm nylon syringe filter. Ascorbic acid was quantified according to Doner and Hicks (1981) using a Varian Pro Star 330 photodiode array detector HPLC system and a Varian Bondesil NH_2 column (250 x 4.6 mm). A 3:1 solution of acetonitrile and potassium phosphate (0.05 M) was used as the mobile phase at 1.5 mL min^{-1}. Detection was measured at 268 nm with a retention time at 2.1 min. Standard solutions of 0, 0.05, 0.10, 0.15 and 0.20 mg mL^{-1} of ascorbic acid were prepared for calibration. Results were expressed in milligram per 100 gram of eggplant on a fresh basis.

Soluble phenolic compounds

Phenolics extract was obtained according to Whitaker and Stommel (2003) with these modifications; 200 mg of freeze dried sample were mixed with 10 mL of methanol containing 0.5% butylated hydroxytoluene (BHT) and sonicated at room temperature for 15 min in an ultrasonic bath (Fisher Scientific Model FS60). After centrifugation at 16,000 *g*, supernatant was recovered and re-extracted twice with 10 mL of solvent. Extracts were combined, filtered through Whatman No. 4 paper and passed through 45 µm PTFE syringe filter.

Total soluble phenolics

Total soluble phenolics were determined using 2 mL aliquots of phenolics extract; the solvent was evaporated at 35 °C by flushing dry N_2 and the residue dissolved in 10 mL of methanol. Total phenols determination was done according to spectrophotometric method based on the colorimetric reaction of the Folin-Ciocalteu reagent (Chun and Kim 2004). Absorbance was measured at 750 nm in a Varian Cary E1 UV-Vis spectrophotometer (Varian, USA). Results were expressed in mg chlorogenic acid equivalents per 100 grams (mg CAE $100g^{-1}$) of eggplant tissue in a dry basis, using a calibration curve with a standard solution of 0, 0.25, 0.50, 0.75 and 1.00 mg mL^{-1} of chlorogenic acid.

Chlorogenic acid

For chlorogenic acid determination, a 1.5 mL aliquot of phenolics extract was taken, solvent was evaporated at 35 °C by flushing dry N_2 and dissolved in 2mM phosphoric acid in methanol-water (1:1); HPLC analysis was done using binary mobile phase gradient of methanol in 1mM aqueous phosphoric acid and a Phenomenex Luna C_{18} column (250 x 4.6 mm) in a Varian Pro Star 330 photodiode array detector HPLC system (Whitaker and Stommel, 2003). Detection was done at 325 nm with a retention time in 19.7 min. Standard solutions of 0, 0.25, 0.50, 0.75 and 1.00 mg mL^{-1} of chlorogenic acid were prepared for calibration curve. Results were expressed in milligram per 100 gram of eggplant tissue on a dry basis. The percentage of chlorogenic acid in comparison with total phenols was calculated, as well.

Total anthocyanins

Total anthocyanin was determined using a spectrophotometric method adapted from Abdel-Aal and Hucl (1999). An eggplant sample of 0.5 g was homogenized with 10 mL of chilled, acidified methanol (95% methanol and 1 N HCl 85:15, v/v). The tube was flushed with nitrogen gas, agitated for 30 min, and centrifuged at 3000 *g* for 10 min, and the supernatant was collected. Solution absorbance was measured immediately at 535 and 700 nm. Anthocyanin content was calculated as the concentration of total anthocyanin, expressed as mg of cyanidin-3-glucoside equivalents per 100 g of sample (mgC$_3$GE 100g^{-1}). The molar absorptivity and molecular weight of cyanidin-3-glucoside are 25965/cm/M and 449.2 g per mole, respectively.

DPPH antioxidant capacity

Eggplant samples (5 g) were homogenized with 20 mL methanol, until reaching uniform consistency and incubated overnight at 4 °C (Heredia and Cisneros-Zeballos, 2009). Homogenates were centrifuged at 29,000 *g* for 15 min at 4 °C. Sample aliquots of 150 μL were taken from the clear supernatants and then diluted with a 2,850 μL DPPH solution, previously prepared with methanol, until reaching 1.1 units of absorbance at 515 nm. Methanol was used as control and to zero the spectrophotometer (Agilent Technologies, Mexico). Further readings of all samples were done at 515 nm. The mixture reaction was allowed to react in a shaker until no significant decrease in absorbance was obtained, compared to the methanol-based control for antioxidant activity. The decrease in absorbance was obtained and the results for antioxidant activity were expressed as % radical scavenging activity, calculated as follows: % Radical scavenging activity = (absorbance$_{control}$-absorbance$_{sample}$/absorbance$_{control}$)*100. Percentage of radical scavenging activity was plotted against the corresponding concentration of the extract to obtain the IC$_{50}$ value, which is defined as the amount of antioxidant material required to scavenge 50% of free radical in the assay system. IC$_{50}$ values are inversely proportional to antioxidant activity.

ORAC antioxidant capacity

The eggplant methanolic extract sample was also used for ORAC measurements against peroxyl radicals (Huang *et al.*, 2002; Prior *et al.*, 2003). Sample extracts, Trolox standard (40 μM) and reagents were dissolved in 75 mM potassium phosphate buffer, pH 7.4, and assayed on a Bio-Tek Synergy HT plate reader with automatic using 96-well plates. The fluorescein stock solution (FLs) was prepared with 0.1125 g of fluorescein (FL) in 50 mL of the buffer. Next, 100 μL of FLs was diluted in 10 mL of the buffer (FL2). A third FL solution (FL3) was prepared by taking 400 μL of FL2 to 25 mL. The assay was performed at 37 °C. Samples were loaded into the 96-well plate (25 μL), with each row containing different samples or dilutions. Two columns were loaded with 25 μL of Trolox and the buffer (blank), respectively. At 15 min, the AAPH peroxyl radical solution (2,2′-azobis 2-amidinopropane) dihydrochloride was prepared right with the incubated buffer. The FL3 and AAPH solutions were then transferred into the incubated autoinjector plastic containers. The plate reader was configured to inject 200 μL of FL3, the shaking plate was set at medium intensity for 3 s, injecting 75 μL of AAPH solution, and taking readings for 50 min every 1:27 min (35 cycles) without shaking during readings. The reading mode was fluorescence kinetic with injection (Ex/Em: 485/528). Data were recorded with the GEN-5 software and exported into Excel. Relative fluorescence (fi = Fi/F1, where Fi is fluorescence at cycle i) was calculated for all wells and kinetic readings, and area under the curve (AUC) were calculated for the 35 cycles as AUC = (f1 + f35)/2 + (f2 + f3...+ f33 + f34). The net area under the curve (NAUC) was calculated as AUC sample - AUC blank. ORAC value was then calculated as (NAUC sample/NAUC Trolox) x 40 and expressed as μmoles Trolox equivalents per gram of sample (μmolTE g^{-1}).

Statistical analysis

ANOVA was used to assess statistical differences among eggplant types with a 5% confidence's level. When significance difference was found, Tukey's multiple range tests were carried out to separate means using MINITAB 14.0 (MINITAB, 2004). Data were expressed as means values of three samples ± standard deviation.

Results and discussion

Proximate composition

In spite of its small fruits, Thai type showed the highest values in crude fibre, ashes and carbohydrates. Although Thai also showed the highest protein content it was only significantly higher than Chinese and as far as dietary fibre only Hindu had values close to Thai (Tab. 1). All eggplants had moisture content greater than 90%, protein content lower than 1% and low fat with 0.04% (Tab. 1). Similar data were reported in different cultivars, with values ranging from 91.8 to 94.2% for moisture, 0.11 to 1.2% for protein and 0.3% to 0.7% for ash (on fresh weight basis) (Flick *et al.*, 1978; Muñoz de Chavez *et al.*, 1996; Maroto, 2002). Vegetables are characterized for the high content of fibre

Tab. 1. Proximate composition of eggplant types grown in Sinaloa, Mexico (g 100g^{-1} fresh weight)

Component	Chinese	Philippine	American	Hindu	Thai
Water	*92.50±0.11^a	92.60±0.09^a	92.70±0.08^a	92.60± 0.04^a	90.10±0.25^b
Crude fibre	0.73±0.01^b	0.78±0.01^b	0.78±0.08^b	0.65±0.02^c	1.54±0.01^a
Ash	0.69±0.05^b	0.72±0.01^b	0.36±0.08^c	0.48±0.02^c	1.04±0.01^a
Protein	0.65 ±0.06^b	0.69±0.09^{ab}	0.67±0.13^{ab}	0.75±0.05^{ab}	0.90±0.07^a
Fat	0.04±0.01^a	0.03±0.00^a	0.04±0.01^a	0.03±0.01^a	0.04±0.01^a
Carbohydrates$^+$	6.12±0.11^{bc}	5.96±0.06^c	6.23±0.27^{bc}	6.14±0.08^b	7.92±0.37^a
Dietary fibre	2.76±0.14^b	1.61 ±0.15^c	2.63±0.18^b	3.59±0.08^a	3.93±0.15^a

Mean values in a row with different letters are significantly different at P≤0.05. *Mean values of three samples±SD
$^+$Carbohydrates (%) =100-(%water+%crude fibre+%ash%+protein+%fat).

and are considered as a primary source for humans. The highest content (>3.5%) in dietary fibre found in Thai and Hindu was significantly different than the rest of the eggplant types. Sánchez-Castillo *et al.* (1999), analyzed the dietary fibre content in different vegetables grown in Mexico reporting a 2.4% values for eggplants, while a study carried out in Chile by Pak (2000), found 1.3% in cooked eggplants without skin and seeds; Takeyama *et al.* (2002) reported a content of around 2.4% in a Japanese cultivar, and Rehman *et al.* (2003) found a value of 1.5% of dietary fibre in eggplants cultivated in Pakistan. An average intake of 25 g day^{-1} of dietary fibre in adults is recommended by the American Dietetic Association (based on a 2000 kcal day^{-1} diet) (Marlett *et al.*, 2002). Based on this value, an intake of 100 g day^{-1} of the eggplant types analyzed in this study could account for 6.5 to 16% of the recommended daily intake (RDI).

Mineral content

Tab. 2 provided results of mineral composition for the five eggplant types. Hindu had the highest values for K, Ca, and Zn; along with American it has the highest concentrations of P, Mg and Mn. Philippine reported the highest Fe content, and no differences among the five types were found for Cu concentration. Although Na content varied widely, it was statistically the same for Hindu, American and Chinese. Ekholm *et al.* (2007) found a similar pattern and mineral content in eggplants cultivated in Finland, as compared to our work, where K showed the highest concentration (175 mg $100g^{-1}$) and Cu showed the lowest result (0.03 $mg100g^{-1}$). On the other hand, a study carried out in eggplants grown in Saudi Arabia by Mohamed *et al.* (2003) showed a different mineral concentration (mg $100g^{-1}$) reporting 274 for Ca, 222 for Na and 212 for K. Those values are higher than those found in our study. Eggplants are a good source of biologically essential minerals like K, Ca, Mg, P, Na and Fe; their contents are similar to those found in tomatoes and higher than those reported in carrots, potatoes and onions, which are the most common vegetables consumed in Mexico.

Ascorbic acid (Vitamin C)

The ascorbic acid content in eggplant types (Tab. 3) showed values ranging between 7.4 and 22 mg $100g^{-1}$ with Thai and Hindu with lower and highest content respectively. All of our results are higher than the 4 mg ascorbic acid $100g^{-1}$ reported in Spain (Esteban *et al.*, 1992) and Turkey (Dürüst *et al.*, 1997). Prohens *et al.* (2007) found 1.4, 1.6 and 1.8 mg ascorbic acid $100g^{-1}$ for a Thai ('Thai Round'), a Chinese ('Kermit') and a European ('Black Beauty') cultivars, respectively. On the other hand, Hanson *et al.* (2006) quantified ascorbic acid contents of

Tab. 2. Mineral content of eggplant types grown in Sinaloa, Mexico (mg $100g^{-1}$ fresh weight)

Element	Chinese	Philippine	American	Hindu	Thai
K	*151.21±17.93^{b}	121.06±4.05^{c}	152.15±8.59^{ab}	191.18±1.49^{a}	176.46±6.11^{b}
Ca	28.00±0.77^{c}	32.80±4.97^{c}	31.36±1.06^{c}	59.63±7.71^{a}	45.08±1.14^{b}
P	21.21±0.56^{b}	13.80±0.86^{c}	29.61±3.28^{a}	33.52±1.34^{a}	30.42±1.25^{a}
Mg	15.29±0.86^{c}	15.74±0.95^{c}	25.35±2.13^{a}	28.96±1.97^{a}	20.88±1.17^{b}
Na	9.40±0.28^{a}	5.76±0.65^{b}	8.49±0.69^{a}	11.54±2.65^{a}	5.61±0.93^{b}
Fe	2.40±0.17^{b}	3.13±0.32^{a}	0.86±0.05^{d}	1.53±0.25^{c}	1.80±0.26^{c}
Zn	0.33±0.04^{d}	0.26±0.03^{d}	0.51±0.01^{b}	0.78±0.04^{a}	0.45±0.00^{c}
Mn	0.36±0.01^{b}	0.30±0.02^{c}	0.41±0.02^{ab}	0.44±0.02^{a}	0.39±0.01^{b}
Cu	0.15±0.02^{a}	0.13±0.03^{a}	0.15±0.01^{a}	0.15±0.01^{a}	0.18±0.03^{a}
Total	228.35	192.98	248.89	327.73	281.27

Mean values in a row with different letters are significantly different at P≤0.05. *Mean values of three samples±SD

Tab. 3. Nutraceutical components and antioxidant capacity of eggplant types grown in Sinaloa, Mexico

Attribute	Chinese	Philippine	American	Hindu	Thai
Ascorbic acid (mg $100g^{-1}$ of fresh sample)	*11.2±2.5^{b}	8.6±0.6^{b}	8.9±2.2^{b}	22.0±4.1^{a}	7.4±2.9^{b}
Total soluble phenols (mg CAE $100g^{-1}$ of dry sample)	1350.0±37.5^{d}	1562.7±57.3^{c}	1512.5±21.7^{c}	1750.0±43.3^{b}	2049.8±77.8^{a}
Chlorogenic acid (mg $100g^{-1}$ of dry sample)	953.3±47.3^{c}	916.7±45.1^{c}	860.0±30.0^{c}	1200.0±0.0^{b}	1700.0±100^{a}
Chlorogenic acid %**	70.61	58.66	56.87	68.57	82.93
Total anthocyanins***	128.7±1.5^{b}	161.1±2.9^{a}	123.6±5.7^{b}	82.8±2.6^{c}	3.9±0.2^{d}
ORAC****	296.2±16.7^{b}	538.9±61.2^{a}	94.8±12.4^{c}	352.1±24.5^{b}	102.8±23.2^{c}
DPPH*****	56.1±3.3^{c}	92.5±0.3^{a}	78.5±2.9^{b}	82.1±4.1^{ab}	77.8±7.8^{b}
IC_{50} (DPPH, µg)	0.0067	0.0033	0.0092	0.0039	0.0046

Mean values in a row with different letters are significantly different at P≤0.05, *Mean values of three samples±SD,
%Chlorogenic acid=(chlorogenic acid/total phenols)X100, * $mgC_3GE\ 100g^{-1}$, **** µmolTE g^{-1} dry basis, *****% Inhibition of DPPH

10.2, 7.3 and 5.2 mg 100g^{-1} for Hindu, Philippine, and Chinese eggplant types respectively, in all cases the same types show higher values in our trial. To put consumption in perspective, Young (1999) suggested a recommended daily intake (RDI) of vitamin C in the range from 60 to 100 mg. Eggplant consumptions of 100 g day^{-1} as analyzed in this study, will account for 7 to 22% of the RDI.

Total soluble phenols

Total soluble phenols in our work (Tab. 3) ranged from 1350 to 2049 mg CAE 100g^{-1} in Chinese and Hindu, respectively a significant gradient was found among the rest of the eggplant types. Our results were higher than those reported by Raigón *et al.* (2008), their report was based on a study of 31 *S. melongena* cultivars from around the world with total phenolic content ranged from 545 to 1048 mg CAE 100g^{-1}, Ninfalli *et al.* (2005) also reported 820 and 925 mg CAE 100g^{-1} for 'Black Beauty' and 'Violetta Lunga' varieties. Using five *S. melongena* cultivars of varied colours, Akanitapichat *et al.* (2010) reported values from 739 to 1116 mg GAE 100g^{-1}, and Hanson *et al.* (2006) found values from 740 to 1430 mg GAE 100g^{-1} in cultivars from around the world. In the other hand, a study by Okmen *et al.* (2009) showed contents from 878 to 1984 mg GAE 100g^{-1} in 26 Turkish cultivars, such data covered the whole range found in our study, except for the highest value we found in Thai.

Chlorogenic acid

Chlorogenic acid is one of the most ubiquitous phenolic acid in plants, particularly in the solanaceas (Clifford, 1999; Niggeweg *et al.*, 2004). It has been reported that following consumption it is absorbed and metabolized by humans having a beneficial effect on health (Olthof *et al*, 2001; Olthof *et al.*, 2003). The chlorogenic acid content found in our work resulted in a range from 860 to 1700 mg 100g^{-1}, American showed the lowest amount, while Thai reach the highest value. Winter and Herrmann (1986) reported 902 and 821 mg 100g^{-1} of chlorogenic acid for Italian and Spanish eggplant cultivars, respectively. Comparing with our data, these results are similar to the ones reported for Philippine and American, respectively. Papanga *et al.* (1999) and Whitaker and Stommel (2003) found 947 and 960 mg 100g^{-1} of chlorogenic acid in eggplant flesh, respectively, which is similar to our data for Chinese. In the other hand, Mattila and Hellström (2007) reported a chlorogenic acid content of 492 mg 100g^{-1} in an eggplant cultivated in Finland and Singh *et al.* (2009) found a chlorogenic acid content of 498 and 805 mg 100g^{-1} for 'Millionaire' and 'Blackbell' cultivars. The levels of these studies are lower than all the eggplant types analyzed in our work. Several studies related to phenolic acids determination in eggplants, concluded that chlorogenic acid is the main phenolic acid in the flesh of this vegetable, but none of them, have reported it in terms of percentage as compared to the total phenols content. Analyzing data from total phenols and chlorogenic acid shows that chlorogenic acid represents more than 56% of total phenols in all analyzed eggplant types (Tab. 3). Data variation in eggplant total phenols and chlorogenic acid among different reports is mainly attributed to solvent and extraction techniques used for extraction of phenolic compounds and such topic is analyzed by Luthria and Mukhopadhyay (2006).

Anthocyanins

Among the purple eggplant types, the highest amount of anthocyanins was observed in Philippine with 161 mgC$_3$GE 100g^{-1}, which coincides with the results for ORAC and DPPH antioxidant activity, while the lowest result was found in Hindu with 83 mg C$_3$GE 100g^{-1} (Tab. 3). No statistical differences between Chinese and American were found with 129 and 124 mg C$_3$G 100g^{-1}, respectively. Anthocyanins data obtained in this study of purple fruits are higher than those reported by Papanga *et al.* (1999) and Sadilova *et al.* (2006) whom reported 55 and 64 mg 100g^{-1}, respectively. In the other hand, our anthocyanins data in Hindu and American are similar to those reported by Lo Scalzo *et al.* (2010) and Koponen *et al.* (2007) who measured 90 and 107 mg 100g^{-1}. Former reports show that Bor *et al.* (2006) sampled *S. melongena* at retails in Taiwan, while Wu *et al.* (2006) in a comprehensive sample at 12 cities around the U.S. reported anthocyanin content values ranging from 167 and 1067 mg 100g^{-1} in commercial eggplants at retail points, which is higher than the amounts found in our five types study. All of these anthocyanin comparative studies used delphinidin or delphinidin derivatives as standard compounds for analysis.

Antioxidant capacity: DPPH radical scavenging activity and ORAC

Antioxidant activity was evaluated using DPPH and AAPH free radicals. AAPH free radical was used in the ORAC assay and results were expressed as μmoles Trolox equivalents per gram of sample (μmolTE g^{-1}), showing significant differences ($P \leq 0.05$) among the five types (Tab. 3). Results ranged from 95 to 539 μmolTE g^{-1}, with American and Philippine at each extreme. Thai samples (103 μmolTE g^{-1}) had no significant differences as compared to the American type, while Hindu and Chinese samples showed no significant differences with 296 and 352 μmolTE g^{-1}, each. IC_{50} is defined as the amount of antioxidant material required to scavenge 50% of free radical in the assay system. Its value are inversely proportional to the antioxidant activity, the smaller the number the highest the scavenging activity. All eggplant types ranged from IC_{50} values of 0.0033 to 0.0092. According to these, Philippine had the highest antioxidant activity with an IC_{50} of 0.0033 μg, while American has the lowest with only 0.0092 μg. Methodological similarities among the ORAC and DPPH methods may be due to the presence and specificity of the diverse bioactive molecules found in the evaluated eggplant types.

Conclusions

The five evaluated eggplant types specially Hindu and Thai, have significant content of some human promoting health components, such as crude and dietary fibre, as well as biologically essential minerals like K, Ca, Mg, P, Na and Fe. They also provide important amounts of antioxidants like ascorbic acid and soluble phenolic compounds (mainly chlorogenic acid), as well as significant antioxidant.

Acknowledgments

Authors would like to acknowledge the scholarship awarded to Guillermo Niño-Medina by CONACYT (Consejo Nacional de Ciencia y Tecnología) and the laboratory assistance by Laura Contreras-Angulo, Rosalba Contreras-Martínez, Verónica Pérez-Rubio and Werner Rubio-Carrasco.

References

Abdel-Aal ES, Hucl P (1999). A rapid method for quantifying total anthocyanins in blue aleurone and purple pericarp wheats. Cereal Chem 76(3):350-354.

Akanitapichat P, Phraibung K, Nuchklang K, Prompitakkul S (2010). Antioxidant and hepatoprotective activities of five eggplant varieties. Food Chem Toxicol 48(10):3017-3021.

AOAC (1998). Official Methods of Analysis. Association of Official Analytical Chemist. Maryland, AOAC International.

Bor J, Chen H, Yen G (2006). Evaluation of antioxidant activity and inhibitory effect on nitric oxide production of some common vegetables. J Agr Food Chem 54(5):1680-1686.

Cao G, Sofic E and Prior RL (1996). Antioxidant capacity of tea and common vegetables. J Agr Food Chem 44(11):3426-3431.

Chun OK, Kim DO (2004). Consideration on equivalent chemicals in total phenolic assay of chlorogenic acid-rich plums. Food Res Int 37(4):337-342.

Clifford MN (1999). Chlorogenic acids and other cinnamates– nature, occurrence and dietary burden. J Sci Food Agr 79(3):362-372.

Doner LW, Hicks KB (1981). High-performance liquid chromatographic separation of ascorbic acid, erythorbic acid, dehydroascorbic acid, dehydroerythorbic acid and diketogluconic acid. Anal Biochem 115(1):225-230.

Dürüst N, Sümengen D, Dürüst Y (1997). Ascorbic acid and element contents of foods of Trabzon (Turkey). J Agr Food Chem 45(6):2085-2087.

Ekholm P, Reinivuo H, Mattila P, Pakkala H, Koponen J, Happonen A, Hellström J, Ovaskainen ML (2007). Changes in the mineral and trace element contents of cereals, fruits and vegetables in Finland. J Food Compos Anal 20(6):487-495.

Esteban RM, Mollá EM, Robredo FJ, López-Andréu FJ (1992). Changes in the chemical composition of eggplant fruits during development and ripening. J Agr Food Chem 40(6):998-1000.

Flick GJ, Burrnette FS, Aung LH, Ory RL, St Angelo AJ (1978). Chemical composition and biochemical properties of mirlitons (*Sechium edule*) and purple, green and white eggplants (*Solanum melongena*). J Agr Food Chem 26(5):1000-1005.

Hanson PM, Yang RY, Tsou SC, Ledesma D, Engle L, Lee TC (2006). Diversity in eggplant (*Solanum melongena*) for superoxide scavenging activity, total phenolics and ascorbic acid. J Food Compos Anal 19(6-7):594-600.

Heredia JB, Cisneros-Zevallos L (2009). The effects of exogenous ethylene and methyl jasmonate on the accumulation of phenolic antioxidants in selected whole and wounded fresh produce. Food Chem 115(4):1500-1508.

Huang D, Ou B, Hampsch-Woodil M, Flanagan J, Prior RL (2002). High-throughput assay of oxygen radical absorbance capacity (ORAC) using a multichannel liquid handling system coupled with a microplate fluorescence reader in 96-well format. J Agr Food Chem 50(16):4437-4444.

Kantharajah AS, Golegaonkar PG (2004). Somatic embryogenesis in eggplant. Sci Hortic-Amsterdam 99(2):107-117.

Kashyap V, Kumar SV, Collonnier C, Fusari F, Haicour R, Rotino GL, Sihachakr D, Rajam MV (2003). Biotechnology of eggplant. Sci Hortic-Amsterdam 97(1):1-25.

Koponen JM, Happonen AM, Mattila PH, Törrönen R (2007). Contents of anthocyanins and ellagitannins in selected foods consumed in Finland. J Agr Food Chem 55(4):1612-1619.

Lo Scalzo R, Fibiani M, Mennella G, Rotino GL, Dal Sasso M, Culici M, Spallino A, Braga PC (2010). Thermal treatment of eggplant (*Solanum melongena* L.) increases the antioxidant content and the inhibitory effect on human neutrophil burst. J Agr Food Chem 58(6):3371-3379.

Luthria DL, Mukhopadhyay S (2006). Influence of sample preparation on assay of phenolic acids from eggplant. J Agr Food Chem 54(1):41-47.

Manach C, Scalbert A, Morand C, Rémésy C, Jiménez L (2004). Polyphenols: food sources and bioavailability. Am J Clin Nutr 79(5):727-747.

Marlett JA, McBurney MI, Slavin JL (2002). Position on the American dietetic association: health implications of dietary fiber. J Am Diet Assoc 102(7):993-100.

Maroto JV (2002). Horticultura Herbácea Especial. 5th Ed. Ediciones Mundi-Prensa. Mexico. pp. 481-495 p.

Mattila P, Hellström J (2007). Phenolic acids in potatoes, vegetables and some of their products. J Food Compos Anal 20(3-4):152-160.

MINITAB (2004). Statistical software. Release number 14. www.minitab.com.

Mohamed AE, Rashed MN, Mofty A (2003). Assessment of essential and toxic elements in some kinds of vegetables. Ecotox Environ Safe 55(3):251-60.

Muñoz de Chávez M, Chávez A, Roldán J, Ledesma J, Mendoza E, Pérez-Gil F (1996). Tablas de valor nutritivo de los alimentos de mayor consumo en Mexico. 2nd Ed. Editorial Pax. Mexico, 54 p.

Ninfalli VP, Mea G, Giorgini S, Rocchi M, Bacchiocca M (2005). Antioxidant capacity of vegetables, spices and dressings relevant to nutrition. Brit J Nutr 93(2):257-266.

Niggeweg R, Michael AJ, Martin C (2004). Engineering plants with increased levels of the antioxidant chlorogenic acid. Nat Biotechnol 22(6):746-754.

Noda Y, Kneyuki T, Igarashi K, Mori A, Packer L (2000). Antioxidant activity of nasunin, an anthocyanin in eggplant peels. Toxicology 148(2-3):119-123.

Okmen B, Sigva HO, Mutlu S, Doganlar S, Yemenicioglu A, Frary A (2009). Total antioxidant activity and total phenolic contents in different Turkish eggplant (*Solanum melongena* L.) cultivars. Int J Food Prop 12(3):616-624.

Olthof MR, Hollman PC, Katan MB (2001). Chlorogenic acid and caffeic acid are absorbed in humans. J Nutr 131(1):66-71.

Olthof MR, Hollman PC, Buijsman NC, Van Amelsvoort JM, Katan MB (2003). Chlorogenic acid, quercetin-3-rutinoside and black tea phenols are extensively metabolized in humans. J Nutr 133(6):1806-1814.

Özgür MÜ, Sungur S (1995). Third order derivative spectrophotometric determination of ascorbic acid in fruits and vegetables. Talanta 42(11):1631-1640.

Pak DN (2000). Fibra dietética en verduras cultivadas en Chile. Arch Latinoam Nutr 50(1):97-101.

Papanga G, Miller N, Rice-Evans CA (1999). The polyphenolic content of fruit and vegetables and their antioxidant activities. What does a serving constitute? Free Radical Res 30(2):153-162.

Prior RL, Hoang H, Gu L, Wu X, Bacchiocca M, Howard L, Hampsch-Woodill M, Huang D, Ou B, Jacob R (2003). Assays for hydrophilic and lipophilic antioxidant capacity (oxygen radical absorbance capacity ($ORAC_{FL}$)) of plasma and other biological and food samples. J Agr Food Chem 51(11):3273-3279.

Prohens J, Rodríguez-Burruezo A, Raigón MD, Nuez F (2007). Total phenolic concentration and browning susceptibility in a collection of different varietal types and hybrids of eggplant: implications for breeding for higher nutritional quality and reduced browning. J Am Soc Hortic Sci 132(5):638-646.

Raigón MD, Prohens J, Muñoz-Falcón JE, Nuez F (2008). Comparison of eggplant landraces and commercial varieties for fruit content of phenolics, minerals, dry matter and protein. J Food Compos Anal 21(5):370-376.

Rehman ZU, Islam M, Shah WH (2003). Effect of microwave and conventional cooking on insoluble dietary fibre components of vegetables. Food Chem 80(2):237-40.

Sadilova E, Stintzing FC, Carle R (2006). Anthocyanins, colour and antioxidant properties of eggplant (*Solanum melongena* L.) and violet pepper (*Capsicum annuum* L.) peel extracts. Z Naturforsch C 61(7-8):527-535.

Sánchez-Castillo CP, Englyst HN, Hudson GJ, Lara JJ, Solano ML, Munguía JL, James WP (1999). The non-starch polysaccharide content of Mexican foods. J Food Compos Anal 12(4):293-314.

Scalbert A, Williamson G (2000). Dietary intake and bioavailability of polyphenols. J Nutr 130(8):2073S-2085S.

Singh AP, Luthria D, Wilson T, Vorsa N, Singh V, Banuelos GS, Pasakdee S (2009). Polyphenols content and antioxidant capacity of eggplant pulp. Food Chem 114(3):955-961.

Takeyama E, Fukushima M, Tanimura A (2002). Effect of heat-treatment on the content and polysaccharide composition of dietary fiber. Food Sci Technol Res 8(2):125-130.

United States Department of Agriculture (USDA) (2014). USDA National Nutrient Database for Standard Reference. http://www.nal.usda.gov/fnic/foodcomp/search.

Whitaker BD, Stommel JR (2003). Distribution of hydroxycinnamic acid conjugates in fruit of eggplant (*Solanum melongena* L.) cultivars. J Agr Food Chem 51(11):3448-3454.

Winter M, Herrmann K (1986). Esters and glucosides of hydroxycinnamic acids in vegetables. J Agr Food Chem 34(4):616-620.

Wu X, Beecher GB, Holden JM, Haytowitz DB, Gebhardt SE, Prior RL (2006). Concentrations of anthocyanins in common foods in the United States and estimation of normal consumption. J Agr Food Chem 54(11):4069-4075.

Young M (1999). How much vitamin C do you need?. JAMA-J Am Med Assoc 281(15):1460.

Assessment of the Potential of 1-Methylcyclopropene Treatments to Maintain Fruit Quality of the Common Fig (*Ficus carica* L. cv. 'Bursa Siyahi') during Refrigerated Storage

Okan OZKAYA*, Songül ÇÖMLEKÇIOĞLU, Hatice DEMIRCIOĞLU

*Cukurova University, Agriculture Faculty, Horticulture Department, Balcali, Adana, Turkey; oozkaya@cu.edu.tr (*corresponding author)*

Abstract

The fig fruit is a unique, climacteric, highly perishable subject to rapid physiological breakdown. Application of 1-methylcyclopropene (1-MCP) was tested to delay ripening of black fig (*Ficus carica* L. cv. 'Bursa Siyahi') during storage over two growing seasons. Fruits were pre-cooled to 1 °C for 6 hours and afterwards treated with 500 or 1000 nl l^{-1} of 1-MCP for 24 hours. Treated fruits were stored for 10 days at 1 °C, 90-95% RH and then evaluated. 1-MCP treatments showed that ethylene production, respiration rate, weight loss and concentrations of glucose, fructose and total soluble solids (TSS) were negatively correlated to the 1-MCP doses during cold storage (with the exception of TSS in the first year of experiment and respiration rate in the second year of the experiment). In contrast, pulp firmness and colour (h°) during cold storage were positively correlated to the 1-MCP applied doses. Results of this study showed that although 1-methylcyclopropene applications slowed down fruit softening during the 10 days of storage, 1-methylcyclopropene appeared to have a relatively limited effect on slowing ripening of 'Bursa Siyahi' figs.

Keywords: fresh fig, quality, 1-MCP, shelf life

Introduction

Ficus carica L., the common fig, is a species of great commercial importance cultivated for its sweet fruits widely used both as a food and as medicine all over the world. Fresh and dried fig fruits are an excellent source of minerals, vitamins and dietary fiber (Solomon *et al.*, 2006; Veberic *et al.*, 2008). In addition, they contain high levels of polyphenols, flavonoids and anthocyanin, hence they are a good source of antioxidants for humans (Oliveira *et al.*, 2009). Figs have been traditionally used for their medicinal benefits as a laxative, for favouring cardiovascular circulation and providing anti-inflammatory properties. Fig fruits and latex have also been shown to reduce the risk of cancer (Rubnov *et al.*, 2001).

The fig fruit is a unique, highly perishable climacteric fruit subject to rapid physiological breakdown. The postharvest life of the fresh fig fruit typically does not exceed beyond 7-10 days even when low temperature storage is used (Chessa, 1997). Maintenance of the cold chain is important to slow down the deterioration of the fresh fig fruits. Basic studies on the processes that occur during ripening are essential for investigating systems in which the biological and physiological processes linked to maturation are involved in postharvest deterioration (Owino *et al.*, 2004) to determine more effective tools for the maintenance of postharvest life of fresh figs.

Ripening and senescence is mainly controlled by ethylene, one of several plant growth regulators involved in the growth and developmental processes, especially in climacteric fruits and vegetables (Abeles *et al.*, 1992). This simple hydrocarbon can diffuse into and out of plant tissues from both endogenous and exogenous sources and can greatly affect the quality of harvested products (Watkins, 2006). These effects can be beneficial or deleterious depending on the product, its ripening stage and its usage (Saltveit, 1999). Exogenous ethylene applications in bananas, for example, initiate uniform ripening. However, most commonly, commercial approaches for horticultural products are based on avoiding exposure to ethylene and/or attempting to minimize ethylene production and action during ripening, harvest, storage, transport and handling by temperature and atmosphere control (Watkins, 2002).

The discovery and commercialization of 1-MCP, an ethylene inhibitor, provided beneficial effects by suppressing the ethylene production and thus slowing down the ripening process and senescence of the fruits, especially climacteric ones. 1-MCP acts by blocking ethylene receptors (even at very low doses), thus ethylene cannot bid with the fruit (Sisler and Serek, 1997). Therefore, 1-MCP can be used as an effective tool for extending shelf-life and improving quality of fruits and vegetables (Blankenship and Dole, 2003). Several studies have successfully demonstrated that 1-MCP can extend shelf-life and quality of fruits and vegetables (Khan *et al.*, 2007; Özkaya and Dündar, 2009; Özkaya *et al.*, 2010; Valero *et al.*, 2003; Yuan *et al.*, 2010). However, there are very few reports on postharvest management of fresh fig cultivars.

The aim of these trials reported here was to evaluate the potential of 1-MCP to maintain the quality and extend the storage life of fresh figs. (*Ficus carica* L. cv. 'Bursa Siyahi'). Two different doses of 500 and 1000 nl l^{-1} of 1-MCP applications were tested over two seasons.

Materials and methods

Plant material

Black figs (*Ficus carica* L. cv. 'Bursa Siyahi') were hand-harvested at commercial ripening stage, from an orchard located at Adana-Turkey, over two seasons. After harvest, fruits were brought to the postharvest physiology laboratory of the Department of Horticulture, Faculty of Agriculture, Cukurova University. Figs were then selected according to their uniformity of size, colour and absence of defects, according to the exporting criteria of packing houses. Fruits were selected and divided randomly into 3 groups as control, 500 and 1000 nl l^{-1} doses of 1-MCP treatments.

1-MCP treatments

Fruits were pre-cooled at 1 °C for 6 h before treatments. 1-MCP was used as powder, which after addition of warm water (40 °C) released the active ingredient as a gas. Different amounts of the powder were weighed and warm water was added to obtain the doses of 500 and 1000 nl l^{-1}. Treatments were performed in 0.1 m^3 hermetically sealed containers through injection of 1-MCP using plastic syringes and air circulation inside the containers were done by a small ventilator that works with a battery. Duration of treatment was 24 h at 1 °C. Control fruits were kept in similar containers for 24 h at 1 °C without 1-MCP application.

Ethylene and respiration rate determination

Ethylene production was measured by enclosing 1 kg of fruits for each treatment in a 2.5 l airtight glass jar hermetically sealed with two rubber septa for 2 hours. The headspace air was mixed with a 20 ml syringe and 1-ml sample was withdrawn from each jar for ethylene analysis. Ethylene (C_2H_4) concentration was determined by using a gas chromatograph fitted with a flame ionization detector (FID) with a stainless steel column filled with Alumina GC (Gapper *et al.*, 2006). Ethylene production rate was expressed as nl g^{-1} h^{-1}.

The CO_2 determination was made by CO_2 analyser by which gas sample was withdrawn from the headspace of the same jars which the samples were taken for ethylene, after two hour closing time and calculated as mg CO_2 kg^{-1} h^{-1}.

Fruit quality assessment

Following 1-MCP applications, treated fig fruits together with untreated control fruits were stored at 1 °C, 90-95% RH for 10 days and fruit weight loss, firmness, total soluble solids concentration (TSS), titratable acidity (TA), colour change, individual sugars were determined after storage to assess fruit quality.

Fruits were weighed at harvest and at the end of the storage period and weight loss was determined. Fruit firmness of 30 fig samples per treatment were determined on opposite sides of the fruit, using a hand penetrometer with an 8-mm-diameter tip and expressed as Newton (N). Fruit juice was extracted and diluted with pure water (3x) to measure total soluble solids concentration (TSS), using a hand-held refractometer. Titratable acidity (TA) was determined by titration of juice (5 ml) with 0.1 N sodium hydroxide and measured with a pH meter and expressed as g malic acid/100 ml fruit juice.

Fruit skin colour was assessed using a CIE colorimeter after calibration with white tile and the results expressed as L, a*, b* and Hue angle (h °) [arctangent (b*/a*)] (Voss, 1992). Thirty fruits from each treatment were used for colour measurement. Each was characterized by using the average of 2 measurements at equidistant points around equatorial circumference of the fruits.

Fruit sugars (g/kg) were analysed by high-performance liquid chromatography equipped with an in-line degasser, pump, manual injection (20 μ injection volume) interfaced to a PC running Class VP chromatography manager software. Fruit juice for each 1-MCP treatment and control samples was prepared by dilution with ultra pure water (18.2 MΩ cm) and filtered (nylon syringe filters, 0.45 μm, 13 mm diameter) for the determination of individual sugars. Separations were performed on a 150 mm 4.6 mm i.d., 5 μm, reverse-phase analytical column at 40 °C oven temperature with a flow rate of 0.6 ml min^{-1}. Elution was isocratic with pure water. Components were identified by comparing the retention times with the retention time of standards run at the same time during the analysis. A 25 minutes equilibrium time was allowed between injections.

Statistical analysis

The experiment was set up as a completely randomized design with six replicates and 30 fruits comprised each replicate. The Statistical Analysis System (SAS) was used to analyse the data and means were compared with Tukey HSD multiple range test at significance level $P<0.05$.

Results

Fruit weight loss, ethylene production and respiration rate

Significantly more weight loss occurred in the 1-MCP treatments compared with the control fruit, except in the 1000 nl treatment in the first year (Fig. 1). At the end of cold storage control figs lost 7.27% in first year, 6.72% in second year, whereas 500 nl l^{-1} 1-MCP treated fruits lost 6.26% in first year, 5.70% in second year and 1000 nl l^{-1} figs lost 5.97% in the first year, 6.48 % in the second year, of their initial weight.

No significantly differences were found in ethylene production after 10 days of storage in both years of storage in the 1-MCP and control treatments (Fig. 2). However, there was a trend in both years for reduced ethylene production in the 1-MCP treatments respectively. The respiration rate was increased by 40% in control fruits after 10 days of storage, whereas it was decreased by 21% and 25% in 500 and 1000 nl l^{-1} 1-MCP treatments respectively, during the first year. The respiration rate was 9.65 mg CO_2 kg^{-1} h^{-1} in the second year at the harvest day and it decreased in control and other treatments 1%, 21% and 22% in control, 500 and 1000 nl l^{-1} 1-MCP treatments respectively (Fig. 3).

Fruit quality parameters

1-MCP treated figs maintained firmness better than the

control after 10 days of storage during both the first year and second year (Fig. 4). Contrarily, in control figs firmness decreased in the first year treatment with 30% than the harvest one measured. 1-MCP treated figs exhibited a sharp decrease in firmness, although levels were higher than in control figs, for both 1-MCP treatments in the second year.

There were no significant differences between control and 1-MCP treatment in regard of TSS in the first year, whereas the 1000 nl l^{-1} 1-MCP treatment produced significantly higher TSS levels in the second year (Fig. 5). There were slight decreases in TA in both years; the changes in TA were not found to be significantly difference between treatments in the first year, however they were significantly different in the second year (Fig. 6).

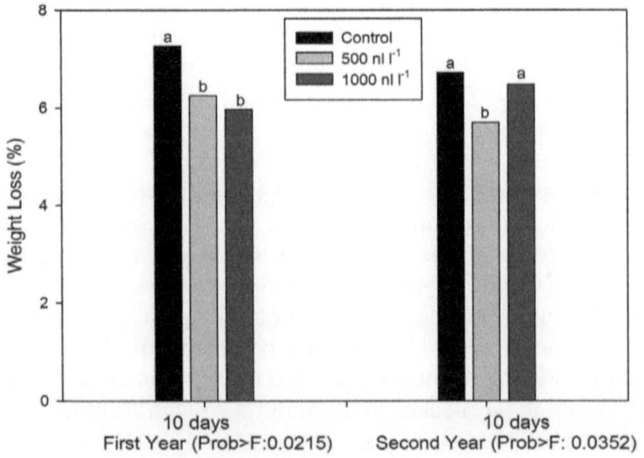

Fig. 1. Weight losses after storage

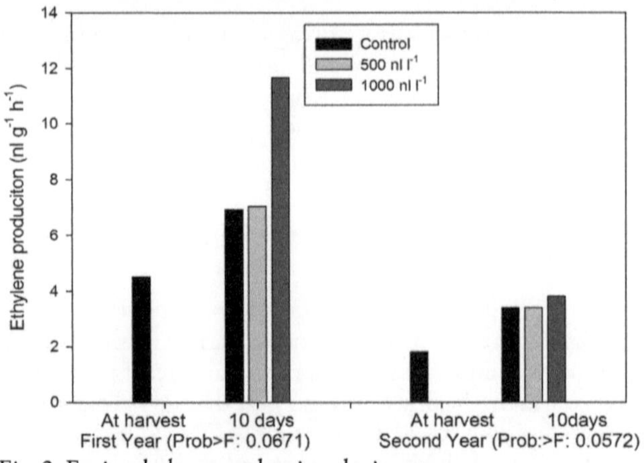

Fig. 2. Fruit ethylene production during storage

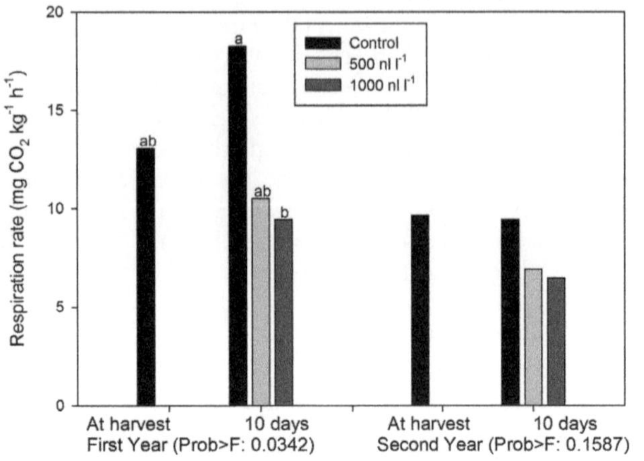

Fig. 3. Respiration rate of fruits during storage

Fig skin colour did not significantly change in control and 1-MCP treatments for both years of the experiment during cold storage. There was more sharp decrease in the first year storage period regarding h° value of fig skins than in the second year storage period (Fig. 7).

Differences between the glucose content of fig flesh between treatments and control were not significantly important. However, it was found significantly important the increase between treatments in the second year of the experiment after 10 days of storage (Fig. 8). The fructose content of figs was increased in both storage period and the differences between control and 1-MCP treatments (Fig. 9) were not significantly different.

To check the effect of the 1-MCP applied dose on fruit quality parameters during cold storage, linear regressions were performed taking into account the experimental data for storage period. Thus, ethylene production (Fig. 10a), respiration rate (Fig. 10b), weight loss (Fig. 10c), glucose (Fig. 10h), fructose (Fig. 10i) and TSS (Fig. 10e) were negatively correlated to the 1-MCP dose during cold storage (with the exception of TSS in the first year respiration rate for stage 2 of ripeness) or shelf life. Contrarily, pulp firmness (Fig. 10d) and colour (h°) (Fig. 10g) during cold storage were positively correlated to the 1-MCP applied dose.

Discussion

Fruit ethylene evolution regulates the ripening process of climacteric fruits by initiating a wide range of physical and chemical changes such as tissue softening, pigment degradation and changes in sugar and organic acid levels in the fruit (Giovannoni, 2001). Although ethylene production is inhibited by 1-MCP application in most climacteric fruits (Blankship and Dole, 2003; Watkins, 2006) it was not inhibit in firm mature figs. When fruit were treated with 1-MCP the ethylene climacteric was delayed by 6 days and reduced in magnitude over 50% in avocado (Jeong *et al.*, 2002) and it was also reduced in plums (Abdi *et al.*, 1997) and apricots (Fan *et al.*, 2000). In contrast, 1-MCP treated pineapple fruit produced more ethylene than the control (Selvarajah *et al.*, 2001). Ethylene production was also increased by 1-MCP treatment in coriander (Jiang *et al.*, 2002). The ethylene production in first year was higher in both 1-MCP doses than in control fruits and 1000 nl l^{-1} dose of 1-MCP also had higher amount than control fruits after 10 days of storage in the second year. These results may indicate that 1-MCP may affect normal feedback regulation in cooled and low temperature application of 1-MCP treated figs.

The respiration rate varied depending on the 1-MCP dose in this study. The respiratory climacteric onset in plums was delayed by 1-MCP (Abdi *et al.*, 1997; Dong *et al.*, 2002). Respiration rate was reduced in apricots treated with 1-MCP (Fan *et al.*, 2000). However, 1-MCP treatment had no effect on nectarine (Dong *et al.*, 2001) and apricot (Dong *et al.*, 2002). The respiration rate of figs were not significantly differenced after 10 days of storage, however the respiration rate was lower in 1-MCP treated fruits. The differing results in climacteric fruits might be due to fruit maturity, cultivar, or some other unknown factors that influence the climacteric onset.

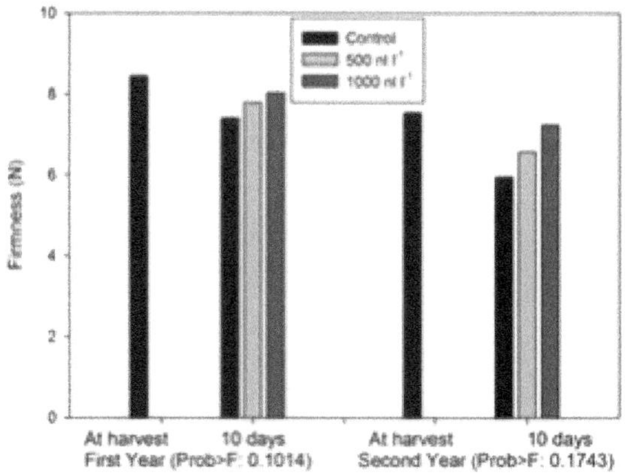

Fig. 4. Fruit firmness changes during storage

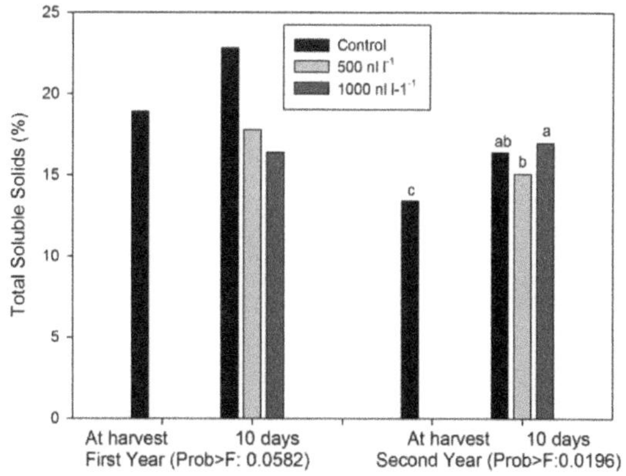

Fig. 5. Total soluble solid changes during storage

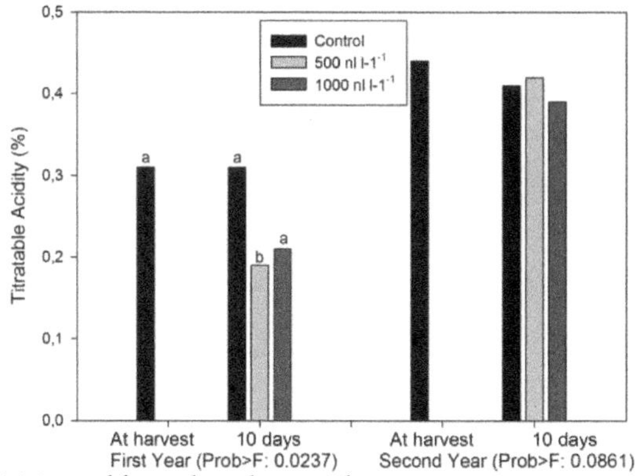

Fig. 6. Titratable acidity changes during storage

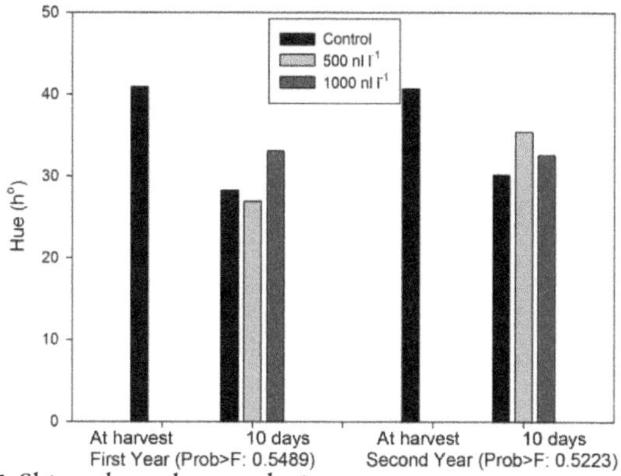

Fig. 7. Skin colour changes during storage

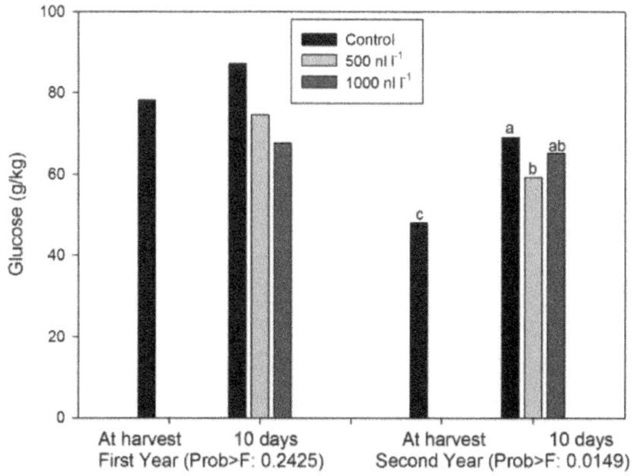

Fig. 8. Glucose content change during storage

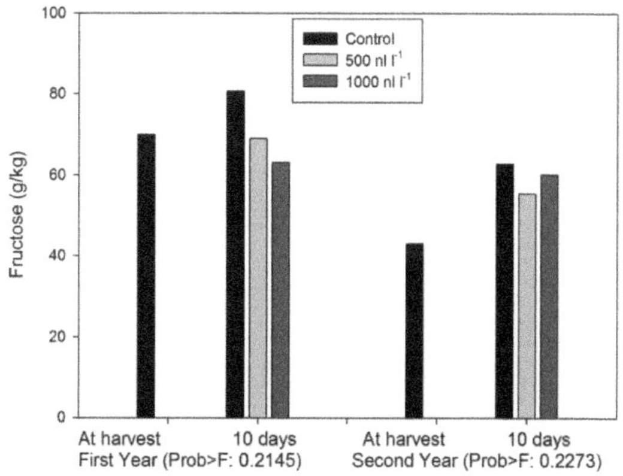

Fig. 9. Fructose content change during storage

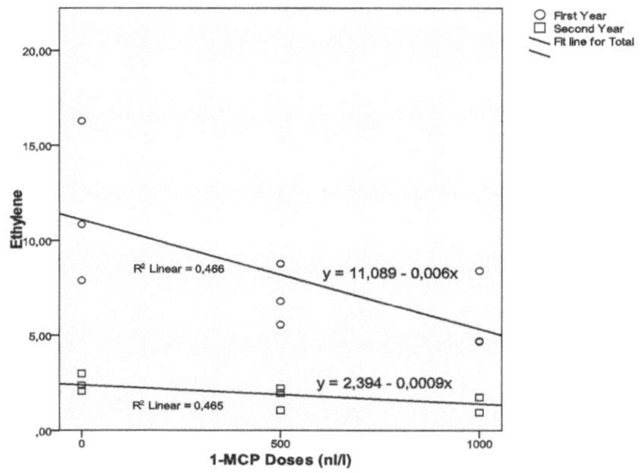

(a)

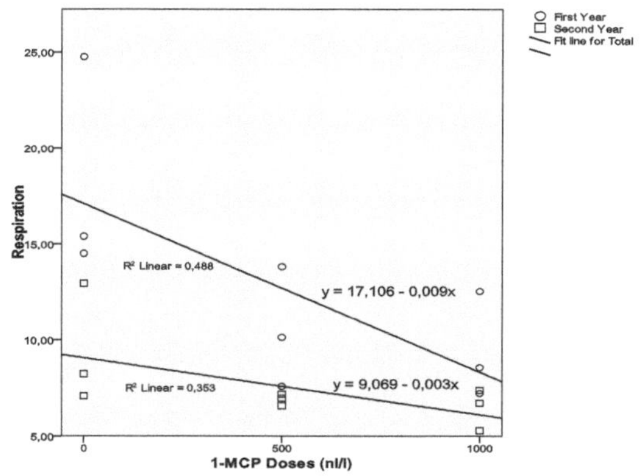

(b)

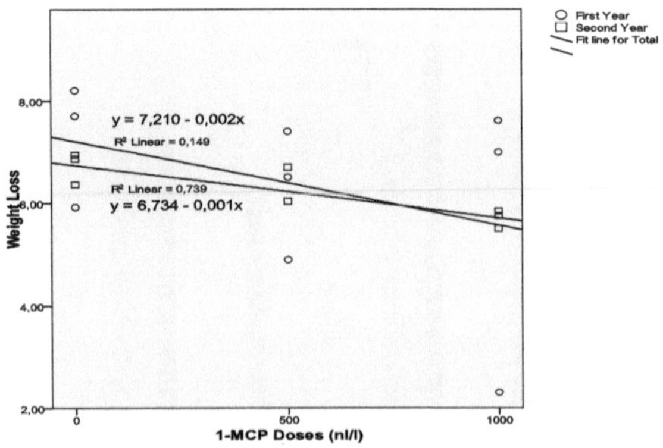

(c)

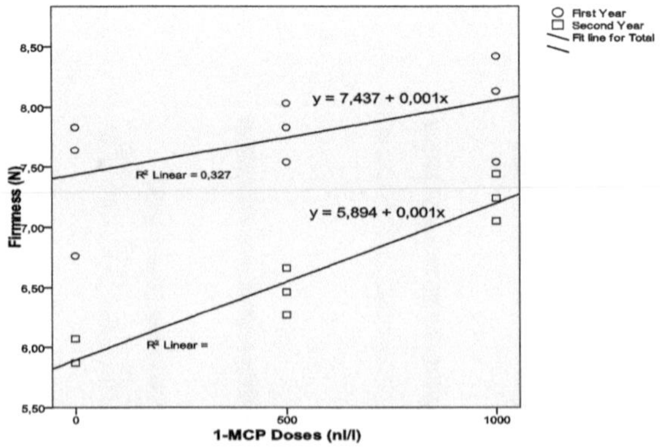

(d)

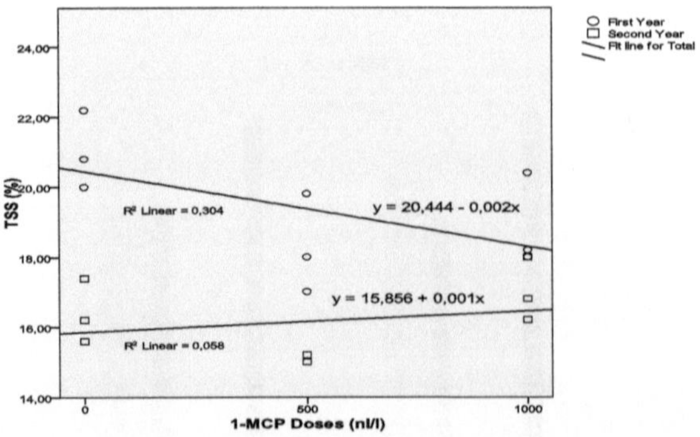

(e)

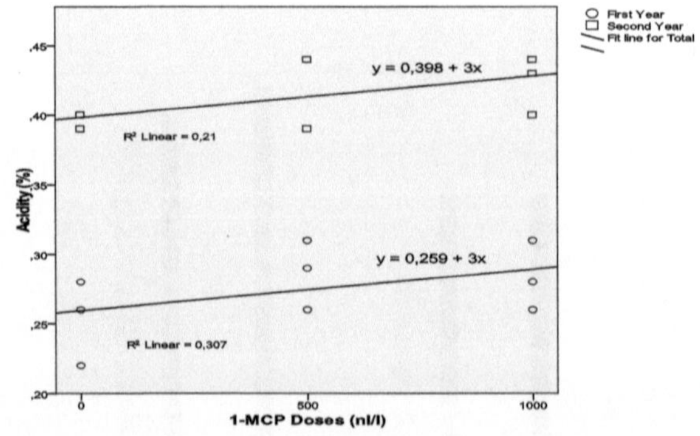

(f)

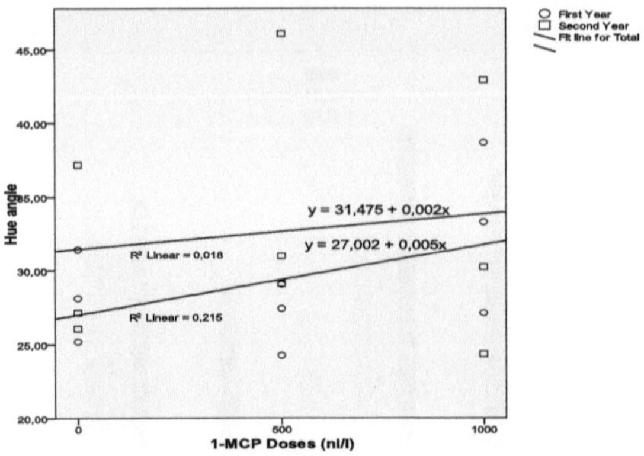

(g)

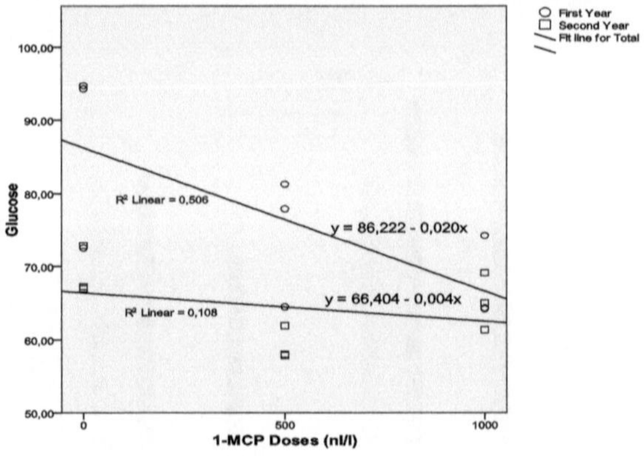

(h)

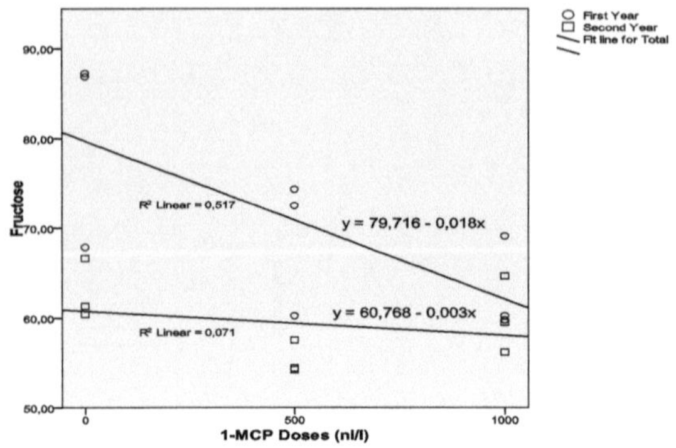

(i)

Fig. 10. Linear regressions figures between fruit quality parameters and 1-MCP dose; (a): Ethylene production; (b): Respiration rate; (c): Weight loss; (d): Fruit firmness; (e): Total soluble solid; (f): Titratable acidity; (g): Skin colour; (h): Glucose content; (i): Fructose content

Weight loss is one of the most important causes responsible for fruit quality deterioration. The findings of the present study revealed that weight loss was slightly reduced in 'Bursa Siyahi' fig treated with both 1-MCP doses, the effect being independent of the applied concentration. As shown in this study, 1-MCP inhibited ethylene evolution and decreased fruit respiration, which likely resulted in less weight loss in the 1-MCP treated fruit. Other studies in apple, kiwi etc. fruits, however, have reported no effect of 1-MCP application on fruit weight loss (Watkins *et al.*, 2006). Fruit weight loss can vary depending on the maturity stage, ventilation of cold store and epicular wax metabolism during ripening and cold storage (Valero *et al.*, 2003).

The fruit firmness observed in fruits subjected to both 1-MCP dose treatments was higher than control fruits in the first year and it was significantly different than control in the second year. Reduction in loss of firmness with 1-MCP treatment has been reported in tomato (Wills and Ku, 2002), apple (Watkins *et al.*, 2006) and other climacteric fruits. 1-MCP treatments have significantly lowered polygalacturonase (PG) and cellulose activities in avocado fruits, which may explain the increased fruit firmness (Feng *et al.*, 2000). Even though ethylene production of figs was not affected by 1-MCP treatment, the retard in softening of 1-MCP treated fruits indicate that 1-MCP might play a role retarding some enzyme activity in cell wall construction that influence fruit softening.

TSS varied between 1-MCP and control treatments in both years. 1-MCP application might not directly affect TSS levels, but it can indirectly, by influencing the maturity stage of the stored fruits. Soluble solids were higher in 1-MCP treated pineapple (Selvarajah *et al.*, 2001), papaya (Hofman *et al.*, 2001) and apples (Fan *et al.*, 1999). However, soluble solids were reduced in 1-MCP treated strawberries regardless of the presence or absence of exogenous ethylene (Tian *et al.*, 2000). Soluble solids were unaffected by 1-MCP in oranges (Porat *et al.*, 1999), apricots and plums (Dong *et al.*, 2002).

In general, fruits lost acidity during cold storage and even more following ripening at room temperature. The effect of 1-MCP on TA is mixed, with some crops being affected and others unaffected. 1-MCP completely inhibited TA loss in tomatoes (Wills and Ku, 2002), delayed TA loss in apples (Özkaya and Dündar, 2009). The results of the present research showed that TA was affected by 1-MCP treatments in the first year whereas the acidity loss was not affected in the second year.

The association between colour changes and ethylene varies with the type of pigment, the species and the tissue which the pigment is being produced (Abdi *et al.*, 1997). 1-MCP resulted in lowered phenylalanine ammonia lyase activity and anthocyanin production in strawberries (Jiang *et al.*, 2001). Nevertheless, the application of up to 1000 nl l^{-1} was insufficient to prevent changes of h° in figs. The inherent variability of samples may partially mask the effect of 1-MCP.

Glucose and fructose content of control and 1-MCP treatments were increased during storage. As it is mentioned, individual sugars are affected by physiological maturation levels. These effects change in different ways, depending on cultivar. The results of this experiment showed that both glucose and fructose level were in an increase trend.

Conclusion

It is very important for both exporters and consumers to purchase a good quality fruit with acceptable price. Reducing fruit losses is one of the main key during postharvest practices to supply the market with lower priced fruits and balance the marginal income. The results of the presented study showed that 1-MCP have a limited effect on fresh fig fruit to slow down the initial ethylene production depending on physiological process.

Acknowledgement

This manuscript was edited by Professor Richard C. Rosecrance, California State University, Chico. Authors would like to thank for his evaluation and editing.

References

Abdi N, Holford P, McGlasson WB, Mizrahi Y (1997). Ripening behavior and responses to propylene in four cultivars of Japanese type plums. Postharvest Biol Tec 12:21-34.

Abeles FB, Morgan PW, Saltveit ME (1992). Ethylene in plant biology. San Diego, California: Academic Press, p. 64-87.

Blankenship SM, Dole JM (2003). 1-Methylcyclopropene: a review. Postharvest Biol Tec 28:1-25.

Chessa I (1997). Fig, p. 245-268. In: Mitra S (Ed.). Postharvest Physiology and Storage of Tropical and Subtropical Fruits. CAB International, Wallingford, Oxon, UK.

Dong L, Zhou H, Sonego L, Lers A, Lurie S (2001). Ethylene involvement in the cold storage disorder of 'Flavortop' nectarine. Postharvest Biol Tec 23:105-115.

Dong L, Lurie S, Zhou H (2002). Effect of 1-methylcyclopropene on ripening of 'Canino' apricots and 'Royal Zee' plums. Postharvest Biol Tec 24:135-145.

Fan X, Mattheis JP (1999). Impact of 1-methylcyclopropene and methyl jasmonate on apple volatile production. J Agric Food Chem 47:2847-2853.

Fan X, Argenta L, Mattheis JP (2000). Inhibition of ethylene action by 1-methylcyclopropene prolongs storage life of apricots. Postharvest Biol Tec 20:135-142.

Feng X, Apelbaum A, Sisler, EC, Goren R (2000). Control of ethylene responses in avocado fruit with 1-methylcyclopropene. Postharvest Biol Tec 20:143-150.

Gapper NE, Bai J, Whiteaker BD (2006). Inhibition of ethyleneinduced α-farnasene synthase gene Pc*AFS1* expression in 'd'Anjou' pears with 1-MCP reduces synthesis and oxidation of α-farnasene and delays development of superficial scald. Postharvest Biol Tec 4:225-233.

Giovannoni J (2001). Molecular biology of fruit maturation and ripening. Annu Rev Plant Phys 52:725-749.

Hofman PJ, Jobin-Décor M, Meiburg GF, Macnish AJ, Joyce DC (2001). Ripening and quality responses of avocado, custard apple, mango and papaya fruit to 1-methylcyclopropene. Aust J Exp Agric 41:567-572.

Jeong J, Huber DJ, Sargent SA (2002). Influence of 1-methylcyclopropene (1-MCP) on ripening and cell-wall matrix polysaccharides of avocado (*Persea americana*) fruit. Postharvest Biol Tec 25:241-364.

Jiang Y, Joyce DC, Terry LA (2001). 1-methylcyclcopropene treatment affects strawberry fruit decay. Postharvest Biol Tec 23:227-232.

Jiang W, Sheng Q, Zhou X, Zhang M, Liu X (2002). Regulation of coriander senescence by 1-methylcyclopropene and ethylene. Postharvest Biol Tec 26:339-345.

Khan AS, Zora S (2007). 1-MCP regulates ethylene biosynthesis and fruit softening during ripening of 'Tegan Blue' plum. Postharvest Biol Tec 43:298-306.

Oliveira AP, Valentão P, Pereira J, Silva BM, Tavares F, Andrade PB (2009). *Ficus carica* L.: Metabolic and biological screening. Food Chem Toxicol 47:2841-2846.

Owino OW, Nakano R, Kubo Y, Akitsugu I (2004). Alterations in cell wall polysaccharides during ripening in distinct anatomical tissue regions of the fig (*Ficus carica* L.) fruit. Postharvest Biol Tec 32:67-77.

Özkaya O, Dündar Ö, Demircioglu H (2010). Ripening and quality response of 'Blackamber' plums to 1-Methylcyclopropene during long-term storage. J Food Agric Environ 8:284-291.

Özkaya O, Dündar Ö (2009). Influence of 1-methylcyclopropene (1-MCP) on 'Fuji' apple quality during long-term storage. J Food Agric Environ 7:146-148.

Porat R, Weiss B, Cohen L, Daus A, Goren R, Droby S (1999). Effects of ethylene and 1-methylcyclopropene on the postharvest qualities of 'Shamouti' oranges. Postharvest Biol Tec 15:155-163.

Rubnov S, Kashman Y, Rabinowitz R, Schlesinger M, Mechoulam R (2001). Suppressors of cancer cell proliferation from fig (*Ficus carica*) resin: isolation and structure elucidation. J Nat Prod 64:993-996.

Saltveit ME (1999). Effect of ethylene on quality of fresh fruits and vegetables. Postharvest Biol Tec 15:279-92.

Selvarajah S, Bauchot AD, John P (2001). Internal browning in cold-stored pineapples is suppressed by a postharvest application of 1-methylcyclopropene. Postharvest Biol Tec 23:167-170.

Sisler EC, Serek M (1997). Inhibitors of ethylene responses in plants at the receptor level. Recent developments. Physiol Plantarum 100:577-582.

Solomon A, Golubowicz S, Yablowicz Z, Grossman S, Bergman M, Gottlieb H, Altman A, Kerem Z, Flaishman MA (2006). Antioxidant activities and anthocyanin content of fresh fruits of common fig (*Ficus carica* L.). J Agric Food Chem 54:7717-7723.

Tian MS, Prakash S, Elgar HJ, Young H, Burmeister DM, Ross GS (2000). Responses of strawberry fruit to 1-methylcyclopropene (1-MCP) and ethylene. Plant Growth Regul 32:83-90.

Valero D, Martínez -Romero D, Valverde JM, Guillen F, Serrano M (2003). Quality improvement and extension of shelf life by 1-methylcyclopropene in plum as affected by ripening stage at harvest. Innov Food Sci Emerg 4:339-348.

Veberic R, Colaric M, Stampar F (2008). Phenolic acids and flavonoids of fig fruit Ficus carica in the northern Mediterranean region. Food Chem 106:153-157.

Voss HD (1992). Relating colourimeter measurement of plant colour to the Royal Horticultural Society Colour Chart. HortScience 27:12.

Watkins CB (2002). Ethylene synthesis, mode of action, consequences and control, p. 180-224. In: Knee M (ed.). Fruit quality and its biological basis. Boca Raton, Florida: Sheffield Academic Press.

Watkins CB (2006). The use of 1-methylcyclopropene (1-MCP) on fruits and vegetables. Biotechnol Adv 24:389-409.

Wills RBH, Ku VVV (2002). Use of 1-MCP to extend the time to ripen of green tomatoes and postharvest life of ripe tomatoes. Postharvest Biol Tec 26:85-90.

Yuan G, Sun B, Yuan J, Wang Q (2010). Effect of 1-methylcyclopropene on shelf life, visual quality, antioxidant enzymes and health-promoting compounds in broccoli florets. Food Chem 118:774-781.

22

Calcium and Calmodulin Involve in Mycorrhizal and Root Development in Trifoliate Orange Colonized by *Rhizophagus intraradices*

Qiang-Sheng WU*, Qiu-Dan NI, Yan-Cong QUE, Wei HUANG

*Yangtze University, College of Horticulture and Gardening/Institute of Root Biology, Jingzhou, Hubei 434025, People's Republic of China; wuqiangsh@163.com (*corresponding author)*

Abstract

A pot experiment was made to study effects of ethylene glycol tetraacetic acid (EGTA, an inhibitor of Ca^{2+}) and trifluoperazine (TFP, an inhibitor of calmodulin (CaM) on mycorrhizal colonization, growth performance, and chlorophyll, sucrose and glucose concentrations of four-month-old trifoliate orange (*Poncirus trifoliata*) seedlings under mycorrhization with *Rhizophagus intraradices*. Exogenous EGTA and TFP notably inhibited root mycorrhizal colonization, and the addition of EGTA also decreased soil hyphal length. In general, EGTA treatment decreased but TFP increased easily extractable glomalin-related soil protein (EE-GRSP) and total GRSP (T-GRSP) concentrations. In addition, EGTA and TFP applications generally significantly inhibited growth performance (height, stem diameter, leaf number, and shoot and root biomass), root traits (total length, surface area, volume, and number of 1st, 2nd and 3rd order lateral root), and chlorophyll *a*, *b* and *a+b* concentrations, the mycorrhizal inoculation generally reversed the negative effects and markedly increased these variables, irrespective of whether the seedlings were applied by inhibitors or not. EGTA and TFP treatments generally inhibited sucrose and glucose levels of leaf and root, except that TFP application notably increased root glucose in AM and non-AM seedlings. AMF inoculation resulted in carbohydrate modification: decrease in leaf sucrose, increase in root sucrose and leaf glucose, as well increase in root glucose under no-inhibitor and EGTA conditions and decrease in root glucose under TFP. It suggests that Ca^{2+} and CaM were involved in mycorrhizal and root development in trifoliate orange.

Keywords: arbuscular mycorrhizal fungi, ethylene glycol tetraacetic acid, roots, trifluoperazine

Abbreviations: AM, arbuscular mycorrhiza; AMF, arbuscular mycorrhizal fungi; CaM, calmodulin; EE-GRSP, easily extractable glomalin-related soil protein; EGTA, ethylene glycol tetraacetic acid; GRSP, glomalin-related soil protein; TFP, trifluoperazine; T-GRSP, total glomalin-related soil protein

Introduction

Arbuscular mycorrhizal fungi (AMF), one of the most widespread endophytic microorganisms in soil, can establish symbiotic association, namely, arbuscular mycorrhizas (AMs), with the roots of most terrestrial plants. AM symbiosis can provide roots of the host with plenty of water and mineral nutrition from the soil (Wu *et al.*, 2011), in requital for up to 30% of the host's photosynthates for AM development (Kivlin *et al.*, 2011). During the formation and development of AM symbiosis, many diffusible symbiotic signals such as a mixture of sulphated and non-sulphated simple lipochitooligosaccharides are involved in the initial communication between AMF and the fungal partner, while the exact signal molecules are not well known (Harrison, 2005; Maillet *et al.*, 2011).

Calcium (Ca^{2+}) is considered as an universal second messenger in the signal transduction networks to induce a series of extracellular stimuli with particular physiological process (Hetherington and Brownlee, 2004; Dodd *et al.*, 2010). Plant cells receive the diffusible signals released by spore and extraradical mycelium of AMF to stimulate a Ca^{2+} dependent pathway (Navazio *et al.*, 2007; Navazio and Mariani, 2008). Spores of *Gigaspora margarita* were able to release some active molecules to arouse a Ca^{2+}-mediated signaling, thus, leading to the AM symbiosis-specific responses (Navazio *et al.*, 2007). As reported by Breuninger and Requena (2004), some genes related to Ca^{2+}-dependent signaling were up-regulated during interaction between AMF and plants, implying that Ca^{2+} as an intracellular messenger plays an important role in early stages of AM formation (Navazio *et al.*, 2007). Calmodulin (CaM) is a small and broadly distributed Ca^{2+}-binding protein, which is related with signal pathways and adventitious rooting (Liao *et al.*, 2012). It is not well known whether the CaM mediated mycorrhizal and root development.

Plant roots generally possess various physiological functionings including water and nutrient uptake, anchorage to the soil, and tolerance capacity of abiotic and biotic stresses (Lopez-Bucio *et al.*, 2003). Emerging evidence implicates that root development is involved in signal transduction (Zamioudis *et al.*, 2013). Wu *et al.* (2012) have shown that AMF mediated the root morphology in citrus plants through the endogenous polyamine metabolism. Even so, the AM-mediated mechanisms of root morphology are not yet fully understood. Liao *et al.* (2012) reported that Ca^{2+} might act as an upstream regulator for CaM in the process of adventitious rooting. In addition, CaM mainly distributed in cortical microtubules of root meristematic zone and in the root tip (Fraichard *et al.*, 1996).

Citrus, one of the most important fruit trees, is widely planted in the world including the Southern regions of China and strongly depends on AMs due to short and poor root hairs (Wu *et al.*, 2013a). The objective of the present work was to investigate the influences of exogenous Ca^{2+}/CaM inhibitors on plant growth, carbohydrate concentrations, and AM and root development in citrus seedlings.

Materials and methods

Plant culture

After sterilized with 70% alcohol for 15 min, seeds of trifoliate orange [*Poncirus trifoliata* (L.) Raf.] were germinated in plastic boxes containing autoclaved sands in a controlled growth chamber, where day/night temperature is 28/20 °C, photosynthetic photon flux density 745 μmol/m^2/s, and relative humidity 80%. The five-leaf-old seedlings with similar size and non-mycorrhization were transferred into a plastic pot (13 cm in depth and 15.5 cm in mouth diameter) containing the autoclaved (0.11 MPa, 121 °C, 40 min) substrates of yellow soil and vermiculite (2:1, v/v) on March 15, 2012. The AMF inoculated pots received 50 g inoculums of *Rhizophagus intraradices* (N.C. Schenck & G.S. Sm.) C. Walker & A. Schüßler at 5 cm soil depth, and the non-AMF pots also received the same quantity sterilized inoculums as the control. The fungal strain was isolated from the rhizosphere of *Lycopersicon esculentum* in Chaoyang District, Beijing. The AM fungus was propagated with *Sorghum vulgare* as the host for 16 weeks under pot culture in terms of the identified spores, and the inocula were comprised of both spores (35 spores/g) and sands (root AM colonization was 89% at harvest). The AM and non-AM seedlings were grown in a plastic greenhouse of Yangtze University (Jingzhou, China) from March to July 2012.

After 16 weeks of acclimation or AMF inoculation, 250 mL 0.5 mM ethylene glycol-bis (β-aminoethyl ether)-N,N,N',N'-tetraacetic acid (EGTA, an inhibitor of Ca^{2+}, from Tianda Chemical Reagent Factory, Tianjin, China) and 150 μM trifluoperazine (TFP, an inhibitor of CaM, from SIGMA-ALDRICH) were exogenously applied to the designed pots at an interval of two days from June 7 to July 23, respectively. Meanwhile, the other pots received 250 mL distilled water as the control.

Experimental design

In a 3^2 completely randomized arrangement, six treatments with three replicates each for a total of 18 pots (two seedlings/pot) were as follows: (i) non-AMF control (–AMF), (ii) AM *R. intraradices* only (+AMF), (iii) EGTA plus *R. intraradices* (EGTA+AMF), (iv) EGTA only (EGTA–AMF), (v) TFP plus *R. intraradices* (TFP+AMF), and (vi) TFP only (TFP–AMF).

Measurements of plant growth and root traits

After twenty times of EGTA and TFP applications, the AM and non-AM seedlings were harvested, and plant height, stem diameter, and leaf number per plant were directly recorded. And then, these seedlings were divided into shoots and roots, and the number of 1st-, 2nd-, and 3rd-order lateral roots was determined by means of manual count. The root systems per seedlings were scanned with an Epson scanner, and total length, surface area and volume were automatically obtained by a professional WinRHIZO root analysis system. The dry biomass of shoots and roots was determined after 75 °C for 48 h.

Measurements of mycorrhizal development

Root mycorrhizas were stained with the procedure reported by Phillips and Hayman (1970), and the root AM colonization was expressed as the percentage of colonized root lengths versus observed root lengths. Soil hyphal length was determined by the protocol described by Bethlenfalvay and Ames (1987).

The easily extractable glomalin-related soil protein (EE-GRSP) and total GRSP (T-GRSP) were done following the method of Zou *et al.* (2014).

Determinations of chlorophyll and sugar concentrations

The concentration of chlorophyll *a* and *b* was determined using the method described by Lichtenthaler and Wellburn (1983).

Glucose and sucrose of leaves and roots were extracted with 80% of ethanol and determined with the colorimetric method of Zhang and Zai (2004).

Statistical analysis

The data were analyzed by one-way ANOVA with SAS software, and the Duncan's multiple range test was used to compare the significant differences between the treatments at the 5% level.

Results

Mycorrhizal development

Root AM colonization varied from 22.81 to 32.05% in the inoculated seedlings, whereas no AMF colonization was found in the non-inoculated seedlings (Tab. 1). Compared to sole AMF treatment, application of EGTA or TFP significantly inhibited root AM colonization by 29 or 25%, respectively. In addition, in the AM seedlings, EGTA treatment significantly decreased soil hyphal length but TFP did not alter the parameter (Tab. 1).

In general, the AM seedlings recorded significantly higher rhizospheric EE-GRSP and T-GRSP concentrations

Tab. 1. Mycorrhizal development of *Rhizophagus intraradices* - colonized trifoliate orange (*Poncirus trifoliata*) seedlings applied by EGTA and TFP

Treatment	Root AM colonization (%)	Soil hyphal length (m/g FW)	Rhizospheric GRSP (mg/g DW)	
			EE-GRSP	T-GRSP
–AMF	0	0	0.69±0.02^{c}	0.97±0.03^{c}
+AMF	32.05±3.5^{3a}	0.21±0.02^{a}	0.79±0.01^{b}	1.19±0.01^{a}
EGTA–AMF	0	0	0.63±0.02^{d}	0.89±0.03^{d}
EGTA+AMF	22.81±1.04^{b}	0.10±0.01^{b}	0.70±0.01^{c}	1.06±0.03^{b}
TFP–AMF	0	0	0.74±0.04^{bc}	1.04±0.04^{b}
TFP+AMF	23.91±1.30^{b}	0.21±0.02^{a}	0.85±0.05^{a}	1.24±0.05^{a}

Same letter within each column indicates no significant difference among treatments (P<0.05)

than the non-AM seedlings, irrespective of exogenous inhibitor types (Tab. 1). Compared to the no-inhibitor treatment, EGTA application significantly decreased rhizospheric EE-GRSP and T-GRSP concentrations in AM or non-AM seedlings, whilst TPF significantly increased rhizospheric EE-GRSP concentration in AM seedlings and rhizospheric T-GRSP concentration in non-AM seedlings.

Growth performance

Mycorrhizal inoculation significantly increased plant height, stem diameter, and shoot and root dry biomass than non-mycorrhizal treatment, irrespective of exogenous inhibitors applied or not (Tab. 2). The AMF treatment and the inhibitor applications did not alter leaf number per seedling. In the non-AM seedlings, EGTA significantly decreased stem diameter and shoot dry biomass, and TPF significantly inhibited shoot and root dry biomass production. In the AM seedlings, both EGTA and TFP significantly decreased plant height, stem diameter, and shoot and root dry biomass, except a similar root dry biomass between TFP+AMF treatment and +AMF treatment.

Root morphological traits

Compared to non-AMF control, AMF inoculation generally significantly increased root total length, surface area, volume, and number of 1st-, 2nd- and 3rd-order lateral roots, irrespective of the inhibitors applied or not, except a similar 1st-order lateral root number between +AMF and –AMF and between TFP+AMF and TFP–AMF (Tab. 3). In the non-AM seedlings, EGTA significantly decreased total length, surface area, volume, and 1st-order lateral root number, and TFP significantly decreased total length, surface area, and 1st-, 2nd- and 3rd-order lateral root number. In the AM seedlings, EGTA significantly inhibited total length, surface area, and first-order lateral root number, and TFP notably inhibited 1st-, 2nd- and 3rd-order lateral root number.

Chlorophyll concentrations

Mycorrhizal seedlings showed significantly higher chlorophyll *a*, *b*, and *a+b* concentrations than non-mycorrhizal seedlings, regardless of exogenous inhibitors or not (Tab. 4), except that chlorophyll *b* concentration was

Tab. 2. Effect of EGTA and TFP on growth performance of trifoliate orange (*Poncirus trifoliata*) seedlings inoculated with and without *Rhizophagus intraradices*

Treatment	Plant height (cm)	Leaf number	Stem diameter (mm)	Dry biomass (g/plant)	
				Shoot	Root
–AMF	20.1±2.0^{bcd}	20.0±1.0^{a}	2.18±0.04^{c}	0.86±0.06^{b}	0.30±0.02^{bc}
+AMF	25.3±1.6^{a}	20.3±2.1^{a}	2.59±0.04^{a}	0.99±0.04^{a}	0.43±0.02^{a}
EGTA–AMF	18.6±1.0^{cd}	19.0±2.0^{a}	1.96±0.06^{d}	0.70±0.04^{c}	0.26±0.02^{cd}
EGTA+AMF	22.3±1.3^{b}	19.7±1.5^{a}	2.34±0.10^{b}	0.81±0.05^{b}	0.32±0.02^{b}
TFP–AMF	18.2±1.1^{d}	18.3±2.1^{a}	2.08±0.06^{cd}	0.41±0.03^{d}	0.24±0.03^{d}
TFP+AMF	20.8±0.8^{bc}	20.7±1.2^{a}	2.45±0.09^{b}	0.77±0.09^{bc}	0.40±0.03^{a}

Same letter within each column indicates no significant difference among treatments (P<0.05)

Tab. 3. Effects of EGTA and TFP on root morphological traits of trifoliate orange (*Poncirus trifoliata*) seedlings inoculated with and without *Rhizophagus intraradices*

Treatment	Total length (cm)	Surface area (cm^2)	Volume (cm^3)	Number of lateral root		
				1st-order	2nd-order	3rd-order
–AMF	239±9^{c}	33.6±2.5^{c}	0.38±0.04^{c}	43±5^{ab}	78±2^{bc}	10±1^{c}
+AMF	455±26^{a}	56.3±3.4^{a}	0.56±0.05^{b}	47±2^{a}	102±9^{a}	16±2^{a}
EGTA–AMF	206±16^{d}	27.6±2.5^{d}	0.29±0.07^{d}	28±1^{e}	63±7^{cd}	9±2^{cd}
EGTA+AMF	378±11^{b}	49.7±2.9^{b}	0.52±0.05^{b}	40±3^{bc}	88±11^{ab}	14±2^{ab}
TFP–AMF	202±17^{d}	29.0±1.4^{d}	0.33±0.02^{cd}	32±3^{de}	60±6^{d}	6±3^{d}
TFP+AMF	428±18^{a}	58.4±0.8^{a}	0.64±0.04^{a}	36±1^{cd}	77±13^{bc}	11±1^{bc}

Same letter within each column indicates no significant difference among treatments (P<0.05)

Tab. 4. Effects of EGTA and TFP on chlorophyll *a*, *b* and *a+b* concentrations and sucrose and glucose concentrations in leaf and root of trifoliate orange (*Poncirus trifoliata*) seedlings inoculated with and without *Rhizophagus intraradices*

Treatment	Chlorophyll (mg/g FW)			Sucrose (mg/g DW)		Glucose (mg/g DW)	
	a	*b*	*a+b*	Leaf	Root	Leaf	Root
−AMF	1.67±0.11[b]	0.62±0.07[b]	2.29±0.14[b]	77.34±3.04[a]	51.92±1.68[a]	88.52±3.82[ab]	36.00±3.14[d]
+AMF	2.01±0.11[a]	0.79±0.10[a]	2.80±0.19[a]	70.75±4.39[b]	57.08±0.37[a]	97.74±5.25[a]	45.16±3.21[c]
EGTA−AMF	0.25±0.03[e]	0.24±0.04[d]	0.49±0.01[e]	44.37±2.58[d]	25.96±2.89[c]	59.82±8.44[e]	28.34±2.31[e]
EGTA+AMF	0.63±0.04[d]	0.45±0.08[c]	1.09±0.09[c]	48.38±1.11[d]	35.09±8.34[b]	69.64±10.47[de]	32.78±0.58[de]
TFP−AMF	0.62±0.02[d]	0.18±0.03[d]	0.80±0.03[d]	56.54±3.59[c]	36.90±1.94[b]	75.92±1.80c[d]	56.77±3.63[a]
TFP+AMF	0.81±0.03[c]	0.30±0.03[d]	1.11±0.06[c]	44.56±2.41[d]	52.26±1.33[a]	83.19±3.53[bc]	50.05±1.80[b]

Same letter within each column indicates no significant difference among treatments ($P<0.05$)

not significant between TFP-AMF and TFP+AMF seedlings. In addition, EGTA and TFP applications notably decreased chlorophyll *a*, *b*, and *a+b* concentrations than non-applied control, irrespective of AM or not.

Sucrose and glucose concentrations

In seedlings inoculated with AMF only, AM seedlings recorded significantly lower leaf sucrose concentration and higher root glucose concentration (Tab. 4). In seedlings applied by EGTA, inoculated seedlings showed significantly higher root sucrose concentration than un-inoculated seedlings. In seedlings applied by TFP, AM seedlings represented significantly higher root sucrose concentration but lower leaf sucrose and root glucose concentrations, as compared with non-AM seedlings. EGTA and TFP applications notably decreased leaf and root sucrose concentration and leaf glucose concentration, irrespective of AM or not, except a similar root sucrose between +AMF and TFP+AMF. In addition, addition of EGTA significantly decreased root glucose concentration but TFP markedly increased root glucose concentration than no-inhibitor treatment.

Discussion

In a plant-AMF system, transient Ca^{2+} changes in plant cells have been released by AMF (Navazio and Mariani, 2008), implying that Ca^{2+} as a symbiotic signal is generated during AM formation (Navazio *et al.*, 2007; Chabaud *et al.*, 2011). Our study showed that exogenous EGTA (an inhibitor of Ca^{2+}) significantly inhibited root AM colonization, soil hyphal length, and rhizospheric EE-GRSP and T-GRSP production, suggesting that Ca^{2+} might participate in AM establishment, soil hyphal growth, and release of GRSP fractions. On the other hand, TFP (an inhibitor of CaM) did not alter soil hyphal length and rhizospheric T-GRSP level, but significantly decreased root AM colonization and increased EE-GRSP level, suggesting that CaM might involve only in the root AM development but not in the soil hyphal development and behaviour. It is well documented that Ca^{2+} acts as an upstream regulator for CaM (Liao *et al.*, 2012). Therefore, CaM as the downstream product may less take part in the AMF-plant-soil system.

The present results showed that the exogenous applications of EGTA and TFP had negative effects on plant height, stem diameter, leaf number, and shoot and root biomass in AM and non-AM seedlings, which agrees with those obtained by Xu *et al.* (2009) and Ozgen *et al.* (2011). Nevertheless, AMF inoculation significantly improved these growth traits under no-inhibitor, EGTA and TFP conditions than non-AMF treatment, indicating that AMF alleviated the inhibition of EGTA and TFP to plant growth in a certain extent. It seems that CaM/Ca^{2+} mediated the functioning of AMF or vice-versa (Conn *et al.*, 2011; Poovaiah *et al.*, 2013).

CaM or Ca^{2+} as a signal transducer acts upon the development of lateral roots (Kanchiswamy *et al.*, 2014). In the present study, the application of EGTA and TFP considerably limited the root total length, surface area, volume, and number of 1st-, 2nd- and 3rd-order lateral roots, which is in agreement with Muto and Hirosawa (1987). The result also suggested that CaM and Ca^{2+} closely connect with root growth, because Ca^{2+}/CaM is involved in auxin-induced adventitious root development (Liao *et al.*, 2012). Presence of AMF, in the present study, had the ability to alleviate or eliminate the negative effects of EGTA and TFP on root morphology. However, it is not clear whether the AM-mediated root improvement is related either with AM-induced Ca^{2+} and CaM levels or with improvement of nutrients by mycorrhization. In roots, CaM mainly distributed in cortical microtubules of root meristematic zone and in the root tip (Fraichard *et al.*, 1996), where the AMF less colonized. The TFP application did not alter root total length and surface area, but increased root volume under mycorrhization and notably decreased root total length and surface area and did not alter root volume under non-mycorrhization. It seems that the presence of AMF has the ability to defend the CaM inhibitor, and therefore, better root development occurs in AM plants.

Our study showed that under no-inhibitor of Ca^{2+} or CaM conditions, AMF significantly enhanced chlorophyll *a*, *b*, and *a+b* concentrations relative to non-AMF control, which is consistent with a previous study reported by Campanelli (2013). In addition, chlorophyll *a*, *b*, and *a+b* concentrations were inhibited by exogenous applications of EGTA and TFP under non-mycorrhization, which agree with Lechowski and Bialczyk (1993). It implies that CaM and Ca^{2+} may be involved in the activity of chlorophyll composition. However, the integration of AMF with EGTA or TFP significantly increased these three chlorophyll concentrations than the EGTA/TFP-AMF, indicating that AMF could alleviate the negative effects of EGTA and TFP on activity of chlorophyll composition.

Mycorrhizal symbiosis would request 4–20% of plant fixed carbohydrates for development, especially hexose (Finlay, 2008). In the present study, AMF inoculation generally significantly decreased leaf sucrose concentration but significantly increased root sucrose concentration, irrespective of inhibitor types, confirming that AMs might shift an allocation of sucrose to pools for cleaving (Wu *et al.*, 2013b). Moreover, EGTA and TFP applications generally inhibited leaf and root sucrose concentrations, whilst AM presence altered the inhibited effects: decrease in leaf and increase in root, suggesting that AM-mediated sucrose modification was dependent on tissue position. Although EGTA and TFP significantly decreased leaf glucose, AMF could eliminate the negative effects of EGTA and TFP. It is well documented that AMF could induce cleavage of sucrose into glucose and fructose into the root (Wu *et al.*, 2013b), thereby, triggering better translocation of glucose, leading to better root and shoot biomass yield. In the present study, root glucose was notably decreased by EGTA treatment but increased by TFP, whilst AMF inoculation did not significantly alter root glucose under EGTA but decreased under TFP and increased under no-inhibitor. These results illuminate that AM-induced root glucose modification might be regulated by Ca^{2+} or CaM. In addition, glucose not only controls root growth and development, but also modulates AM development (Mishra *et al.*, 2009). As a result, higher glucose in AM leaf would benefit AM and root development. On the other hand, increasing concentration of root glucose under no-inhibitor and EGTA conditions and decreasing concentration of root glucose under TFP caused by mycorrhization may be due to the integration of AMs and Ca^{2+} or CaM, which needs to be further studied.

Conclusions

The present study indicated that Ca^{2+} involved in root mycorrhizal colonization, soil hyphal development and GRSP production, and CaM involved in root colonization and GRSP production. AMF inoculation could reverse the negative effects of EGTA and TFP on plant growth performance, root traits, and chlorophyll concentrations, suggesting that AMF might mediate endogenous Ca^{2+} or CaM for normal physiology. Plant carbohydrates would be altered by the integration with EGTA and TFP or AMF inoculation, which might benefit root and shoot biomass production.

Acknowledgements

This study was supported by the Excellent Young Teacher Research Support Program of Yangtze University (cyq201324) of China.

References

Bethlenfalvay GJ, Ames RN (1987). Comparison of two methods for quantifying extraradical mycelium of vesicular arbuscular mycorrhizal fungi. Soil Sci Soc Am J 51:834-837.

Breuninger M, Requena N (2004). Recognition events in AM symbiosis: analysis of fungal gene expression at early appressorium stage. Fungal Genet Biol 41:794-804.

Campanelli A, Ruta C, de Mastro G, Morone-Fortunato I (2013). The role of arbuscular mycorrhizal fungi in alleviating salt stress in *Medicago sativa* L. var. icon. Symbiosis 59:65-76.

Chabaud M, Genre A, Sieberer BJ, Faccio A, Fournier J, Novero M, Barker DG, Bonfante P (2011). Arbuscular mycorrhizal hyphopodia and germinated spore exudates trigger Ca^{2+} spiking in the legume and nonlegume root epidermis. New Phytol 189:347-355.

Conn S, Gilliham M, Athman A, Schreiber AW, Baumann U, Moller I, Cheng NH, Stancombe MA, Hirschi KD, Webb AA, Burton R, Kaiser BN, Tyerman SD, Leigh RA (2011). Cell-specific vacuolar calcium storage mediated by CAX1 regulates apoplastic calcium concentration, gas exchange, and plant productivity in *Arabidopsis*. Plant Cell 23:240-257.

Dodd AN, Kudla J, Sanders D (2010). The language of calcium signaling. Annu Rev Plant Biol 61:593-620.

Finlay RD (2008). Ecological aspects of mycorrhizal symbiosis: with special emphasis on the functional diversity of interactions involving the extraradical mycelium. J Exp Bot 59:1115-1126.

Fraichard A, Perotti E, Gavin O, Chanson A (1996). Subcellular localization, distribution and expression of calmodulin in *Zea mays* roots. Plant Sci 118:157-165.

Harrison MJ (2005). Signaling in the arbuscular mycorrhizal symbiosis. Annu Rev Microbiol 59:19-42.

Hetherington AM, Browlee C (2004). The generation of Ca^{2+} signals in plants. Annu Rev Plant Biol 55:401-427.

Kanchiswamy CN, Malnoy M, Occhipinti A, Maffei ME (2014). Calcium imaging perspectives in plants. Int J Mol Sci 15:3842-3859.

Kivlin SN, Hawkes CV, Treseder KK (2011). Global diversity and distribution of arbuscular mycorrhizal fungi. Soil Biol Biochem 43:2294-2303.

Lechowski Z, Bialczyk J (1993). Calcium mediated cytokinin action on chlorophyll synthesis in isolated embryo of Scots pine. Biol Plant 35:53-62.

Liao WB, Zhang ML, Huang GB, Yu JH (2012). Ca^{2+} and CaM are involved in NO- and H_2O_2-induced adventitious root development in marigold. J Plant Growth Regul 31:253-264.

Lichtenthaler HK, Wellburn AR (1983). Determination of total carotenoids and chlorophyll a and b of leaf extracts in different solvents. Biochem Soc Trans 603:591-603.

Lopez-Bucio J, Cruz-Ramirez A, Herrera-Estrella L (2003). The role of nutrient availability in regulating root architecture. Curr Opin Plant Biol 6:280-287.

Maillet F, Poinsot V, Andre O, Puech-Pages V, Haouy A, Gueunier M, Cromer L, Giraudet D, Formey D, Niebel A, Martinez EA, Driguez H, Becard G, Denarie J (2011). Fungal lipochitooligosaccharide symbiotic signals in arbuscular mycorrhiza. Nature 469:58-63.

Mishra BS, Singh M, Laxmi A (2009). Glucose and auxin signaling interaction in controlling *Arabidopsis thaliana* seedlings root growth and development. PLoS ONE 4:e4502.

Muto S, Hirosawa T (1987). Inhibition of adventitious root growth in tradescantia by calmodulin antagonists and calcium inhibitors. Plant Cell Physiol 28:1569-1574.

Navazio L, Mariani P (2008). Calcium opens the dialogue between plants and arbuscular mycorrhizal fungi. Plant Signal Behav 3:229-230.

Navazio L, Moscatiello R, Genre A, Novero M, Baldan B, Bonfante P, Mariani P (2007). A diffusible signal from arbuscular mycorrhizal fungi elicits a transient cytosolic calcium elevation in host plant cells. Plant Physiol 144:673-681.

Ozgen S, Busse JS, Palta JP (2011). Influence of root zone calcium on shoot tip necrosis and apical dominance of potato shoot: simulation of this disorder by ethylene glycol tetra acetic acid and prevention by strontium. HortScience 46(10):1358-1362.

Phillips JM, Hayman DS (1970). Improved procedures for clearing roots and staining parasitic and vesicular-arbuscular mycorrhizal fungi for rapid assessment of infection. Trans Br Mycol Soc 55:158-161.

Poovaiah BW, Du LQ, Wang HZ, Yang TB (2013). Recent advances in calcium/calmodulin-mediated signaling with an emphasis on plant-microbe interactions. Plant Physiol 163: 531-542.

Wu QS, Li GH, Zou YN (2011). Roles of arbuscular mycorrhizal fungi on growth and nutrient acquisition of peach (*Prunus persica* L. batsch) seedlings. J Anim Plant Sci 21:746-750.

Wu QS, Srivastava AK, Zou YN (2013a). AMF-induced tolerance to drought stress in citrus: A review. Sci Hortic 164:77-87.

Wu QS, He XH, Zou YN, Liu CY, Xiao J, Li Y (2012). Arbuscular mycorrhizas alter root system architecture of *Citrus tangerine* through regulating metabolism of endogenous polyamines. Plant Growth Regul 68:27-35.

Wu QS, Zou YN, Huang YM, Li Y, He XH (2013b). Arbuscular mycorrhizal fungi induce sucrose cleavage for carbon supply of arbuscular mycorrhizas in citrus genotypes. Sci Hortic 160:320-325.

Xu T, Li TL, Qi MF (2009). Calcium effects on mediating polygalacturonan activity by mRNA expression and protein accumulation during tomato pedicel explants abscission. Plant Growth Regul 60:255-263.

Zamioudis C, Mastranesti P, Dhonukshe P, Blilou I, Pieterse CMJ (2013). Unraveling root developmental programs initiated by beneficial *Pseudomonas* spp. bacteria. Plant Physiol 162:304-318.

Zhang ZL, Zai LJ (2004). Experimental Instruction of Plant Physiology. 3rd ed. Higher Education Press, Beijing, 328 p.

Zou YN, Srivastava AK, Wu QS, Huang YM (2014). Glomalin-related soil protein and water relations in mycorrhizal citrus (*Citrus tangerina*) during soil water deficit. Arch Agron Soil Sci 60:1103-1114.

23

Cutting Type and IBA Treatment Duration Affect *Teucrium fruticans* Adventitious Root Quality

Leo SABATINO, Fabio D'ANNA, Giovanni IAPICHINO*

*University of Palermo, Department of Agricultural and Forest Sciences, Viale delle Scienze, 90128, Palermo, Italy; leo.sabatino@unipa.it; fabio.danna@unipa.it; giovanni.iapichino@unipa.it (*corresponding author)*

Abstract

Root development of stem cuttings of Silver Germander (*Teucrium fruticans*) was investigated in relation to cutting type and indole-3-butyric acid (IBA) treatment. Terminal cuttings of a clone grown in Sicily were trimmed to three types: terminal cuttings with apex (TWA), terminal cuttings without apex (TWOA) or sub-terminal cuttings (ST). To verify the cutting response to exogenous auxin, cuttings were dipped to a 2.0 cm depth in a 0.5% indole-3-butyric acid solution for 0, 5 or 7 minutes. Overall percent survival was 97 to 98%. Rooting percent, root number and root length were affected by cutting type and indole-3-butyric acid treatment. In general, TWA cuttings demonstrated a higher capacity to form roots than cuttings without apex (TWOA and ST cuttings). In absence of indole-3-butyric acid treatment, TWA and ST cuttings gave higher rooting percentages than TWOA cuttings. Exposing cuttings to indole-3-butyric acid improved percent rooting, number of roots per cutting and root length. The best results in terms of rooting percentage and root number per cuttings were obtained with TWA cuttings in combination with 7 min indole-3-butyric acid basal dip. However, cuttings taken farther down the stem, such as sub-terminal cuttings gave satisfactory rooting performance as well. We suggest that the use of all cutting types tested associated to indole-3-butyric acid basal dip for 5 or 7 min may be beneficial to propagators wishing to produce *T. fruticans* rooted cuttings with well-developed root system.

Keywords: basal dip, growth regulators, Labiatae, native plants, rooting

Introduction

The genus *Teucrium* belongs to the family Labiatae and contains more than 300 perennial herbs, shrubs and sub-shrubs mainly native to Mediterranean region and south-western Asia (Bryant, 2003). *Teucrium fruticans*, also known as Silver Germander, is a shrub native to Southern Europe and North Africa (Graham, 1992). Cultivated for its attractive blue/violet flowers and evergreen foliage, gray-green above and silver-white beneath, *T. fruticans* is high drought tolerant and suitable for landscaping in areas with Mediterranean-type environment (Cassaniti *et al.*, 2009; La Mantia *et al.*, 2012; Lopez *et al.*, 2006). Four-year old plants (sold in 22-cm-diameter pots) obtained by rooted cuttings are generally used for landscaping. *Teucrium fruticans* propagation by seed is difficult because of poor germination (Bryant, 2003) and high percentage of male sterility (Ojeda and Dìez, 1992); as a result, plant propagation via seed is unsatisfactory for the commercial production (Frabetti *et al.*, 2009). The species has been also characterized as difficult to propagate by cutting in certain periods of the year (Frangi and Nicola, 2004). To circumvent these problems micropropagation has been recommended (Frabetti *et al.*, 2009). Vegetative propagation by stem cuttings has been successfully applied to other *Teucrium* species (*T. divaricatum*, *T. polium*, *T. chamaedrys*, *T. flavum*, *T. scorodomia*) (Dirr and Heuser, 2006; Iapichino *et al.*, 2006; MacDonald, 1987; Maloupa *et al.*, 2008). Although the effects of exogenous auxin on adventitious root formation in *Teucrium* species has been documented (Dirr and Heuser 1987; Frangi and Nicola 2004; Maloupa *et al.*, 2008) no research on the influence of the morphological characteristics of the cuttings on rooting of *T. fruticans* has been reported. The aim of our study was to examine rooting of stem cuttings in relation to cutting type and IBA treatment duration in a *T. fruticans* clone grown in Sicily.

Materials and methods

Plant material

The research was conducted at the University of Palermo, Experimental Station in the Northern coast of Sicily (Italy) (longitude 13° 19' E, latitude 38° 9' N). Terminal 30 cm long stem cuttings were harvested on 23 April 2012 from actively growing stock plants. Plants were 5 years old and from a single Sicilian unnamed clone. Cuttings were stored overnight at 6 °C. The next day, the cuttings were trimmed to three types:

softwood stem-tip cuttings in which the tip was retained (terminal cuttings with apex, TWA), softwood stem cuttings in which the tip was removed (terminal cuttings without apex, TWOA) or semi-hardwood cuttings taken farther down the stem (sub-terminal cuttings, ST) (Fig. 1).

Fig. 1. *Teucrium frutescens* cutting types tested; from left, terminal with apex, terminal without apex and sub-terminal

All cuttings were trinodal with the lower cut below a node at the base of each cutting. Leaves were removed from the base of the stem; length and diameter of the cuttings were 12-14 cm and 0.4 cm, respectively.

Auxin treatment

To verify the cutting response to exogenous auxin, cuttings were dipped to a 2.0 cm depth in a 0.5% indole-3-butyric acid (IBA) solution (Frangi and Nicola, 2004) for five or seven minutes. Reagent grade IBA (Sigma n. I 5386) was dissolved in 90% ethanol, then brought to final concentration of 6% ethanol 94% distilled water (v/v). Untreated cuttings were dipped in distilled water (control= 0 minutes).

Fig. 2. *Teucrium frutescens* cuttings in the rooting bench

Rooting conditions

Propagation was performed on 24 April 2012 in an unheated greenhouse covered with clear polyethylene (PE) and external 50% shade-cloth. Air temperature in the greenhouse was 18-22 °C during the day and 14-16 °C during the night. Cuttings were inserted 4 cm deep in a bottom heated bench containing perlite. Basal heat was provided at constant temperature of 25 ± 2 °C. Benches were covered with clear PE to maintain high relative humidity. Intermittent mist operated daily 30 sec every 2 hours from 8:30 AM to 6:00 PM. Ventilation of the cuttings was increased with time by increasing size of the holes made in the plastic.

Experimental design, data collection and analysis

The design was a randomized complete block (RCB) with three replications, nine treatments [three cutting types x three IBA exposure time duration (0, 5 or 7 minutes)] and 20 cuttings per replication. An IBA cutting exposition of a few seconds (quick dip method) was not included in our experimental design because in a preliminary experiment (performed in the same season) we did not find any significant increase in rooting capacity as compared to the untreated control. After 4 and 6 weeks, cuttings were evaluated for percent survival, percent rooting, number of roots, length of the six longest roots. Cuttings with non-withered stems (with and without roots) were considered to have survived. Cuttings with necrotic tissue without roots were classified as dead. Percentage data were subjected to arcsin transformation before ANOVA analysis. Mean separation was performed by Fisher least-significant-difference test at $p \leq 0.01$ (Petersen, 1985).

Plant establishment

Single plants were transferred to each plastic pot (diameter 16 cm) containing a peat-perlite mixture 1:1 (v/v) and kept for five weeks in a lath house covered with 70% shade-cloth. Acclimatized plants were thereafter transferred in the open field.

The experiment was repeated a second year using the same stock plants and obtaining similar results.

Results

Cutting survival and rooting percentage

Overall percent survival was 97 to 98%; no significant effect of cutting type and IBA treatment duration was found on cutting survival (Tab. 1). Rooting percent was affected by cutting type and IBA treatment (Tab. 1). In absence of IBA treatment, TWA and ST cuttings gave higher rooting percentages than TWOA cuttings. Percent rooting in TWA cuttings increased from 70% in absence of IBA treatment to 100 and 96% in the presence of IBA for 5 and 7 min, respectively; percent rooting in TWOA cuttings also increased from 56% in untreated control to 100% in cuttings receiving either 5 or 7 min IBA basal dips. Percent data recorded on ST cuttings paralleled data recorded on TWA and TWOA cuttings exposed to IBA giving higher performance than control.

Root number

The number of roots was affected by cutting type and IBA treatment. In general, TWA cuttings demonstrated a higher capacity to form roots than cuttings without apex (TWOA and ST cuttings). Exposing cuttings to IBA improved the number of roots per cutting (Fig. 2). Data collected 4 weeks after planting revealed that formed roots on TWA cuttings increased from 1.0 in untreated control to 14 and 18 roots on stem cuttings receiving 5 and 7 min IBA basal dips, respectively. Root count in TWOA cuttings also increased from 1.5 roots in absence of IBA to 11 roots at 5 min duration and to 20 roots at 7 min duration. Data collected on ST cuttings supported the trend established for the other two cutting types with a maximum of 13 roots for the cuttings receiving 7 min IBA basal dip. Data collected after 6 weeks again revealed that rooting TWA, TWOA and ST cuttings in the presence of IBA substantially increased the number of roots per cutting as compared to untreated cuttings. All cutting types generally produced higher number of roots per cutting with 7 min IBA basal dip as compared with 5 min IBA basal dip. Among the treatments tested, TWA cuttings in combination with 7 min IBA basal dip was the most effective in terms of number of adventitious roots produced.

Tab. 1. Effects of cutting type and IBA treatment on *Teucrium fruticans* rooting

Treatments		Roots per cutting (n)		Root length (mm)		Rooting (%)	Survival (%)
Cutting type	IBA (min.)	Weeks after planting					
		4	6	4	6	6	6
TWA	0	1.0 ± 0.8 e	11.4 ± 4.2 d	3.5 ± 1.6 de	6.8 ± 2.5 e	70 ± 0.3d	97.0 ± 1.3 a
TWA	5	14.7 ± 3.8 b	18.4 ±3.9 bc	5.9 ± 0.5 ab	10.1 ± 3.9 abc	100 a	98.0 ± 2.8 a
TWA	7	18.5 ± 3.3 a	24.0 ± 3.6 a	4.7 ± 1.8 bcd	11.5 ± 4.1a	96 ± 2.8ab	97.1 ± 2.5 a
TWOA	0	1.5 ± 0.7 e	4.0 ± 2.4 e	2.5 ± 0.8 e	5.3 ± 2.8 f	56 ± 2.1e	98.0 ± 2.1 a
TWOA	5	11.0 ± 4.2 cd	14.3 ±2.5 d	7.0 ± 1.0 a	10.5 ± 4.1 ab	100 a	98.0 ± 1.8 a
TWOA	7	20.1 ± 1.5 a	20.2 ± 2.1ab	5.0 ± 2.1 abc	9.3 ± 2.1bcd	100 a	97.6 ± 2.8 a
ST	0	1.9 ± 0.6 e	6.1 ± 2.6 e	2.6 ± 1.2 e	5.0 ± 2.7 f	81 ± 3.4cd	98.0 ± 3.6 a
ST	5	9.7 ± 2.9 d	13.1± 4.7 d	3.2 ± 0.9 de	7.8 ± 3.2 e	100 a	98.0 ± 2.0 a
ST	7	13.0 ± 1.8 bc	15.0 ± 2.2 d	4.2 ± 1.6 cd	7.7 ± 3.1 e	92 ±1.7abc	97.6 ± 2.5 a

Values represent mean ± standard error. In each column, means followed by the same letters are not significantly different at the $p \leq 0.01\%$ level by Fisher's protected least significant difference test.
TWA = terminal cuttings with apex; TWOA = terminal cuttings without apex; ST = sub-terminal cuttings.

Fig. 3. *Teucrium frutescens* sub-terminal cuttings, exposed to 7 min IBA basal dip, showing a well developed rooting system

Root length

Root length also varied by cutting type and IBA treatment. Generally, TWA cuttings formed longer roots than cuttings without apex (TWOA and ST cuttings). Data collected four and six weeks after planting, revealed that root length increased from the untreated control to stem cuttings receiving IBA basal dips. However, no significant differences were found between 5 and 7 min IBA basal dip.

Plant establishment and performance

In the present study, propagated plants grew vigorously, were morphological normal and began to flower during spring 2013.

Discussion

Our results are similar to those of Krisantini *et al.* (2003) who report tip cuttings having a higher capacity to form roots than decapitated cuttings in *Grevillea* and are consistent with those of Marks (1996) who found inhibition of rooting following tip removal in *Quercus* and *Daphne* cuttings. The higher rooting percentage obtained using tip cuttings rather than stem (decapitated) cuttings has been hypothesized as due to production of a factor (presumably, IAA basipetally transferred from the meristems) that from the apex has an important role in adventitious root formation even though exogenous auxin is applied. In our study, the slightly higher, but not statistically different percentage rooting observed in ST cuttings as compared to TWA cuttings could be attributed to the fact that IAA was supplied from quiescent axillary buds. However, our study also demonstrated that the number of roots formed in cuttings with apices was superior to that observed in those without apices even in the presence of exogenous auxin.

In our study, the application of IBA improved rooting percentage and root number as compared to untreated control. IBA has been reported to markedly increase adventitious root formation in many species (Blazich, 1988; De Klerk *et al.*, 1999; Hartmann *et al.*, 1997). Our results are only partially in agreement with those obtained by Frangi and Nicola *et al.* (2004) who found IBA treatments enhancing rooting in *T. fruticans* stem tip cuttings. These authors report a low rooting response during spring (40% rooting) even applying exogenous auxin. Our higher response in the same season could be related to diverse genotype studied or different environmental conditions under which the stock plants were grown. Our results are in accord with the results obtained by Carpenter and Cornell (1992) who, investigating the adventitious root proliferation response of three *Hibiscus rosa-sinensis* cultivars, found that rooting is highly influenced by IBA application duration and concentration. Our results are also consistent with those of Telgen *et al.* (2007) who reported that diluted IBA-solutions and longer dipping times resulted in more uniform cuttings and higher root number than quick dips in *Rosa hybrida* 'First Red'.

In the present study, the best results in terms of rooting percentage and root number per cuttings were obtained with TWA cuttings. However, this propagation method yields fewer cuttings per plant compared with methods using cuttings taken farther down the stem, such as sub-terminal cuttings which, in our study, also performed well. Therefore, we suggest that all cutting types tested and exposed to an IBA basal dip of 5 or 7 min may be successfully used by propagators to produce *T. fruticans* rooted cuttings with well-developed root system.

This is an important information for nursery growers since there is little knowledge about cultural techniques of this species of increasing importance in the landscape and nursery trade.

References

Blazich FA (1988). Chemicals and formulations used to promote adventitious rooting, p.132-149. In: Davis TD, Haissig BE, Sankhla N (Eds.). Adventitious root formation in cuttings. Dioscorides Press, Portland, Oregon.

Bryant G (2003). Plant propagation A to Z. Firefly Books (U.S.) Inc. Buffalo, New York, 224 pp.

Carpenter WJ, Cornell JA (1992). Auxin application duration and concentration govern rooting of *Hibiscus* stem cuttings. J Amer Soc Hort Sci 117:68-74.

Cassaniti C, Li Rosi A, Romano D (2009). Salt tolerance of ornamental shrubs mainly used in the Mediterranean landscape. Acta Hort 807:675-680.

De Klerk GJ (1999). The formation of adventitious roots: new concepts, new possibilities. In Vitro Cell Dev Biol Plant 35:189-199.

Dirr MA, Heuser CW (2006). The reference manual of woody plant propagation. Varsity Press, Athens, 410 pp.

Frabetti M, Gutiérrez-Pesce P, Mendoza de-Gyves E, Rugini E (2009). Micropropagation of *Teucrium fruticans* L., an ornamental and medicinal plant. In Vitro Cell Dev Biol Plant 45:129-134.

Frangi P, Nicola S (2004). Study of propagation by cuttings of Mediterranean native species with ornamental potential (in Italian). Italus Hortus 4:191-193.

Graham ST (1992). Ornamental shrubs, climbers and bamboos. Sagapress, Inc./Timber Press, Inc, Portland, Oregon, 583 pp.

Hartmann HD, Kester DE, Davies FJ, Geneve RL (1997). Plant propagation: principles and practices. Prentice Hall, New Jersey, 770 pp.

Iapichino G, Amico Roxas U, Bertolino M, Accardo Palumbo S, Moncada A (2006). Propagation techniques for three Mediterranean native shrubs with potential as ornamental outdoor plants. Acta Hort 723:433-436.

Krisantini S, Johnston M, Williams RR, Beveridge C (2003). Propagation of Grevillea. Comb Proc Inter Plant Prop Soc 53:154-158.

La Mantia T, Messana G, Billeci V, Dimarca A, Del Signore M, Leanza M, Livreri S, Console G, Maraventano G, Nicolini G, Prazzi E, Quatrini P, Sanguedolce F, Sorrentino G, Pasta S. (2012). Combining bioengineering and plant conservation on a Mediterranean islet. iForest-Biogeosciences 12:296-305.

Lopez J, Gonzalez A, Fernandez JA, Banon S (2006). Ornamental use of Labiates for xeriscape in Mediterranean area. Acta Hort 723:459-464.

MacDonald B (1987). Introducing new and recommended plants into the nursery industry of British Columbia. Comb Proc Inter Plant Prop Soc 37:336-343.

Maloupa E, Grigoriadou K, Papanastassi K, Krigas N (2008). Conservation, propagation, development and utilization of xerophytic species of the native greek flora towards commercial floriculture. Acta Hort 766:205-214.

Marks TR (1996). The role of the shoot apex in controlling rhizogenesis *in vitro*. Plant Growth Reg 20:57-60.

Ojeda F, Dìez MJ (1992). Pollen dimorphism in three *Teucrium* species (*Lamiaceae*). Plant Syst Evolution 183:43-49.

Petersen RG (1985). Design and analysis of experiments. Marcell Dekker, New York, 429 pp.

Telgen HJ van, Eveleens B, Garcia N (2007). Improving rooting uniformity in rose cuttings. Propagation of Ornamental Plants 4:190-194.

The Effect of Different Irrigation Water Levels on Grain Yield, Yield Components and Some Quality Parameters of Silage Maize (*Zea mays indentata* Sturt.) in Marmara Region of Turkey

Abdullah KARASU[1]*, Hayrettin KUŞCU[1], Mehmet ÖZ[1], Gamze BAYRAM[2]

[1]*Uludag University, Mustafakemalpasa Vocational School, Department of Plant Production, 16500 Bursa, Turkey; akarasu@uludag.edu.tr (*corresponding author); kuscu@uludag.edu.tr; momer@uludag.edu.tr*
[2]*Uludag University, Faculty of Agriculture, Department of Field Crops, 16059 Bursa, Turkey; gbayram@uludag.edu.tr*

Abstract

This research was conducted in Bursa, Marmara region, Turkey, in order to determine grain yield and some morphological traits which affect the silage maize response to different irrigation water amounts. The grains' percentage of crude oil and of crude protein were determined. Field experiments were planned following randomized complete block design with three replications and included six irrigation treatments. Irrigation treatments were created as water levels of pan evaporation (E_{pan}) applied via drip irrigation [1.25 × E_{pan} (I_{125}), 1.00 × E_{pan} (I_{100}), 0.75 × E_{pan} (I_{75}), 0.50 × E_{pan} (I_{50}), 0.25 × E_{pan} (I_{25}) and 0 × E_{pan} (I_0)]. The highest value of grain yield was found to be 18,268 kg ha^{-1} in the I_{125} treatment, which represents excessive water. A quadratic relationship between grain yield and irrigation water applied was obtained. Deficit irrigation decreased grain yield and yield components except the percentage of crude oil and crude protein of grain, but improved the efficient use of irrigation water. Relationships between the grain yield and each yield component were positively significant. The highest correlation coefficient in the research gave the relationship between grain yield and plant height ($r=0.957^{**}$). The results revealed that 1.25 × E_{pan} and 1.00 × E_{pan} treatments are preferable for higher yield. The results of this study also suggest that if water is limited, the application of 0.75 × E_{pan} can be recommended as optimal treatment, because the best compromise among yield, yield components, quality and irrigation water use efficiency for maize was achieved with this application.

Keywords: crude oil, crude protein, irrigation water productivity, morphological properties, seed quality, water stress

Introduction

Maize is one of the most important plants worldwide, because of its high grain and forage yield. It has extensive adaptation characters and is intensely cultivated in Turkey. The total production of maize is not sufficient to meet the continuous increase of consumption. Although the surfaces cultivated with maize are increasing, the total production of the maize is not sufficient in Turkey. Because of this, the attempts to increase maize production are of great importance. To increase the maize production, there are needs for the adequate supply of irrigation water, N, P, K fertilizers, high yielding cultivars, agronomical practices like optimum plant density, timing of different treatments and interventions etc. affecting directly the growth and productivity.

Maize is very sensitive to water stress (Rhoads and Bennett, 1990; Pandey *et al.*, 2000; Cakir, 2004; Kuscu and Demir, 2013). Payero *et al.* (2009) reported that water stress can effect growth, development and physiological processes of maize plants, which reduce biomass yield. Farré and Faci (2009) noted that the maize needs for the highest water amount is during the flowering period. Because of this, one of the most important factors that can limit crop production is availability of water. If water stress can be avoided during silking and early ear development, high yield could be expected.

Irrigation water supplies are decreasing in Turkey, likewise many areas of the world, in recent years. Because of the climate change, predictions of increase in temperature and decrease in rainfall mean, water will become increasingly scarce. Many farmers in Turkey are currently facing some irrigation water problems because of shortage irrigation water supplies. This water shortage has motivated some researchers and farmers to find ways to produce maize with less irrigation water and changing from fully-irrigated to deficit irrigated cropping system.

Water is usually the most important natural factor limiting expansion and development of agriculture in Marmara region, a sub-humid climate of Turkey. The total precipitation does not meet the water requirements of maize crop. Bursa has approximately 126 mm precipitation from the May to September, the growing season of maize, this being not enough for high maize yields. On the other hand, for the high yields, the seasonal water requirements of maize vary from 500 to 1,000 mm depending on the climate and the total length of the growing period. Therefore, irrigation is necessary for optimal vegetative and reproductive development in the periods of insufficient precipitation during the growing season in the Marmara region. The furrow and sprinkler irrigation methods are usually used for irrigation of maize in Turkey and Marmara region, because of the traditional habits. However, the use of drip irrigation has increased rapidly in the Turkish maize cultivation. The major reasons for the conversion from furrow and sprinkler irrigation are the improved water application efficiency provided by drip irrigation (Kuscu *et al.*, 2014). Additionally, the limited fresh water resources and the cost of irrigation water are the most important reasons that induced many farmers to decrease irrigation in the Marmara region. Farmers in the region and researchers are motivated to find ways to produce crops with less irrigation water, such as using more efficient irrigation systems and changing from fully irrigated to deficit irrigated cropping systems (Kuscu *et al.*, 2010). Some farmers think that when the amount of irrigation water is increased, the yield increases too, and because of this, they make excessive irrigation within the expectation of higher yield. This excessive irrigation can be seen especially in field areas with cheap irrigation water. However, excessive irrigation can cause some social problems such as the increase in frequency of water distribution conflicts. Furthermore, excessive irrigation increases the cost of production and cause environmental and agronomical problems such as salinity, erosion and drainage.

There are large populations of cattle in Marmara region. In some years, farmers who cultivate the maize for silage as a main crop or second crop, when the prices of silage are comparatively lower and there is more production than the need of farmers, leave the remaining product for grain productions. Namely, the maize grown for silage is kept waiting to obtained maize grain. Under these circumstances what is the grain yield of maize is important, like for silage.

The main aim of this study was 1) to determine the effect of different irrigation water amounts applied via drip irrigation on grain yield, some yield components and the content of crude protein and crude oil of silage maize grown in a sub-humid climate of Turkey, 2) to define correlation coefficients among yield and yield-related components in maize, 3) to conclude which irrigation appropriate for silage maize where growing sub-humid region of Turkey.

Materials and methods

Field experiments were conducted at the experimental station of the Mustafakemalpasa Vocational School, Uludag University located in Bursa, Turkey (40°02'N, 28°23'E, 25 m above sea level) for two consecutive summer seasons. The climate in this part of the country is classified as sub-humid according to mean rainfall amount (from 600 to 700 mm of annual precipitation) (Jensen, 1980). Long term annual total precipitation was 680 mm, the mean temperature was 14.0 °C and the relative humidity was 68% (means for the vegetation season of maize, along May to September, are 126.3 mm precipitation, 21.1 °C and 62.6% relative humidity). In the vegetation seasons of the experiment, these values were 77.1 mm, 23.7 °C and 55.4% in the first year and 122.8 mm, 22.0 °C and 65.8% in the second year, respectively. The soil in the experimental field was clay loam with low organic matter content and slightly alkaline nature with pH 7.8.

Treatments were arranged in the field according to a completely randomized block design with three replications, in both seasons. The treatments consisted of six irrigation levels of pan evaporation (E_{pan}) replenishment [$1.25 \times E_{pan}$ (I_{125}), $1.00 \times E_{pan}$ (I_{100}), $0.75 \times E_{pan}$ (I_{75}), $0.50 \times E_{pan}$ (I_{50}), $0.25 \times E_{pan}$ (I_{25}), and $0 \times E_{pan}$ (I_0)]. The United States Weather Bureau (USWB) Class A evaporation pan was used to measure daily evaporation. Irrigation management was based on the common practice in the area for maize, which consists of irrigation at 7-day intervals. The area of each plot was 13.0 m^2 (long: 5.00 m, wide: 2.60 m). A buffer zone of 2 m was provided between the plots. The variety hybrid 'Ada-523' was planted at a spacing of 0.10 m × 0.65 m. There were 200 plants in each plot (15. 4 plants per m^2). In the first year of experiment, maize was planted on May 10 and harvested on October 16. In the second year, maize was planted on May 17 and harvested on October 22. Two - fold seeds were sown at every 10 cm and 5 cm depth and hand - thinned to target densities with the first hoeing. Weeding was performed with hand twice during the growing season. Maize plots were fertilized with 180 kg ha^{-1} N and 120 kg ha^{-1} P_2O_5 in both years, before planting. In the field, all crops were irrigated by drip irrigation.

The drip irrigation system was designed and installed to meet the objectives of the project. The irrigation water was pumped directly from the Mustafakemalpasa Aquifer to the drip system. One screen filter was installed on the mainline to minimize dripper blockage. Low density polyethylene pipes of 32 mm diameter were used for sub-main. The lateral lines were laid adjacent to each crop row. The thick-walled dripper lines had inline compensating emitter pressure and the discharge rate of the emitters was 3.0 L h^{-1} at an operating pressure of 100 kPa. The emitter spacing was chosen as 0.30 m based on the soil characteristics.

The amount of first irrigation water for all the plots was based on the moisture deficit that would be needed to bring the 0-90 cm layer of soil to field capacity and it was applied using the drip irrigation system (Cetin *et al.*, 2002). It was given 20 mm and 24 mm water for germination in the two years respectively, before seedling. The seasonal water applied in treatments I_{125}, I_{100}, I_{75}, I_{50}, I_{25} and I_0 were 1.047, 842, 636, 431, 226 and 20 mm in the first season of the experiment, and 1.007, 810, 613, 417, 220 and 24 mm for treatments I_{125}, I_{100}, I_{75}, I_{50}, I_{25} and I_0 respectively, in the second year of the experiment. The differences between the seasonal water applied were simply due to the number of irrigation events in each season.

For morphological measurements, ten plants were randomly selected from each plot and taken just before harvesting for grain. Plant height, ear length, the number of row per ear, number of grain per row, the number of grain per ear, the number of ear per plant, 1,000 grain weight, hectoliter weight, grain yield, crude protein content and crude oil content were analyzed. Composition of crude protein and crude oil of maize were determined using the standard methods, as described in Association of Official Analytical Chemists (AOAC, 1990). Seed samples of plots were grounded to pass through a sieve of 0.8 mm diameter. Ground seed samples were analyzed by Kjeldahl method to determine the total nitrogen. Crude protein content was calculated by multiplying total nitrogen value with a coefficient of 6.25. Crude oil percentage was determined by the Soxhlet extraction technique. Approximately 3 g of sample were extracted with di-ethyl ether for a period of 4 h in Soxhlet device.

Two rows of the each plot were harvested for grain yield when the humidity of grain was 18%.

The irrigation water use efficiency was calculated (Howell *et al.* 1995):

$$IWUE = (GY_t - GY_{ni}) / (IR_t - IR_{ni})$$

where:

IWUE = of treatment t ($kg\,m^{-3}$)
GY_t = grain yield of treatment t ($kg\,m^{-2}$)
GY_{ni} = grain yield of a nonirrigated treatment (I_0, $g\,m^{-2}$)
IR_t = applied irrigation water to treatment t (mm)
IR_{ni} = applied irrigation water for germination (mm)

The data were subjected to analyses of variance using MINITAB software. LSD test was used to group the means per year, irrigation and their interactions when the F-test was significant. The correlation coefficient between grain yield and yield components was calculated using the MINITAB software.

Results and discussions

Results of variance analysis

The analysis of variance indicated that growing season significantly affected all characters except ear length, the number of row per ear, the number of ear per plant, crude protein and crude oil percentage (Tables 1 and 2).

Results indicated that the effect of irrigation water amount was significant at 99% probability level for plant height, ear length, the number of grain per row, the number of grain in the ear, 1,000 grain weight, hectoliter weight, grain yield, and that was significant at 95% probability level for the number of row per ear and the number of ear per plant in average data of two years. The irrigation levels did not have effect on the percentage of crude protein and crude oil at averaged over two growing seasons. Additionally, the effects of the growing season x irrigation interactions on grain yield, number of grain per ear and plant height were significant at the 5% and 1% levels (Table 1 and Fig. 1). This result indicated that irrigation levels responses were various according to the seasons. Values related to these parameters increased as the irrigation water levels increased. However, the values obtained from the same irrigation water levels at different growing seasons formed different statistical groups and this situation leaded to interaction.

Yield components

Yield components such as plant height, ear length, the number of row per ear, the number of grain per row, the number of grain per ear and the number of ear per plant of maize grown under different levels of irrigation are shown in Table 1 and quality parameters including 1,000 grain weight, hectoliter weight, crude protein and crude oil in Table 2.

The effect of different irrigation water amounts was statistically significant at plant height of silage maize. Results of combined data indicated that all applied water treatments yielded plants having higher and similar heights than unirrigated plants (Table 1). In similar experiments (Bozkurt *et al.*, 2006; Cakir, 2004; Istanbulluoglu *et al.*, 2002; Otegui *et al.*, 1995; Pandey *et al.*, 2000), plant heights were reported to be higher with full irrigation (100% ETc or Epan = 1) and slightly deficit irrigation throughout the crop growing season, which is in agreement with the results of the current study.

As shown in Table 1, data obtained from two years study showed that ear length was significantly affected by irrigation levels ($p<0.01$). The higher ear length values were obtained from treatments I_{125} and I_{100} (20.6 cm in both irrigation levels), while the shortest ear length (16.0 cm) was obtained from treatment I_0. Because the ear length affects the number of grain per ear, it is accepted as one of the most important yield components that affects the grain yield. Our findings showed that when the irrigation levels decreased, the ear length decreased too. This result is consistent with the results of Ertek and Kara (2013). They showed that ear length was affected by different irrigation water levels (12.8-18.8 cm) and reported that the ear length decreased with decreasing water application. Kuscu (2010) noted that full irrigation during total crop growing season increased the ear length, but deficit irrigation at different phenological stages decreased it. In a similar study, Istanbulluoglu *et al.* (2002) reported values varying between 16.4 and 20.5 for ear length in relation to irrigation water levels.

The number of rows per ear was statistically affected by different irrigation water amounts (Table 1). Results revealed that the higher number of row per ear was found in treatments of I_{125}, I_{100}, I_{75}, I_{50} and I_{25} (15.3, 15.3, 15.1, 14.8,and 14.7 respectively), whereas the lowest was found in non-irrigated plants (14.2). Our results are similar with those of Sampathkumar *at al.* (2013). In a study carried out by Esmailian *et al.* (2011) under the ecological conditions of Iran, the number of row per ear varied between 12.4 and 14.1. This result was lower than ours because of different cultivar and ecological conditions. Similar findings are reported by Kuscu (2010), who stated that the lowest number of row per ear was obtained from non-irrigated plots.

Differences between irrigation treatments were significant for the number of grain per row (Table 1). The highest number of grain per row was obtained from treatments of I_{125}, I_{100}, and I_{75} (42.4, 41.0 and 41.1, respectively), while the lowest was obtained from treatment I_0 (33.9). Results showed that the number of grain per row increased as irrigation water amount increased up to I_{75} level. Our results were similar with those of Claassen and Shaw (1970), Frey (1982), Fischer and Palmer (1984) and Kuscu (2010), who reported that the water stress decreased the number of grain per row at silking stage. The values of number of grain per row obtained in this study are in agreement with those of other researches on maize (Carpici, 2009; Okay, 2006; Turgut *et al.*, 2005).

Significant differences between irrigation treatments were found in terms of the number of grain per ear for combined data of two years. The highest numbers of grain per ear were obtained from I_{125} and I_{100} irrigation water amounts (555.0 and 549.3, respectively), whereas the lowest number of grain per ear was obtained from non-irrigated treatment (413.3). The number of grain per ear is related with the ear length, the number of row per ear and the number of grain per row. It was seen that the number of grain per ear increased like the number of grain per row, as the amount of irrigation water was increased. The ear length, the number of row per ear and the number of grain per row gave the highest values when there was no deficit irrigation (I_{125} and I_{100} irrigation levels).

At the applications of deficit irrigation, the values considerably decreased (Table 1). Ertek and Kara (2013) also reported that deficit irrigation decreased the number of grain per ear, which was in agreement with our findings.

Table 1. Effect of irrigation water amounts on some morphological traits of maize and the result of F testing and statistical grouping irrigation water amount

Irrigation treatments	Plant height (cm)	Ear length (cm)	The number of row per ear	The number of grain per row	The number of grain per ear	The number of ear per plant
I_0	267.3 c	16.0 d	14.2 b	33.9 d	413.3 e	0.65 b
I_{25}	302.1 b	17.6 c	14.7 ab	37.4 c	481.3 d	0.83 a
I_{50}	310.0 ab	19.7 ab	14.8 ab	39.4 bc	504.8 c	0.86 a
I_{75}	316.6 ab	19.4 b	15.1 ab	41.1 ab	530.0 b	0.88 a
I_{100}	317.7 ab	20.6 a	15.3 a	41.0 ab	549.3 a	0.88 a
I_{125}	323.3 a	20.6 a	15.3 a	42.4 a	555.0 a	0.89 a
Analysis of variance						
Growing season	**	ns	ns	**	**	ns
Irrigation	**	**	*	**	**	*
Growing season × irrigation	*	ns	ns	ns	**	ns

Note: Means followed by the same letter are not significantly different at the p = 0.05 level using LSD test;
*,**, ns F- test significant at $p \leq 0.05$, $p \leq 0.01$, respectively, and not significant

Table 2. Effect of irrigation water amounts on some quality characters of maize and the result of F testing and statistical grouping irrigation water amount

Treatments	1,000 grain weight (g)	Hectoliter weight (kg hl^{-1})	Crude protein (%)	Crude oil (%)
I_0	261.2 e	63.1 c	8.6	3.3
I_{25}	270.7 d	63.4 c	8.3	3.2
I_{50}	280.5 c	66.4 b	8.2	3.4
I_{75}	287.3 b	67.8 ab	8.2	3.6
I_{100}	308.0 a	69.1 a	8.2	3.2
I_{125}	309.3 a	69.2 a	7.8	3.0
Growing season	**	**	ns	ns
Irrigation	**	**	ns	ns
Growing season × irrigation	ns	ns	ns	ns

Note: Means followed by the same letter are not significantly different at the p = 0.05 level using LSD test;
*,**, ns F- test significant at $p \leq 0.05$, $p \leq 0.01$, respectively, and not significant

Table 3. Seasonal irrigation water applied, grain yield, irrigation water use efficiency (IWUE) and saved irrigation water for maize under different irrigation treatments in combined growing seasons

Treatment	Seasonal irrigation water applied (mm)	Grain yield (kg ha^{-1})	IWUE (kg m^{-3})	Grain yield losses (%)	Irrigation water saving (%)
I_0	22[x]	7,123 f	-	-59.9	97.4
I_{25}	223	10,580 e	1.72	-40.5	73.0
I_{50}	424	13,383 d	1.56	-24.7	48.7
I_{75}	625	15,702 c	1.42	-11.7	24.4
I_{100} (Control)	826	17,777 b	1.32	0	0
I_{125}	1027	18,268 a	1.11	+2.76	-24.4

[x] Irrigation water applied for germination in sowing time. The first irrigation water was applied for each treatment. Grain yields' means followed by the same letter are not significantly different at the p = 0.05 level using LSD test

Table 4. Correlation coefficients among yield and yield-related components in maize (averages of overall data)

Parameters	Ear length (cm)	The number of row per ear	The number of grain per row	The number of grain per ear	The number of ear per plant	1000 grain weight (g)	Hectoliter weight (kg hl^{-1})	Grain yield (kg ha^{-1})
Plant height (cm)	0.830**	0.772**	0.578**	0.883**	0.875**	0.890**	0.761**	0.957**
Ear length (cm)		0.592**	0.598**	0.643**	0.754**	0.717**	0.513**	0.844**
The number of row per ear			0.489**	0.807**	0.720**	0.738**	0.737**	0.729**
The number of grain per row				0.517**	0.581**	0.556**	0.331**	0.622**
The number of grain per ear					0.866**	0.840**	0.767**	0.857**
The number of ear per plant						0.805**	0.647**	0.894**
1000 grain weight (g)							0.713**	0.938**
Hectoliter weight (kg)								0.675**

** F-test significant at $p \leq 0.01$

The effect of different irrigation water amounts was statistically important for the number of ear per plant of silage maize. The numbers of ear per plant varied between 0.89 and 0.65 number per plant. The higher number of ear per plant were obtained from I_{125}, I_{100}, I_{75}, I_{50} and I_{25} treatments, which were statistically similar (0.89, 0.88, 0.88, 0.86 and 0.83, respectively), while the lowest value was determined at non-irrigated plots (0.65). The soil water stress affected the ear number per plant. Pandey *et al.* (2000) reported that water deficit decreased the ear number per unit area. Our results are similar with those of Cakir (2004) who reported the stress of water affected the ear number of per plant. Because of the higher plant density in our study (15.4 plant per m^2) ear number per plant is less than those of some early researches (Carpici, 2009; Geren *et al.*, 2003; Kusaksiz and Kaya, 2005; Turgut *et al.*, 2005).

Quality parameters

The effect of different irrigation water amount was statistically important for the 1,000 grain weight of silage maize for combined data of two years. As shown Table 2, the highest 1,000 grain weights were obtained from treatments of I_{125} and I_{100} (309.3 and 308.0 g, respectively), while the lowest 1,000 grain weight was obtained from non-irrigated plots (261.2 g). As a result, 1,000 grain weight increased as the amount of irrigation water increased. Our results were in agreement with the results of Mansouri *at al.* (2010). They reported that when the amount of water decreased, both the 1,000 grain weight and grain yield were decreased. Similarly, Ogretir (1993) reported that the application of deficit irrigation on maize at the flowering period decreased the 1,000 grain. Cakir (2004) also stated that the irrigations during milk maturation period increased the 1,000 grain weight.

Hectoliter weight was statistically affected by different irrigation water amounts (Table 2). The highest hectoliter weights were obtained from treatments of I_{125}, I_{100} and I_{75} (69.2, 69.1 and 67.8 kg hl^{-1}, respectively), while lower hectoliter weights were obtained from treatments of I_0 and I_{25} (63.1 and 63.4 kg hl^{-1}, respectively). Vartanli and Emeklier (2007) reported that hectoliter weights for different maize cultivars were between 65.43-73.53 kg hl^{-1}. In general, deficit irrigation levels adversely affected hectoliter weight (Kuscu, 2010).

The effect of different irrigation water amounts was not statistically important for crude protein content. As shown Table 2, crude protein content ranged from 7.8 to 8.6%. Similarly, Vartanli and Emeklier (2007) reported that crude protein contents of some maize cultivars were between 6.21 and 8.65%. Conversely, Ertek and Kara (2013) who worked with the similar subjects on sugar maize reported that deficit irrigation levels affected the content of crude protein, which vary in their study between 10. 63 - 11.25%. Esmailian *et al.* (2011), Farhad *et al.* (2013) and Aydinsakir *et al.* (2013) who worked on different irrigation water levels and different maize cultivars also reported that the grain protein contents were significantly influenced by different irrigation levels. The observed differences are possibly due to variations in experimental conditions and/or cultivar, soil and ecological differences.

Irrigation treatments had no statistically significant effect on crude oil content, for which the values ranged between 3.0-3.6% (Table 2). In a similar study, Vartanli and Emeklier (2007) found that crude oil contents of some maize cultivars were between 2.04-6.90%. However, in contrast to our findings, Farhad *et al.* (2013) reported that grain oil contents of maize hybrids were significantly affected with different irrigation water levels and the content of oil was differed among maize cultivars, varying between 1.95-4.55 %. In a similar study, Esmailian *et al.* (2011) also reported that the grain oil contents were significantly influenced with different irrigation levels. Cultivar and experimental conditions may be the cause of these differences.

Grain yield and water-yield relationships

Grain yield results according to irrigation treatments are shown Fig. 1. The grain yield was different between the two growing seasons at 99% probability level. Similarly, Mengü and Özgürel (2008) reported that the effects of years were important on grain yield. Additionally, the effect of the growing season × irrigation interactions on grain yield was significant at 1% level (Fig. 1). This result indicated that irrigation levels responded variously according to the growing seasons. This may be attributed to the differences in climatic conditions among growing seasons. Irrigation water amounts applied in the first growing season were higher than the amounts in the second growing season due to the higher temperature and lower rainfall.

The effect of different irrigation water amounts was statistically important for grain yield per hectare for combined data of two years (Table 3). In general, there was a close relationship between irrigation and grain yield. The relationship between grain yield and irrigation water applied is shown in Fig. 2. The relationship was mainly quadratic due to excessive irrigation (1.25 E_{pan}). The quadratic equation for the relationship between grain yield and seasonal irrigation water applied implies that a grain yield of approximately 6,669 kg ha^{-1} can be achieved even without the application of irrigation water after germination. This is only possible if there is residual moisture in the soil to support such a yield (Igbadun *et al.*, 2008). Quadratic relationships between grain yield and irrigation were also reported by Gencoglan and Yazar (1999), Kipkorir *et al.* (2002), Bozkurt *et al.* (2006) and Farré and Faci (2009). However, Payero *et al.* (2006) found that there was a linear relationship between grain yield and seasonal irrigation water amount. The differences among the relationships reported by different researchers are due to different experimental conditions, seasonal rainfall amounts and distribution (Howell *et al.*, 1995).

In this study, the highest grain yield (18,268 kg ha^{-1}) was obtained in the treatment I_{125} due to satisfactory soil moisture during the growing period, while the lowest yield (7,123 kg ha^{-1}) was obtained from treatment I_0. The second highest grain yield (17,777 kg ha^{-1}) was obtained in the I_{100} treatment. Furthermore, deficit irrigation applications decreased the grain yield. The results for the two years can be summarized by stating that a producer would have obtained the highest grain yield using full irrigation (1.00 × E_{pan}) or slightly excessive irrigation (1.25 × E_{pan}). This results are consistent with findings of Anac and Ul (1992), Yildirim *at al.* (1996), Stan and Naescu (1997), Pandey *et al.* (2000), Istanbulluoglu *et al.* (2002), Farre and Faci (2009), Ko and Piccinni (2009) and Mansouri *et al.* (2010), who showed that grain yield was affected by irrigation water amount.

As shown Table 3, irrigation water use efficiency (IWUE) values varied from 1.11 to 1.72 kg m^{-3}, which are similar to reported values from 1.51 to 2.48 kg m^{-3} by Howell *et al.* (1995) and up to 1.62 kg m^{-3} reported by Kuscu and Demir (2013). Differences in the rainfall during growing seasons could be the cause of small differences in the results of IWUE values because the amount of rainfall affects the amount of irrigation water applied. In this study, irrigation water use efficiency increased with decreasing irrigation water applied. The highest IWUE was obtained from treatment I_{25}. The relationship between IWUE and irrigation water applied is shown in Fig. 2.

In regions where water scarcity exists, irrigation managers should adopt the deficit irrigation approach to achieve sustainable crop production. In this study, compared to I_{100} (control) the least yield loss of 11.7% was seen in I_{75} treatment with 24.4% saving of irrigation water, and IWUE was also improved. Even though highest IWUE value was obtained from I_{25}, grain yield was dramatically reduced (-40.5%) (Table 3).

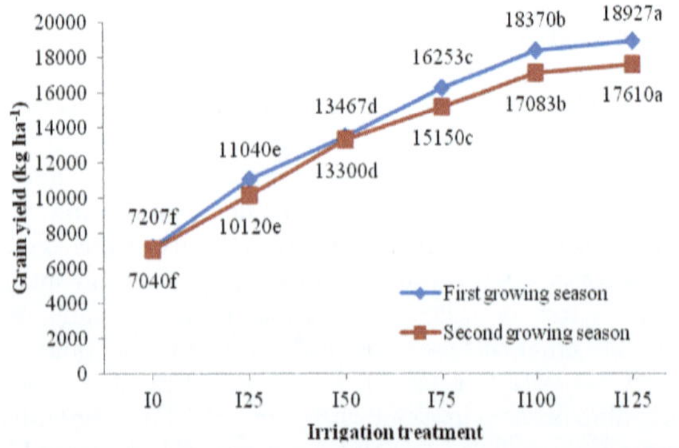

Fig. 1. Grain yield results according to irrigation treatments

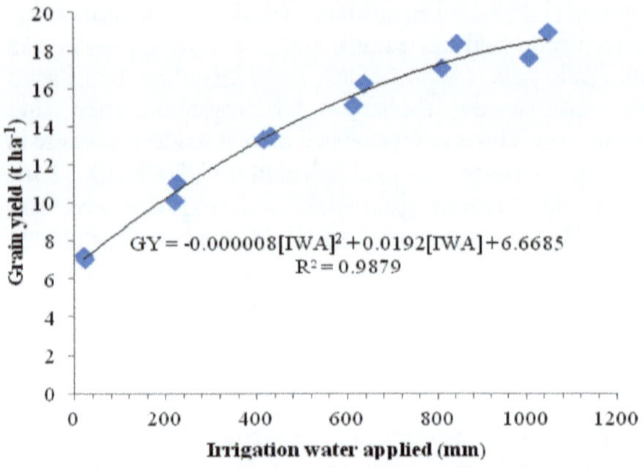

Fig. 2. Relationship between seasonal irrigation water applied (IWA) and grain yield (GY) (pooled data of the two growing seasons)

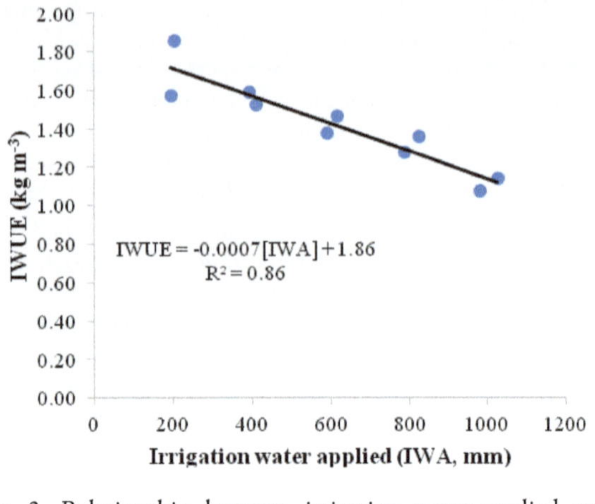

Fig. 3. Relationship between irrigation water applied and irrigation water use efficiency (IWUE)

Correlations between grain yield and yield components

The simple correlation coefficients averaged over all treatments are shown in Table 4. As can be seen, all correlation coefficients between yield and yield components were statistically significant ($p<0.01$). Positive and statistically significant relationships existed between the grain yield and all of yield components. The grain yield was positively and significantly correlated with plant height ($r = 0.957^{**}$), 1,000 grain weight ($r = 0.938^{**}$), the number of ear per plant ($r = 0.894^{**}$), the number of grain per ear ($r = 0.857^{**}$), ear length ($r = 0.844^{**}$), the number of row per ear ($r = 0.729^{**}$), hectoliter weight ($r = 0.675^{**}$) and the number of grain per row ($r = 0.622^{**}$). In addition, all the reciprocal relationships among the variable all yield components were positively significant. For instance, the relationships between plant height and, 1,000 grain weight ($r = 0.890^{**}$), the number of grain per ear ($r = 0.883^{**}$), the number of ear per plant ($r= 0.875^{**}$), ear length ($r = 0.830^{**}$), the number of row per ear ($r = 0.772^{**}$), hectoliter weight ($r = 0.761^{**}$), the number of grain per row ($r = 0.578^{**}$) were positive and significant. Also, correlations between ear length and the number of ear per plant ($r = 0.754^{**}$), 1,000 grain weight ($r = 0.717^{**}$), the number of grain per ear ($r = 0.643^{**}$), the number of grain per row ($r = 0.598^{**}$), the number of row per ear ($r = 0.592^{**}$), hectoliter weight ($r = 0.513^{**}$) were positively significant. The highest correlation coefficient in the research gave relationship between grain yield and plant height ($r=+0.957^{**}$) followed by correlations between grain yield and 1,000 grain weight ($r=+0.938^{**}$). Our findings on this subject are in agreement with those of Ilker (2011), Hasyan *et al.* (2012) and Kumar *et al.* (2014) who reported that correlations were significant between yield and yield-related components of maize.

Conclusions

Irrigation levels significantly affected the maize grain yield, all morphological and quality parameters observed, except crude protein and crude oil. Deficit irrigation decreased the grain yield and yield components. The results of present research show that $1.25 \times E_{pan}$ and $1.00 \times E_{pan}$ irrigation applications at 7 day intervals during the total growing period can be considered as maximum grain yield in maize grown in ecological conditions of Marmara Region, Turkey. However, irrigation at $0.75 \times E_{pan}$ could be recommended where water resources are scarce such as that of present experiment. Because this DI regime achieved irrigation water savings up to 40% compared with $1.25 \times E_{pan}$. Because the grain yield of maize grown for silage was not lower than those grown for grain, this can be a solution for gaining profit if the prices of silage are comparatively lower or the production is too high compared with the demand at some moment.

References

Anac S, Ul MA (1992). Deficit irrigation studies on corn. Presented at the FAO/IAEA research co-ordination meeting on the use of nuclear and related techniques in assessment irrigation schedules of field crops to increase effective use of water. Irrigation Projects, 3 to 7 February 1992, Vienna, Austria.

AOAC (1990). Official methods of analysis. Association of Official Agricultural Chemists. Washington, DC.

Aydinsakir K, Erdal S, Buyuktas D, Bastug R, Toker R (2013). The influence of regular deficit irrigation applications on water use, yield, and quality components of two corn (*Zea mays* L.) genotypes. Agricultural Water Management 128:65-71.

Bozkurt Y, Yazar A, Gencel B, Sezen MS (2006). Optimum lateral spacing for drip-irrigated corn in the Mediterranean Region of Turkey. Agricultural Water Management 85:113-120.

Cakir R (2004). Effect of water stress at different development stages on vegetative and reproductive growth of corn. Field Crops Research 89:1-16.

Carpici EB (2009). Evaluation of the effects of plant densities and nitrogen rates on stress physiology traits in silage corn (*Zea mays* L.) production. PhD thesis, Field Crops, Graduate School of Natural and Applied Sciences. Uludag University, Bursa, Turkey.

Cetin Ö, Yildirim O, Uygan D, Boyaci H (2002). Irrigation scheduling of drip-irrigated tomatoes using class a pan evaporation. Turkish Journal of Agriculture and Forestry 26:171-178.

Claassen MM, Shaw RH (1970). Water deficit effects on corn. I. Grain components. Agronomy Journal 62:652-655.

Ertek A, Kara B (2013).Yield and quality of sweet corn under deficit irrigation. Agricultural Water Management 129:138-144.

Esmailian Y, Ghanbari A, Babaeian M, Tavassoli A (2011). Influence of organic and inorganic fertilizers and wastewater irrigation on yield and quality traits of corn. American-Eurasian Journal of Agricultural & Environmental Sciences 10(4):658-666.

Farhad W, Cheema MA, Saleem MF, Radovich T, Abbas F, Hammad HM, Wahid MA (2013). Yield and quality response of maize hybrids to composted poultry manure at three irrigation levels. International Journal of Agriculture & Biology 15:181-190.

Farré I, Faci JM (2009). Deficit irrigation in maize for reducing agricultural water use in a Mediterranean environment. Agricultural Water Management 96(3):383-394.

Fischer KK, Palmer FE (1984). Tropical maize. In: Goldsworthy PR, Fischer NM (Eds). The Physiology of Tropical Field Crops. Wiley, New York pp 213-248.

Frey NM (1982). Dry matter accumulation in kernels of maize. Crop Science 21:118-122.

Gencoglan C, Yazar A (1999). The effects of deficit irrigations on corn yield and water use efficiency. Turkish Journal of Agriculture and Forestry 23:233-242.

Geren H, Avcioğlu R, Kir B, Demiroğlu G, Yilmaz M, Cevheri AC (2003). Effect of different sowing dates on the yield and quality characteristics of some maize cultivars grown as second crop for silage. The Journal of Ege University Faculty of Agriculture 40(3):57-64.

Hasyan RM, Moualla YM, Ahmad AAS (2012). Potence ratio and path coefficient analysis for some quantitative traits of maize (*Zea mays* L.) hybrids developed in Syria. Jordan Journal of Agricultural Sciences 8(4):557-565.

Howell TA, Yazar A, Schneider AD, Dusek DA, Copeland KS (1995). Yield and water use efficiency of corn in response to lepa irrigation. Transactions of the ASAE 38(6):1737-1747.

Igbadun HE, Salim BA, Tarimo AKPR, Mahoo HF (2008). Effects of deficit irrigation scheduling on yields and soil water balance of irrigated maize. Irrigation Science 27(1):11-23.

Ilker E (2011). Correlation and path coefficient analyses in sweet corn. Turkish Journal of Field Crops 16(2):105-107.

Istanbulluoglu A, Kocaman I, Konukcu F (2002). Water use-production relationship of maize under Tekirdag conditions in Turkey. Pakistan Journal of Biological Science 5(3):287-291.

Jensen ME (1980). Design and operation of farm irrigation systems. An ASAE Monography, Number 3 in a Series Published by American Society of Agricultural Engineers, 2950 Niles Road, Michigan 49085, USA.

Kipkorir EC, Raes D, Massawe B (2002). Seasonal water production functions and yield response factors for maize and onion in Perkerra, Kenya. Agricultural Water Management 56:229-240.

Ko J, Piccinni G (2009). Corn yield responses under crop evapotranspiration-based irrigation management. Agricultural Water Management 96:799-808.

Kumar GP, Prashanth Y, Reddy VN, Sudheer S, Rao VP (2014). Character association and path coefficient analysis in maize (*Zea mays* L.). International Journal of Applied & Biology and Pharmaceutical Technology 5(1):257-260.

Kusaksız T, Kaya C (2005). An investigation on the herbage yield of different maize cultivars (*Zea mays* L.) grown under Manisa conditions. Turkey VI. Field Crops Congress, September 5-9, Antalya 2:1021-1026.

Kuscu H (2010). Effects of deficit irrigation on yield and yield components of maize grown under Bursa conditions. PhD thesis, Irrigation and Agricultural Structures, Graduate School of Natural and Applied Sciences, Uludag University, Bursa, Turkey.

Kuscu H, Karasu A, Öz M (2010). Effects of drip irrigation on corn yield in Bursa conditions. I. National Symposium on Irrigation and Agricultural Structures. Kahramanmaras, Turkey, May 27-29, 1:156-167.

Kuscu H, Demir AO (2013). Yield and water use efficiency of maize under deficit irrigation regimes in a sub-humid climate. Philippine Agricultural Scientist 96(1):32-41.

Kuscu H, Karasu A, Öz M, Demir A, Turgut I (2013). Effect of irrigation amounts applied with drip irrigation on maize evapotranspiration, yield, water use efficiency, and net return in a sub-humid climate. Turkish Journal of Field Crops 18:13-19.

Kuscu H, Turhan A, Demir AO (2014). The response of processing tomato to deficit irrigation at various phenological stages in a sub-humid environment. Agricultural Water Management 133:92-103.

Mansouri-Far CSA, Sanavy MM, Saberali SF (2010). Maize yield response to deficit irrigation during low sensitive growth stages and nitrogen rate under semi-arid climatic conditions. Agricultural Water Management 97(1):12-22.

Mengü GP, Özgürel M (2008). An evaluation water-yield relations in maize (*Zea mays* L.) in Turkey. Pakistan Journal of Biological Sciences 11(4):517-524.

Ogretir K (1993). The water-yield relationships of corn in Eskişehir conditions. PhD thesis, Irrigation and Agricultural Structures, Graduate School of Natural and Applied Sciences. Cukurova University, Adana.

Okay D (2006). Determination of corn plant water-yield relationships with CERES-maize model in Bursa conditions. PhD thesis, Irrigation and Agricultural Structures, Graduate School of Natural and Applied Sciences. Uludag University, Bursa, Turkey.

Otegui ME, Andrade FH, Suero EE (1995). Growth, water use and kernel abortion of maize subjected to drought at silking. Field Crops Research 40(2):87-94.

Pandey RK, Maranville JW, Admou A (2000). Deficit irrigation and nitrogen effects on maize in Sahelian environment. II. Shoot growth. Agricultural Water Management 46(1):1-13.

Payero JO, Steven M, Irmak S, Tarkalson DD (2006). Yield response of corn to deficit irrigation in a semiarid climate. Agric Water Manage 84:895-908.

Payero JO, Tarkalson DD, Irmak S, Davison D, Petersen JL (2009). Effect of timing of a deficit–irrigation allocation on corn evapotranspiration, yield, water use efficiency and dry mass. Agricultural Water Management 96:1387-1397.

Rhoads FM, Bennett JM (1990). Corn. In: Stewart BA, Nielsen DR (Eds). Irrigation of agricultural crops. Madison, WI, ASA, CSSA, and SSSA, Agronomy Monograph 30:569-596.

Sampathkumar T, Pandian BJ, Rangaswamy MV, Manickasundaram P, Jeyakumar P (2013). Influence of deficit irrigation on growth, yield and yield parameters of cotton-maize cropping sequence. Agricultural Water Management 130:90-102.

Stan I, Naescu V (1997). Maize response to water deficit. Romanian Agriculture Research 7-8:77-90.

Turgut I, Duman A, Bilgili U, Açikgoz E (2005). Alternate row spacing and plant density effects on forage and dry matter yield of corn hybrids (*Zea mays* L.) Journal of Agronomy and Crop Science 191:146-151.

Vartanlı S, Emeklier HY (2007). Determination of the yield and quality characteristics of hybrid maize varieties under Ankara conditions. Journal of Agricultural Sciences 13(3):195-202.

Yildirim O, Kodal S, Selenay F, Yildirim YE, Oztürk A (1996). Corn grain yield response to adequate and deficit irrigation. Turkish Journal of Agriculture and Forestry 20(4):283-288.

25

Phenolics Content, Antioxidant Activity and Color of Green Walnut Extracts for Preparing Walnut Liquor

Sina COSMULESCU[1*], Ion TRANDAFIR[2], Violeta NOUR[1], Mira IONICA[1], Felicia TUTULESCU[1]

[1]*University of Craiova, Agriculture & Horticulture Faculty, Department of Horticulture & Food Science, A.I.Cuza Street, 13, Craiova, Romania; sinacosmulescu@hotmail.com (*corresponding author), vionour@central.ucv.ro, miraelena29@yahoo.com, felixdragomir@yahoo.com*
[2]*University of Craiova, Sciences Faculty, Department of Chemistry, Calea Bucuresti Street, 107, Craiova, Romania; nelutrandafir@yahoo.com*

Abstract

Different extracts prepared from green fruits of 'Sibisel 44' walnut cultivar were investigated for their total phenolics and flavonoids, antioxidant activity, individual phenolics content and colour component. Extracts were prepared using three variants, to analyze the effect of solvent concentration and sugar addition on properties of green walnuts extracts. The antioxidant activity of alcoholic extracts was investigated by the 2,2'-diphenyl-1-picrylhydrazyl (DPPH) radical scavenging method while total phenolic content was determined by using the Folin-Ciocalteu assay. Seventeen individual phenolics were detected and quantified by using HPLC in walnut green fruit extracts. For total phenolics and flavonoids extraction from green walnut fruits 70% ethanol was more efficient. Highest concentration was found in samples with 70% ethanol for gallic, vanillic, chlorogenic, caffeic, syringic, salicylic, ellagic acids, juglone, catechin, epicatechin, myricetin, and quercetin. The use of 40% ethanol led to a significant increase of ferulic acid and rutin content, while the greatest amount of rutin was extracted using 40% ethanol and sugar.

Keywords: antioxidant activity, color component, polyphenols, walnut liqueur

Introduction

In Romanian folk medicine, walnut has many uses. It is said that green walnuts are a symbol of life and fecundity. Fruits (dry and green), green husk, kernels, barks and leaves are a source of phytochemicals that increase the value of walnut production, as well as offer utilisation for by-products (Akbari *et al.,* 2012; Cosmulescu and Trandafir 2012). In food preparation, jam and liqueur made of green walnut are among the most used recipes. Walnut liqueur is subject of research papers in literature. Stampar *et al.* (2006) consider traditional walnut liqueur a cocktail of phenolics. Leaves of walnut and green walnuts are particularly rich in phenolic compounds (Cosmulescu *et al.*, 2014; Cosmulescu and Trandafir, 2012). Traditional walnut liqueur is made of green walnuts just prior to hardening of the endocarp and this drink is commonly used in folk medicine (Stampar *et al.*, 2007). A strong influence of cultivar choice and picking on phenolics content in walnut liqueur was observed by Jakopic *et al.* (2007). Stampar *et al.* (2006) found that concentrations of individual phenolics in liqueur were quite low compared to the contents in green husks, due to the traditional way of making the liqueur. Processing variables are influencing the characteristics of walnut liqueur. Alamprese and Pompei (2005a) found that the highest content of phenolic substances and the highest antioxidant power was obtained with the least ripe batch of walnuts and that temperature and length of steeping have little effect on phenolic composition of nocino. Jakopic *et al.* (2008) found that the content of total phenolics and of some individual phenolic compounds (protocatechuic, sinapic and *p* - coumaric acids as well as 1,4-naphthoquinone) increased in walnut liqueur with increasing ethanol concentration. Stampar *et al.* (2006) identified fourteen phenolic compounds in walnut liqueur (chlorogenic acid, caffeic acid, ferulic acid, sinapic acid, gallic acid, ellagic acid, protocatechuic acid, syringic acid, vanillic acid, catechin; epicatechin, myricetin, 1,4-naphthoquinone, juglone) while Jakopic *et al.* (2007) quantified ten phenolic compounds (gallic, protocatechuic, ellagic, chlorogenic, syringic, *p*-coumaric, sinapic acids, catechin, 1,4-naphthoquinone and juglone).

In relation to the antioxidant properties of walnut extract, Negi *et al.* (2011) believe that walnut may act as a cancer chemopreventive agent, exhibiting high phenolic content, antioxidant activity, and potent antiproliferative activity. In a study of nocino liqueur samples aged for different lengths of time, Alamprese *et al.* (2005b) showed that total phenols, total tannins, non-tannin phenolics and antioxidant activity have remained unchanged during the aging process in the case of home-made nocino compared to the commercial products.

The aim of this work was to analyze the effect of ethanol concentration and sugar addition on the bioactive content and antioxidant activity of green walnuts hydroalcoholic extracts. Extracts were compared with respect to their total phenolics content, total flavonoids, individual phenolics content, antioxidant activity, and color.

Materials and methods

Green walnuts were collected in the second half of June 2013 from the experimental walnut orchard belonging to the Research Station of University of Craiova (44°20′ N, 23°49′ E). Walnuts of 'Sibisel 44' cultivar were utilized in preparing the extracts. 200 g of knife crushed green walnuts were mixed with 400 mL of liquid, then placed in a sealed glass vessel and left to soak for 6 weeks. Walnut extracts were prepared in three solvents: (a) 70% ethanol; (b) 40% ethanol and sugar (ratio 1:1) - traditional recipe/method; (c) 40% ethanol in three repetitions. The extracts were filtered through Whatman cellulose nitrate membranes (0.8 μm pore size) and stored at 4°C until analysis.

Total phenolics content

Chemicals and reagents: Folin-Ciocalteu reagent (2N, Sigma-Aldrich), gallic acid (Sigma-Aldrich), anhydrous sodium carbonate (Sigma-Aldrich), methanol were used in this experiment.

Method: Total phenolics content in the extracts was determined colorimetrically with Folin-Ciocalteu reagent by using the method described by Singleton and Rossi (1965) with some modifications (Cosmulescu and Trandafir, 2012). 5 mL Folin-Ciocalteu reagent (diluted 1:10 with ultrapure water) were added to 1 mL extract (diluted 1:20 with methanol) and stirred for one minute. After 2 min, 4 mL of sodium carbonate aqueous solution 7.5% were added and incubated for 2 hours at the room temperature (approx 25 °C). The absorbance of the mixture was measured at 765 nm using an Evolution 600 UV/VIS spectrophotometer (Thermo Scientific, USA). Gallic acid was used to prepare the standard curve (0 - 250 mg/L). Gallic acid was used as reference standard and results were expressed as gallic acid equivalents in milligrams per liter of extract (mg GAE/L).

Total flavonoids content

Chemicals and reagents: Methanol (Merck), aluminium nitrate (Sigma-Aldrich), potassium acetate (Sigma-Aldrich), quercetin (Sigma-Aldrich) were used in this experiment.

Methods: Determination of flavonoids was done by aluminium nitrate colorimetric method described by Rahimipanah *et al.* (2010) and Mohammadzadeh *et al.* (2007). Briefly, 0.5 mL of extract was diluted with methanol (1:10) and mixed in a test tube with 0.1 mL of 10% aluminium nitrate, 0.1 mL of 1 M aqueous potassium acetate and 4.3 mL methanol. After keeping for 40 min at room temperature, the absorbance of the reaction mixture was measured at 415 nm. Quercetin was used for preparing the standard curve (0-100 mg/L). The results were expressed as milligrams of quercetin equivalents per liter (mg QE/L).

Antioxidant Activity

Chemicals and reagents: Methanol (Merck), 2,2-diphenyl-1-picrylhydrazyl (DPPH; Merck), ascorbic acid (Merck), were used in this experiment.

Method: The capacity of scavenging the 2,2-diphenyl-1-picrylhydrazyl free radical was monitored according to the method reported by Cosmulescu and Trandafir (2012). Briefly, 50 μL of sample extracts were mixed with 3 mL of methanolic solution which is containing DPPH radicals (40 mg/L). The mixture was kept in dark for 30 minutes, and the absorbance was measured at 517 nm. All assays were conducted in triplicate. Standards of ascorbic acid with various concentrations (0-200 mg/L) were used. Antioxidant capacity was expressed in mg ascorbic acid equivalents per liter.

Individual phenolics and juglone

Chemicals and reagents: Standards of phenolic acids (gallic, vanillic, chlorogenic, caffeic, syringic, *p*-coumaric, ferulic, sinapic, salicylic, ellagic and *trans*-cinnamic), flavonoids (catechin, epicatechin, rutin, myricetin and quercetin) and juglone (Sigma-Aldrich, Germany), methanol (HPLC grade, Baker, Netherlands), acetic acid (HPLC grade, Merck, Germany) and BHT (Sigma-Aldrich, Germany) were used in this experiment.

Method: Phenolics were extracted from a 1000 mg sample in a DK 102 p Bandelin ultrasonic bath with 20 ml methanol and 1% BHT (2,6-di-tert-butyl-4-methylphenol) at 25 °C for 40 min. The extracts were centrifuged at 1200 g and the supernatants were filtered through a 0.2 μm polyamide membrane and then stored at -20 °C. HPLC analyses were performed on a Finningan Surveyor Plus HPLC system (Thermo Electron Corporation, San Jose, CA). Separation was achieved by a reversed-phase Hypersil Gold C18 column (5 μm particle size, 250 x 4.6 mm) provided by Thermo Electron Corporation. According to Nour *et al.* (2013), the mobile phase consisted of 1% aqueous acetic acid solution (A) and methanol (B). Samples were eluted with the following gradient: 90% A from 0 to 27 min, from 90 to 60% A in 28 min, 60% A for 5 min, from 60 to 56% A in 2 min, 56% A for 8 min, from 56 to 90% A in 1 min and 4 min 90% A to re-establish the initial conditions, before the injection of another sample. All gradients were linear. The flow rate was 1 mL/min and injection volume was 5 μL. Column temperature was maintained at 20 °C. Each compound was identified by its retention time and by spiking with standards under the same conditions. Identities of constituents were also confirmed with a photodiode array (PDA) detector by comparison with ultraviolet (UV) spectra of standards in the wavelength range of 220-450 nm. Each compound was quantified according to peak area measurements, which were reported in calibration curves of corresponding standards. All samples were extracted and analyzed in triplicate. The contents of investigated phenolic compounds were expressed in mg/L fresh weight (FW) as mean values ± standard deviations.

*Measurement of color by the CIEL*a*b* system*

The CIEL*a*b* system (International Commission on Illumination, Vienna) has been embraced by the US food industry for measuring color of food products. Absorbance of undiluted samples was measured in quartz cuvettes with a path length of 0.1 cm from 350 to 850 nm using an Evolution 600 spectrophotometer. L* (lightness), a* (red/green colour component), b* (yellow/blue colour component) were determined using the color analysis program VL-ColorCalc. Chroma (C) or saturation and hue angle (H) or tones were determined considering McGuire (1992): $C = (a^{*2}+b^{*2})^{1/2}$; H = arctangent of b*/a*.

Statistical Analysis

Data were evaluated by one-way analysis of variance (ANOVA) using Statgraphics Centurion XVI software (StatPoint Technologies, Warrenton, VA). Differences in content levels among the variants were estimated with a multiple range test based on Fisher's least significant difference (LSD) procedure at $P<0.05$.

Results and discussions

Total flavonoid, total phenolics content and antioxidant capacity in green walnut extracts

Starting from the idea of using walnut green fruit extract for preparing liqueur, the present study was focused on the hydroalcoholic extraction of phenolics. Walnut green fruits were extracted with 70% ethanol (a); 40% ethanol and sugar (b); 40% ethanol (c). Results of total flavonoids, total phenolics content and antioxidant capacity in green walnut extracts are presented in Tab. 1. In liqueur, phenolics contribute to astringent but pleasant taste (Stampar *et al.*, 2006) and low risk of some chronic diseases may be associated at higher dietary flavonoid intakes (Knekt *et al.*, 2002). In the present study, in walnut extracts, total phenolics content ranged from 2365 mg gallic acid/L to 4610 mg gallic acid/L and total flavonoids content from 171.02 mg quercetin/L to 423.97 mg quercetin/L.

There were significant differences ($P<0.05$) among phenolic contents and total flavonoids of extracts using the variants mentioned (Tab. 1).

The highest phenolic levels and also flavonoids were obtained at extractions in 70% ethanol, followed by extractions in 40% ethanol and sugar (ratio 1:1) and in 40% ethanol. It shows that high concentration of ethanol (70%) is causing high content in total flavonoids and total phenolics, and at the same ethanol concentration (40%), the addition of sugar increases their amount. The result is similar to that previously obtained by other researchers. Other studies have established that extraction yield of phenols is greatly dependent on solvent and solvent concentration (Stampar *et al.*, 2006). In walnut alcoholic drink, total phenolic content increased with increasing ethanol concentration (Jakopic *et al.*, 2008). On the effectiveness of solvent, Jakopic *et al.* (2009) concluded that the amounts of total phenolics were higher when methanol was used for extraction of phenolic compounds from green walnut fruits. In order to optimise extraction process and the quality of extracts, Fernández-Agulló *et al.* (2013) propose the factors to consider in future studies would be temperature, time and solid to liquid ratio. The maximum flavonoid / phenolic ratio (0.116) was found in green walnut extracts obtained in variant (b) - 40% ethanol and sugar - thus indicating that more flavonoids are extracted in this variant. In terms of antioxidant capacity the variation was similar, the highest antioxidant capacity was found in extracts made in 70% ethanol (Tab. 1). The antioxidant capacity of green walnut extracts ranged from 4175 mg Trolox equivalents/L to 7850 mg Trolox equivalents/L (Tab. 1). Alamprese *et al.* (2005b) studied the antioxidative potential of walnut liqueur and showed that antioxidant activity was directly correlated with total phenolics content and this characteristic did not change during storage, even for many years. The highest value of antioxidant activities and total phenolic contents were determined in bilberry, blackberry and black mulberry wines by Kalkan Yildirim (2006). Mrvcic *et al.* (2013) examined a number of 46 spirit drinks and liqueurs produced in Croatia and concluded that walnut or sour cherry liquors contained the most polyphenols and had very high antioxidant activity.

Phenolic profile of green walnut extracts

The contents and retention time of individual phenolic compounds in green walnut extracts are presented in Tab. 2. There were identified and quantified eleven phenolic acids (gallic, vanillic, chlorogenic, caffeic, syringic, *p*-coumaric, ferulic, sinapic, salicylic, ellagic and *trans*-cinnamic), five flavonoids (catechin, epicatechin, rutin, myricetin and quercetin) and juglone in green walnut extracts. Regarding gallic acid, this study found an average content between 66.72 and 55.12 mgGAE/L. Gallic acid was the major phenolic compound in walnut liqueur found by Stampar *et al.* (2006).

Previous experimental results have indicated that ellagic acid is the dominating phenolic acid of walnut leaves, followed by *trans*-cinnamic, chlorogenic and caffeic acids (Nour *et al.*, 2013). There were significant differences among samples in all individual phenolic acids. Highest concentration was found in 70% ethanol samples (a) for gallic, vanillic, chlorogenic, caffeic, syringic, salicylic and ellagic acid. The use of 40% ethanol (c) led to significantly increase in ferulic acid. Determinations revealed also high concentrations of flavonoids. Catechin, a strong antioxidant that scavenges radicals (Alshatwi, 2010) was established as

Tab. 1. Total flavonoid, total phenolics content and antioxidant capacity in green walnut extracts (mg/L)

Solvents	Total phenolics content*	Total flavonoid content**	Antioxidant capacity***	Flavonoid/ Phenolics
a	$4610{\pm}262.73^{c}$	$423.97{\pm}10.37^{c}$	$7850{\pm}337.55^{c}$	$0.092{\pm}0.003^{b}$
b	$3182{\pm}108.24^{b}$	$370.42{\pm}13.22^{b}$	$5330{\pm}245.18^{b}$	$0.116{\pm}0.002^{c}$
c	$2365{\pm}78.53^{a}$	$171.02{\pm}6.84^{a}$	$4175{\pm}212.92^{a}$	$0.072{\pm}0.005^{a}$

Data expressed as means ± SE of three samples analyzed separately. * gallic acid equivalents; ** quercetin equivalents; ***Trolox equivalents. Different superscript letters within the same column indicate significant differences ($P < 0.05$) among solvents

Tab. 2. Phenolic compounds in green walnut extracts (mgGAE/L)

Phenolics	Retention time (min)	Solvent a	Solvent b	Solvent c
Gallic acid	5.25	66.72±3.07^b	65.18±2.34^b	55.12±3.36^a
(+)-Catechin	18.42	530.80±15.39^b	504.60±22.21^b	257.33±10.33^a
Vanillic acid	28.09	9.53±0.43^b	7.65±0.29^a	8.07±0.45^a
Chlorogenic acid	28.90	16.20±0.68^c	3.77±0.21^a	6.47±0.32^b
Caffeic acid	30.95	2.20±0.09^c	0.33±0.02^a	0.44±0.02^b
Acid syringic	37.58	23.00±1.06^b	18.37±0.69^a	17.28±0.98^a
(−)-Epicatechin	40.39	350.33±11.91^b	337.31±13.49^b	167.41±6.62^a
p-Coumaric acid	45.95	15.21±0.72^a	18.27±0.96^b	14.59±0.87^a
Ferulic acid	49.53	12.80±0.43^a	11.70±0.38^a	24.82±0.96^b
Sinapic acid	51.63	23.84±1.22^a	78.63±3.61^c	34.90±1.81^b
Salicylic acid	52.04	186.58±6.34^c	50.96±2.47^b	26.39±1.26^a
Rutin	57.24	44.57±2.22^b	74.70±3.43^c	33.56±1.87^a
Ellagic acid	58.51	32.19±1.65^c	19.08±1.08^b	12.86±0.66^a
Myricetin	61.56	20.76±0.98^c	11.80±0.45^b	5.34±0,31^a
Juglone	62.73	34.40±1.33^c	22.42±0.88^b	12.47±0.54^a
Trans-cinnamic acid	68.63	0.32±0.02^b	2.74±0.14^c	0.12±0.01^a
Quercetin	70.68	8.16±0.43^c	5.82±0.26^b	5.04±0.31^a

Data expressed as means ± SE of three samples analyzed separately. Different superscript letters between the same column indicate significant differences ($P < 0.05$) among solvents

Tab. 3. CIEL*a*b* parameters for three differently extracts in green walnut

Parameters	Solvents		
	a	b	c
L*	81.4±1.95^b	68.6±1.57^a	91.3±2.16^c
a*	16.4±0.76^c	13.6±0.58^b	7.3±0.33^a
b*	38.4±1.36^b	28.4±0.85^a	26.9±1.07^a
$C = (a^{*2}+b^{*2})^{1/2}$	41.75±0.95^c	31.48±0.51^b	27.87±0.94^a
H° = arctangent b*/a*	66.873±1.69^a	64.411±1.61^a	74.816±1.23^b

All data are expressed as average value±standard deviation. Different superscript letters between columns indicate significant differences ($P < 0.05$) among solvents

the dominating flavonoid of green walnut extracts, followed by epicatechin, rutin, myricetin and quercitin. Catechin was found in concentrations between 257.33 and 530.8 mgGAE/L, while quercetin was found in concentrations between 5.04 and 8.16 mgGAE/L. Myricetin, one of the most common flavonols with potent anticarcinogen and antimutagen activities (Ong and Khoo, 1997), was found in concentrations between 20.76 and 5.34 mgGAE/L. High content of flavonoids was found in variant (a), except rutin, while the greatest amount of rutin was found in variant (b).

Juglone is known as the characteristic compound of *Juglans spp.* and is reported to occur in different parts of walnut (Jakopič *et al.*, 2009). The highest juglone levels (Tab. 2) were obtained at extractions in 70% ethanol (a), followed by extractions in 40% ethanol and sugar (b) and in 40% ethanol (c). As regards the amounts of juglone content in extracts from green walnut fruits, Jakopic *et al.* (2009) concluded that content is higher when methanol was used for extraction, compared to ethanol.

*Measurement of color by the CIEL*a*b* system*

To investigate color quality of the three extracts, lightness and color components were measured. Tab. 3 presents the results of color evaluation of the green walnut extracts using the CIEL*a*b* color system. Significant differences of yellow/blue color component (b*), red/ green (a*) color component, values of lightness (L*), hue angle and chroma (C) were obtained for extracts from green walnut.

Tab. 3 shows that lightness (L*) decreased at lower ethanol concentration (40%) and sugar addition (variant b) and increased at variant (c) - 40% ethanol. Indicators a* and b* decreased at variants (b) and (c) compared to variant (a), along with the decrease of alcohol concentration and sugar addition. L* is the indicator of lightness-darkness and the higher it is, the lighter the extracts. In this case, 40% ethanol extraction solution was the most transparent. Parameters a* and b* have recorded positive values, thus pointing out a greater contribution of red and yellow components. Intensity or saturation (C) is higher in the extract (a) 70% ethanol and hue angle (H°) in the 40% ethanol. Hue angle presented values between 64.41-74.8° from red purple (0°) to yellow (90°).

Conclusions

We can conclude that the extraction efficiency of flavonoid and phenolics from green walnut fruits and the color quality of the extract depend on the methods used. For total phenolics and flavonoid extraction from green walnut fruits, 70% ethanol was more efficient. Highest concentrations were found in 70% ethanol samples for gallic, vanillic, chlorogenic, caffeic, syringic, salicylic, ellagic acid, juglone, catechin, epicatechin, myricetin, and quercetin. The use of 40% ethanol led to a significant increase of ferulic acid and rutin, while the greatest amount of rutin was found in the extract made in 40% ethanol and sugar. Hue angle (H°) of the extracts presented values between 64.41-74.81° from red purple (0°) to yellow (90°). The results suggest that, for food value purpose, green walnut fruit is a valuable raw material for obtaining quality liquor and the 70% ethanol extract provides the highest amount of phenolics and flavonoids. Considering the fact that concentration of alcohol in the liqueur is up to 45%, the traditional version (40 % ethanol and sugar) of preparing the extract may be a viable alternative. Further research is needed on the quality of the liquor.

Acknowledgement

This work was partially supported by the grant number 11C/2014, awarded in the internal grant competition of the University of Craiova, Romania.

References

Akbari V, Jamei R, Heidari R, Esfahlan A J (2012). Antiradical activity of different parts of walnut (*Juglans regia* L.) fruit as a function of genotype. Food Chem 135(4):2404-2410.

Alamprese C, Pompei C (2005a). Influence of processing variables on some characteristics of nocino liqueur. Food Chem 92(2):203-209.

Alamprese C, Pompei C, Scaramuzzi F (2005b). Characterization and antioxidant activity of nocino liqueur. Food Chem 90(4):495-502.

Alshatwi A (2010). Catechin hydrate suppresses MCF-7 proliferation through TP53/Caspase-mediated apoptosis. J Exp Clin Canc Res 29:167.

Cosmulescu S, Trandafir I (2012). Anti-oxidant activities and total phenolics contents of leaf extracts from 14 cultivars of walnut (*Juglans regia* L.). J Hortic Sci Biotech 87(5):504-508.

Cosmulescu S, Trandafir I, Nour V (2014). Seasonal variation of the main individual phenolics and juglone in walnut (*Juglans regia*) leaves. Pharm Biol 52(5):575-580.

Fernández-Agulló A, Pereira E, Freire MS, Valentão P, Andrade PB, González-Álvarez J, Pereira JA (2013). Influence of solvent on the antioxidant and antimicrobial properties of walnut (*Juglans regia* L.) green husk extracts. Ind Crops Prod 42:126-132.

Jakopic J, Colaric M, Veberic R, Hudina M, Solar A, Stampar F (2007). How much do cultivar and preparation time influence on phenolics content in walnut liqueur? Food Chem 104(1):100-105.

Jakopic J, Solar A, Colaric M, Hudina M, Veberic R, Stampar F (2008). The influence of ethanol concentration on content of total and individual phenolics in walnut alcoholic drink. Acta Aliment 37(2):233-239.

Jakopic J, Veberic R, Stampar F (2009). Extraction of phenolic compounds from green walnut fruits in different solvents. Acta Agr Slov 93(1):11-15.

Kalkan Yildirim H (2006). Evaluation of colour parameters and antioxidant activities of fruit wines. Int J Food Sci Nutr 57(1-2):47-63.

Knekt P, Kumpulainen J, Järvinen R, Rissanen H, Heliövaara M, Reunanen A, Hakulinen T, Aromaa A (2002). Flavonoid intake and risk of chronic diseases. Am J Clin Nutr 76(3):560-568.

McGuire RG (1992). Reporting of objective color measurements. HortScience 27:1254-1255.

Mohammadzadeh S, Sharriatpanahi M, Hamedi M, Amanzadeh Y, Sadat Ebrahimi SE, Ostad SN (2007). Antioxidant power of Iranian propolis extract. Food Chem 103:729-733.

Mrvčić J, Posavec S, Kazazić S, Stanzer D, Peša A, Stehlik-Tomas V (2013). Spirit drinks: a source of dietary polyphenols. Croat J Food Sci Technol 4(2):102-111.

Negi AS, Luqman S, Srivastava S, Krishna V, Gupta N, Darokar MP (2011). Antiproliferative and antioxidant activities of *Juglans regia* fruit extracts. Pharm Biol 49(6):669-673.

Nour V, Trandafir I, Cosmulescu S (2013). HPLC determination of phenolic acids, flavonoids and juglone in walnut leaves. J Chromatogr Sci 51(9):883-890.

Ong KC, Khoo HE (1997). Biological effects of myricetin. Gen Pharmacol Vasc Syst 29(2):121-126.

Rahimipanah M, Hamedi M, Mirzapour M (2010). Antioxidant activity and phenolic contents of Persian walnut (*Juglans regia* L.) green husk extract. Afr J Food Sci Technol 1(4):105-111.

Singleton VL, Rossi JA (1965). Colorimetry of total phenolics with phosphomolybdic - phosphotungstic acid reagent. Am J Enol Viticult 16:144-158.

Stampar F, Solar A, Hudina M, Veberic R, Colaric M (2006). Traditional walnut liqueur - cocktail of phenolics. Food Chem 95(4):627-631.

Stampar F, Solar A, Hudina M, Veberic R, Colaric M, Fabcic J (2007). Phenolics in walnut liqueur. Acta Hort 744:451-454.

26

Phenolic Compounds and Antifungal Activity of *Hedera helix* L. (Ivy) Flowers and Fruits

Marcel PÂRVU[1*], Laurian VLASE[2], Alina E. PÂRVU[3], Oana ROSCA-CASIAN[4], Ana-Maria GHELDIU[2], Ovidiu PÂRVU[5]

[1]"Babes-Bolyai" University, Faculty of Biology and Geology, Department of Biology, 42 Republicii Street, 400015 Cluj-Napoca, Romania; marcel.parvu@ubbcluj.ro (*corresponding author)
[2]"Iuliu Hatieganu" University of Medicine and Pharmacy, Faculty of Pharmacy, Department of Pharmaceutical Technology and Biopharmaceutics, 12 Ion Creanga Street, 400010 Cluj-Napoca, Romania; laurian.vlase@yahoo.com
[3]"Iuliu Hatieganu" University of Medicine and Pharmacy, Faculty of Medicine, Department of Pathophysiology, 3 Victor Babes Street, 400012 Cluj-Napoca, Romania; parvualinaelena@yahoo.com
[4]"Babes-Bolyai" University, "A. Borza" Botanical Garden, 42 Republicii Street, 400015 Cluj-Napoca, Romania; casioana@yahoo.com
[5]"Babes-Bolyai" University, Faculty of Mathematics and Computer Science, 1 Kogălniceanu Street, 400084 Cluj-Napoca, Romania; ovidiu.parvu@gmail.com

Abstract

Identification and quantitative analysis of the phenolic compounds from *Hedera helix* L. (ivy) flower and fruit ethanol extracts by LC/MS, *in vitro* germination and growth inhibition effects on *Aspergillus niger, Botrytis cinerea, Fusarium oxysporum* f.sp. *tulipae, Penicillium gladioli* and *Sclerotinia sclerotiorum* were performed. In the non-hydrolyzed samples of flower and fruit extracts were determined, in different amounts, five polyphenols (p-coumaric acid, ferulic acid, rutoside, quercetol and kaempferol) while quercitrin was identified only in the ivy flower extract. The hydrolyzed samples of the same ivy extracts indicated four phenolic compounds (p-coumaric acid, ferulic acid, quercetol and kaempferol), in different concentrations, whereas sinapic acid was only detected in the ivy fruit extract. The antifungal activity of the fresh flower extract was stronger than that of the fresh fruit extract and was compared to that of an antimycotic drug.

Keywords: agar-dilution assay, concentration, ethanol extracts, LC/MS, phytopathogenic fungi, quercetol, rutoside

Introduction

Hedera helix L. (ivy) is an evergreen woody liana of Araliaceae family which presents alternate and evergreen leaves, with accentuated polymorphism. The ivy flowers produced from summer until late autumn are small, greenish-yellow and the fruits are small black berries, ripening in winter. Common ivy naturally grows in the Western, Central and Southern Europe, but has also been introduced to North America and Asia (Gruenwald *et al.*, 2000). *H. helix* is a popular ornamental plant in many countries and it is used as a medicinal plant (Ferrara *et al.*, 2013; Gruenwald *et al.*, 2000; Lutsenko *et al.*, 2010).

Regarding the chemical composition and biological activity of *H. helix* leaves literature mentions different information (Ferrara *et al.*, 2013; Lutsenko *et al.*, 2010; Mandade *et al.*, 2010), but there are little data about fruits (Bedir *et al.*, 2000; Lutsenko *et al.*, 2010) and especially ivy flowers.

The *H. helix* leaves contain saponins (Lutsenko *et al.*, 2010; Sieben *et al.*, 2009; Song *et al.*, 2014), flavonoids (Trute and Nahrstedt, 1997), phenolic acids (Wichtl, 2004), emetine alkaloid (Mahran *et al.*, 1975), aminoacids (Ferrara *et al.*, 2013), sterols (Wichtl, 2004), proteins (Ferrara *et al.*, 2013), vitamins (Wichtl, 2004), polyacetylenes (Gafner *et al.*, 1989; Wichtl, 2004) etc. Other studies revealed active constituents from the *H. helix* fruits (Lutsenko *et al.*, 2010) like triterpene saponins (Bedir *et al.*, 2000), fatty acids (Grosbois, 1971, 1976), polyacetylenes (Christensen *et al.*, 1991) and β-lectins (Gleeson and Jermyn, 1979).

The main objectives of this study were to evaluate the fungicide potential of the ivy flower and fruit extracts against five economically important plant pathogens, selected from different taxonomic groups, and to determine polyphenols from these extracts known as antifungal compounds (Arif *et al.*, 2011; Del Río *et al.*, 2000; Lattanzio *et al.*, 2006), since different studies present antibacterial activity of the *H. helix* leaf extract against *Erwinia amylovora* on apple (Baysal *et al.*, 2002; Baysal and Zeller, 2004) and phytopathogenic fungi like *Phytophthora infestans, Pseudoperonospora cubensis* (Röhner *et al.*, 2004), *Alternaria solani* (Yanar *et al.*, 2011) etc.

Materials and methods

Plant material

Ivy (*Hedera helix* L.) was collected from the A. Borza Botanical Garden in Cluj-Napoca (46°45′36″N and 23°35′13″E) and was identified by Dr. M. Parvu, Babes-Bolyai University of Cluj-Napoca. A voucher specimen (CL 664210) was deposited at the Herbarium of Babes-Bolyai University, Cluj-Napoca, Romania.

Plant extract preparation

Fragments (0.5-1 cm) from flowers and fruits, respectively, were extracted with 70% ethanol in the Mycology Laboratory of Babes-Bolyai University, Cluj-Napoca, Romania, by cold repercolation method (Mishra and Verma, 2009; Sundaram and Gurumoorthi, 2012) at room temperature, for 3 days (Sundaram and Gurumoorthi, 2012) to obtain the two *H. helix* extracts. The ivy flowers were harvested in 23.09.2011 and the fruits in 29.12.2011. The content of plant extracts (w/v; g/ml) was 1/1.1 for flower extract and 1/1 for fruit extract.

Preparation of fungal colonies

Aspergillus niger Tiegh. isolated from *Allium cepa* L. bulbs, *Botrytis cinerea* Pers. isolated from 'Kordes Perfecta' rose flowers, *Fusarium oxysporum* f. sp. *tulipae* W.C. Snyder and H.N. Hansen isolated from *Tulipa gesneriana* L. flowers, *Penicillium gladioli* Machacek isolated from *Gladiolus* x *hybridus* C. Morr. corms, *Sclerotinia sclerotiorum* (Lib.) de Bary isolated from *Daucus carota* L. roots, were included in this study. All the fungal species were identified in the Mycology Laboratory, Babes-Bolyai University, Cluj-Napoca, Romania, by Dr. M. Parvu. Colonies were obtained in Petri dishes on Czapek-agar medium (BD Difco, Budapest, Hungary), by inoculation in the central point and incubation at 22 °C for 5 days.

Identification and quantitative determinations of the polyphenols

A high-performance liquid chromatography method coupled with mass spectrometry (LC/MS) was used to analyze the polyphenolic compounds in the *H. helix* plant extracts. The method used was a previously published HPLC method with minor changes (Compaore *et al.*, 2012; Mocan *et al.*, 2014; Vlase *et al.*, 2013). The method is suitable for qualitative (18 compounds) and quantitative (14 compounds) analyses. In this study, 18 standards of the polyphenolic compounds were used, namely caftaric acid, gentisic acid, caffeic acid, chlorogenic acid, paracoumaric acid, ferulic acid, sinapic acid, hyperoside, isoquercitrin, rutoside, myricetol, fisetin, quercitrin, quercetol, patuletine, luteolin, kaempferol, and apigenin (Mocan *et al.*, 2014; Vlase *et al.*, 2013).

The experiments were performed using an Agilent 1100 HPLC Series system equipped with a degasser, binary gradient pump, column thermostat, autosampler and UV detector. The HPLC system was coupled with an Agilent 1100 mass spectrometer (LC/MSD ion trap VL). For the separation, a reverse-phase analytical column was employed (Zorbax SB-C18 100 x 3.0 mm i.d., 3.5 μm particle); the temperature was 48 °C. The compounds were detected in both the UV and MS mode. The MS signal was used only for qualitative analysis based on the specific mass spectra of each polyphenol (Mocan *et al.*, 2014; Vlase *et al.*, 2013).

Four polyphenols could not be quantified under the chromatographic conditions because of overlapping (caftaric acid with gentisic acid and caffeic acid with chlorogenic acid). However, all 4 compounds were selectively identified using MS detection (qualitative analysis) based on differences in their molecular mass and MS spectra (Vlase *et al.*, 2013).

Determination of antifungal activity

The antifungal activity of *H. helix* flower and fruit extracts, expressed as minimum inhibitory concentration (MIC), was determined by the agar-dilution assay, and was compared to the antimycotic drug fluconazole (2 mg·mL^{-1}, Krka, Novo Mesto, Slovenia) and a control (nutritive medium and 70% ethanol). The percentage of fungal growth inhibition (*P*) at each concentration was calculated using the formula $P = (C\text{-}T) \times 100/C$, where C is the diameter of the control colony and T is the diameter of the treated colony (Nidiry and Babu, 2005). MIC was considered the lowest concentration that completely inhibited the pathogen growth according to a visual evaluation (Pinto *et al.*, 2010).

Statistical analysis

Statistical analyses were performed using the program R environment, version 2.14.1. The results for each group were expressed as mean ± standard deviation. Data were evaluated by analysis of variance (ANOVA). A P value of ≤ 0.05 was considered statistically significant. The correlation analysis was performed by the Pearson test.

Results and discussions

Quantitative analysis of polyphenolic compounds

The non-hydrolyzed sample (Fig. 1) of *H. helix* flower extract contains p-coumaric acid (1.043 μg·mL^{-1}), ferulic acid (0.506 μg·mL^{-1}), rutoside (129.711 μg·mL^{-1}), quercitrin (1.300 μg·mL^{-1}), quercetol (7.111 μg·mL^{-1}), kaempferol (7.911 μg·mL^{-1}), whereas the hydrolyzed sample (Fig. 2) of the same extract revealed p-coumaric acid (1.404 μg·mL^{-1}), ferulic acid (0.961 μg·mL^{-1}), quercetol (37.777 μg·mL^{-1}) and kaempferol (20.581 μg·mL^{-1}).

The analysis of polyphenols from the non-hydrolyzed sample (Fig. 3) of the *H. helix* fruit extract revealed the presence of p-coumaric acid (0.682 μg·mL^{-1}), ferulic acid (3.237 μg·mL^{-1}), rutoside (169.643 μg·mL^{-1}), quercetol (2.156 μg·mL^{-1}) and kaempferol (1.146 μg·mL^{-1}), while the hydrolyzed sample (Fig. 4) of the same extract indicated p-coumaric acid (0.862 μg·mL^{-1}), ferulic acid (2.933 μg·mL^{-1}), sinapic acid (0.660 μg·mL^{-1}), quercetol (31.886 μg·mL^{-1}) and kaempferol (3.865 μg·mL^{-1}).

Antifungal activity

The results of the antifungal assays of *H. helix* extracts are presented in Table 1. *H. helix* flower extract MIC was 12% for *A. niger*, 8% for *B. cinerea* and *S. sclerotiorum*, 10% for *F. oxysporum* f. sp. *tulipae* and *P. gladioli*.

H. helix fruit extract MIC was 14% for *A. niger*, 10% for *B. cinerea* and *S. sclerotiorum*, 12% for *F. oxysporum* f. sp. *tulipae* and for *P. gladioli*.

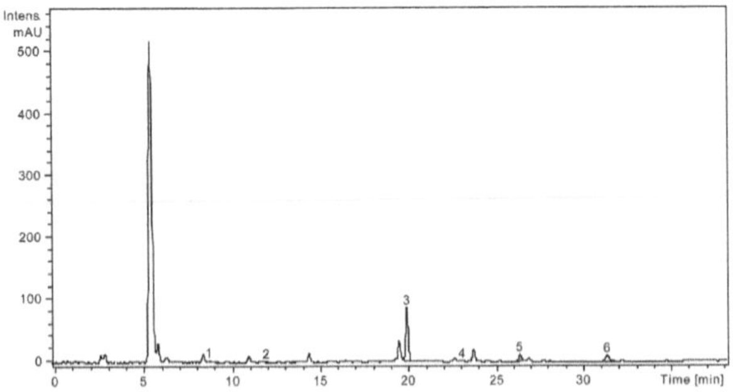

Fig. 1. Chromatogram of polyphenols from the non-hydrolyzed sample of *Hedera helix* flower extract. The peaks are marked: "1" p-coumaric acid; "2" Ferulic acid; "3" Rutoside; "4" Quercitrin; "5" Quercetol; "6" Kaempferol

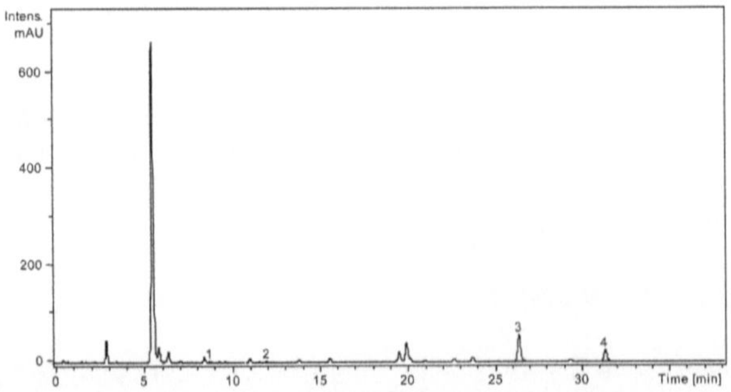

Fig. 2. Chromatogram of polyphenols from the hydrolyzed sample of *Hedera helix* flower extract. The peaks are marked: "1" p-coumaric acid; "2" Ferulic acid; "3" Quercetol; "4" Kaempferol

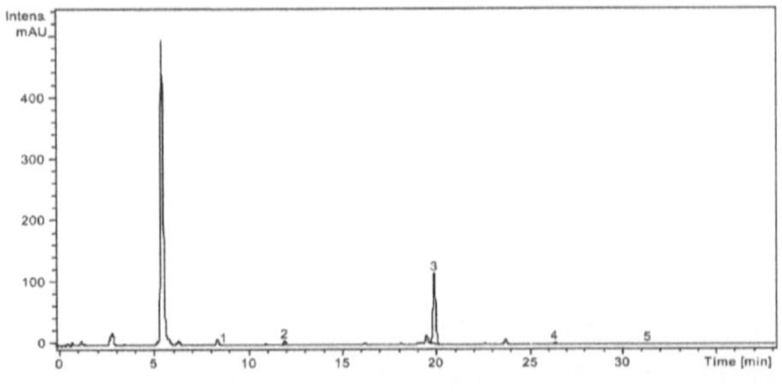

Fig. 3. Chromatogram of polyphenols from the non-hydrolyzed sample of *Hedera helix* fruits extract. The peaks are marked: "1" p-coumaric acid; "2" Ferulic acid; "3" Rutoside; "4" Quercetol; "5" Kaempferol

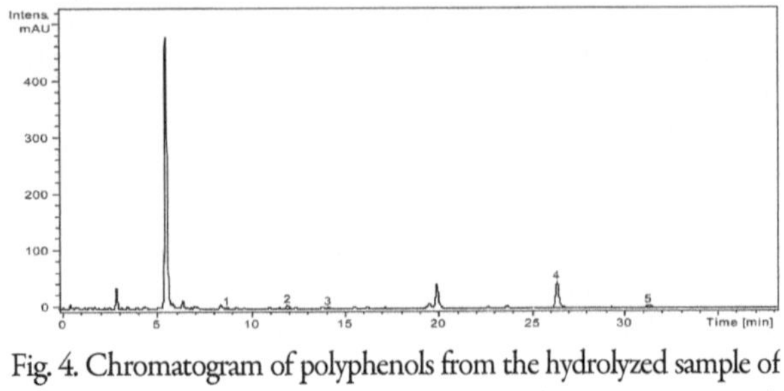

Fig. 4. Chromatogram of polyphenols from the hydrolyzed sample of *Hedera helix* fruits extract. The peaks are marked: "1" p-coumaric acid; "2" Ferulic acid; "3" Sinapic acid; "4" Quercetol; "5" Kaempferol

Fluconazole MIC was 30% for *A. niger* and *P. gladioli*, 12% for *B. cinerea*, 10% for *F. oxysporum* f. sp. *tulipae* and 8% for *S. sclerotiorum* (Table 1).

The inhibitory effect of *H. helix* flower extract was stronger than that of the fruit extract for all concentrations and for all tested fungi ($P<0.001$).

Since generally the active antifungal compounds of most plant extracts are phenolic compounds (El-Khateeb *et al.*, 2013), the analysis of polyhenols from flowers and fruits (Figs. 1-4) is necessary for studying the antifungal effects of ivy extracts.

The identification and quantitative determinations of the polyphenols from ivy flowers and fruits are in agreement with other authors who mention that there are differences regarding these compounds in relation to the plant species (El-Khateeb *et al.*, 2013; Mocan *et al.*, 2014; Wojdyło *et al.*, 2007), plant organ (Singh *et al.*, 2011; Wojdyło *et al.*, 2007) and each analyzed polyphenol (Ferrara *et al.*, 2013; Mocan *et al.*, 2014; Vlase *et al.*, 2013; Wojdyło *et al.*, 2007).

The ivy extracts obtained from leaves (Ferrara *et al.*, 2013), flowers (Fig. 1) and fruits (Fig. 2) contain polyphenols in different concentrations. So, *H. helix* leaves contain rutin, quercetin, kaempferol and apigenin in different quantities. In the non-hydrolyzed samples of the ivy flower and fruit extracts were determined five polyphenols (p-coumaric acid, ferulic acid, rutoside, quercetol and kaempferol) in different amounts, in relation to analysed phenolic compound, while quercitrin was determined only in the ivy flower extract. The hydrolyzed samples of the ivy extracts indicated four phenolic compounds (p-coumaric acid, ferulic acid, quercetol and kaempferol) in different concentrations, whereas sinapic acid was identified only in the ivy fruit extract. In higher quantities, the phenolic compounds p-coumaric acid, quercetol and kaempferol were determined in the non-hydrolyzed and hydrolyzed flower extracts (Figs. 1 and 2), while polyphenols ferulic acid and rutoside in the ivy fruit extract (Figs. 3 and 4). Rutoside was the dominant phenolic compound in both non-hydrolyzed ivy extracts and quercetol was in hydrolyzed samples of the same extracts.

Some polyphenols identified in *H. helix* flowers and fruits extracts as p-coumaric acid, kaempferol, rutoside, showed important antibacterial and antifungal effects (Del Río *et al.*, 2000; Singh *et al.*, 2011; Stojković *et al.*, 2013). The phenolic compound kaempferol obtained from *Cassia renigera* Wall. presented significant antifungal effects against *Aspergillus flavus*, *A. niger*, *Fusarium moniliforme* and *Rhizoctonia bataticola* phytopathogenic fungi (Singh *et al.*, 2011). Moreover, all phenolic compounds identified in *Olea europaea* and *Citrus* spp. acted as antifungal agents and were capable of inhibiting the growth of phytopathogenic fungi (Del Río *et al.*, 2000). The antimicrobial activity of phenolic extracts obtained from wild edible species against postharvest fungal pathogens had been little explored (Gatto *et al.*, 2011). Moreover, the phenolic compounds verbascoside and isoverbascoside isolated from *Orobanche crenata* proved to be effective for biological control of important phytopathogenic fungi of fresh fruit and vegetables (Gatto *et al.*, 2013). The antifungal activity of polyphenols verbascoside and isoverbascoside depends on pathogenic species. So, verbascoside showed significant activity against *Penicillium italicum* and little activity against *Botrytis cinerea*, *Monilinia fructicola* and *Penicillium digitatum*, while isoverbascoside

Table 1. Antifungal activity of *Hedera helix* flower and fruit extracts on *in vitro* germination and growth of some phytopathogenic fungi

Phytopathogenic fungi	Flower extract conc. (%)	Colony [a] diameter (mm)	P [a]	Fruit extract conc. (%)	Colony [b] diameter (mm)	P [b]	Fluconazole conc. (%)	Colony [c] diameter (mm)	P [c]
Aspergillus niger	C^c	22	0	C^c	22	0	C^c	22	0
	2	19	13.63±0.81	4	18	18.18±0.81	10	12	45.45±0.81
	4	14	36.36±0.81	6	13.5	38.63±0.57	20	7	68.18±0.81
	6	10.25	53.40±0.50	8	10	54.54±0.81	25	2.75	87.50±0.5
	8	6.5	70.45±0.57	10	5	77.27±0.81	30	0	100
	10	3	86.36±0.81	12	2.5	88.63±0.57			
	12	0	100	14	0	100			
Botrytis cinerea	C^c	65	0	C^c	65	0	C^c	65	0
	2	55	15.38±0.81	2	60	7.69±0.81	2	40.5	37.69±0.57
	4	31	52.30±0.81	4	40.5	37.69±0.57	6	20.25	68.84±0.5
	6	3.75	94.23±0.50	6	22	66.15±0.81	10	3	95.38±0.81
	8	0	100	8	5	92.30±0.81	12	0	100
				10	0	100			
Fusarium oxysporum f.sp. *tulipae*	C^c	34	0	C^c	34	0	C^c	34	0
	2	28	17.64±0.81	4	32	5.88±0.81	2	20	41.17±0.81
	4	19	44.11±0.81	6	21	38.23±0.81	6	8	76.47±0.81
	6	10	70.58±0.81	8	10.75	68.38±0.50	8	2	94.11±0.81
	8	2.25	93.38±0.50	10	2.5	92.64±0.57	10	0	100
	10	0	100	12	0	100			
Penicillium gladioli	C^c	15	0	C^c	15	0	C^c	15	0
	2	13	13.33±0.81	4	12	20±0.81	10	11	26.66±0.81
	4	8	46.66±0.81	6	8	46.66±0.81	16	10	33.33±0.81
	6	5.50	63.33±0.57	8	5	66.66±0.81	20	7.5	50.0±0.57
	8	2	86.66±0.81	10	2.25	85.0±0.50	25	3.75	75.0±0.50
	10	0	100	12	0	100	30	0	100
Sclerotinia sclerotiorum	C^c	64	0	C^c	64	0	C^c	64	0
	2	52	18.75±0.81	2	58	9.37±0.81	2	30	53.12±0.81
	4	31	51.56±0.81	4	37	42.18±0.81	4	14.50	77.34±0.57
	6	2.75	95.70±0.50	6	13	79.68±0.81	6	4.75	92.57±0.50
	8	0	100	8	2.25	96.48±0.50	8	0	100
				10	0	100			

Legend: [a] = the effect of *H. helix* flower extract; [b] = the effect of *H. helix* fruit extract; [c] = the effect of Fluconazole; C = control (50% aq. EtOH); P = mycelial growth inhibition - results are the mean ± SD of 4 experiments. The same doses of *H. helix* flower and fruit extracts were tested against all fungal species

completely inhibited *Botrytis cinerea, Penicillium digitatum, P. italicum* and *P. expansum* and was less effective against *M. fructicola* (Gatto *et al.,* 2013).

The mechanisms of action thought to be responsible for phenolic toxicity involve enzyme inhibition by the oxidized compounds, possibly through reaction with sulfhydryl groups or nonspecific interactions with the proteins (Arif *et al.,* 2009; Mason and Wasserman, 1987).

The *Hedera helix* L. (ivy) extracts possess antifungal activities due to polyphenols p-coumaric acid, kaempferol, rutoside etc. (Del Río *et al.,* 2000; Singh *et al.,* 2011; Stojković *et al.,* 2013) and the inhibitory effect of flower extract was stronger than that of the fruit extract, against all tested fungi (Table 1). The MIC of *H. helix* flower and fruit extracts depended on the pathogenic species and on the biologically active compounds content of the extracts (Al-Reza *et al.,* 2010; El-Khateeb *et al.,* 2013; Dissanayake, 2014). In general, there is a positive relation between the concentration of the plant extracts and the inhibition rate of mycelia growth in all tested fungi (El-Khateeb *et al.,* 2013).

Our results (Table 1) completed the literature data which mention different antimicrobial effects of *H. helix* extracts against phytopathogenic bacteria and fungi: antibacterial activity against *Erwinia amylovora* (Baysal *et al.,* 2002; Baysal and Zeller, 2004); a high degree of antifungal activity against *Phytophthora infestans* from tomato and *Pseudoperonospora cubensis* from cucumber (Röhner *et al.,* 2004); an insignificant antifungal effect against *Alternaria solani* from potato (Yanar *et al.,* 2011); an ineffective antifungal action against *in vitro* growth of the fungus *Colletotrichum lindemuthianum* from common bean (Pinto *et al.,* 2010).

The antifungal effects of the ivy extracts recommend them as good candidates for the *in vivo* biological control of phytopathogenic fungi similar to other fungicide plant extracts known for their antifungal activity (Al-Reza *et al.*, 2010; Dissanayake, 2014; El-Khateeb *et al.*, 2013; Ribera and Zuñiga, 2012), limiting the abuse of chemical fungicides (Gatto *et al.*, 2013; Lattanzio *et al.*, 2006; Pârvu and Pârvu, 2011).

Conclusions

The quantitative and qualitative analysis of polyphenols revealed differences among these compounds in *H. helix* extracts and completed the literature regarding the chemical composition of ivy flower and fruit extracts. The present study clearly indicated that the ivy extracts obtained from flowers and fruits possess antifungal constituents and can be further explored as useful source of antimicrobial agents for *in vivo* control of plant pathogenic fungi.

Acknowledgments

This paper was published under the frame of European Social Found, Human Resources Development Operational Programme 2007–2013, project no. POSDRU/159/1.5/S/136893.

References

Al-Reza SM, Rahman A, Ahmed Y, Kang SC (2010). Inhibition of plant pathogens *in vitro* and *in vivo* with essential oil and organic extracts of *Cestrum nocturnum* L. Pestic Biochem Phys 96(2):86-92.

Arif T, Bhosale JD, Kumar N, Mandal TK, Bendre RS, Lavekar GS, Dabur R (2009). Natural products – antifungal agents derived from plants. J Asian Nat Products Res 11:621-638.

Arif T, Mandal TK, Dabur R (2011). Natural products: Antifungal agents derived from plants, pp 283-311. In: Tiwari VK (Ed). Opportunity, challenge and scope of natural products in medicinal chemistry. Research Singpost, Kerala, India.

Baysal O, Laux P, Zeller W (2002). Further studies on the induced resistance (IR) effect of plant extract from *Hedera helix* against fire blight (*Erwinia amylovora*) (Conference Paper). Acta Hort 590:273-277.

Baysal O, Zeller W (2004). Extract of *Hedera helix* induces resistance on apple rootstock M26 similar to Acibenzolar-S-methyl against fire blight (*Erwinia amylovora*). Physiol Mol Plant P 65(6):305-315.

Bedir E, Kirmizipekmez H, Sticher O, Calis I (2000). Triterpene saponins from the fruits of *Hedera helix*. Phytochemistry 53(8):905-909.

Christensen LP, Lam J, Thomasen T (1991). Polyacetylenes from the fruits of *Hedera helix*. Phytochemistry 30(12):4151-4152.

Compaore M, Lamien CE, Lamien-Meda A, Vlase L, Kiendrebeogo M, Ionescu C, Nacoulma OG (2012). Antioxidant, xanthine oxidase and lipoxygenase inhibitory activities and phenolics of *Bauhinia rufescens* Lam. (Caesalpiniaceae). Nat Prod Res 26:1069-1074.

Del Río JA, Arcas MC, Botía JM, Báidez AG, Fuster MD, Ortuno AM (2000). Involvement of phenolic compounds in the antifungal defense mechanisms of *Olea europaea* L. and *Citrus* sp., pp 331-341. In: Pandalai SG (Ed). Recent Research Developments in Agricultural & Food Chemistry. Research Signpost, Trivandrum.

Dissanayake MLMC (2014). Inhibitory effect of selected medicinal plant extracts on phytopathogenic fungus *Fusarium oxysporum* (Nectriaceae) Schlecht. Emend. Snyder and Hansen. Annu Res Rev Biol 4(1):133-142.

El-Khateeb AY, Elsherbiny EA, Tadros LK, Ali SM, Hamed HB (2013). Phytochemical analysis and antifungal activity of fruit leaves extracts on the mycelial growth of fungal plant pathogens. J Plant Pathol Microb 4(9):1-6.

Ferrara L, Naviglio D, Faralli S (2013). Identification of active principles of *Hedera helix* L. in aqueous extracts. The Journal of Phytochemistry. Photon 114:170-175.

Gafner F, Reynolds GW, Rodriguez E (1989). The diacetylene 11, 12-dehydrofalcarinol from *Hedera helix*. Phytochemistry 28(4):1256-1257.

Gatto MA, Ippolito A, Linsalata V, Cascarano NA, Nigro F, Vanadia S, Di Venere D (2011). Activity of extracts from wild edible herbs against postharvest fungal diseases of fruit and vegetables. Postharvest Biol Technol 61(1):72-82.

Gatto MA, Sanzani SM, Tardia P, Linsalata V, Pieralice M, Sergio L, Di Venere D (2013). Antifungal activity of total and fractionated phenolic extracts from two wild edible herbs. Natural Science 5(8):895-902.

Gleeson PA, Jermyn MA (1979). Alteration in the composition of β-lectins caused by chemical and enzymic attack. Austral J Plant Physiol 6(1):25-38.

Grosbois M (1971). Biosynthesis of fatty acids during the development of ivy fruit and seed (in French). Phytochemistry 10(6):1261-1273.

Grosbois M (1976). Changes in the amount of complex lipids in the seeds and in the pericarp during the development of ivy fruit *Hedera helix*. Physiol Plantarum 36(2):145-149.

Gruenwald J, Brendler T, Jaenicke C (2000). PDR for Herbal Medicines. Medical Economics Company, Montvale, 284 p.

Lattanzio V, Lattanzio VMT, Cardinali A (2006). Role of phenolics in the resistance mechanisms of plants against fungal pathogens and insects, pp 23-67. In: Imperato F (Ed). Phytochemistry: advances in research. Trivandrum, Kerala, India.

Lutsenko Y, Bylka W, Matlawska I, Darmohray R (2010). *Hedera helix* as a medicinal plant. Herba Pol 56(1):83-96.

Mahran GH, Hilal SH, El-Alfy TS (1975). The isolation and characterisation of emetine alkaloid from *Hedera helix*. Planta Med 27(2):127-132.

Mandade RJ, Choudhuri A, Mashirkar V, Sakarkar D (2010). Analgesic and anti-inflammatory activities of *Hedera helix* leaf extract. International Journal of Comprehensive Pharmacy 1(4):1-3.

Mason TL, Wasserman BP (1987). Inactivation of red beet betaglucan synthase by native oxidized phenolic compounds. Phytochemistry 26:2197-2202.

Mishra R, Verma DL (2009). Antifungal activity and flavonoid composition of *Wiesnerella denudata*. Steph Academia Arena 1:42-45.

Mocan A, Vlase L, Vodnar DC, Bischin C, Hanganu D, Gheldiu A-M, Oprean R, Silaghi-Dumitrescu R, Crisan G (2014). Polyphenolic content, antioxidant and antimicrobial activities of *Lycium barbarum* L. and *Lycium chinense* Mill. leaves. Molecules 19:10056-10073.

Nidiry ESJ, Babu CSB (2005). Antifungal activity of tuberose absolute and some of its constituents. Phytother Res 19:447-449.

Pârvu M, Pârvu AE (2011). Antifungal plant extracts, pp 1055-1062. In: Méndez-Vilas A (Ed). Science against microbial pathogens: communicating current research and technological advances. Formatex Research Center, Badajoz, Spain.

Pinto JMA, Souza EA, Oliveira DF (2010). Use of plant extracts in the control of common bean anthracnose. Crop Protection 29(8):838-842.

Ribera AE, Zuñiga G (2012). Induced plant secondary metabolites for phytopatogenic fungi control: a review. J Soil Sci Plant Nutr 12(4):893-911.

Röhner E, Carabet A, Buchenauer H (2004). Effectiveness of plant extracts of *Paeonia suffruticosa* and *Hedera helix* against diseases caused by *Phytophthora infestans* in tomato and *Pseudoperonospora cubensis* in cucumber. J Plant Dis Prot 111(1):83-95.

Sieben A, Prenner L, Sorkalla T, Wolf A, Jakobs D, Runkel F, Häberlein H (2009). Alpha-hederin, but not hederacoside C and hederagenin from *Hedera helix*, affects the binding behavior, dynamics, and regulation of beta 2-adrenergic receptors. Biochemistry 48(15):3477-3482.

Singh D, Sharma SK, Rani R, Mishra S, Sharma RA (2011). Kaempferol-*7-O*-glucoside and their antimicrobial screening isolate from *Cassia renigera* Wall. Int J Pharm Clin Res 3(2):30-34.

Stojković D, Petrović J, Soković M, Glamočlija J, Kukić-Marković J, Petrović S (2013). *In situ* antioxidant and antimicrobial activities of naturally occurring caffeic acid, p-coumaric acid and rutin, using food systems. Journal of the Science of Food and Agriculture 93(13):3205-3208.

Sundaram U, Gurumoorthi P (2012). Validation of HPTLC method for quantitative estimation of L-Dopa from *Mucuna pruriens*. Int Res J Pharm 3:300-304.

Trute A, Nahrstedt A (1997). Identification and quantitative analysis of phenolic compounds from the dry extract of *Hedera helix*. Planta Med 63(2):177-179.

Vlase L, Pârvu M, Pârvu EA, Toiu A (2013). Chemical constituents of three *Allium* species from Romania. Molecules 18(1):114-127.

Wichtl M (2004). Herbal drugs and phytopharmaceuticals. A handbook for practice on a scientific basis. 3rd Ed. Stuttgart, 274 p.

Wojdyło A, Oszmiański J, Czemerys R (2007). Antioxidant activity and phenolic compounds in 32 selected herbs. Food Chem 105:940-949.

Yanar Y, Gökçe A, Kadioglu I, Çam H, Whalon M (2011). *In vitro* antifungal evaluation of various plant extracts against early blight disease (*Alternaria solani*) of potato. Afr J Biotechnol 10(42):8291-8295.

Impact of Climate on Vegetation Change in a Mountain Grassland – Succession and Fluctuation

Florin PĂCURAR[1], Ioan ROTAR[1]*, Albert REIF[2], Roxana VIDICAN[1], Vlad STOIAN[1], Stefanie M. GÄRTNER[2], Robert B. ALLEN[3]

[1]*University of Agricultural Sciences and Veterinary Medicine Cluj-Napoca, Faculty of Agriculture, Calea Mănăştur 3-5, 400372 Cluj-Napoca, Romania; fpacurar@gmail.com; ioan.rotar@usamvcluj.ro (*corresponding author); roxana.vidican@usamvcluj.ro; vladh_stoian@yahoo.com*
[2]*University of Freiburg, Faculty of Environment and Natural Resources, Tennenbacherstr. 4, 79085 Freiburg, Germany; albert.reif@waldbau.uni-freiburg.de; stefanie.gaertner@waldbau.uni-freiburg.de*
[3]*Landcare Research, P.O. Box 69040, Lincoln 7640, New Zealand; AllenR@landcareresearch.co.nz*

Abstract

Traditionally managed Central European mountain grasslands have high nature conservation value because of their high species diversity. Whether these grasslands and their diversity can be preserved will depend on many factors, including how plant species composition responds to changes in climate conditions. To differentiate between fluctuations and directional succession in the herbaceous layer composition of a Romanian *Festuca rubra* L. and *Agrostis capillaris* L. grassland in Apuseni and whether any compositional changes can be related to climate. The vegetation of permanent plots was recorded annually between 2004 and 2012. Temperature and precipitation were measured by an automatic weather station at the study site. Cluster analysis, Indicator Species Analysis and the co-dominance ratio between *F. rubra* L. - *A. capillaris* were analysed. The compositional data was related to the climate variables. Thresholds of relevant climate variables differentiating between clusters of plots with similar vegetation composition were determined using classification tree methods. The vegetation composition in our plots within the years 2004, 2005 and 2008 were different from each other. From 2004 to 2006 directional succession could be identified; however the major patterns to emerge were fluctuations which occurred over the whole study period. Compositional shifts included *A. capillaris* L. and *F. rubra* L exchanging co-dominance with each other. The most important variables differentiating clusters were temperature during the dormant and vegetation periods and water balance during the vegetation period. It can be concluded that compositional shifts among years were largely a consequence of year to year climatic fluctuations; however, there is some evidence for a directional shift during the early years of the study.

Keywords: climate fluctuation, vegetation dynamic, *Festuca rubra* L. - *Agrostis capillaris* L. meadow, Romania

Introduction

Habitats with a long history have reached an equilibrium which is not static and fluctuates permanently by undirected changes of lesser magnitude (Frelich, 2002). Climate fluctuations also generate floristic, structural and functional changes in plant communities including grassland (Petrychenko *et al.*, 2012). These climate introduced fluctuations were shown to occur in grasslands of North America (Collins *et al.*, 1987), the Czech Republic (Dostálek and Frantík, 2011), Poland (Zarzycki and Szewczyk, 2013) and other areas (Cui, 2009; Egan and Crandall, 2008; Ihm *et al.*, 2007; Lal, 2012, Miller, 2008; Schmer *et al.*, 2010).

Successional trends are difficult to separate from fluctuations (Kasperek 1998, Matesanz *et al.*, 2009). Changes of temperature and precipitation have caused well documented shifts in species composition, community structure and turnover of grassland vegetation (Deutsch *et al.*, 2011; Heisler-White *et al.*, 2009; Niu and Wan, 2008). The duration of these changes and their impact at the ecosystem level are still largely unknown (Zurek 2011, 2012). The existing knowledge is based on phytosociological studies (Brito *et al.*, 2014; Leonard, 2014; Worthington, 2014), but experimental studies are lacking.

In an experimental approach site factors can be quantified and management can be simulated. The climate effect on the behaviour of plant species is less known (Niu and Wan, 2008). Understanding and quantifying the impacts of climate remains a key-challenge for ecologists (Heisler-White *et al.*, 2009). Knowledge about the relationships between climate and vegetation is essential for planning land-use, retaining and enhancing biodiversity, and climatic change modelling (Morecroft *et al.*, 2004;

Plantureux *et al.*, 2012; Sala *et al.*, 2000; Schneider *et al.*, 2009; Schaumberger, 2012; Trnka 2011).

Only long-term monitoring in multifactorial experiments exploring the interacting effects of land use change and climatic fluctuations can provide the necessary data to distinguish these processes, e.g., the impact of changing precipitation patterns (Morecroft *et al.*, 2004). Only few studies exist and they are largely confined to systems with a capacity for rapid compositional change (e.g. Grime *et al.*, 2008).

Mountain grasslands contain some of the most important high nature conservation value areas in Europe (Veen *et al.*, 2009). Highly diverse mountain farmlands can increase tourism incomes and provide a potential seed source for local biodiversity restoration (Hopkins, 2011).

The interactions between climate and management in grasslands dominated by *Festuca rubra* L. and *Agrostis capillaris* L. seem to produce fluctuations in species abundance and cause changes in succession affecting the species composition, diversity and biomass production (Pleşa, 2012). Phytosociological studies rarely define robust community dynamics (Cristea *et al.*, 2004).

The objectives of this study were to distinguish the successional trends and fluctuations in a *Festuca rubra* L. - *Agrostis capillaris* L. grassland in the Apuseni Mountains, Romania in an experimental approach. Permanent sample plot data from Romanian mountain grassland were measured annually over nine years. Our aim was: (1) to quantify the long-term directional successional floristic changes under the impact of climate and management through mowing (2) to identify the most relevant climatic parameters affecting the species composition in the short-term; (3) to develop a model of the grassland community response on the species-level, including the changes in *F. rubra* L. and *A. capillaris* L. co-dominance patterns.

Materials and methods

Study area and experimental design

The research was conducted near the village of Gheţari, Gîrda de Sus commune, situated in the central part of Apuseni Mountains, Romania (coordinates: latitude 46°29'24.88" and longitude 022°48'52.40"). The regional geology is dominated by limestone (Orăşeanu, 2005). The area has a mountainous climate with an average temperature of 5.2 °C and an average annual precipitation of 1122.65 mm (Pleşa, 2012). The experiment was established in 2001 in a *F. rubra* L. – *A. capillaris* L. grassland, at 1130 m elevation on a red preluvosol soil with an average nitrogen content (0,212 %) but poor in phosphorus (3 ppm) and potassium (25 ppm) (Parichi and Stănilă, 2005). The vegetation was recorded on 16 plots, placed on a uniform phytocoenosis with a total area of 3000 m^2, each plot having a size of 10 m^2. Starting from 2001 plots were mowed each year when the grasses were in bloom. As a historical management, land surface was used in a mixed farming system during 1990-2000, through mowing and occasionally autumn grazing with cattle and horses. Therefore, the mowing regime was slightly adjusted as grazing did not occur every year.

Assessment of climatic variables and calculation of indices

An automatic weather station was set up in the research area, which recorded temperature and rainfall every 15 minutes. The climatic data was split into two periods: the vegetation period (May – October) and the dormant period between 20 of November and April (outside of vegetation period). For each period the monthly maximum and minimum temperatures were determined and the monthly precipitation totals were calculated. The combined effect of temperature and precipitation was incorporated into water balance and humidity indices. The soil water balance index was calculated as the difference between rainfall and evapotranspiration on a monthly basis (Vicente-Serrano *et al.*, 2010). The potential evapotranspiration was estimated based on the simplified model proposed by Thornthwaite (1948) using R (R Core Team 2013) package "spei" (Begueria and Vicente-Serrano, 2013). The humidity index is a simplified expression of the ratio between monthly precipitation and potential evapotranspiration (Samaras, 2012).

Additionally, the climate conditions of the three previous years were analyzed regarding their impact on the vegetation composition change. In the following the climate variables were coded as follows: t – mean temperature °C, p – rainfall sum, w – water balance, h – humidity index, m – minimum value in a month, M – maximum value in a month, v – vegetation period, o – outside the vegetation period or dormant period, 1, 2, 3 – the number of years ago.

Vegetation data collection

Vegetation data were collected through floristic relevés based on the cover abundance of each species using the Braun-Blanquét scale (Braun-Blanquet, 1964). For greater precision, the scale's intermediate values were used (Gârda, 2010; Pleşa *et al.*, 2010). Vegetation data from the permanent plots was collected from 2004 to 2012; on the period when the grasses were flowering. Therefore, for a better understanding, plots were placed in 2001, and vegetation data were collected annually from 2004.

Statistical analysis

In the first step of the analysis sequence the plots were classified based on their species composition in order to understand and describe their similarity within and between the years. The hierarchical classification was performed using Soerensen index and the group average algorithm (agnes in the R "cluster" package, 1.14.4, Maechler *et al.*, 2013) recommended by several authors (McCune *et al.*, 2002; McCune and Mefford, 2011; Peck, 2010). Additionally, different dissimilarity indices and several algorithms were applied to ensure that the final solution was stable (Legendre and Legendre, 1998). The optimum number of clusters was confirmed by a multi-response permutation procedure (MRPP). This analysis (performed with PC-ORD ver. 6, McCune and Mefford, 2011) confirmed that there was a significant difference in species composition between the clusters.

In the next step the Indicator Species Analysis was used (Dufrêne and Legendre, 1997) to assess the degree to which a species indicates a compositional group based on its constancy and abundance (Tremp, 2005; Peck, 2010; Leyer and Wesche, 2007; Kent, 2012). This analysis was performed based on Soerensen index - using PC-ORD. For the interpretation, the Ellenberg indicator values adapted by Kovacs (1979) for Romania were used.

Additionally, the turnover between the two dominant species (*F. rubra* L. and *A. capillaris* L.) was analysed with the co-dominance ratio (using Statistica 8.0, Statsoft, 2012). To analyse in detail the effect of the different climate variables on the abundance of the dominant species a comparative Post-hoc Fisher LSD type analysis was performed. This test emphasizes the significant differences among groups (Pleşa, 2012).

To examine variation in species composition we used unconstrained ordination (non-metric multidimensional scaling (NMDS) based on Soerensen index using *metaMDS* function ("vegan" package, Oksanen *et al.*, 2013) and PC ORD. The latter was used to visualize vectors between the clusters in order to find out if a fluctuation or a succession happened and to determine the directional change in vegetation (McCune and Mefford, 2011). To quantify the general influence of the climatic variables (basic and synthetic) on the vegetation they were included in the non-metric multidimensional scaling (NMDS). The corresponding aligning of climatic factors to the ordination was performed using the function *envfit* ("*vegan*" package). Pearson-type correlations among species and climate variables, that were found amongst the climatic variables were analysed ("*ecodist*" package, Goslee and Urban, 2007). Prior to the NMDS, from each pair of strongly correlated climate variables, one was removed (r>0.80, Chan, 2003) to reduce collinearity in the following analysis (Peralta *et al.*, 2012).

To identify the climate thresholds between the different vegetation clusters the dataset was subdivided recursively into subsets using classification tree methods. These were increasingly homogeneous with respect to the defined groups, providing a tree-like classification and an associated dichotomous key to classify unknown samples into groups (Urban 2002). The calculations were performed using the "*rpart*" package (Therneau *et al.*, 2013).

Results and discussions

Vegetation classification

The vegetation plots were divided into five clusters (Fig. 1). The results over nine years show that some years were floristically very homogenous while others were quite different. Especially the first and second year – 2004 (cluster 1) and 2005 (cluster 2) were different. Within the remaining large group, 2008 (cluster 4) forms another floristically different group while the other two comprise data from several years (clusters 3 and 5). This shows that the vegetation plots in some years were compositionally very similar to the same plots measured in other years, like 2006 to 2007 and 2010 (cluster 3) and 2009 to 2011 and 2012 (cluster 5).

During the nine year study period the sample plots fell into one of five distinctive composition (clusters) which largely represent temporal changes in the *Festuca rubra* L. - *Agrostis capillaris* L. grassland. The NMDS ordination showed that the five compositional groups formed through time are related to variability in temperature and precipitation as well as water balance and humidity indices.

The co-dominance ratio between *F. rubra* L. and *A. capillaris* L. changed during the nine years (Tab. 1). The cover of *Agrostis capillaris* L. decreased significantly during the study period from 24.27% (2004) to 10.47% (2009, 2011 and 2012). Contrariwise *F. rubra* L. showed an increasing trend from 15.72% cover (2004) to 23.17% (2009, 2011 and 2012).

Fluctuation and successional trajectory

The successional vectors in the ordination (Fig. 2) show that the plant community was transformed during the nine years due to climatic conditions. It seems that from '2004' (cluster 1) to '2005' (cluster 2) all four plots move along axis 1, with no change on axis 2. Then from '2005' to '2006, 2007 and 2010' (cluster 3) all four plots move to lower scores on axis 1 but to higher scores on axis 2. Then from '2006, 2007 and 2010' to '2008' (cluster 4) plots move to slightly higher scores on axis 1 but barely move on axis 2. The plots of cluster 5

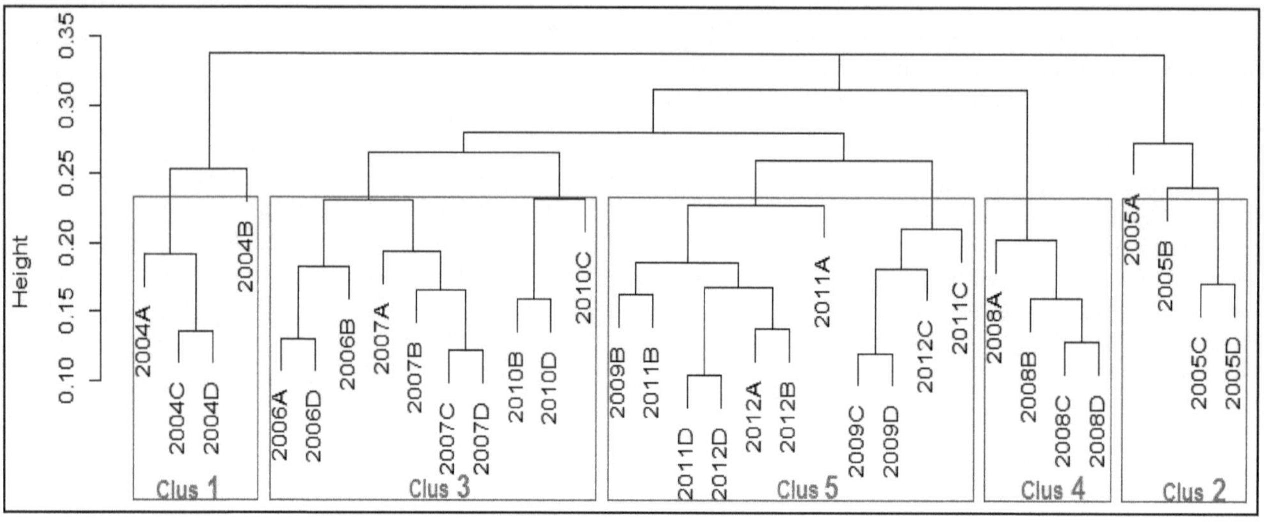

Fig. 1. Dendrogram showing the vegetation classification based on Soerensen index and the group average algorithm. A-D indicate the individual plots and 2004-2012 the year of assessment; Clus – cluster

Tab. 1. Modification of co-dominance relationship between *F. rubra* L. and *A. capillaris* L. during 2004-2012, under the influence of climatic factors

		Agrostis capillaris L.				
Cluster		1	2	3	4	5
	Cover %	24.27	12.99	16.36	11.58	10.47
1	24.27		$p<0.001$ ooo	0.003oo	$p<0.001$ ooo	$p<0.001$ ooo
2	12.99			0.120	0.574	0.197
3	16.36				0.063	0.004oo
4	11.58					0.632
5	10.47					
		Festuca rubra L.				
Cluster		1	2	3	4	5
	Cover %	15.72	16.44	19.67	24.17	23.17
1	15.72		0.859	0.338	0.075	0.043 *
2	16.44			0.358	0.067	0.027*
3	19.67				0.276	0.206
4	24.17					0.905
5	23.17					

Cluster: 1 – 2004; 2 – 2005; 3 – 2006, 2007, 2010; 4 – 2008; 5 – 2009, 2011, 2012

$p < 0.05$ * o $p < 0.01$ ** oo $p < 0.001$ *** ooo

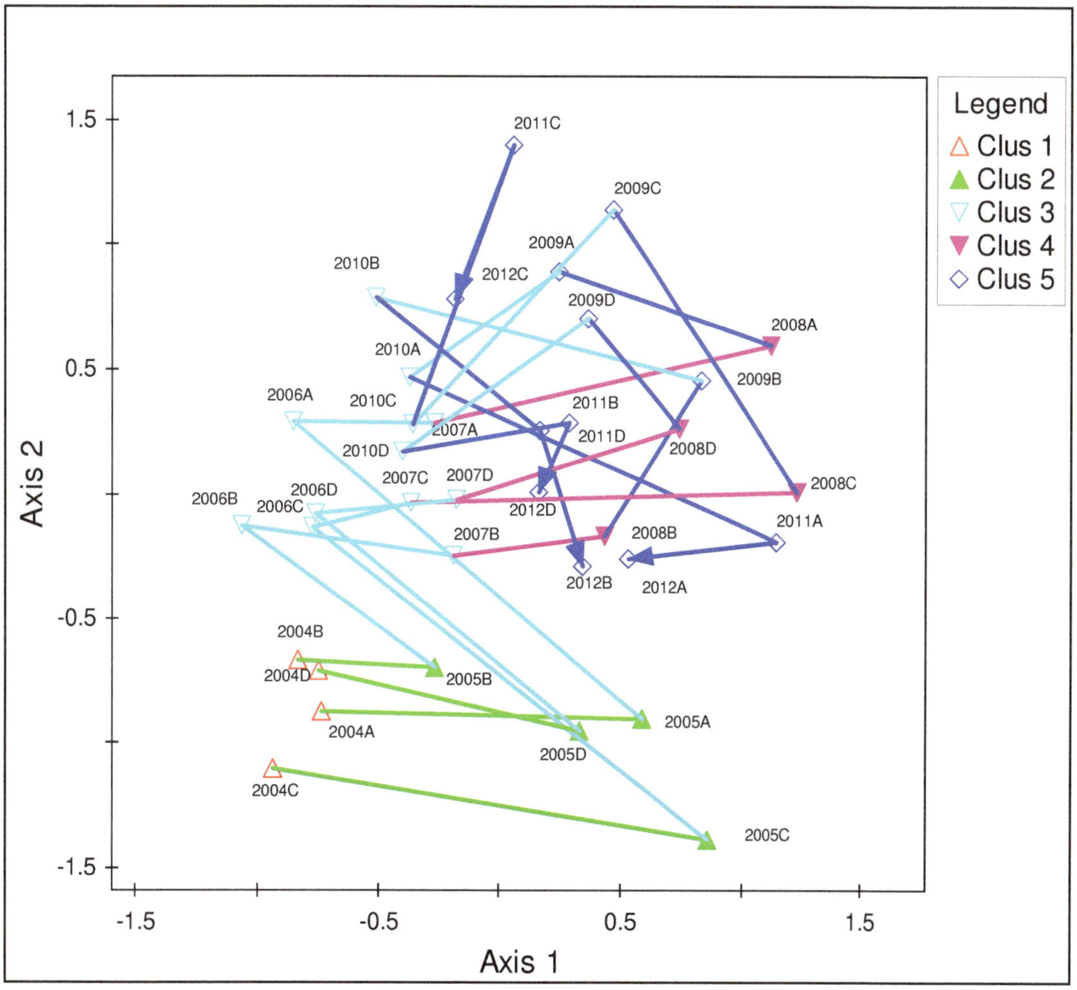

Fig. 2. Succession (trajectory) of the floristic composition during 2004-2012 plotted by succession vectors (clus1 - cluster 1, clus2 – cluster 2, clus3 – cluster 3, clus4 – cluster 4, clus5 – cluster 5)

('2009, 2011 and 2012') are quite scattered in the NMDS ordination space. For this reason we will follow the cluster in segments and can say that from '2008' (cluster 4) to 2009 the plots move to a lower score on axis 1, but to a higher score on axis 2. The majority of the plots of clusters'2009, 2011 and 2012' remain almost at the same position on axis 1 and at a much lower position on axis 2.

Climate caused obvious fluctuations and slight directional succession. During this floristic development the co-dominance ratio between *F. rubra* L. and *A. capillaris* L. often shifted. But the directional succession cannot be explained only by climate. Also, since 2001 (the year that the plots were established) the grassland management consisted in one mown/year. The combination of climate and management

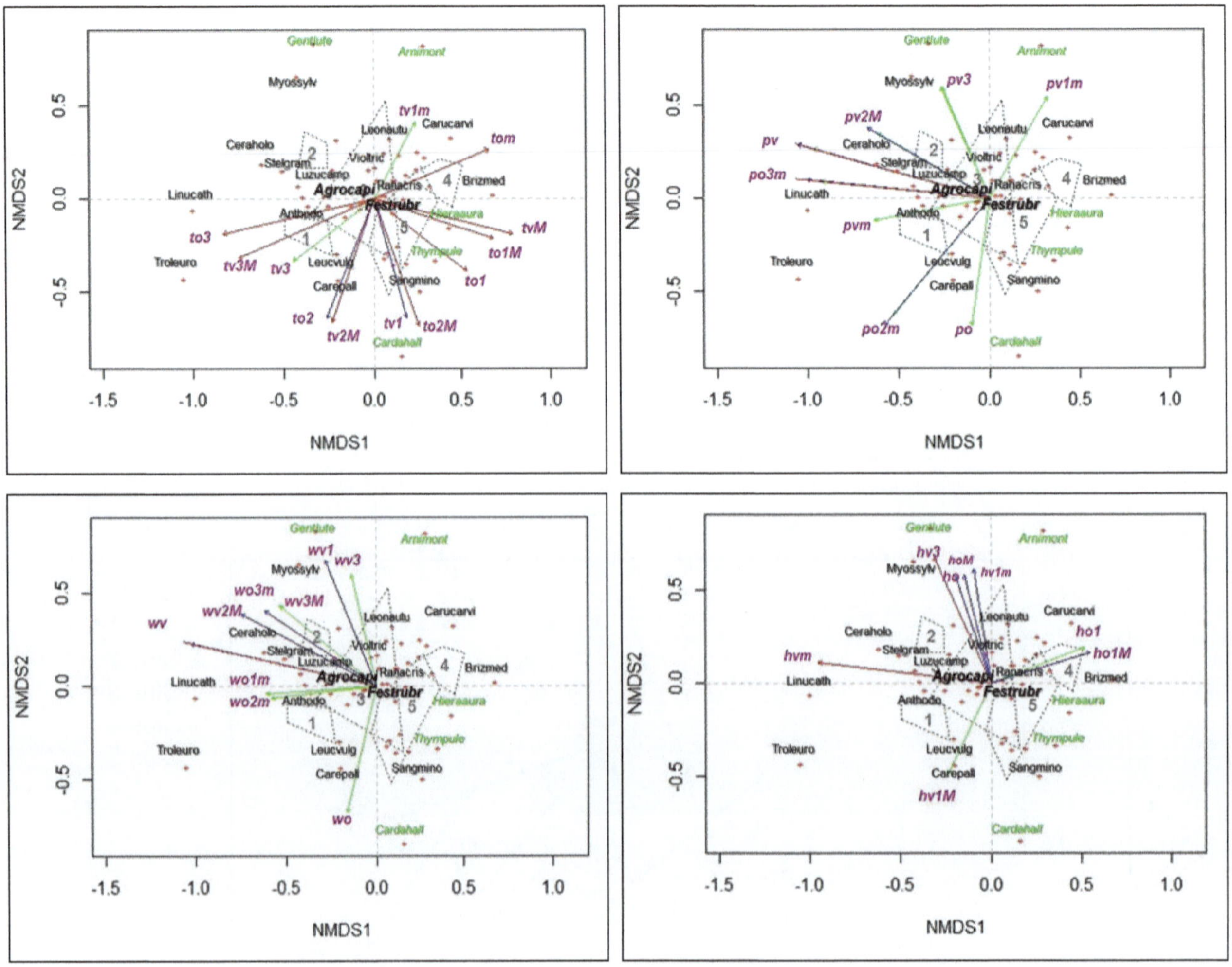

Fig. 3 (a - upper left; b - upper right; c - bottom left; d - bottom right). Clusters ordination and influence of climatic parameters on vegetation: a) mean temperature - t, b) sum of precipitations - p, c) water balance - w, d) humidity index - h (m - minimum value in a month, M - maximum value in a month, v - vegetation period, o - outside the vegetation period, 1, 2, 3 - the number of previous years (their names' combinations codify the climate variables in the figure). The variables drawn in green show the $p<0.05$, in blue $p<0.01$, in red $p<0.001$. The clusters are designed by figures from 1 to 5 drawn in grey. Dominant species: Agrocapi - *Agrostis capillaris* L., Festrubr - *Festuca rubra* L. Indicator species: Anthodo - *Anthoxanthum odoratum* L., Brizmed - *Briza media* L., Carepall - *Carex pallescens* L., Carucarvi - *Carum carvi* L., Ceraholo - *Cerastium holesteoides* Fr.em.Hyl., Leonautu - *Leontodon autumnalis* L., Leucvulg - *Leucanthemum vulgare* Lam., Linucath - *Linum catharticum* L., Luzucamp - *Luzula campestris* L., Myossylv - *Myosotis sylvatica* Ehrh., Ranacris - *Ranunculus acris* L., Sangmino - *Sanguisorba minor* Scop., Stelgram - *Stellaria graminea* L., Troleuro - *Trollius europaeus* L., Violtric - *Viola tricolor* L. Margin species: Arnimont - *Arnica montana* L., Cardahall - *Cardaminopsis halleri* L, Gentlute - *Gentianella lutescens* Velen, Hieraaura - *Hieracium aurantiacum* L., Thympule - *Thymus pulegioides L.*

could have produced a change in vegetation or perhaps it was only one strong factor. In order to separate the effect of these factors, additional data are required which will be analyzed and discussed in a following paper. Collins *et al.* (1987) showed that stopping intensive grazing by goats can lead to an obvious successional trend. Based on our data from 2004-2012, a "dynamic equilibrium" (Cristea *et al.*, 2004) in the *F. rubra* L. – *A. capillaris* L. grassland appears to develop around the '2006-2007-2010' cluster (3) and the '2009-2011-2012'- cluster (5, except 2009). In these periods *F. rubra* L. occur as dominant and *A. capillaris* L. as subordinate. Coldea *et al.* (2012) and Frink (2010) have described the instability of the co-dominance of *F. rubra* L. and *A. capillaris* L. and our study confirms this. Coldea *et al.* (2012) concluded that the instability of the system is mainly due to the high nutrient content of the soil when *A. capillaris* L. prevails. Our study suggests these changes might be driven by climate fluctuations.

The influence of climatic factors on the vegetation change

The species composition of '2004' was positively related to higher temperatures occurring three years earlier (NMDS, Fig. 3a) while '2005' was influenced by higher water availability (Fig. 3bcd). '2008' is positioned opposite to the first two years and therefore also negatively related to their prevailing conditions.

We could not identify species which had a specific ecological indicator value for 'temperature' or 'humidity', this suggests that the constancy and abundance of species reflects the complex action of the different climatic factors. For example, it is the combination of temperature and rainfall that is important for plant species' spreading (Wittig, 2012). In the mountains, differences among plant communities are the result of a complex of factors called "elevation complex" (Dierschke and Briemle (2002) and *Höhenkomplex* by Klapp (1971). This complex includes low temperature, long term snow cover, a short vegetative period, high precipitation, massive leaching through the soil profile, pronounced soil erosion, low biological activity in the soil and humus bioaccumulation (Dierschke and Briemle, 2002).

The first classification tree (Fig. 4) suggests that temperature variables play the main role in the floristic differentiation of the vegetation. An average temperature of less than -1.3 °C two years earlier during the dormant period (to2) separates the vegetation composition of cluster 1 and cluster 5 from the other vegetation clusters. On the next split (right side of the diagram) is indicated that if the minimum temperature during the vegetation period is less than 5.49°C, the same probability exists to maintain the vegetation of cluster 1 or cluster 5. On the left side, if the minimum temperature during the vegetation period in the previous year (tv1m) was less than 5.49°C, the vegetation as observed in the years '2006, 2007 and 2010' (cluster 3) occurred by taking into account the temperature during the growing season of less than 4.6 the vegetation of clusters 2 or 4 will occur.

Influential for '2004' were the temperatures three years before as well as the precipitation in the current and the previous year which were higher than average. A high influence had also the temperature outside the vegetation period (recorded three years earlier; winter 2001/2002) which was much higher than the average during the rest of the study period. Niu and Shiqiang (2008), studying the effect of global warming on natural meadows in the temperate steppe of China, found that *A. capillaris* L. became dominant under higher temperatures. The current year's minimum rainfall during the study period also plays an important role, and was almost twice the average in 2001-2012 (Annex 1). The findings of the present study are partly consistent with those presented by Stampfli and Zeiter (2004), who found that in wet years *A. capillaris* L. and *Anthoxanthum odoratum* L. abundance increased in similar grassland types. The minimum water balance's influence in the dormant period was two years before 2004, and it had a value in 2002 (wv2M) that was similar to the average during the study period, unlike the value registered in 2003 (wv2M) which was almost double the average for vegetation type observed in 2005. Maximum humidity during the vegetative period in '2003' influenced '2004' when it was almost triple than the average of 2001-2012 (Annex 1). Vegetation responses to extremes in the moisture regime are rapid and independent of annual rainfall (Heisler-White *et al.*, 2009).

Considering the ecological indicator value for temperature and humidity (Tab. 2) only *Trollius europaeus* L. (according to the indicator value of Ellenberg et al., 1992) and *Linum catharticum* L. to a lesser degree showed the expected ecological indication in the present study. The presence of a generalist could be explained by climatic fluctuations which happened in the past. However, the ecological indication value might have to be adjusted based on measurements. For example, *Leucanthemum vulgare* L. prefers flooded conditions (Hölzel and Otte, 2003) and is diminished by cold winters (Nissinen, 2004). The vegetation composition in '2005' was influenced by rainfall during the vegetation period (pv = 791.9 mm), which was 20% higher than the average during 2001-2012. Dostalek and Frantik (2011) recorded that the abundance of dicotyledonous species increased with rainfall in the Salabka Natural Reservation, Czech Republic. This result is consistent with our results for the Apuseni Mountains. Additionally, important for '2005' was the water balance during the vegetation period (wv) which was 80% higher than the average of the study period maximum water balance during the vegetation period in 2003 (wv2M), 62% higher than the average during 2001-2012 (Annex 1). The minimum value of the humidity index during the vegetation period in 2005 (hvm) was double the average during 2001-2012, and the humidity index during the vegetation period in 2003 was 20% higher than average and had an important influence on the vegetation composition in cluster 2 (Annex 1).

Tab. 2. Groups' indicator species (x – indifferent species)

Species	*Indicator Value*	*Significance*	*Temperature Index*	*Moisture Index*
		Cluster 1 (2004)		
Anthoxanthum odoratum L.	49.3	0.0002 ***	x	x
Carex pallescens L.	45.0	0.0002 ***	4	x (6)*
Leucanthemum vulgare L.	58.5	0.0004 ***	x	4
Linum catharticum L.	44.9	0.0168 *	x	x
Trollius europaeus L.	75.0	0.0020 **	3	7
		Cluster 2 (2005)		
Luzula campestris L.	43.4	0.0004 ***	x	4
Cerastium holosteoides Fr. em. Hyl.	56.6	0.0106 *	x	5
Myosotis sylvatica Ehrh. ex Hoffm	81.0	0.0008 ***	x	5
Stellaria graminea L.	58.2	0.0006 ***	x	4
Viola tricolor L.	45.3	0.0260 *	5	x
		Cluster 3 (2006, 2007, 2010)		
Leontodon autumnalis L.	57.6	0.0004 ***	x	5
		Cluster 4 (2008)		
Briza media L.	44.5	0.0272 *	x	x
Carum carvi L.	60.1	0.0120 *	4	5
Ranunculus acris L.	81.0	0.0394 *	x	x (6)*
		Cluster 5 (2009, 2011, 2012)		
Sanguisorba minor Scop.	41.7	0.0176 *	6	4

* x – by Kovacz, 1979 and 6 – by Ellenberg, 1997

Based only on the synthetic climatic factors the second classification tree (Fig. 5) reveals that when the water balance was higher or equal to 135.2 mm during the vegetation period in the current year (WV) it was the most important factor, dividing roughly cluster 4 and 5 from the rest of clusters.

The minimum temperature during the dormant period (tom) had a strong influence for cluster 4 (2008), which was lower than the average obtained for 2001-2012 (Annex 1). The minimum rainfall in 2007 (pv1m) was particularly important as it was almost double the average (Annex 1). The value of the humidity index in the dormant period of 2007 (ho1) had a strong influence and was 410.3% higher than the average during 2001-2012 while the maximum value of the humidity index during the dormant period in 2007 (ho1M) was 418.9% times higher than the average (Annex 1). Nowak and Sculz (2002) considered that *F. rubra* L. increased markedly after several rainy years, a result similar to our own. The preferences of the indicator species for temperature and rainfall do not correspond to the described situation and it can be asserted that the species' presence is determined mostly by the interaction between the previous year's temperature and humidity (humidity index). Stampfli and Zeiter (2004) have found that *F. rubra* L. increased in abundance with plentiful rainfall in previous years but average or even lower than average rainfall fell in the study year. They also found that *Briza media* L. decreased during drought, a result similar to the present study.

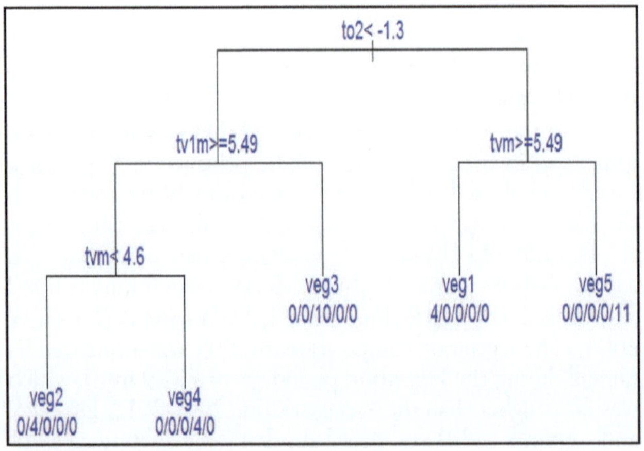

Fig. 4. Climatic factors with major importance to herbaceous layer changes (to2 - temperature of the vegetation period two years before, tvm - the temperature during the vegetation period in the current year, tv1m - the temperature during the vegetation period one year before, veg1 - cluster 1 (2004), veg2 - cluster 2 (2005), veg3 - cluster 3 (2006, 2007, 2010), veg4 - cluster 4 (2008), veg5 - cluster 5 (2009, 2011, 2012)

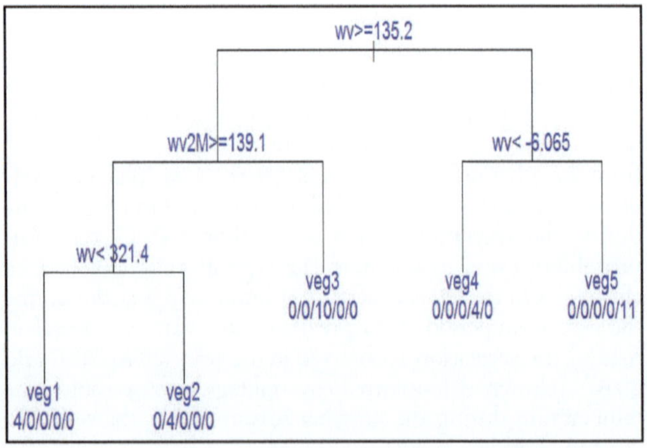

Fig. 5. The synthetic climatic indexes with major importance in herbaceous layer changes (wv - water balance in the current year, wv2M - maximum water balance two years before, veg1 - cluster 1 (2004), veg2 - cluster 2 (2005), veg3 - cluster 3 (2006, 2007, 2010), veg4 - cluster 4 (2008), veg5 - cluster 5 (2009, 2011, 2012)

Cluster 5 (2009, 2011, 2012) was influenced by the maximum temperatures during the vegetation period (tvM) and the maximum temperatures in the dormant period one year before (to1M) and two years before (to2M, Annex 1). Rainfall in the dormant period (po) which, in general, was lower than the average during 2001-2012 also had an influence (Annex 1). Water balance in the dormant period, especially the current year's (wo) balance, which was generally lower than the average during the study period and had a similar influence (Annex 1). Overall, it is *F. rubra* L. as the dominant, *A. capillaris* L. as the co-dominant and *Sanguisorba minor* Scop. as an indicator that were determined first of all by the maximum temperatures both in the current year as well as in previous years and by lower rainfall in the dormant period of the current year. Our results are consistent with Stampfli and Zeiter (2004) who showed that *F. rubra* L. increased its abundance in drought years when compared with *A. capillaris* L. which decreased in abundance. Those authors did not identify changes in *Sanguisorba minor* Scop. abundance after alternating periods of drought and rain.

The classification tree results emphasized that temperature in the dormant period two years before (to2) had the highest influence on changing the herbaceous layer when it was lower than −1.3 °C. This, combined with the minimum temperature during the vegetation period one year before (higher or equal to 5.49 °C), resulted in a specific species composition.

Dostalek and Frantik (2011) studying the dry meadows in the Czech Republic stated that temperature had a lower effect on species abundance and richness compared to precipitation. It is somehow incongruous considering that the limiting factor in their study area is water. However, they found that the annual average temperature in the previous year had no effect upon the species abundance and species richness and only the temperature variation from one year to another played a significant role in the grass canopy changes. Similar to the findings of the present study they reported that the winter temperature considerably influenced the vegetation composition.

Discussion

The impact of climate during 12 years on the *F. rubra* L. – *A. capillaris* L. grassland has led to several alterations in the herbaceous composition which have occurred due to fluctuations and minimal succession. The co-dominance ratio between *F. rubra* L. and *A. capillaris* L. changes towards the dominance of *F. rubra* L at the end of the study period. The high amount of precipitation in the current year and the years before led to the dominance of *A. capillaris* L., whereas high temperatures and lower rainfall caused *F. rubra* L. to become the dominant species.

These results emphasize that precipitation and temperature during the last three years play an even greater role than the current year's precipitation and temperature regarding the grassland vegetation compositional changes. Extreme temperatures affect the herbaceous vegetation not only during the vegetation period but also in the dormant period.

The classification tree result has emphasized that the most important factor in fluctuations and succession on the herbaceous species composition was the temperature. The second most important factor in the fluctuations and succession of *F. rubra* L. – *A. capillaris* L. plant community is the water balance during the current year. Experiments like ours, with long-term observations and equipped with weather stations can provide new information regarding the individual reactions of plant species as well as plant communities' responses to climatic fluctuations or climate changes. Furthermore, it is recommend to verify and to readapt the Ellenberg indicator values (originally adapted by Kovacs (1979) to the Romanian conditions) using up-to-date technologies.

Acknowledgments

This paper was funded by the University of Agricultural Sciences and Veterinary Medicine Cluj-Napoca, the financing agreement 1363/8.02.2013.

References

Beguería S, Vicente-Serrano SM, (2013). SPEI: Calculation of the Standardised Precipitation-Evapotranspiration Index. R package version 1.6. http://CRAN.R-project.org/package=SPEI.

Braun-Blanquet J (1964). Pflanzensoziologie – Grundzüge der Vegetationskunde. Springer Verlag, Wien.

Brito JC, Godinho R, Martínez-Freiría F, Pleguezuelos JM, Rebelo H, Santos X, Vale CG, Velo-Antón G, Boratyński Z, Carvalho SB, Ferreira S, Gonçalves DV, Silva TL, Tarroso P, Campos JC, Leite JV, Nogueira J, Alvares F, Sillero N, Sow AS, Fahd S, Crochet PA, Carranza S (2014). Unravelling biodiversity, evolution and threats to conservation in the Sahara-Sahel. Biol Rev Camb Philos Soc 89(1):215-31.

Chan YH (2003). Biostatistics 104: correlational analysis. Singapore Med J 44(12):614-619.

Coldea G, Oprea A, Sârbu I, Sârbu C, Ștefan N (2012). Les associations vegetal de Romaine. Presa Universitară Clujeană, Cluj-Napoca.

Collins S, Bradford J, Sims P (1987). Succession and fluctuation in *Artemisia* dominanted grassland. Vegetatio 73:89-99.

Cristea V, Gafta D, Pedrotti F (2004). Fitosociologie. Ed. Presa Universitară Clujeană, Cluj-Napoca.

Cui X, Graf HF (2009). Recent land cover changes on the Tibetan Plateau: a review. Climatic Change 94(1-2):47-61.

Deutsch ES, Bork EW, Cahill JF, Chang SX (2011). Short-Term Plant Community Responses to Warming and Defoliation in a Northern Temperate Grassland. ISRN Ecology doi:10.5402/2011/9260612011.

Dierschke H, Briemle G, Kratochwil A (2002). Kulturgrasland: Wiesen, Weiden und verwandte Staudenfluren. E. Ulmer.

Dostálek J, Frantík T (2011). Response of dry grassland vegetation to fluctuations in weather conditions: a 9-year case study in Prague (Czech Republic). Biologia 66(5):837-847.

Dufrêne M, Legendre P (1997). Species assemblages and indicator species: the need for a flexible asymmetrical approach. Ecological Monographs 67(3):345-366.

Egan AN, Crandall KA (2008). Divergence and diversification in North American Psoraleeae (Fabaceae) due to climate change. BMC Biol 6:55.

Ellenberg H, Weber HE, Düll R, Wirth V, Werner W, Paulissen D (1992). Zeigerwerte von Pflanzen in Mitteleuropa. Verlag Erich Goltze KG, Göttingen.

Frelich LE (2002). Forest Dynamics and Disturbance Regimes: Studies from Temperate Evergreen-Deciduous Forests. Cambridge University Press, Cambridge.

Frink JP (2010). Pajiștile de pe valea Arieșului între Lupșa și Turda. Presa Universitară Clujeană, Cluj-Napoca.

Gârda N (2010). Studiul unor elemente de landșaft montan (cu privire specială asupra ecosistemelor de pajiști din comuna Gârda de Sus, Munții Apuseni). PhD thesis USAMV Cluj-Napoca.

Goslee SC, Urban DL (2007). The ecodist package for dissimilarity-based analysis of ecological data. Journal of Statistical Software 22(7):1-19.

Grime JP, Fridley JD, Askew AP, Thompson K, Hodgson JG, Bennett CR (2008). Long-term resistance to simulated climate change in an infertile grassland. Proceedings of the National Academy of Sciences 105(29):10028-10032.

Heisler-White JL, Blair JM, Kelly EF, Harmoney K, Knapp AK (2009). Contingent productivity responses to more extreme rainfall regimes across a grassland biome. Global Change Biology 15(12):2894-2904.

Hölzel N, Otte A (2003). Restoration of a species - rich flood meadow by topsoil removal and diaspore transfer with plant material. Applied Vegetation Science 6(2):131-140.

Hopkins A (2011). Mountainous farming in Europe. In: Pötsch EM, Krautzer B, Hopkins A. (Eds.). Grassland Farming and Land Management Systems in Mountainous Regions. Grassland Science in Europe 16:3-12.

Ihm BS, Lee JS, Kim JW, Kim JH (2007). Relationship between global warming and species richness of vascular plants. J Plant Biol 50:321-324.

Kasperek G (1998). Pflanzenökologische Untersuchungen im mittleren Rur-Tal (Nordrhein-Westfalen): Vegetation und Vegetationsdynamik unter besonderer Berücksichtigung von Fluktuationen in Dauerflächen. 344 S. (Archiv Naturwissenschaftlicher Dissertationen, Band 6.) Martina-Galunder-Verlag, Wiehl.

Kent M (2012). Vegetation Description and Data Analysis: A Practical Approach (2nd Edition). John Wiley & Sons.

Klapp E (1971). Wiesen und Weiden, eine Grünlndlehre. 4.Aufl., Parey, Berlin, Hamburg.

Kovacs JA (1979). Indicatorii biologici, ecologici şi economici ai 563 florei pajiştilor. Ministerul Agriculturii şi Industriei Alimentare. Bucureşti.

Lal R (2012). Climate change and soil degradation mitigation by sustainable management of soils and other natural resources. Agric Res 1(3):199-212.

Legendre P Legendre L (1998). Numerical Ecology. Elsevier, Amsterdam.

Leonard M, Westra S, Phatak A, Lambert M, Hurk B, McInnes K, Risbey J, Schuster S, Jakob D, Stafford-Smith M (2014). A compound event framework for understanding extreme impacts. Wires Clim Change 5:113-128.

Leyer I, Wesche K (2007). Multivariate Statistik in der Ökologie: Eine Einführung. Springer-Verlag Berlin Heidelberg.

Maechler M, Rousseeuw P, Struyf A, Hubert M, Hornik K (2013). Cluster: Cluster Analysis Basics and Extensions. R package version 1.14.4. http://cran.rstudio.com/ web/packages/cluster.

Matesanz S, Brooker R, Valladares F, Klotz S (2009). Temporal dynamics of marginal steppic vegetation over a 26-year period of substantial environmental change., Journal of Vegetation Science 20:299-310.

McCune B, Grace JB, Urban DL (2002). Analysis of ecological communities. 28. Gleneden Beach. OR: MjM software design.

McCune B, Mefford MJ (2011). PC-ORD. Multivariate Analysis of Ecological Data. Version 6. MJM Software. Gleneden Beach OR.

Morecroft MD, Masters GJ, Brown VK, Clarke IP, Taylor ME, Whitehouse AT (2004). Changing precipitation patterns alter plant community dynamics and succession in an ex-arable grassland. Funct Ecol 18:648-655.

Miller W (2008). The Hierarchical Structure of Ecosystems: Connections to Evolution. Evo Edu Outreach 1:16-24.

Nissinen O (2004). Wild flower seed yields in northern Finland. Grassland Science in Europe 9:249-251.

Niu S, Shiqiang W (2008). Warming changes plant competitive hierarchy in a temperate steppe in northern China. Journal of Plant Ecology 1(2):103-110.

Niu S, Wan S (2008). Warming changes plant competitive hierarchy in a temperate steppe in northern China. Journal of Plant Ecology 2:103-110.

Nowak B, Schulz B (2002). Wiese, Nutzung, Vegetation, Biologie und Naturschutz am Beispiel der Wiesen des Südschwarzwaldes und Hochrheingebietes, Verlag Regionalkultur Heidelberg.

Oksanen J, Guillaume Blanchet F, Kindt Roeland, Legendre P, Minchin PR, O'Hara RB, Simpson GL, Solymos P, Stevens MHH, Helene Wagner (2013). Vegan: Community Ecology Package. R package version 2.0-9. http://CRAN.R606 project.org/package=vegan.

Orăşeanu I (2005). Geomorphologie und Geologie des Höhenzuges zwieschen den Tälern von Gârda Seacă – Ordâncusa. In: Rusdea E, Reif A, Povară I, Konold W. Perspektive für eine traditionelle Kulturlandschaft in Osteuropa, Culterra 35.

Parichi M, Stănilă L (2005). Böden der Gemarkung von Gheţari und angrenzender Gebiete. In: Rusdea E, Reif A, Povară I, Konold W. Perspektive für eine traditionelle Kulturlandschaft in Osteuropa, Culterra 35.

Peck JE (2010). Multivariate analysis for community ecologists: step-by-step using PC-ORD. MjM Software Design.

Peralta AL, Jeffrey WM, Diana NF, Angela DK (2012). Environmental Factors at Dissimilar Spatial Scales Influence Plant and Microbial Communities in Restored Wetlands. Wetlands 32(6):1125-1134.

Petrychenko V, Bohovin A, Kurhak V (2012). More efficient use of grassland under climate warming. Grassland Science in Europe 17.

Plantureux S, Michaud A, Baumont R (2012). Influence of pedoclimatic and management factors on botanical and functional composition of grasslands. Grassland Science in Europe 17.

Pleşa A, Păcurar F, Rotar I (2010). New direction with low-input of fertilization in grasslands. Bulletin of University of Agricultural Sciences and Veterinary Medicine Cluj-Napoca 67(1):311.

Pleşa A (2012). Cercetări privind folosirea şi menşinerea pajiştilor montane cu low input. PhD thesis USAMV Cluj-Napoca.

R Core Team (2013). R: A language and environment for statistical computing. R Foundation for Statistical Computing, Vienna, Austria. URL http://www.Rproject.org/.

Sala OE, Chapin FS, Armesto JJ, Berlow E, Bloomfield J, Dirzo R, Huber-Sanwald E, Huenneke LF, Jackson RB, Kinzig A, Leemans R, Lodge DM, Mooney HA, Oesterheld M, Poff NL, Sykes MT, Walker BH, Walker M, Wall DH (2000). Global biodiversity scenarios for the year 2100. Science 287:1770-1774.

Samaras DA (2012). The vegetation of Greek fir (*Abies cephalonica* Loudon) forests on the Oxia-North Vardousia mountain system, central Greece, in relation to drought. Diss. Universität Freiburg, 2012.

Schaumberger A, Pötsch EM, Formayer H (2012). GIS-based analysis of spatio temporal variation of climatological growing season for Austria. Grassland Science in Europe 17.

Schmer MR, Mitchell RB, Vogel KP, Schacht WH, Marx DB (2010). Spatial and Temporal Effects on Switchgrass Stands and Yield in the Great Plains. Bioenerg Res 3:159-171.

Schneider RR, Hamann A, Farr D, Wang X, Boutin S (2009). Potential effects of climate change on ecosystem distribution in Alberta. Canadian Journal of Forest Research 39:1001-1010.

Stampfli A, Zeiter M (2004). Plant regeneration directs changes in grassland composition after extreme drought: a 13-year study in southern Switzerland. J Ecol 92:568-576.

Statsoft Inc. (2012). Electronic Statistics Textbook. Tulsa, ok: Statsoft. Web: http://www.statsoft.com/textbook/.

Stevenson M, Peel S, Martin D (2005). Agri-environment Schemes in England: identifying and targeting semi-natural grasslands for management and restoration. Grassland Science in Europe 10:158-172.

Therneau T, Atkinson B, Ripley B (2013). rpart: Recursive Partitioning. R packageversion 4.1-3. http://CRAN.R-project.org/package=rpart.

Thornthwaite CW (1948). An Approach toward a Rational Classification of Climate, Geographical Review 38:55-94.

Tremp H (2005). Aufnahme und Ansalyse vegetation sökologischer Daten. Verlag Ulmer Stuttgart Eugen.

Trnka M, Bartošová L, Schaumberger A, Ruget F, Eitzinger J, Formayer H, Seguin B Olesen JE (2011). Climate change and impact on European grasslands. Grassland Science in Europe 16.

Urban DL (2002). Classification and regression trees, p 222-232. In: McCune B, Grace JB (Eds.). Analysis of ecological communities. MJM Software Design, Gleneden Beach, OR.

Veen P, Jefferson R, De Smidt J, Van Der Straaten J (2009). Grasslands in Europe of high nature value. KNNV Publishing, Zeist.

Vicente-Serrano SM, Beguería S, López-Moreno JI (2010). A multiscalar drought index sensitive to global warming: the standardized precipitation evapotranspiration index. Journal of Climate 23(7):1696-1718.

Wittig R (2012). Geobotanik. UTB Band-Nr.: 3753. Printed in Germany.

Worthington TA, Shannon KB, Timothy BG, Mueller J (2014). Backcasting the decline of a vulnerable Great Plains reproductive ecotype: identifying threats and conservation priorities. Global Change Biology 20:89-102.

Zarzycki J, Szewczyk W (2013). Impact of abandonment on the floristic composition of permanent grassland and grassland created on former arable land. Grassland Science in Europe 18:421-423.

Zurek G (2011). Effect of climate change on phenology of forage grass species. Journal of Life Sciences 5(9):754-758.

Zurek G (2012). Use of amenity grasses – status quo and innovations. Grassland Science in Europe 17.

Annex 1

Individual and combined effect of climatic factors on grasslands phytocoenosys

Cluster		Year	tvm	tvM	tv1	tv1m	tv2m	tv2M	tv3	tv3M	tom	to1	to1M	to2	to2M	to3	to3M
Temperature	1	2004	5.58	14.7	12.83	4.1	6.5	17.1	11.93	15.6	-10.5	-2.12	3.1	0.63	4.6	-0.08	4.1
	2	2005	3.6	14.8	10.59	5.58	4.1	16.3	12.33	17.1	-9.4	-1.63	4.7	-2.12	3.1	0.63	4.6
	3	2006	5.3	15.5	11.03	3.6	5.58	14.7	12.83	16.3	-9.5	-3.88	2.9	-1.63	4.7	-2.12	3.1
		2007	5.9	15.9	11.03	5.3	3.6	14.8	10.59	14.7	-2	-3.72	4	-3.88	2.9	-1.63	4.7
		2010	3.4	16.3	12.2	5.4	5.6	17.4	12.05	15.9	-6	-1.15	6.7	-1.45	8.2	0.32	5.8
	4	2008	5.6	17.4	12.05	5.9	5.3	15.5	11.03	14.8	-5.1	0.32	5.8	-3.72	4	-3.88	2.9
	5	2009	5.4	16.6	12.33	5.6	5.9	15.9	11.03	15.5	-4.8	-1.45	8.2	0.32	5.8	-3.72	4
		2011	4.3	15.4	11.55	3.4	5.4	16.6	12.33	17.4	-4.9	-0.77	4.8	-1.15	6.7	-1.45	8.2
		2012	5.32	17.6	11.77	4.3	3.4	16.3	12.2	16.6	-8.6	-0.77	5.3	-0.77	4.8	-1.15	6.7
		Mean	*4.93*	*16.02*	*11.71*	*4.80*	*5.04*	*16.07*	*11.81*	*15.99*	*-6.76*	*-1.69*	*5.06*	*-1.53*	*4.98*	*-1.45*	*4.90*
Cluster		**Year**	**pv**	**pvm**	**pvM**	**pv1**	**pv1m**	**pv1M**	**pv2m**	**pv2M**	**pv3**	**pv3m**	**po**	**po1m**	**po2m**	**po3m**	
Precipitations	1	2004	720.8	70.9	144.8	637.1	1.8	211.3	64	239	756.4	35.4	668.7	40.6	41.3	60.1	
	2	2005	791.9	25	223.6	720.8	70.9	144.8	1.8	211.3	828.6	64	246.5	55.2	40.6	41.3	
	3	2006	750.6	40.3	185.9	791.9	25	223.6	70.9	144.8	637.1	1.8	256.2	0	55.2	40.6	
		2007	669.2	79	158	750.6	40.3	185.9	25	223.6	720.8	70.9	372.4	5.4	0	55.2	
		2010	803	70.9	194.5	564.4	31.6	152.4	7.2	130.4	669.2	79	658.6	27.2	52.2	3.3	
	4	2008	476.6	7.2	130.4	669.2	79	158	40.3	185.9	791.9	25	432.6	3.3	5.4	0	
	5	2009	564.4	31.6	152.4	476.6	7.2	130.4	79	158	750.6	40.3	433	52.2	3.3	5.4	
		2011	468.6	14	235.6	803	70.9	194.5	31.6	152.4	476.6	7.2	271	72.8	27.2	52.2	
		2012	515.87	35.2	118.8	468.6	14	235.6	70.9	194.5	564.4	31.6	353.94	0.6	72.8	27.2	
		Mean	*640.11*	*41.57*	*171.56*	*653.58*	*37.86*	*181.83*	*43.41*	*182.21*	*688.40*	*39.47*	*410.33*	*28.59*	*33.11*	*31.70*	
Cluster		**Year**	**wv**	**wvM**	**wv1**	**wv2m**	**wv2M**	**wv3**	**wv3M**	**wo**	**wo1m**	**wo2m**	**wo3m**				
Water balance	1	2004	294.94	80.14	127.74	-36.59	146.52	287.14	139.21	607.14	33.21	15.06	60.1				
	2	2005	347.94	131.59	294.94	-103.37	180.63	339.01	146.52	266.27	45.1	33.21	15.06				
	3	2006	308.43	95.92	347.94	12.12	80.14	127.74	180.63	258.35	0	45.1	33.21				
		2007	188.6	92.79	308.43	3.99	131.59	294.94	80.14	222.23	5.4	0	45.1				
		2010	339.89	142.44	81.77	-111.78	57.71	188.6	92.79	531.53	-16.97	20.15	-35.57				
	4	2008	-13.76	57.71	188.6	-19.57	95.92	347.94	131.59	408.45	-35.57	5.4	0				
	5	2009	81.77	81.11	-13.76	-32.41	92.79	308.43	95.92	292.72	20.15	-35.57	5.4				
		2011	1.63	131.69	339.89	-49.09	81.11	-13.76	57.71	427.86	57.24	-16.97	20.15				
		2012	23.59	58.67	1.63	-18.92	142.44	81.77	81.11	263.75	-4.48	57.24	-16.97				
		Mean	*174.78*	*96.90*	*186.35*	*-39.51*	*112.09*	*217.98*	*111.74*	*364.26*	*11.56*	*13.74*	*14.05*				
Cluster		**Year**	**hvm**	**hvM**	**hv1m**	**hv1M**	**hv2m**	**hv2M**	**hv3**	**hv3m**	**hv3M**	**ho**	**hoM**	**ho1**	**ho1M**	**ho2**	**ho2M**
Humidity index	1	2004	1.13	2.46	0.02	8.67	0.64	3.37	1.56	0.69	2.9	1.84	6.22	1.38	5.76	2.47	10.94
	2	2005	1.05	2.44	1.13	2.46	0.02	8.67	1.97	0.64	3.37	6.64	34.13	1.84	6.22	1.38	5.76
	3	2006	0.67	2.42	1.05	2.44	1.13	2.46	2.19	0.02	8.67	0.58	3.48	6.64	34.13	1.84	6.22
		2007	0.71	2.82	0.67	2.42	1.05	2.44	1.86	1.13	2.46	15.8	83.24	0.58	3.48	6.64	34.13
		2010	0.81	8.14	0.46	3.7	0.06	2.82	1.67	0.71	2.82	2.07	9.66	4.35	25.51	0.23	1.38
	4	2008	0.06	2.82	0.71	2.82	0.67	2.42	1.66	1.05	2.44	0.23	1.38	15.8	83.24	0.58	3.48
	5	2009	0.46	3.7	0.06	2.82	0.71	2.82	1.74	0.67	2.42	4.35	25.51	0.23	1.38	15.8	83.24
		2011	0.19	2.27	0.81	8.14	0.46	3.7	1.33	0.06	2.82	1.72	9.45	2.07	9.66	4.35	25.51
		2012	0.34	2.98	0.19	2.27	0.81	8.14	1.43	0.46	3.7	0.57	3.42	1.72	9.45	2.07	9.66
		Mean	*0.60*	*3.34*	*0.57*	*3.97*	*0.62*	*4.09*	*1.71*	*0.60*	*3.51*	*3.76*	*19.61*	*3.85*	*19.87*	*3.93*	*20.04*
Influence		***		**		*		ns.		°		°°		°°°		*Combined effect*	

Legend: t – average temperature °C ; p – sum of precipitation; w – water balance; h – humidity index; m – minimum value from a month; M-maximal value from a month; v – vegetation period; o – outside vegetation period; 1, 2, 3 – number of years prior.

Effect of Light Intensity and Temperature on Growth and Quality Parameters of Grafted Vines

Bülent KÖSE

Ondokuz Mayis University, Faculty of Agriculture, Department of Horticulture, 55139 Atakum, Samsun, Turkey; bulentk@omu.edu.tr

Abstract

Light and temperature are the most important environmental conditions that impact the growth and quality of grafted vines. In this research, the most suitable light and temperature conditions were tested to determine the best environment for grafted vines grown in pots. This study was conducted to determine the influence of different growth conditions, such as open field, shaded greenhouse and unshaded greenhouse, on grafted vine growth and quality. The length and diameter of shoot, trunk diameter, mean and total leaf area, total soluble sugars, starch and total carbohydrate contents of 'Narince' and 'Trakya Ilkeren' vines, grafted on 5C rootstock, were investigated in this study. Shaded conditions yielded significantly higher shoot lengths, mean and total leaf areas than other conditions ($p<0.05$). Among the growth parameters, the highest correlation ($r = 0.819$) was observed between shoot length and plant total leaf area ($p<0.01$). The total soluble sugars, starch and total carbohydrates content of vines under shaded and unshaded conditions were higher ($p<0.05$) than those under open field conditions. Grafted vines had the highest carbohydrate content in unshaded conditions. Under different light intensity, temperature and relative humidity conditions studied, the better results were obtained in the case of unshaded conditions.

Keywords: carbohydrates, grapevine, light, quality, temperature

Introduction

Grafting *Vitis vinifera* varieties onto resistant rootstocks represents the most effective way of protecting crops against phylloxera (Vršič *et al.*, 2004). Especially in phylloxera-contaminated areas, where the establishment of suitable vines on their own roots is virtually impossible. The only successful means of controlling phylloxera has been the grafting of phylloxera resistant rootstock (Troncoso *et al.*, 1999). Because of this, grafted vines should be be used during the establishment of new vineyards regardless of the phylloxera infection. Quality grafted vines allow for the successful growth and development of both root and vegetative organs after planted in a vineyard. Healthy vineyards are only possible with healthy and quality grafted vines. Nursery soil characteristics and climate conditions have significant impacts on the success of grafted vine operations especially during the initial 2-3 weeks after planting grafted cuttings (Çelik *et al.*, 1992). For successful production, control of the nursery's climatic conditions should be taken. A temperature of 25 °C and 85-95% relative humidity should be provided within the growing conditions for good callus formation (Karakır *et al.,* 1988). Several other environmental factors also affect the growth and development of the plants (Arnold and Mauseth; 1999; Buttrose, 1969; Bindi *et al.*, 1997; Oztürk and Serdar, 2011). The most important of these factors are light, temperature, water (humidity) and nutrition. Light and temperature are two important parameters that affect the physiology of the grapevine. Plants are need light and temperature for basic physiological processes such as photosynthesis, respiration transpiration and carbohydrate assimilation (Agaoglu *et al.*, 1995; Uzun and Demir, 1996). Sivritepe and Türkben (2001) was obtained rooting grafted cuttings between 30.47% and 73.75% on five different rootstocks belonging to the 'Müşküle' grape variety at 25±2°C temperature, 70-75% humidity and 16 hours photoperiod. Yılma (1996) stated that 5BB grafting cuttings of 'Abalıkoca' and 'Kazova' grapes grown under white transparent polythene plastic mini tunnel systems gave the highest rooting rate. Light is another crucial environmental factor determining the morphology and physiology of leaves (Neri *et al.*, 2003). It plays a significant role in physiological processes in plants such as photosynthesis and transpiration (Fluhr *et al.*, 1986; Wang *et al.*, 2014). Sultanina grapes growing under greenhouse conditions required light intensity and temperature at respective rates of 60.000 lux and 25°C for optimum assimilation rates (Kriedemann, 1968). Kısmalı (1984) reported certain increases in photosynthesis capacities with increasing light intensities of greenhouses but also noted that photosynthesis reached full capacity at a certain point even with increasing light intensities.

Carbohydrates, which are stored in plants' roots, trunks, cordons and canes, play major roles in the growth and development of plants. During the bud burst period, carbohydrates are moved from the roots and woody organs to the growing shoot, providing new leaves and other organs with essential reserves needed for growth. Young shoots and leaves utilize carbohydrates for growth and metabolism until they reach approximately a third of their final size (Hale and Weaver, 1962). Buttrose (1968) stated that dry weights of roots, canes, stems, and leaves increased with increasing light intensity and were greater at 25 °C than at 20 °C or 30 °C. Richards (1976) reported that grafted vines produced very good growth results in an environment of around 25 °C. Researchers suggested the optimum temperature for the photosynthesis of grapevines as between 25 °C and 35 °C (Greer and Weedon, 2012; Kriedemann, 1968; Mullins *et al.,* 1992; Schultz, 2000). However, temperatures above 35 °C, generally reduced the photosynthesis rates of grape leaves (Ferrini *et al.,* 1995; Kriedemann, 1968; Luo *et al.,* 2011; Yu *et al.,* 2009; Zsófi *et al.,* 2009). The lowest net photosynthesis values were recorded in vines grown at 35 °C and were closely correlated to chlorophyll content. The effect of this temperature on stomatal conductance was less remarkable and the dry matter accumulation was correlated to average photosynthesis rates (Ferrini *et al.,* 1995).

From dig out till grafting, various factors effect vine growth and quality. The quality of grafted vines diminishes after planting vineyards. Having the best quality vines is a crucial factor for successful grapevine growing. The aim of this study was to determine the influence of different growth conditions (open field, shaded greenhouse and unshaded greenhouse) on the growth and carbohydrate content of grafted vines.

Materials and methods

Experiments were conducted in the experimental fields and plastic greenhouses of Ondokuz Mayis University's Agricultural Faculty. 'Narince' and 'Trakya Ilkeren' scions (*Vitis vinifera* L. cvs.) were grafted on 5C rootstock. 'Narince' is the most important white wine grape in Turkey and is used both as a table food and for wine in the country. 'Trakya Ilkeren' ('Alphonse Lavallée' x 'Perlette') is a very early maturing red table grape. In addition, this grape is quite suitable as a wrapping leaf for foods and for greenhouse cultivation. Material supply, grafting and callusing processes were performed at Tekirdağ Viticultural Research Station. The scions and rootstocks were stored at +2 °C until two months until grafting. The omega grafting was done at the beginning of April and grafted cuttings were stored for 3 weeks at 28 °C and one week at 26 °C in a callusing room. Callused cuttings were planted in 10-L plastic pots with a soil, sand and manure (1:1:1) potting mix (substrate content not analyzed). Plants were irrigated everyday with 1.5 L of water in the early morning. In order to provide different light and temperature conditions, vines were grown in shaded and unshaded greenhouse and open field conditions. Similarly, to render equal soil conditions, the vines were grown in pots. The pots were placed under different growth conditions at the beginning of May. Among the rooted ones, uniformly growing vines were selected for the calculation of certain growth parameters. Shading was provided through netting with a light transmission of 50%. The temperature and relative humidity were measured with a data logger and resultant values are provided in Figs. 1 and 2 for both experimental years. Light intensity (PAR) was measured in 15-day periods between the months of June and September of both years with a canopy analyzer device and expressed in $\mu mol m^{-2} s^{-1}$ (Tab. 1 and Fig. 3). In the study, some growth parameters such as total and mean leaf area, shoot length, trunk and shoot diameter were measured in August. Mean leaf area was determined by dividing the total leaf area of the shoot with the number of leaves over the main shoot. Leaf areas were calculated non-destructively $[-1.41 + 0.527(W^2) + 0.254(L^2)]$ according to Elsner and Jubb (1988). All measurements were repeated in the second year on annual cuttings.

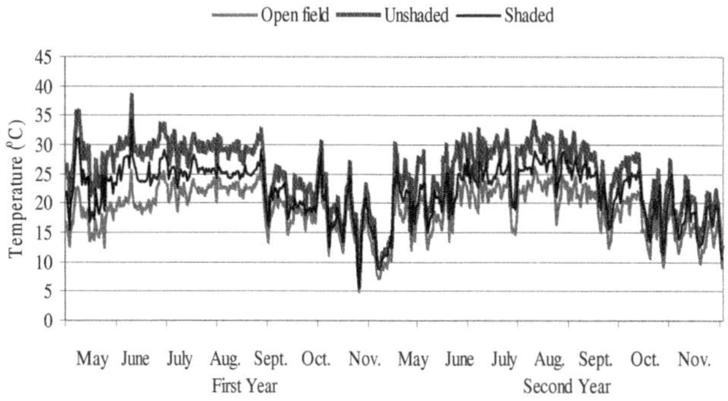

Fig. 1. Course of temperature throughout the experimental years

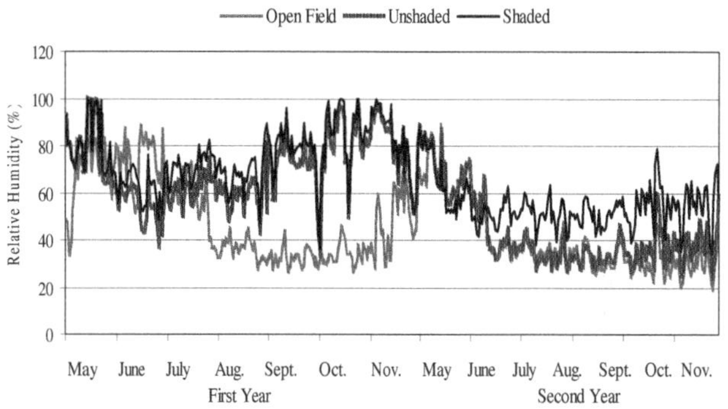

Fig. 2. Course of relative humidity throughout the experimental years

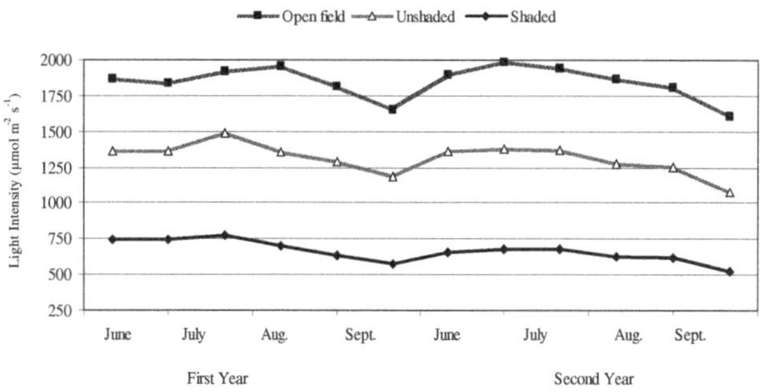

Fig. 3. Course of light intensity throughout the experimental years

Carbohydrate analysis

During the dormant periods of the plants, a 2 cm piece between the 2nd and 3rd nodes from the bottom of the canes; a 2 cm piece from the mid-section of the stem; a piece between 2.5-4 mm from the older sections of the roots; and a 1-2 mm piece from the younger sections of the roots were cut out and grinded together (Bates *et al.,* 2002). Plant tissue samples were pulverized in a mill with a 40-mesh screen for analysis and 200 mg of dust was used for the extraction. Soluble sugars were extracted twice with 8 ml of 80% ethanol at 60 °C for 30 minutes (Candolfi and Koblet, 1990). The content of soluble sugars was determined with the anthrone method as described by Scott and Melvin (1953). Starch was extracted twice with 8 ml of 1 M perchloric acid, one hour each time at 60 °C and measured by the same method. Absorbance readings were made at 620 nm with a spectrophotometer. Glucose was used as a standard for the analysis of soluble sugars and starch.

Experimental design

The randomized complete block design, with three replications, was used both years. Every replication consisted of four uniformly grown grafted vines. Growing conditions, cultivars, years and these interactions were studied by three-factored ANOVA. Data analysis was performed using SPSS 16.0 for Windows. Results studied are presented as means and a pooled standard error of mean (SEM). Differences among means were detected using Duncan's multiple range tests, which were used for statistical analyses ($p \leq 0.05$). Correlation analysis was applied to assess the relationship between growth parameters and sugar, starch and total carbohydrates.

Results and discussion

Changes in temperature, humidity and light intensity

Data related to average temperature (°C), relative humidity (%) and light intensity (μmolm^{-2}s^{-1}) determined daily after the planting of grafted cuttings during the research are shown in Tab. 1 and Fig. 1, 2 and 3. Average temperature, relative humidity and light intensity values under different growth conditions throughout the growth seasons of experimental years are provided in Tab. 1.

Under unshaded, shaded and open field conditions, average temperatures were respectively calculated as 24.9 °C, 22.1 °C and 18.8 °C. Under the same conditions, average light intensities were respectively calculated as 1315.4, 662.2 and 1846.1 μmol m^{-2}s^{-1}.

Vine growth and development

Shoot lengths and diameters, trunk diameters, total and mean leaf areas of grape cultivars grafted on 5C rootstocks under different conditions (shaded and unshaded greenhouses and open field) for both years are presented in Tab. 2. Shoot lengths were significantly affected by growth conditions, year and growth conditions x year interaction ($p<0.05$). The shoot lengths were higher under shaded

Tab. 1. Temperature, relative humidity and light intensity values under different nursery conditions throughout the growth seasons of experimental years

Years	Parameters	Conditions		
		Unshaded	Shaded	Open Field
First Year	Light Intensity (μmol m^{-2}s^{-1})	1343.1	695.0	1841.1
	Temperature (°C)	25.1	22.0	18.7
	Relative Humidity (%)	69.8	75.9	49.4
Second Year	Light Intensity (μmol m^{-2}s^{-1})	1287.7	629.3	1851.1
	Temperature (°C)	24.7	22.1	18.9
	Relative Humidity (%)	44.9	56.2	41.3

Tab. 2. Shoot lengths (cm) and diameters (mm), trunk diameters (mm), total and mean leaf areas (cm^2) of different grafted vine cultivars under different conditions

Growth conditions	Cultivars	Trunk Diameter	Shoot		Leaf Area	
			Length	Diameter	Total	Mean
First Year						
Open field	'Narince'	14.3	75.0	9.9	3420.0	65.2
	'T.Ilkeren'	13.6	92.3	8.9	3786.0	42.6
Unshaded	'Narince'	15.0	118.0	9.3	3840.0	74.6
	'T.Ilkeren'	13.7	114.8	9.4	5729.0	57.5
Shaded	'Narince'	14.4	268.5	10.0	6435.0	116.9
	'T.Ilkeren'	15.0	250.0	10.3	8164.0	88.7
Second Year						
Open field	'Narince'	11.8	46.5	6.5	1865.0	55.7
	'T.Ilkeren'	12.3	47.3	6.3	1843.0	39.0
Unshaded	'Narince'	10.9	118.5	6.9	2439.0	63.0
	'T.Ilkeren'	12.1	117.5	6.8	3794.0	54.4
Shaded	'Narince'	12.3	133.0	6.8	3130.0	97.6
	'T.Ilkeren'	12.6	155.0	6.4	4145.0	69.8
Pooled SEM*		0.134	4.615	0.117	90.234	1.357
Conditions		N.S	*	N.S	*	*
Cultivars		N.S	N.S	N.S	*	*
Year		*	*	*	*	*
Conditions x Cultivars		N.S	N.S	N.S	*	N.S
Conditions x Year		N.S	*	N.S	*	N.S
Cultivars x Year		*	N.S	N.S	N.S	N.S
Conditions x Cultivars x Year		N.S	N.S	N.S	*	N.S

*Pooled SEM: Pooled standard error of the mean (NS. $p>0.05$; *. $p<0.05$).

conditions than in other growth conditions. In the study, decreasing light intensities and increasing temperatures conditions affected shoot lengths. Low light intensities and high temperatures (shaded and unshaded) lead to the elongation of plant heights (Fig. 4). Mean shoot length was 201.6 cm under shaded greenhouse conditions, 117.2 cm under unshaded greenhouse conditions and 62.3 cm under open field conditions (Tab. 3).

Year also significantly affected growth characteristics (Tab. 2). Strong shoot and leaf development of the first year also resulted in larger shoot diameter the same year. Similarly, trunk diameter of vines in the first year was also larger than in the second year. This was due to weak growth in the second year. Although similar environmental conditions and cultural practices were applied, all growth parameters in the first year were higher than in the second year. These differences might have been from a difference in the organic composition of the soil mix and relative humidity. There was no effect of cultivars on growth. The cultivars were only significant on leaf areas. Larger leaf areas of growing vines were obtained under shaded conditions. Despite the fact that shoot and trunk diameters were not affected by growing conditions (Tab. 2, 3), there was a trend for the shoot and trunk diameters under the shaded greenhouse conditions in both years to be higher (8.4 mm and 13.6 mm, respectively) (Tab. 3).

Tab. 3. Mean shoot length (cm) and diameter (mm), trunk diameter (mm) and leaf area (cm^2) values of grafted vines under different conditions

Growth conditions	Trunk Diameter	Shoot Length	Shoot Diameter	Leaf Area Total	Leaf Area Average
Open field	13.0	62.3c	7.9	2728c	50.6c
Unshaded	12.9	117.2b	8.1	3950b	62.4b
Shaded	13.6	201.6a	8.4	5468a	93.2a

Means with different letters in the same column were significantly different (*p<0.05). **(Means represent the averages for cultivars and years).

Shoots lengths, mean and total leaf areas per plant were affected by the growth conditions (Tab. 2). Light intensity measured under shaded conditions was about one-third of that in open field conditions (Tab. 1). Shoot length increased with decreasing light intensity. Leaves growing at different light levels differ in their anatomy and physiology. While increasing light intensities yield improved photosynthesis rates of vine leaves, decreasing light intensities cause longer internode spacing, weak growth, embrittelement and larger leaf areas. Aside from this, such decreasing lighting intensities also result in decreased photosynthesis rates, etiolating and defoliation (Eris, 1990). The results of the study indicated that different light and temperature conditions had significant effects on shoot growth, total leaf area and mean leaf area, especially in the shaded conditions. According to TS 3981 (Anonymous, 1995), first class vines have to have at least 20 cm of shoot length and a trunk diameter of 7 mm. In the study, the average shoot length was above 60 cm, shoot diameter was 7.9 mm and trunk diameter was 12.9 mm. All growth combinations of the first class of grafted vines were according to growth parameters studied in the research (Fig. 4).

Temperature is the most important environmental factor for canopy development and leaf photosynthesis (Baldocchi and Amthor, 2001; Koblet, 1984; Matloobi, 2012). Field *et al.* (2009) reported that the shoot and trunk dry biomass of potted Shiraz grapevines grown at 23 °C in a glasshouse were significantly greater than those grown at 13 °C. Skelton (2007) reported that the soil surface requires a cover of black polythene to keep in moisture and raise the soil temperature to aid in vine development, a strong graft union and a good root system.

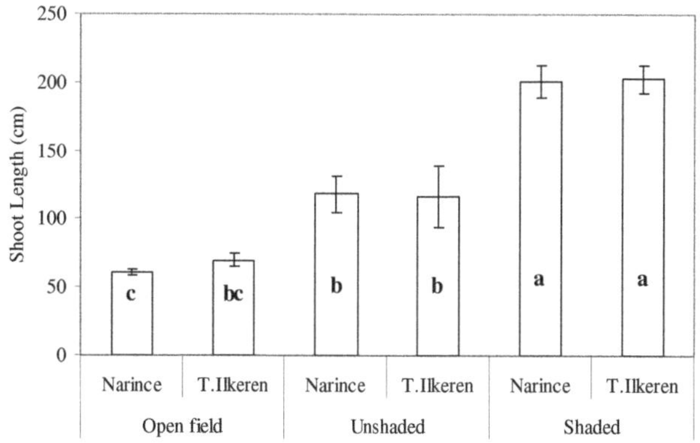

Fig. 4. Shoot lengths of grape cultivars grown in shaded and unshaded greenhouses and in the open field conditions in two different years; means with different letters in the same column were significantly different at P < 0.05

Leaf growth and development

The total and mean leaf areas of grape cultivars grown under different conditions (shaded and unshaded greenhouses and open field conditions) with respect to years and cultivars are presented in Tab. 2. Growth conditions significantly affected the leaf areas of different varieties. Tab. 2 shows that the grape cultivar has a significant effect upon mean and total leaf area and there were significant effects of the year and nursery conditions on leaf area. The mean and total leaf areas of grapes which were grown under shaded greenhouse conditions were significantly higher than those grown under unshaded greenhouse and open field conditions in both years (Tab. 3). On the other hand, growth conditions x cultivar x year interaction had a significant effect on total leaf area but not on mean leaf area. Significant differences were observed in leaf areas of cultivars (Fig. 5). The average total leaf area of vines was calculated as 5229 cm^2 in the first year and as 2869.3 cm^2 in the second year. Similarly, mean leaf areas of the first and second years were respectively observed as 74.3 and 63.3 cm^2 (Tab. 2). The highest total leaf area was at shaded 'Trakya Ilkeren' (6154 cm^2) whereas the lowest at open field 'Narince' cultivar (2643 cm^2). However, the highest mean leaf area was at shaded 'Narince' (107.2 cm^2), whereas the least mean leaf area was at unshaded 'Trakya Ilkeren' (40.8 cm^2) cultivar. Reducing light intensity led to the increased leaf area of cultivars. Bonan (2002) was impressed that leaves growing in sunny environments were smaller and more deeply lobed than leaves growing in shaded environments. Furthermore, the leaf area ratio was 2.7 times greater and leaf weight ratio was about 20% greater in shaded plants compared with sun plants. The reported acclimation to low light was not restricted to dry-weight allocation among shoot organs because the specific leaf area was increased to about 100% by the low-light environment. Increased shading resulted in increased leaf areas and wider leaves. Gutschick and Wiegel (1988) stated that deeper shade allowed plants to construct a greater leaf surface area for light interception at a certain biomass investment in foliage. Excess sun energy may have harmful effects on the photosynthetic devices, and plants arrange morphological, biochemical and physiological strategies to minimize these effects (Lichtenthaler and Burkart, 1999; Lichtenthaler *et al.*, 2007). Leaf area plays a key role in determining crop productivity by controlling the interception of solar radiation (Monteith, 1977). In grapevines, the rate of photosynthesis production of fully expanded leaves depends on air temperature (Buttrose, 1969). In addition, there is a direct connection between intercepted solar radiation and plant leaf area to shoot development and accumulated assimilates in plant organs. For example, Cartechini and Palliotti (1995) reported that grapevines acclimate to modified light intensity and produce a given amount of carbohydrates. Shade-grown vines (30% PPF) had lower total leaf area and soluble carbohydrates per plant and lower starch content in leaves compared to light adapted grapevines. Buttrose (1968) reported that cuttings of grapevines stem lengths and leaf area values grown in growth cabinets were in most cases greatest at 25 °C followed in turn by 30 °C and 20 °C values. Miller *et al.* (1996) indicated a linear relationship between leaf area and shoot lengths of vines. In the research, a relationship was observed between leaf area and shoot length. Shoot length, total and mean leaf areas of the grafted vines grown under shaded greenhouse conditions were higher than that grown under other conditions. There was a highest correlation observed between shoot length, total and mean leaf area (Tab. 6). Total leaf area was correlated between trunk diameter (r=0.577, p<0.01), shoot diameter (r=0.646, p<0.01) and shoot length (r=0.819, p<0.01) per plant.

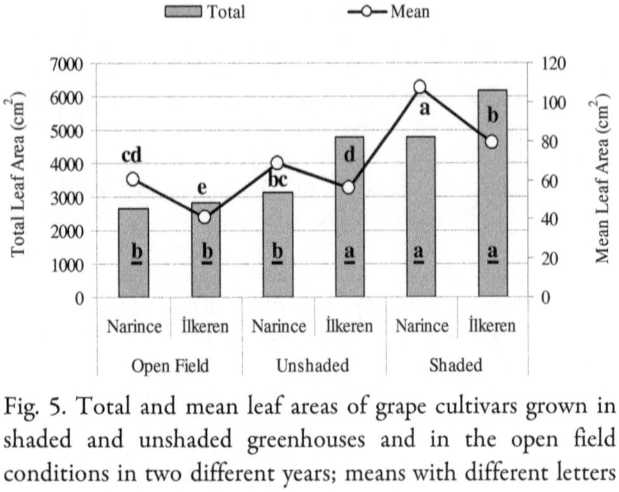

Fig. 5. Total and mean leaf areas of grape cultivars grown in shaded and unshaded greenhouses and in the open field conditions in two different years; means with different letters were significantly different at P<0.05

Plant carbohydrate contents

The total soluble sugars, starch and total carbohydrate contents of different grafted vine cultivars under different growth conditions (shaded and unshaded greenhouses and open field) are presented in Tab. 4. Growth conditions had significant impacts on sugar, starch and carbohydrate contents of grafted vines (Tab. 5).

Data showing significant effects was obtained concerning interactions such as growth conditions x cultivar, cultivar x year and growth conditions x cultivars x year (p<0.05). While significant differences were observed in sugar and starch contents of the years, effects of years on carbohydrate contents were not significant (Tab. 4). While the first year's sugar content was high and starch was low, the second year's sugar content was low and starch was high. So in both years, total carbohydrate content was not significant.

In the study, the total soluble sugars, starch and total carbohydrate content of the vines grown under unshaded greenhouse conditions were higher than that of the other conditions (Fig. 6). Effects of cultivars x conditions on starch and carbohydrate content were significant except for sugar only in the open field condition (Tab. 4). In spite of that, the mean sugar content of grafted vines grown under shaded greenhouse conditions (14.6 mg L^{-1}) and unshaded greenhouse conditions (14.3 mg L^{-1}) were found to be a bit higher than the ones grown under open field conditions (12.4 mg L^{-1}) (Tab. 5).

Carbohydrate contents of grapevines in this study were significantly affected by growth conditions. Starch and carbohydrate contents of plants grown under unshaded greenhouse conditions were higher than the plants grown under shaded greenhouse and open field conditions (Tab. 5, Fig. 6). Growth conditions x year, cultivar x year and growth conditions x cultivar x year interactions were found to be significant with regard to sugar, starch and carbohydrate content. Mean sugar content was 15.8 mg L^{-1} in the first year and 11.8 mg L^{-1} in the second year; starch content was 38.5 mg L^{-1} in the first year and 43.4 mg L^{-1} in the second year; carbohydrate content was 54.3 mg L^{-1} in the first year and 55.2 mg L^{-1} in the second year (Tab. 4). In brief, the sugar, starch and carbohydrate content of plants grown under high light intensity (1846.1 μmolm^{-2}s^{-1}) and low temperature (18.8 °C) conditions (open field) were lower than the ones grown under lower light but higher temperature conditions (662.2 μmolm^{-2}s^{-1}, 22.1 °C for shaded greenhouses and 1315.4 μmolm^{-2}s^{-1}, 24.9 °C for unshaded greenhouses). However, the highest sugar, starch and carbohydrate content was obtained around 25 °C and 30% lower light intensity (unshaded) in all the conditions (Fig. 6). Such a case seemed to be related to decreased photosynthesis efficiencies under high light but low temperature conditions. Among growth conditions, average relative humidity varied. The highest relative humidity (66%) was observed under shaded greenhouse conditions and the lowest (45.4%) under open field conditions (Tab. 1). Relative humidity of the first year was higher than that of the second year under all growth conditions. According to Jang *et al.* (2011) generally, 75-85% relative humidity levels accelerate the photosynthesis rate. In grapevines, carbohydrates in roots and wood portions play important roles for vine longevity and quality. Indeed, carbohydrates are not only involved in the protection against frost (Keller, 2010), but they also take part in leaf area development, shoot growth and flower induction (Keller and Koblet, 1994; Murisier and Aerny, 1994; Yang and Hori, 1979). Carbohydrate accumulation correlated with photosynthesis efficiency in the plants. In addition, photosynthesis is highly dependent on temperature, light intensity and relative humidity (Gomes-Laranjo *et al.,* 2006; Koorneef *et al.,* 2002). It was reported that the optimal temperature range for photosynthesis was between 26-30 °C, with a decrease of about 50% when the temperature increases to 32-34 °C; the decrease being higher than 80% when temperatures reach 38 °C in chestnut (Gomes-Laranjo *et al.,* 2006). Schultz and Matthews (1993) stated that in potted-grown White Riesling grapes dry-weight allocation of leaves of shaded plants was 9% greater, whereas allocation to stem tissue was 10% less than in sun plants. Photosynthesis rates decrease at temperatures above the optimum levels (Spayd, 2000). In the case of temperatures up to 40 °C, the photosynthesis of grapevine leaves are significantly reduced (Ferrini *et al.,* 1995; Kriedemann, 1968; Yu *et al.,* 2009; Zsofi *et al.,* 2009). Greer and Weedon (2012) reported remarkable decreases in the photosynthesis of Semillon grape leaves at temperatures above 35 °C. In part, these reductions in photosynthesis are related to stomatal limitation due to high temperature (Ferrini *et al.,* 1995; Sepúlveda and Kliewer, 1986; Soar *et al.,* 2009).

Most of the physiological processes of plants are related to carbohydrates. Carbohydrates are the most important products of photosynthesis and they are the most significant component of various physiological processes. In general, carbohydrates constitute about 50-80% of the total dry weight of most plants (Eris, 1990). Goldschmidth (1997) indicated the role of carbohydrates in various physiological processes of plants as the source of energy. Monteith (1977) reported linear relationships between dry matter accumulation and total PAR (Photosynthetic Active Radiation) up to a certain point. However, since high light intensities may damage plant chlorophylls, photosynthetic efficiencies of plants may decrease because of photoinhibition. Bertamini and Nedunchezhian (2004) reported net photosynthesis under low and high light intensities respectively as 42 and 76% lower than the normal light conditions. Thus, photosynthesis rates increase at optimum temperature levels. Since the average temperatures in shaded and unshaded greenhouses were within the optimum range of temperature (25-30 °C) for photosynthesis, the plants had higher carbohydrate contents under shaded and unshaded greenhouse conditions than those under open field conditions. Higher total and mean leaf areas of the plants under shaded and unshaded greenhouse conditions also resulted in higher carbohydrate contents for those plants than the ones grown under open field conditions (high light, low average temperature). Accordingly, there was also a significant correlation between total leaf area and sugar content (Tab. 6; r=0.695, $p<0.01$).

Tab. 4. The sugar, starch and total carbohydrate content (mg L^{-1}) of grafted vines grown under shaded and unshaded greenhouses and open field conditions

Growth conditions	Cultivars	Sugar	Starch	Carbohydrate
		First Year		
Open field	'Narince'	14.0	30.7	44.7
	'T.Ilkeren'	13.8	34.2	47.9
Unshaded	'Narince'	16.3	35.5	51.8
	'T.Ilkeren'	18.1	48.7	66.8
Shaded	'Narince'	14.6	36.0	50.5
	'T.Ilkeren'	18.0	46.1	64.1
		Second Year		
Open field	'Narince'	11.0	39.2	50.2
	'T.Ilkeren'	10.9	38.9	49.8
Unshaded	'Narince'	13.1	57.5	70.6
	'T.Ilkeren'	10.9	44.3	55.2
Shaded	'Narince'	13.6	41.5	55.1
	'T.Ilkeren'	11.2	39.1	50.3
Pooled SEM*		0.089	0.235	0.236
Conditions		*	*	*
Cultivars		N.S	*	*
Year		*	*	N.S
Conditions x Cultivars		N.S	*	*
Conditions x Year		*	*	*
Cultivar x Year		*	*	*
Conditions x Cultivars x Year		*	*	*

*Pooled SEM: Pooled standard error of the mean (N.S. $p>0.05$; *. $p<0.05$).

Tab. 5. Mean sugar, starch and carbohydrate contents (mg L^{-1}) of grafted vines grown under shaded and unshaded greenhouses and open field conditions

Growth conditions	Sugar	Starch	Carbohydrate
Open field	12.4b	35.8c	48.1c
Unshaded	14.3a	46.5a	61.1a
Shaded	14.6a	40.7b	55.0b

Means with different letters in the same column were significantly different (*$p<0.05$).

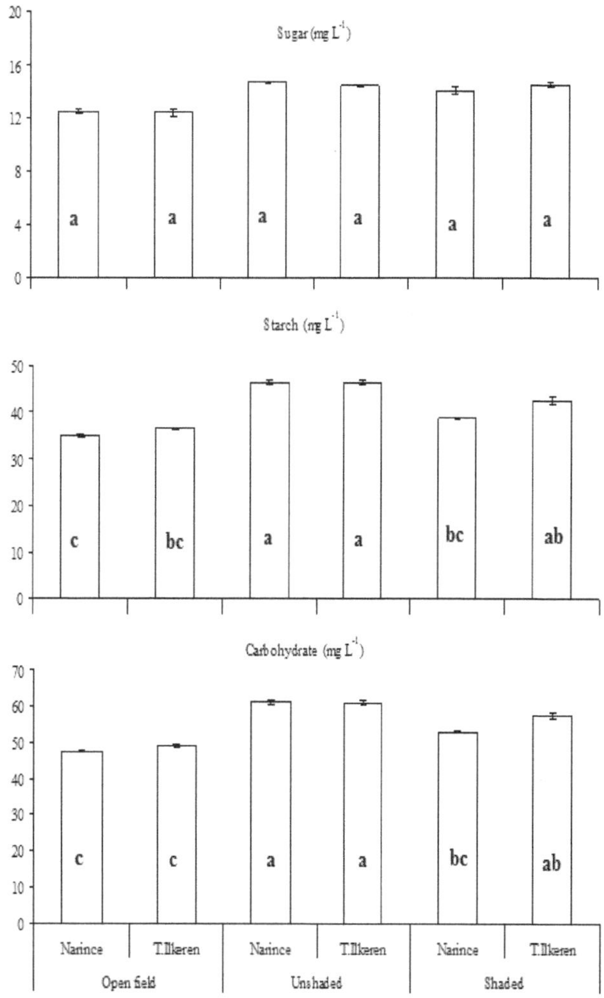

Fig. 6. Sugar, starch and total carbohydrate content of grape cultivars grown in shaded and unshaded greenhouses and in the open field conditions in two different years; means with different letters were significantly different at P<0.05

Tab. 6. Correlations amongst the plant growth parameters and carbohydrates

Parameters	Sugar	Starch	Carbohydrate	Total Leaf Area	Trunk Diameter	Mean Leaf Area	Shoot Diameter
Starch	0.134						
Carbohydrate	0.444**	0.948**					
Total Leaf Area	0.695**	0.068	0.286*				
Trunk Diameter	0.544**	-0.403**	-0.189	0.577**			
Mean Leaf Area	0.356*	-0.032	0.086	0.555**	0.339*		
Shoot Diameter	0.742**	-0.232	0.030	0.646**	0.706**	0.319*	
Shoot Length	0.447**	0.129	0.261	0.819**	0.354*	0.754**	0.395**

**Correlation is significant at the $p < 0.01$ level (2-tailed).
*Correlation is significant at the $p < 0.05$ level (2-tailed).

Conclusions

According to the results of this study, vines' growth and quality characteristics were affected by different growth conditions. The vines grown under unshaded and shaded conditions exhibited higher performances than the ones grown under open field conditions. Shoot lengths and leaf areas were especially affected by decreasing light intensities and increasing temperatures. Additionally, the carbohydrate content of vines was affected by growth conditions. Higher carbohydrate levels were observed under increasing temperature and decreasing light intensity conditions as opposed to higher light intensity and lower temperature conditions. However, all three growth conditions produced top quality vines. Providing optimum temperature or the establishment of nurseries obtaining these temperature regions is gaining importance in the area of quality grapevine production. To reduce the negative impact of high light intensity on plants, canopy shading throughout the growth cycle should be provided. In this study, better growth and quality results were obtained under high temperatures (25 °C) and low light intensity conditions (1300 μmol $m^{-2}s^{-1}$).

References

Agaoglu SY, Ayfer M, Fidan F, Köksal I, Çelik M, Abak K, Celik H, Kaynak L, Gülsen Y (1995). General Horticulture. Ankara University Agricultural Faculty Press, Volume No: 1009, Ankara, Turkey, 283 p.

Anonymous (1995). Turkish Standard of Grape Vine (TS 3981). Necatibey Street 112. Ministries Ankara (in Turkish).

Arnold DH, Mauseth JD (1999). Effects of environmental factors on development of wood. Americ J of Botany 86(3):367-371.

Baldocchi, DD, Amthor JS (2001). Canopy Photosynthesis: History, measurements and models, p. 9-31. In: J. Roy, B. Saugier and H. Mooney (Eds.). Terrestrial Global Productivity: Past, Present and Future. Academic Press, San Diego.

Bates TR, Dunst RM, Joy P (2002). Seasonal dry matter, starch, and nutrient distribution in 'Concord' grapevine roots. HortScience 37(2):313-316.

Bertamini M, Nedunchezhian N (2004). Phoptosynthetic responses for *Vitis vinifera* plants grown at different photon flux densities under field conditions. Biologia Plantarum (48):149-152.

Bindi M, Miglietta F, Gozzini B, Orlandini S, Seghi L (1997). A simple model for simulation of growth and development in grapevine (*Vitis vinifera* L.) I. Model description. Vitis 36(2):67-71.

Bonan GB (2002). Ecological climatology: concepts and applications. Cambridge University Press, 678 p.

Buttrose MS (1968). Some Effects of Light Intensity and Temperature on Dry Weight and Shoot Growth of Grape-Vine. Ann Bot 32(4):753-765.

Buttrose MS (1969). Fruitfulness in grapevines: Effects of light intensity and temperature. Botanical gazette 130(3):166-173.

Cangi R, Kelen M, Doğan A (1999). The posibility of grapevine production in cool climate conditions. Turkey 3rd National Horticulture Congress. 14-17 September 1999, Ankara, Turkey. p. 430-435 (in Turkish).

Candolfi-Vasconcelos MC, Koblet W (1990). Yield, fruit quality, bud fertility and starch reserves of the wood as a function of leaf removal in *Vitis vinifera* - Evidence of compensation and stress recovering. Vitis (29):199-221.

Cartechini A, Palliotti A (1995). Effect of shading on vine morphology and productivity and leaf gas exchange characteristics in grapevines in the field. Am J Enol Vitic 46(2):227-234.

Çelik S, Delice A, Arın L (1992). The grafted grapevine production in the nursery condition. Tr J Agric Forestry 16:507-518.

Eris A (1990). Physiology of Horticulture. Uludag University Lecture Notes No:11, Bursa, Turkey (in Turkish).

Elsner EA, Jubb GL (1988). Leaf Area Estimation of Concord Grape Leaves from Simple Linear Measurements. Am J Enol Vitic 39(1):95-97.

Ferrini F, Mattii GB, Nicese FP (1995). Effect of temperature on key physiological responses of grapevine leaf. Am J Enol Vitic 46(3):375-379.

Field SK, Smith JP, Holzapfel BP, Hardie WJ, Emery RN (2009). Grapevine response to soil temperature: xylem cytokinins and carbohydrate reserve mobilization from budbreak to anthesis. Am J Enol Vitic 60(2):164-172.

Fluhr R, Moses P, Morelli G, Coruzzi G, Chua NH (1986). Expression dynamics of the pea rbcS multigene family and organ distribution of the transcripts. EMBO J 5(9):2063-2071.

Greer DH, Weedon MM (2012). Modelling photosynthetic responses to temperature of grapevine (*Vitis vinifera* cv. Semillon) leaves on vines grown in a hot climate. Plant, Cell and Environment 35(6):1050-1064.

Goldschmidth EE (1997). Carbohydrate supply as a limiting factor for citrus fruit growth and productivity. HortScience 32(3): 550-551.

Gomes-Laranjo J, Peixoto F, Wong Fong Sang HW, Torres-Pereira J (2006). Study of the temperature effect in three chestnut (*Castanea sativa* Mill.) cultivars' behavior. J of Plant Phys 163(9): 945-955.

Gutschick VP, Wiegel FW (1988). Optimizing the canopy photosynthetic rate by patterns of investment in specific leaf mass. The American Naturalist 132(1):67-86.

Hale CR, Weaver RJ (1962). The efect of developmental stage on direction of translocation of photosynthate in *Vitis vinifera*. Hilgardia 33(3):89-131.

Jang Y, Goto E, Ishigami Y, Mun B, Chun C (2011). Effects of light intensity and relative humidity on photosynthesis, growth and graft-take of grafted cucumber seedlings during healing and acclimatization. Horticulture, Environment, and Biotechnology 52(4): 331-338.

Karakır MN, Uzun HI, İlter E (1988). The research of Yuvarlak Çekirdeksiz grape variety grafted on 5BB and 99 R rootstocks; graft success, rooting and anatomical investigation of graft union. Turkey 3rd Viticultural Symposium, pp. 88-92, Bursa, Turkey (in Turkish).

Keller M (2010). The Science of Grapevines. Anatomy and Physiology. Academic Press London, 400 p.

Keller M, Koblet W (1994). Is carbon starvation rather than excessive nitrogen supply the cause of inflorescence necrosis in *Vitis viniferous*? Vitis 33:81-86.

Kısmalı I (1984). The research of some of the table grape varieties on winter bud fertility. 2rd Turkish National Viticulture and Enology Symposium, 14-17 November 1983, pp. 35-48, Manisa, Turkey (in Turkish).

Koblet W (1984). Influence of light and temperature on vine performance in cool climates and applications to vineyard management, p. 139-157. In: Heatherbell DA, Lombard, PB, Bodyfelt FW, Price SF (Eds.). Proc Int Symp on Cool Climate. Vitic Enol Oregon State University, Oregon.

Koornneef M, Bentsink L, Hilhorst H (2002). Seed dormancy and germination. Current Opinion Plant Biol 5: 33-36.

Kriedemann PE (1968). Photosynthesis in vine leaves as a function of light intensity, temperature, and leaf age. Vitis 7:213-220.

Lichtenthaler H, Burkart S (1999). Photosynthesis and high light stress. Bulgarian J Plant Physiology 25(3-4):3-16.

Lichtenthaler H, Marek A, Kalina J, Urban O (2007). Differences in pigment composition, photosynthetic rates and chlorophyll fluorescence images of sun and shade leaves of four tree species. Plant Physiology and Biochemistry 45(8):577-588.

Luo HB, Ma L, Xi HF, Duan W, Li SH, Loescher W, Wang JF, Wang LJ (2011). Photosynthetic responses to heat treatments at different temperatures and following recovery in grapevine (*Vitis amurensis* L.) leaves. PLoS One 6(8):e23033.

Miller DP, Howell GS, Flore AJ (1996). Effect of Shoot Number on Potted Grapevines: I. Canopy Development and Morphology. Am J Vitic 47(3):247-250.

Matloobi, M (2012). Light harvesting and photosynthesis by the canopy. Advances in photosynthesis - fundamental aspects. InTech: Janeza Trdine 9(51000):235-256.

Monteith JL (1977). Climate and the efficiency of crop production in Britain. Physiological Translocations of the Royal Society of London 281:277-294.

Mullins MG, Bouquet A, Williams LE (1992). Biology of grapevine. Cambridge Univ Pres, Cambridge, UK, P. 252.

Murisier F, Aerny J (1994). Influence du niveau de rendement de la vigne sur les réserves de la plante et sur la chlorose. Rôle du porte-greffe. Rev Suisse Vitic Arboric. Hortic 26(5):281-287.

Neri D, Battistelli R, Albertini G (2003). Effects of low-light intensity and temperature on photosynthesis and transpiration of *Vigna sinensis* L. J Fruit Ornamental Plant Res 11:17-24.

Oztürk A, Serdar U (2011). Effects of different nursery conditions on the plant development and some leaf characteristics in Chestnuts (*Castanea sativa Mill.*). Aust J Crop Sci 5(10):1218-1223.

Richards M (1976). Propagation of grapes by grafting. Plant Propagator 22(1):9-10.

Sepúlveda G, Kliewer WM (1986). Effect of high temperature on grapevines (*Vitis vinifera* L.). II. Distribution of soluble sugars. American Journal of Enology and Viticulture 37(1):20-25.

Schultz H (2000). Climate change and viticulture: A European perspective on climatology, carbon dioxide and UV-B effects. Aust J Grape and Wine Res 6(1):2-12.

Schultz HR, Matthews MA (1993). Xylem development and hydraulic conductance in sun and shade shoots of grapevine (*Vitis vinifera* L.): evidence that low light uncouples water transport capacity from leaf area. Planta 190(3):393-406.

Scott TA, Melvin EH (1953). Determination of Dextran with Anthrone. Anal Chem 25(11):1656-1661.

Sivritepe N, Türkben C (2001). The Effects of Different Rootstocks on Grafting Success and Ratio of Grafted Vine in Müşküle Grapevine Cultivar. J Uludag Uni Agr Fac 15:47-58 (in Turkish).

Skelton S (2007). Viticulture - An introduction to commercial grape growing for wine production. Lulu Com London p. 146.

Spayd S (2000). The casual cellar rat explores the importance of leaves. Grape Research News. Washington State University Press 11(2):1-11.

Soar CJ, Collins MJ, Sadras VO (2009). Irrigated Shiraz vines up-regulate gas exchange and maintain berry growth under short spells of high maximum temperature in the field. Funct Plant Biol 36(9):801-814.

Troncoso A, Atte CM, Cantos M (1999). Evaluation of salt tolerance of in vitro-grown grapevine rootstock varieties. Vitis 38:55-60.

Uzun S, Demir Y (1996). The effect of temperature on plant growth, development and productivity (II. Growth). J Agr Fac OMU 11(3):201-212 (in Turkish).

Vršič S, Valdhuber J, Pulko B (2004). Compatibility of the rootstock Börner with various scion varieties. Vitis 43:155-156.

Wang WH, Chen J, Liu TW, Han AD, Simon M, Dong XJ, He HX, Zheng HL (2014). Regulation of the calcium-sensing receptor in both stomatal movement and photosynthetic electron transport is crucial for water use efficiency and drought tolerance in *Arabidopsis.* J Experimental Botany 65(1):223-234.

Yang YS, Hori Y (1979). Studies on retranslocation of accumulated assimilates in Delaware grapevines. I. Retranslocation of ^{14}C assimilates in the following spring after ^{14}C feeding in summer and autumn. Tohoku J Agric Res 30:43-56.

Yılma P (1996). The effect of different grafting time and nursery systems on grafted grapevine production by grafted cuttings directly planted in the nursery. Ondokuz Mayis University, MSD Thesis. Applied and Natural Sciences Institute. Samsun, Turkey, p. 126 (in Turkish).

Yu DJ, Kim SJ, Lee HJ (2009). Stomatal and non-stomatal limitations to photosynthesis in field-grown grapevine cultivars. Biol Plant 53(1):133-137.

Zsófi Z, Gál L, Szilágy Z, Szücs E, Marschall M, Nagy Z, Bálo B (2009). Use of stomatal conductance and pre-dawn water potential to classify terroir for the grape variety Kékfrankos. Aust J Grape Wine Research 15(1):36-47.

Molecular Role of Nitric Oxide in Secondary Products Production in *Ginkgo biloba* Cell Suspension Culture

Hossam Saad EL-BELTAGI[1]*, Osama K. AHMED[1], Adel E. HEGAZY[2]

[1]*Cairo University, Biochemistry Department, Faculty of Agriculture, Gamma st, Giza, Cairo, Egypt; lbltg@yahoo.com (*corresponding author)*

[2]*University of Sadat City, Genetic Engineering and Biotechnology Institute (GEBRI), Plant Biotechnology Department, Egypt*

Abstract

Effects of sodium nitroprusside (SNP; nitric oxide donor) treatment on the enhancement of secondary metabolites production, oxidative stress mediators ($O_2^{\cdot-}$) accumulation and antioxidant defense enzymes of *Ginkgo biloba* callus culture was investigated. On one hand, the obtained data showed a highly metabolic modification of chemical constituents, PAL activity and various antioxidant defense enzymes (APX, SOD), which gradually increased in response to SNP treatments. On the other hands the high NO levels significantly increased the accumulation of various oxidative burst of $O_2^{\cdot-}$. MS basal medium supplemented with casein hydrolase (500 mg/L), NAA and BA at equal concentration (0.5 mg/L) recorded the highest number of regenerated shoots (4.81 cm) and shoot height (4.96 cm) as well as root number (2.25 cm) and root length (4.5 cm). The highest survival (40 %) was shown in acclimatization on the mixture containing sand, peat moss and vermiculite (1: 1: 1, v/v/v), which significantly confirmed and reflected the variation in survival percentage. Meanwhile, higher treatment (500 µM) of NO positively enhanced secondary products accumulation of total tannins, saponins, phenols and total flavonoids in *G. biloba* callus culture.

Keywords: antioxidant enzymes, *Ginkgo biloba*, nitric oxide, oxidative burst, secondary products, tissue culture

Abbreviations: NO: Nitric oxide; SNP: sodium nitroprusside; $O_2^{\cdot-}$: superoxide anion radical; MDA: malonaldehyde; H_2O_2: hydrogen peroxide; GSH: glutathione; AsA: Ascorbate; PAL: Phenyl alanine ammonia lyase; SOD: Superoxide dismutase; APX: ascorbate peroxidase; NAA: Naphthaleneacetic acid; BA: Benzyl amino purine; MS: Murashige and Skoog medium.

Introduction

Ginkgo biloba has been known as a popular remedy in traditional Chinese medicine for over 4000 years, and it has been a common herbal medicine in Europe since the 1730's. The *Ginkgo* tree, the only existing tree in the family *Ginkgoaceae*, is the world's oldest living tree and is thus sometimes referred to as a "living fossil". *Ginkgo* trees are now widely planted in China, Japan, Korea, France, Germany and the United States for both ornamental and medicinal purposes. Kuo *et al.* (2004) demonstrated that *G. biloba* extract competitively inhibited rat hepatic microsomal CYP1A-mediated enzyme activity, as determined by experiments performed on extracts containing terpene trilactones and ginkgo flavonol glycosides. Despite the fact that the plant appears to possess antioxidant activity, the mechanism of action of *G. biloba* is not known. Plant cells and tissue cultures have been used as alternative sources to the whole plant for the production of valuable phytochemicals such as flavors, fragrances, pharmaceuticals and nutraceuticals. Numerous strategies have been developed to improve the productivity of plant cell culture such as medium optimization, cell line selection, cell immobilization, precursor addition, elicitation, genetic transformation, organ or hairy root cultures, metabolic and integrated bioreactor engineering (Abdullah *et al.*, 2005; El-Beltagi *et al.*, 2011).

Nitric oxide as a bioactive molecule shows prooxidant as well as antioxidant properties in plants (Beligni and Lamattina, 1999; Delledonne *et al.*, 2001; Beligni and Lamattina, 2002; Delledonne *et al.*, 2002). The biosynthetic origin of NO during plant pathogen interactions involves L-arginine conversion into L citrulline. Nitric oxide induces a complementary set of plant defense genes, including two key enzymes of the phenylpropanoid pathway, so-called Phenyl alanine ammonia lyase (PAL) and chalcone synthase. Furthermore, NO-treated tobacco (*Nicotiana tabacum*) cells were shown to induce the accumulation of cyclic acid, a key molecule for the expression of systemic acquired resistance (Durner *et al.*, 1998; Crawford and Guo, 2005). The free radical NO has a half-life of just a few seconds and can also react with the free radical superoxide to form the reactive molecule peroxynitrite (ONOO), which can lead to the formation of NO_2 and the potent oxidant hydroxyl radical, which is a very strong oxidizing specie that can rapidly attack biological membranes and all types of bio-molecules such as DNA and proteins leading to irreparable damage, metabolic dysfunction and cell death (Del-Rio *et al.*, 2003). ONOO is responsible for tyrosine nitration and oxidation of thiol (Stamler *et al.*, 1992; Tamir *et al.*, 1993; Lamattina *et al.*, 2003). Sodium nitroprusside (SNP) was used as the donor of nitric oxide (NO). SNP at high concentrations stimulated catharanthine formation of *Catharantuse roseus* cells, but inhibited the growth of the cells (Xu *et al.*, 2009). It stimulated production of hypericin by *Hypericum perforatum*. The cells were significantly increased after elicitor treatment by NO, achieving 4.2-fold higher over the control (Xu *et al.*, 2005).

Antioxidant defense enzymes such as superoxide distumase (SOD), catalase (CAT), ascorbate peroxidase (APX), glutathione reductase (GR) and monodehydroascorbate reductase (MDAR) are the systems designed to minimize the concentrations of superoxide and hydrogen peroxide. A demonstrated strong inhibitory effect of NO on the catalase (CAT) peroxidase (POD) activity of cytc/CL complex combined with its increased amount in mitochondria suggest that NO may regulate lipid peroxidation and apoptotic activity of CL membrane bound cytc during apoptosis (Vlasova *et al.*, 2006).

Recently, a function of nitric oxide (NO) in the protection of plants against oxidative stress under various adverse conditions was reported (Beligni and Lamattina, 2002; Shi *et al.*, 2005). Many previous studies had reported the presence of NO in the plant kingdom and its involvement in growth, development and defense responses (Beligni and Lamattina 1999). In addition, Tu *et al.* (2003) found that 0.1 mM SNP delayed the senescence of wheat leaves by inhibition of the degradation of chlorophyll and soluble proteins, especially Rubisco enzyme. Therefore, this study was conducted to investigate the molecular role of nitric oxide in secondary products production in *G. biloba* cell suspension culture.

Materials and methods

Plant materials and surface sterilization

All excised tissues of young stems of *G. biloba* were collected from Orman botanical garden, Giza governorate and stored temporarily in an antioxidant solution (150 mg/l citric acid and 100 mg/l ascorbic acid), prior to surface-sterilization in 1% chlorox for 20 min, 2 drops/ 100 ml solution of Tween 20 as wetting agent then rinsed three times with sterile distilled water.

Callus initiation and maintenance

Callus was initiated from the sterilized young stems of *G. biloba*. The callus was initiated in a MS-Medium (Murashige and Skoog, 1962) supplemented casein hydrolisate (500 mg/L) and different combinations of Naphthaleneacetic acid (NAA) and kinetin (kin) each with concentrations of 1, 2, 3 and 4 mg/L and sucrose (30 g/L) solidified with agar (6 g/L). Three jars (replicates) were used for each treatment. Cultures were incubated at 25±1 °C, and re-cultured 3 times in the same medium with four week intervals. Data of explants callus formation responses were recorded.

Growth dynamic

After several subcultures of callus (5 g F.W.) they were transferred into 250-ml flasks containing 50 ml of the same medium without agar and maintained on a rotary shaker at 100 rpm. Microscopic observations were carried out on cell suspension cultures during the growth cycle, which revealed the presence of single cells and small cell aggregates. The number of cell per unit can be determined using a hemocytometer. Cell growth was measured every two days for a total period of 12 days; starting, 2, 4, 6, 8, 10 and 12 days.

Cells viability

Cells viability was determined using Evan's blue staining test. A 2 mL sample from tank was incubated into 0.25% Evan's blue stain for 5 minutes and then 700 cells were counted. The viability was calculated taking into account the non-viable cells (Evans *et al.*, 2003).

Experimental design and cell culture treatment

G. biloba cell suspension cultures were treated with various concentrations of sodium nitroprusside (SNP, 50, 100, 250 and 500 μM) after sub-culturing and followed by cells harvest at time intervals. All parameters and enzymes activity were monitored after 0, 12, 48 and 72 h post NO donors treatments.

Determination of phytochemicals in *Ginkgo biloba*

Total tannins, saponins, flavonoids, and phenolic compounds were determined according to the methods described by: Balbaa, 1974; Ebrahimzadeh and Niknam, 1998; Zhuang *et al.*, 1992; Swain and Hillis, 1959.

The production rate of superoxide anion ($O_2^{\cdot-}$)

The production rate of superoxide anion ($O_2^{\cdot-}$) was measured by the modified method described by Elstner and Heupel (1976). Fresh mass (200 mg) from culture was homogenized in 1 ml of 50 mM phosphate buffer (pH 7.8), and the homogenate was centrifuged at 10,000 g for 10 min. Then 0.5 ml of the supernatant was added to 0.5 ml 50 mM phosphate buffer (pH 7.8) and 0.1 ml of 10 mM hydroxylamine hydrochloride. After 1 h reaction at 25 1C, 1 ml of 17 mM sulfanilamide and 1 ml 7 mM a-naphthylamine were added to the mixture at 25 1C, and after 20 min, the specific absorbance at 530 nm was determined. Sodium nitrite was used as standard solution to calculate the production rate of $O_2^{\cdot-}$.

Assay of total soluble protein

Soluble proteins were measured by the Bio-Rad micro assay modification of the Bradford (1976) procedure using crystalline bovine serum albumin as a reference.

Preparation of enzyme extracts and assay of protein content

The callus samples (1.0 g) were crushed into fine powder using liquid nitrogen. Soluble protein was extracted by homogenizing the powder in 5 mL 50 mM phosphate buffer (pH 7.8) containing ethylenediamine-N,N,N,N-tetraacetic acid (EDTA, 1 mM) and polyvinylpyrrilidone (PVP, 1%) in addition to ASC

(1mM) in the case of APX assay. The samples homogenate was centrifuged at 15,000 x g for 20 minutes and the supernatant was used for further enzyme activity assays.

Assay of superoxide dismutase (SOD) and ascorbate peroxidase (APX) activity

The activity of SOD was assayed by measuring its ability to inhibit the photochemical reduction of NBT using the method of Beauchamp and Fridovich (1971). APX activity was determined (13.7 mM^{-1} cm^{-1}) using the method of Nakano and Asada (1981).

Assay of phenylalanine ammonia lyase (PAL)

PAL activity was determined as the rate of conversion of L phenylalanine to trans-cinnamic acid at 290 nm as described by Dickerson *et al.* (1984).

Callus formation responses

It was determined according to the rate of scaling by Pottino (1981), which included: - No response, + Poor response, ++ Average response, +++ High response and ++++ very high response.

Callus regeneration and development

Callus was cultured on MS basic medium supplemented with casein hydrolisate (500 mg/L) and auxin (NAA) Naphthalene acetic acid in combination with cytokinins (BA) 6- Benzyl amino purine was used (mg/L); with different concentrations, as follows: 0.2, 0.5, 1.0 and 2.0 (mg/L) each. Incubation was carried out in a growth room at 25±1 °C with 16 h photoperiod (1500 lux). After 6 weeks, data including the number of regenerated shoots, shoot height (cm) as well as no. of roots and root length (cm) were recorded.

Acclimatization

In order to realize a proper acclimatization, plantlets were removed from rooting medium, rinsed under tap water and then, planted in plastic pots (5 cm) filled with a soil mixture as follows: sand, peat moss and vermiculite (1:1:1, v/v/v). Plants were covered with transparent polyethylene sheet and irrigated if needed. Potted plants were incubated for 30 days in phytotron at 25 ± 1 °C, relative humidity (80-90 %) and 16 h photoperiod with 3000 lux light intensity. Acclimatization was realized through gradual remove of the plastic sheet, every day, until it became removed after 30 days. Survival percent, the number of leaves/plant and leaf length (cm) were recorded.

Statistical analysis

All statistical analyses were carried out using SPSS 10 software. Mean and standard error were descriptive measures of quantitative data using the analysis of variance test (ANOVA) for independent samples. All determinations have been done in triplicate, *p*-Values < 0.05 were considered significant.

Results and discussions

Establishment of Ginkgo callus culture

Callus formation occurred on young leaves excised from *G. biloba* tree. On one hand, explants were individually cultured on MS medium supplemented with different combinations of NAA (1, 2, 3 and 4 mg/l) and kinetin with concentrations of 1, 2, 3 and 4 mg/l. Data in Tab. 1 represents the effect of plant growth regulators (NAA and kin) combinations on callus formation responses. The highest response has been recorded with NAA and Kin with equal concentrations (3 mg/L) each followed by (2 mg/L) and (1 mg/L). On the other hand the application of NAA (1 mg/L) and kinetin (3 mg/L) did not represent any callus formation response of *Ginkgo* explants *in vitro*. Suspension culture of the cell line was initiated from *G. biloba* callus on a liquid MS medium supplemented with NAA (2mg/L), kinetin (2 mg/L) and sucrose 30 g/L. These data were in accordance with Wang *et al.* (2006), who found that hypocotyls and cotyledons of *Arctium lappa* were induced to form callus by culturing on MS medium supplemented with 2, 4-D (2.0 mg/L) and BA (0.5–2.0 mg /L).

Data in Tab. 2 showed that, the growth rate and biomass were gradually enhanced and reached their optimum on the 8th day of cultivation. These data are in agreement with Bulgakov *et al.* (2002), Chong *et al.* (2005) and Tanveer *et al.* (2012); they found a significant decrease in the culture growth.

Chemical analysis of different types of G. biloba explants

While data in Tab. 3 show that the percentage of tannins and saponins compounds were 2.02%, 0.59% and 1.15%, 0.79% based on the dry weight, respectively for the *G. biloba* leaves and cell suspension cultures. While the percentage of flavonoids and phenolic compounds were 1.75%, 1.09% and 2.95%, 1.68%. The highest concentration of tannins, saponins, flavonoids and phenolic compounds were found in the *G. biloba* leaves compared to its derived cell suspension culture.

As shown in Tab. 3, there were a high accumulation of total soluble phenols under various NO treatments. The highest

Tab. 1. Effect of plant growth regulators (NAA and kin) in combination with different concentrations on callus formation and growth of *Ginkgo biloba* explants grown *in vitro*

Treatments (mg/L)	Callus formation responses			
NAA \ Kin	1	2	3	4
1	+	-	-	-
2	+	+++	++	++
3	-	++	++++	++
4	-	-	+	+

Rating Scale: - No response +Poor response ++Average response +++High response ++++very high response.

Tab. 2. Growth dynamic of suspension culture of *Ginkgo biloba* cultured for 12 days *in vitro*

Time (day)	Cells No./l
Starting	x10^5 564
2 days	x10^5 2288
4 days	x10^5 3521
6 days	x10^5 3800
8 days	x10^5 4287
10 days	x10^5 3987
12 days	x10^5 3870

Cells/ml = average count per square *dilution factor* 1000
Total cells = cells/ml * original volume of fluid from which the cells sample was taken.

Tab. 3. Effect of various NO treatments on total tannins, saponins, flavonoids and Phenolic compounds contents (% Dry weight) in *Ginkgo biloba* leaves and cell suspension culture

Treatments	Total % of			
	Tannins	Saponins	Flavonoids	Phenolic compounds
Ginkgo biloba Leaves	2.02±0.14^{c}	1.15±0.05^{c}	1.75±0.08^{b}	2.95±0.12^{c}
Ginkgo biloba Cell culture	0.59±0.03^{f}	0.79±0.02^{f}	1.09±0.05^{e}	1.68±0.04^{e}
SNP (50 μM)	0.95±0.08^{e}	0.90±0.04^{e}	1.26±0.11^{d}	2.45±0.06^{d}
SNP (100 μM)	1.87±0.10^{d}	1.01±0.06^{d}	1.43±0.07^{c}	2.92±0.08^{c}
SNP (250 μM)	2.49±0.13^{b}	1.45±0.12^{b}	1.79±0.10^{b}	3.36±0.14^{b}
SNP (500 μM)	2.61±0.14^{a}	1.74±0.13^{a}	1.93±0.03^{a}	3.59±0.11^{a}

- Means within the same column followed by different letters are significantly different at $P < 0.05$. Values are means of three replicates (±SE).

Tab. 4. Effect of administration of different SNP doses (0, 50, 100 and 250 μM) on superoxide anion ($O_2^{\cdot-}$) contents (nmol/h g FW) in *Ginkgo biloba* elicited suspension culture

SNP Treatments (μM)	Superoxide anion determination			
	0 h	12 h	48 h	72 h
Control (0)	0.36±0.02^{c}	1.03±0.04^{c}	1.25±0.05^{b}	1.75±0.05^{e}
50	0.47±0.02^{c}	1.16±0.03^{b}	1.36±0.04^{ab}	1.99±0.05^{d}
100	0.76±0.02^{b}	1.21±0.03^{b}	1.57±0.02^{ab}	2.31±0.06^{c}
250	0.96±0.03^{ab}	1.25±0.02^{b}	1.80±0.03^{ab}	2.46±0.04^{b}
500	1.10±0.10^{a}	1.44±0.01^{a}	2.31±0.03^{a}	2.57±0.05^{a}

- Means within the same column followed by different letters are significantly different at $P < 0.05$. Values are means of three replicates (±SE).

accumulation has been observed with SNP (500 μM). All NO treatments significantly increased the total phenolic compounds, tannins, flavonoids and total saponins contents. These data were in accordance with Goodman *et al.* (1967), who found a multifold increase of phenols in inoculated plant cells after challenging with SNP. In the present study the increase of phenols it can be due to the excess production of H_2O_2 in elicited plant cells through increased respiration (Farkas and Kiraly, 1962) or due to the activation of hexose-monophosphate pathway, acetate pathway and release of bound phenols by hydrolytic enzymes. Moreover, callus cultures can be employed and exploited for secondary metabolites production (Jedinák *et al.*, 2004; Radfar *et al.*, 2012). In tissue culture procedures, this used frequently for production of flavonoids. The results also demonstrate that NO signaling is involved in ROS induced secondary products production of *G. biloba* cell suspension culture. In addition to NO, many other signal molecules such as hydrogen peroxide (H_2O_2), jasmonicacid (JA), and salicylic acid (SA) have also been reported to play important roles in secondary metabolite production (Hahlbrock *et al.*, 2003). Whether these signal molecules are involved in NO-induced secondary metabolites production and their relationship and/or interaction with ROS in mediating the NO-stimulated secondary metabolite synthesis still needs further investigations. The role of nitric oxide in the biosynthesis of secondary metabolites in plant cells is not well established yet. Nitric oxide released from its donor sodium nitroprusside (SNP) enhanced the production of terpenoid indole alkaloids of *Catharanthus roseus* cells. Furthermore, one or more protein kinases were demonstrated to act downstream of nitric oxide and transduce the signal to indole alkaloid biosynthesis (Delledonne *et al.*, 2002).

Data in Tab. 4 show that, all NO treatments increased the $O_2^{\cdot-}$ content in a positive correlation with time duration. NO treatments increased the accumulation of defense related to enzymes such as PAL, APX, catalase and then increased the accumulation of phenols, secondary metabolites and reactive oxygen species (ROS). Data are in agreement with Lu *et al.* (2011), who found that Nitric oxide (NO) has recently emerged as an important signaling molecule in plants. It has been well documented that NO is widely involved in plant adaptive responses to biotic and abiotic stresses. Accumulating evidences indicate that this free radical, which was first identified as a second messenger in animals, plays an important role in stress-induced secondary metabolite accumulation in plants.

Data in Tab. 5 reveal that, the increase in phenylalanine ammonia lyase (PAL) activity has been triggered under all NO treatments. Furthermore, there was a positive correlation between administrated NO and affected PAL activity. In accordance to Goodman *et al.* (1967); Kobeasy *et al.* (2011), the administration of NO increase the PAL activity which might enhance the phenol content in challenge plant cells. In addition, NO treatments also increased the PAL activity which enhanced the phenolic compound and flavonoid accumulation (Durner and Klessig, 1999). These data indicate that many studies have implicated nitric oxide (NO) as a key regulator for many different physiological processes in plants.

Evaluation of elicitation induces antioxidant defense enzymes activity

Data in Tab. 5 show that, SOD and APX activities increased after elicitation of different SNP doses. Similar results were obtained by Qiao and Fan (2008) which confirms that, NO can protect oxidative stresses by decreasing the carbonyl group and H_2O_2 contents. The decrease of H_2O_2 concentration was probably produced by the direct NO and H_2O_2 interactions. Also, it is highly possible that the protective effect of NO may result from the increased expression of genes encoding active oxygen scavenging enzymes. Pretreatment of plants with NO also enhanced the antioxidant enzyme activities in concentrations (Mackernessa *et al.* 2001) compared with its corresponded control. In addition, the NO elicitation processes trigger a specific activation of peroxidase activity that interferes with the process of hypersensitive reactions against infection example: lignifications and antibacterial toxic compounds formation during and after the elicitation process. The application with 500 μM of SNP reduced the SOD and APX activities compared with its corresponded control. These data were in agreement with Zhu *et al.* (2009) and Cui *et al.* (2010) confirming that, NO treatment has led to inhibition in POD and SOD activity after fresh-cut of peach and copper toxicity of tomato plants. It has also been reported that NO inhibited the activity of haem-containing enzymes, such as catalase and peroxidases, the important H_2O_2-scavenging enzymes during pathogen attack (Clark *et al.*, 2000).

Regeneration and development

Data in Tab. 6 demonstrate that MS medium supplemented with NAA and BA with equal concentration (0.5 mg/L) recorded the highest number of regenerated shoots

Tab. 5. Effect of different SNP administrated doses (0, 50, 100, 250 and 500µM) on the activities (Unit /mg protein^{-1}/h) of phenylalanine ammonia lyase (PAL), ascorbate peroxidase (APX) and superoxide dismutase (SOD) in *Ginkgo biloba* suspension culture

SNP Treatments (µM)	PAL				APX				SOD			
	0	12	48	72	0	12	48	72	0	12	48	72
Control (0)	101.9 ± 5.2^{e}	163.9 ± 4.7^{e}	218.7 ± 6.1^{b}	235.8 ± 6.7^{d}	2.65 ± 0.3^{e}	4.89 ± 0.3^{c}	6.83 ± 0.3^{d}	7.02 ± 0.2^{d}	143.8 ± 4.5^{c}	170.3 ± 4.8^{d}	200.2 ± 5.4^{c}	208.6 ± 6.94^{c}
50	116.6 ± 5.5^{d}	192.2 ± 6.0^{d}	229.5 ± 7.5^{b}	262.6 ± 6.6^{c}	3.72 ± 0.2^{d}	7.03 ± 0.3^{b}	7.43 ± 0.4^{c}	8.43 ± 0.3^{c}	154.1 ± 2.7^{b}	183.7 ± 5.5^{cd}	217.3 ± 6.7^{b}	231.6 ± 5.4^{b}
100	125.5 ± 5.8^{c}	228.3 ± 7.1^{c}	253.0 ± 6.9^{a}	271.6 ± 7.1^{b}	6.09 ± 0.4^{b}	7.04 ± 0.2^{b}	8.43 ± 0.2^{a}	8.88 ± 0.5^{b}	162.1 ± 3.1^{d}	190.5 ± 3.7^{bc}	231.3 ± 5.8^{ab}	245.8 ± 3.6^{ab}
250	141.0 ± 6.3^{b}	241.1 ± 5.7^{b}	258.2 ± 5.8^{a}	285.3 ± 7.3^{a}	6.78 ± 0.3^{a}	7.51 ± 0.5^{a}	8.32 ± 0.4^{a}	9.43 ± 0.4^{a}	166.0 ± 5.8^{a}	211.6 ± 4.8^{ab}	237.7 ± 6.2^{a}	260.2 ± 5.9^{a}
500	163.0 ± 5.8^{a}	253.6 ± 4.3^{a}	267.8 ± 4.4^{a}	290.7 ± 4.9^{a}	5.87 ± 0.2^{c}	7.00 ± 0.4^{b}	8.00 ± 0.5^{b}	8.84 ± 0.3^{b}	150.3 ± 0.9^{a}	201.9 ± 1.0^{a}	221.8 ± 0.9^{ab}	256.0 ± 1.5^{a}

- Means within the same column followed by different letters are significantly different at $P < 0.05$. Values are means of three replicates (±SE).

Tab. 6. Effect of plant growth regulators (NAA and BA) in combination with different concentrations on regeneration response of *Gingko* callus grown for 6 weeks *in vitro*

Treatments (mg/L)		No. of regenerated shoots	Shoot height (cm)	No. of roots	Root Length (cm)
NAA	BA				
0.2	0.2	3.73^{a}	4.73^{a}	1.30^{a}	3.20^{a}
0.5	0.5	4.81^{a}	4.96^{a}	2.25^{a}	4.50^{a}
1.0	1.0	2.21^{b}	3.11^{b}	0.00^{b}	0.00^{b}
2.0	2.0	2.18^{b}	3.12^{b}	0.00^{b}	0.00^{b}

- Means within each column followed by the same letter are not significantly different at $P < 0.05$.

Tab. 7. Effect of NAA and IBA on in vitro rooting and their subsequent relation to plantlets survival %, after transplanted ex vitro for 3 months for acclimatization

Treatments (mg/L)		Soil mixture type treatment	Survival % in acclimatization
NAA	BA		
0.2	0.2		33.33^{b}
0.5	0.5	Sand + Peat moss +	40.00^{a}
1.0	1.0	Vermiculite (1: 1: 1, v/v/v)	0.00^{c}
2.0	2.0		0.00^{c}

- Means within each column followed by the same letter are not significantly different at $P < 0.05$.

(4.81 cm) and shoot height (4.96 cm) as well as the number of roots (2.25 cm) and root length (4.5 cm) as compared to the other studied concentration treatments tested. In this concern, *Ginkgo* embryos culture initiated on MS medium with 2, 4-D plus NAA for 5 weeks produced shoots and roots when were transferred to media with 4.5 µM 2, 4-D alone for an additional 5 weeks. Plants were transferred from the 2, 4-D media to pots and maintained in the greenhouse (Camper *et al.* 1997).

High growth regulators (NAA and BA) with equal concentrations and combination (1.0 and 2.0 mg/L) gradually decreased the number of regenerated shoots and shoot height as well as the completely inhibited root formation.

Acclimatization

Data presented in Tab. 7 show that soil mixture containing sand, peat moss and vermiculite (1: 1: 1, v/v/v) significantly confirmed and reflected survival rate variation which is mainly ascribed to the strong relationship between plantlets and their received *in vitro* auxin concentrations during growth and development on rooting phase before *ex vitro* transfer. In this concern, Al-Salih *et al.* (1986) suggested that, success or failure of transferred plantlets to greenhouse is dependent primarily on the quality and type of materials produced in the previous stages of *in vitro* propagation.

Conclusion

This study shows that the NO treatments in *G. biloba* cell suspension cultures induced significant increase in phenolic compounds and glycosides accompanied with a significant increase in oxidative burst of $O_2^{\cdot-}$ which enhance specific enzymes activities of PAL and antioxidant defense enzymes SOD and APX. In addition, the optimal media for induced *G. biloba* embryo callus and subculture of callus were MS basaic medium supplemented with casein hydrolase (500 mg/L), NAA and BA with equal concentrations (0.5 mg/L).

Acknowledgments

The authors extend their appreciation to the Deanship of Scientific Research at King Saud University for funding the work through the research group project no RGP-VPP-130.

References

Abdullah R, Alizah Z, Wee YH, Leaw CL, Yeap CB, Yap WSP, Yeun LH, Rashdan MM, Joseph JL, Sitiazma J, Lee MP, Salwa AS (2005). Immature Embryo: A useful tool for oil palm genetic transformation studies. Electr J Biotechnol 8(1):24-34.

Al-Salih AA, Bader SM, Jarrah AZ, Al-Qadi MT (1986). A comparative morphological and anatomical study of seed and embryo culture-derived seedling of *Phoenix dactylifera* L. Date Palm J 4:153-161.

Balbaa SI (1974). Chemistry of crude drugs - laboratory manual. Al – Shaab Printing House. Cairo, Egypt, 194 p.

Beauchamp C, Fridovich I (1971). Superoxide dismutase: improved assays and assay applicable to acrylamide gels. Anal Biochem 44:276-287.

Beligni MV, Lamattina L (1999). Nitric oxide counteracts cytotoxic processes mediated by reactive oxygen species in plant tissues. Planta 208:337-344.

Beligni MV, Lamattina L (2002). Nitric oxide interferes with plant photo-oxidative stress by detoxifying reactive oxygen species. Plant Cell Environ 25:737-748.

Bradford MM (1976). A rapid and sensitive method for the quantification of microgram quantities of protein utilizing the principle of protein-dye binding. Anal Biochem 72:248-254.

Bulgakov VP, Tchernoded GK, Mischenko NP, Khodakovskaya MV, Glazunov VP, Radchenko SV, Zvereva EV, Fedoreyev

SA, Zhuravlev YN (2002). Effect of salicylic acid, methyl jasmonate, ethephon and cantharidin on anthraquinone production by *Rubia cordifolia* callus cultures transformed with the rolB and rolC genes. J Biotechnol 97:213-221.

Camper ND, Coker PS, Wedge DE, Keese RJ (1997). In vitro culture of *Ginkgo*. *In Vitro* Cellular and Developmental Biology Plant 33:125-127.

Chong TM, Abdullah MA, Fadzillah NM, Lai OM, Lajis NH. (2005). Jasmonic acid elicitation of anthraquinones with some associated enzymic and non-enzymic antioxidant responses in *Morinda elliptica*. Enzyme Microb Technol 36:469-477.

Clark D, Durner J, Navarre DA, Klessig DF (2000). Nitric oxide inhibition of tobacco catalase and ascorbate peroxidase. Mol Plant Microbe Interact 13(12):1380-1384.

Crawford NM, Guo FQ (2005). New insights into nitric oxide metabolism and regulatory functions. Trends Plant Sci 4:195-200.

Cui XM, Zhang, YK, Wu XB, Liu CS (2010). The investigation of the alleviated effect of copper toxicity by exogenous nitric oxide in tomato plants. Plant Soil Environ 56(6):274-281.

Delledonne M, Zeier J, Marocco A, Lamb C (2001). Signal interactions between nitric oxide and reactive oxygen intermediates in the plant hypersensitive disease resistance response. P Natl Acad Sci USA 23:13454-13459.

Delledonne M, Murgia I, Ederle D, Sbicego PF Biondani A, Polverari A, Lamb C (2002). Reactive oxygen intermediates modulate nitric oxide signaling in the plant hypersensitive disease-resistance response. Plant Physiol Biochem 40:605-610.

Del-Rio, LA, Corpas FJ, Sandalio LM, Palma JM, Barroso JB (2003). Plant peroxisomes, reactive oxygen metabolism and nitric oxide. IUBMB Life 2:71-81.

Dickerson DP, Pascholati SF, Hagerman AE, Butler LG, Nicholson RL (1984). Phenylalanine ammonia-lyase and hydroxyl cinnamate: CoA ligase in maize mesocotyls inoculated with *Helminthosporium maydis* or *Helminthosporium carbonum*. Physiol. Plant Pathol 25:111-123.

Durner J, Klessig DF (1999). Nitric oxide as a signal in plants. Current Opinion in Plant Biology 2(5):369-374.

Durner J, Wendehenne D, Klessig DF (1998). Defence gene induction in tobacco by nitric oxide, cyclic GMP, and cyclic ADP-ribose. P Natl Acad Sci USA 95:10328-10333.

Ebrahimzadeh H, Niknam V (1998). A revised spectrophotometric method for determination of triterpenoid saponinis. Indian Drugs 35:379-381.

El-Beltagi HS, Ahmed OK, El-Desouky W (2011). Effect of low doses γ-irradiation on oxidative stress and secondary metabolites production of rosemary *(Rosmarinus officinalis* L.) callus culture. Radiat Phys Chem 80:968-976.

Evans DE, Coleman JOD, Kearns A (2003). Plant Cell Culture. BIOS Scientific Publ. Taylor and Francis, London.

Elstner EF, Heupel A (1976). Inhibition of nitrite formation from hydroxylaminonium- chloride: a simple assay for superoxide dismutase. Anal Biochem 70:616-620.

Farkas GL, Kiraly Z (1962). Role of phenolic compounds in the physiology of plant disease resistance. Phytopathol Z 44:105-150.

Goodman RN, Kiraly Z, Zaitlin M (1967). The Biochemistry and Physiology of Infectious Plant Disease. Van Nostrand Co Inc. Princeton, New Jersey 354 pp.

Hahlbrock K, Bednarek P, Ciolkowski I, Hamberger B, Heise A, Liedgens H, Logemann E, Nürnberger T, Schmelzer E, Somssich IE, Tan J (2003). Non-self-recognition, transcriptional reprogramming, and secondary metabolite accumulation during plant/pathogen interactions. P Natl Acad Sci USA 100:14569-14576.

Jedinák A, Faragó J, Pšenáková I, Maliar T (2004). Approaches to flavonoid production in plant tissue cultures. Biologia Bratislava 59:697-710.

Kobeasy MI, El-Beltagi HS, El-Shazly MA, Khattab EAH (2011). Induction of resistance in *Arachis hypogaea* L. against *Peanut Mottle Virus* by nitric oxide and salicylic acid. Physiol Mol Plant Pathol 76:112-118.

Kuo IF, Jie C, Thomas KH (2004). Chang Effect of *Ginkgo biloba* extract on rat hepatic microsomal CYP1A activity: role of ginkgolides, bilobalide, and flavonols. Can J Physiol Pharmacol 82:57-64.

Lamattina L, Garcia-Mata C, Graziano M, Pagnussat G (2003). Nitric oxide: the versatility of an extensive signal molecule. Ann Rev Plant Physiol 54:109-136.

Lu D, Dong J, Jin H, Sun L, Xu X, Zhou T, Zhu Y, Xu M (2011). Nitrate reductase-mediated nitric oxide generation is essential for fungal elicitor induced camptothec in accumulation of *Camptotheca acuminata* suspension cell cultures. Appl Microbiol Biotechnol 90:1073-1081.

Mackernessa SAH, John CF, Jordan B, Tomas B (2001). Early signaling components in ultraviolet-B responses: distinct roles for different reactive oxygen species and nitric oxide. FEBS Lett 489:237-242.

Murashige T, Skoog F (1962). A revised medium for rapid growth and bioassays with tobacco tissue culture. Physiol Plant 15:473-497.

Nakano Y, Asada K (1981). Hydrogen peroxide is scavenged by ascorbate specific peroxidase in spinach chloroplasts. Plant Cell Physiol 22:867-880.

Pottino BG (1981). Methods in Plant Tissue Culture. Dept Hort Agric. Maryland Univ, College Park, Maryland, USA p. 8-29.

Qiao W, Fan L (2008). Nitric oxide signaling in plant responses to abiotic stresses. J Integr Plant Biol 50(10):1238-1246.

Radfar M, Sudarshana MS, Niranjan MH (2012). Betalains from stem callus cultures of *Zaleya decandra* L. N. Burm. f. - A medicinal herb. Med Plants Res 6(12):2443-2447.

Shi SY, Wang G, Wang YD, Zhang LG, Zhang LX (2005). Protective effect of nitric oxide against oxidative stress under ultraviolet-B radiation. Nitric Oxide 13:1-9.

Stamler JS, Singel DJ, Loscalzo J (1992). Biochemistry of nitric oxide and its redox-activated forms. Science 258:1898-1902.

Swain T, Hillis WE (1959). The phenolic constituents of *Prunus domestica*. I. The quantitative analysis of phenolic constituents. J Sci Food Agric 10(1):63-68.

Tamir S, Lewis RS, De-Rojas WT, Deen WM, Wishnok JS, Tannenbaum SR (1993). The influence of delivery rate on the chemistry and biological effects of nitric oxide. Chem Res Toxicol 6:895-899.

Tanveer H, Ali S, Asi MR (2012). Appraisal of an important flavonoid, quercetin in callus cultures of *Citrullus colocynthis*. Int J Agric Biol 14:528-532.

Tu J, Shen WB, Xu LL (2003). Regulation of nitric acid on the aging process of wheat leaves. Acta Bot Sin 45:1055-1062.

Vlasova II, Tyurin VA, Kapralov AA, Kurnikov IV, Osipov AN, Potapovich MV, Stoyanovsky DA, Kagan VE (2006). Nitric Oxide Inhibits Peroxidase Activity of Cytochrome c - Cardiolipin Complex and Blocks Cardiolipin Oxidation. Journal of Biological Chemistry 281(21):14554-14562.

Wang CY, He WT, Hou SW (2006). Callus induction and high-frequency plant regeneration from hypocotyl and cotyledon explants of *Arctium lappa*. In vitro Cellular and Developmental Biology - Plant 39(5):468-474.

Xu MJ, Dong JF, Zhu MY (2005). Nitric oxide mediates the fungal elicitor-induced hypericin production of *Hypericum perforatum* cell suspension cultures through a jasmonic-acid-dependent signal pathway. Plant Physiology 139(2):991-998.

Xu MJ, Dong J, Wang H, Huang L (2009). Complementary action of jasmonic acid on salicylic acid in mediating fungal elicitor-induced flavonol glycoside accumulation of *Ginkgo biloba* cells. Plant Cell Environ 32:960-967.

Zhu LQ, Zhou J, Zhu SH, Guo LH (2009). Inhibition of browning on the surface of peach slices by short-term exposure to nitric oxide and ascorbic acid. Food Chem 114:174-179.

Zhuang XP, Lu YY, Yang GS (1992). Extraction and determination of flavonoids in ginkgo. Chin Herbal Med 23:122-124.

Revegetation of Abandoned Sulphur Flotation Waste in Keçiborlu, Isparta/Turkey: Heavy Metals Concentrations of Growing Media and *Agropyron elongatum* Grass

Huseyin KALKAN, Mustafa KAPLAN, Şule ORMAN*

*Akdeniz University, Faculty of Agriculture, Department of Soil Science and Plant Nutrition, Antalya, Turkey; hkalkan@akdeniz.edu.tr, mkaplan@akdeniz.edu.tr, suleorman@akdeniz.edu.tr (*corresponding author)*

Abstract

The main objectives of the study are to evaluate the effectiveness of using amendment materials (calcareous soil and farmyard manure) and to grow *Agropyron elongatum* (tall wheat grass) for the revegetation of sulphur flotation waste in tailing ponds. Calcareous soil (CS) at 10, 20, 30, 40% and farmyard manure (FYM) at 4, 8% were applied to the flotation waste (FW). pH, electrical conductivity (EC), DTPA-extractable Fe, Zn, Mn, Cu, Ni, Cr, Pd, Cd, Co concentrations of growing media, also, total Fe, Zn, Mn, Cu, Ni, Cr, Pd, Cd, Co concentrations of plants were analysed together with the dry weight yield of the plant which were determined at end of the experiment. The results have shown that pH, shoot Zn, Mn, Cu and Pb concentrations increased by CS-alone but this increment tended to decrease by FYM. The EC, DTPA-extractable Fe, Zn, Mn, Ni, Cr, Co and shoot Fe, Ni, Cr, Co concentrations were decreased by CS-alone, but this decline tended to increase by FYM. There was no change in DTPA-extractable Pb and Cd concentrations between CS and CS with FYM. The dry weight of the plant shoots increased CS with FYM more than CS-alone. The plant grew well and had high cover in CS with FYM treatments compared to CS-alone. However this work indicates that *A. elongatum* can be used to revegetate sulphur flotation waste, further studies with different plant species are needed in order to obtain better plant cover in revegetation efforts.

Keywords: calcareous soil, farmyard manure, grass, inorganic waste, phytoremediation, reclamation, tailings

Introduction

Abandoned tailings which are the minerals processing residues around former mine plants constitute a major source of environmental contamination throughout the world. Tailings are often almost completely devoid of vegetation due to toxic levels of heavy metals and unfavourable edaphic conditions. The bare surfaces of tailings are susceptible to wind and water erosion and act as continuous source of metal contaminators to the surroundings (Shu *et al*., 2005). The release of metals from mine sites takes place mainly through acid mine drainage and erosion of waste dumps and tailings deposits (Salomons, 1995). These pollutant effects can reach local and, in some cases, regional scales (Rybicka, 1996). In a study, investigating environmental effects of sulphur flotation waste ponds in Keçiborlu, Isparta/Turkey, an increasing concentration gradient for total and DTPA-extractable heavy metal in surface soil of both agricultural lands and centre of Keçiborlu district was detected toward the waste ponds (Yazar, 2010). Heavy metals in the soil of old mining areas, besides affecting the productivity of their ecosystems, could also affect animal and human health (Gutierrez-Gines *et al*., 2010).

Many abandoned mines and their tailing ponds are barren or have minimal vegetation to slow natural colonization. Common physical and chemical limitations to plant growth on mine tailings include: low pH, high salt content, lack of organic matter and plant nutrients, metal toxicities, high bulk density, lack of soil structure, slow water infiltration, low water retention, and low air permeability (Henriquez and Fernandez, 1991; Wong *et al*., 1998; Wong, 2003). Inorganic and organic amendments are useful for stabilization of mine soil prior to revegetation (Alvarenga *et al*., 2008). Revegetation can be used for in situ reclamation of mine wastes, and this technique is also considered as the most suitable to achieve long term reclamation (Tordoff *et al*., 2000). Vegetation can provide effective protection against wind carried polluted particles, reducing runoff and the overland flow of water and sediments. Vegetation may also improve nutrient conditions in the soil and form the basis for the establishment of a self-sustaining vegetative cover (Norland and Veith, 1995). Revegetation would require the addition of lime and/or organic matter to facilitate plant growth and increase the diversity of plant species (Conesa *et al*., 2007).

Organic residues (e.g. manure, compost, sewage sludge) are increasingly being considered in land rehabilitation (Walker *et al.*, 2004; Brown *et al.*, 2005; Pérez-de-Mora *et al.*, 2006). The addition of organic matter to mine soils can significantly improve the physical nature of the root medium, especially by improving water and nutrient holding capacity, supply of plant nutrients in a slow release form, facilitating plant establishment and in situ chemical immobilization of metals, reducing leachability and phytotoxicity (Tordoff *et al.*, 2000).

Therefore, soil acidity is the limiting factor for the establishment of vegetation in mine wastes, it is common practice to apply liming materials to overcome some of the problems associated with soil acidity (Wong, 2003; Sutton and Dick, 1987; Ye *et al.*, 1999). Alkaline amendments reduce the concentration of metals in soil solution by raising soil pH, thereby allowing the formation of insoluble metal precipitates, complexes and secondary minerals (Basta *et al.*, 2001). Soils polluted with sulphur-mine wastes should be remediated by using inorganic materials containing $CaCO_3$ (higher than 40 g $CaCO_3$/kg dry soil) (Del Moral *et al.*, 2010). The applications of calcareous soil and farmyard manure increased the soil's pH, decreased electrical conductivity (EC) and DTPA-extractable concentrations of Fe, Zn, Mn, Cu, Ni, Cr, Pb and Cd in sulphur mine tailings, and also increased shoot dry weight of *Zea mays* L. plant (Orman and Kaplan, 2007).

Choosing the plant material is very important for revegetation of mine wastelands. Plants should have to develop an extensive root system and a large amount of biomass in the presence of high concentrations of heavy metals, in order to keep translocation of metals from roots to shoots as low as possible (Wong, 2003; Rizzi *et al.*, 2004; Mendez and Maier, 2008). Some *Agropyron* species, such as *A. desertorum* and *A. intermedium* have been used for mine reclamation (Williamson *et al.*, 1982). *A. elongatum* has also been used to grow on salt-affected lands (Roundy, 1985).

The main objective of the present study was to evaluate the effects of a farmyard manure and calcareous soil in the remediation of a sulphur flotation waste as, also, called sulphur mine tailings and to determine the possibility of using *A. elongatum* (tall wheat grass) in its revegetation.

Materials and methods

Experimental materials

The abandoned sulphur flotation waste ponds of Keçiborlu Sulphur Factory were located in Keçiborlu/Isparta/Turkey (37° 57' N, 30° 18' E) with a mean annual rainfall 615 mm, which falls in winter and spring months. The wastes from the flotation of sulphur ores were stored permanently in the tailing ponds, approximately up to 1 million tons (Fig. 1). The sulphur flotation waste ponds requires revegetation to reduce the environmental impact, which is a major source of heavy metal pollution in the local environment, owing to dust below and from leaching of the products of mineral weathering into nearby watercourses. The sulphur flotation waste was collected from the tailing ponds form 0-20 cm depths. The extremely calcareous soil was taken from the campus area of Akdeniz University in Antalya, Turkey. The soils were classified as Lithic Xerorthent. The farmyard manure was supplied by a local farm. All samples (flotation waste, soil and farmyard manure) were air dried and screened through a 4 mm mesh sieve before use. *A. elongatum* (tall wheat grass) was grown as experimental plant.

Experimental design

The greenhouse pot trial was designed to test the growth of *A. elongatum* in sulphur flotation waste amended with calcareous soil and farmyard manure. A total of 6 kg of experimental materials (Flotation Waste (FW) + Calcareous Soil (CS) + Farmyard Manure (FYM) = Growing Media) at predetermined ratios (Tab. 1) were thoroughly mixed and then N:P:K fertilizer (15:15:15) was added at a rate equivalent to 150 kg da^{-1} per pot. The pots (510x165x125 mm) were filled up with growing mediums. There were prepared four replicates for each treatment.

Each pot was hand broadcasted with seeds *A. elongatum*, at rate of 100 g m^{-2}. The pots were arranged in a randomized factorial design and were equally watered in greenhouse and allowed to grow for 3 months (from the beginning of June - until the end of September). The above-ground parts (shoots) of the plants were harvested 5 times during the experiment. After harvest, plant samples were washed with tap water, rinsed three times with deionized water and then oven-dried at 65 °C to a constant weight. Dry weight yields were recorded. The dried plants resulted from the 5 harvests were homogeneously combined together to obtain a composite plant sample for each treatment. After this process, plant samples were reduced to powder by an electric mill and analysed to determine the concentrations of Fe, Zn, Mn, Cu, Ni, Cr, Pb, Cd, Co. The growing media samples were taken from each pot at the end of the experiment, air-dried, grounded to pass through a 2 mm sieve, and determined the pH, EC (Electrical Conductivity) values and concentrations of DTPA-extractable Fe, Zn, Mn, Cu, Ni, Cr, Pb, Cd and Co.

Tab. 1. Experimental design for pot trial

Treatments	A total of 6 kg per pot (w/w)
$FWCS_{10}$	90% FW + 10% CS
$FWCS_{20}$	80% FW + 20% CS
$FWCS_{30}$	70% FW + 30% CS
$FWCS_{40}$	60% FW + 40% CS
$FWCS_{10}FYM_4$	86% FW + 10% CS + 4% FYM
$FWCS_{10}FYM_8$	82% FW + 10% CS + 8% FYM
$FWCS_{20}FYM_4$	76% FW + 20 %CS + 4% FYM
$FWCS_{20}FYM_8$	72% FW + 20% CS + 8% FYM
$FWCS_{30}FYM_4$	66% FW + 30 %CS + 4% FYM
$FWCS_{30}FYM_8$	62% FW + 30% CS + 8% FYM

Also, air-dried flotation waste, calcareous soil and farmyard manure were grounded to pass through a 2 mm sieve and some physico-chemical parameters were analysed prior to the plant growth experiment (Tab. 2).

Analytical chemistry

The pH and EC values were measured in H_2O (1-2.5 soil: deionized water). The $CaCO_3$ content was determined by using a Scheibler calcimeter. The total N was done by using modified Kjeldahl procedure (Kacar, 1994). The available P was extracted by sodium bicarbonate (Olsen and Sommers, 1982), exchangeable K^+, Ca^{+2}, Mg^{+2}, Na^+ were extracted by ammonium acetate (Kacar, 1994), extractable Fe, Zn, Mn, Cu, Ni, Cr, Pb, Cd and Co were extracted by diethylenetriaminepentaacetic

Fig. 1. General views of the sulphur flotation waste (sulphur mine tailings) in Keçiborlu, Isparta, Turkey

acid-DTPA (Lindsay and Norvel, 1978) and then concentrations were determined by Induced-Coupled Plasma with Optical Emission Spectrophotometer (ICP-OES).

Plant samples (shoots) were digested with HNO_3 and $HClO_4$ (4:1) (Kacar and Inal, 2008). The concentrations of Fe, Zn, Mn, Cu, Ni, Cr, Pb, Cd and Co in the digests were determined by ICP-OES.

The data were analysed by standard ANOVA procedures and their significances were based on the $p<0.05$ level using the Duncan's Multiple Range Test.

Tab. 2. General properties of sulphur flotation waste, calcareous soil and farmyard manure

Parameters	Flotation Waste	Farmyard Manure	Calcareous Soil
pH	3.12	7.62	7.97
EC dS m^{-1}	9.99	4.73	0.16
$CaCO_3$ %	1.61	9.66	48.25
N %	0.038	2.13	0.035
P mg kg^{-1}	0.17	3876.16	0.49
K mg kg^{-1}	0.01	6767.84	23.87
Ca mg kg^{-1}	1442.00	36362.10	1442.00
Mg mg kg^{-1}	406.00	5116.83	130.00
Na mg kg^{-1}	4.92	0.15	15.01
Fe mg kg^{-1}	6250.00	3582.50	3.53
Zn mg kg^{-1}	8.17	152.58	1.18
Mn mg kg^{-1}	53.21	75.58	3.00
Cu mg kg^{-1}	10.30	2.56	0.59
Ni mg kg^{-1}	64.84	28.09	0.11
Cr mg kg^{-1}	14.86	57.56	0.02
Pb mg kg^{-1}	0.53	5.35	0.16
Cd mg kg^{-1}	0.13	0.56	0.01
Co mg kg^{-1}	5.82	0.27	0.04

Results and discussions

The pH, EC values and DTPA-extractable metal concentrations in growing media

The results of pH, EC and DTPA-extractable metal concentrations of growing media were given in Tab. 3 and Fig. 2. The pH of growing media increased by CS-alone, but this increment tended to decrease by FYM. Also, the EC was decreased by CS-alone, but this decline tended to increase by FYM. When contaminated soil is moist and exposed to air, oxidation and hydrolysis of sulphides (S^{2-}) in the waste leads to the formation of sulphuric acid, thus lowering pH and greatly increasing heavy metal solubility (Williamson and Johnson, 1981; Simón *et al.*, 1999; Clemente *et al.*, 2003). The application of both lime and organic matter to contaminated soil by mineral sulphides caused increasing pH and decreasing EC in soil (Clemente *et al.*, 2003).

The concentrations of DTPA-extractable Fe, Zn, Mn, Ni, Cr and Co concentrations were decreased by CS-alone, but this decline tended to increase by FYM. There was no change in Pb and Cd concentrations between CS and CS with FYM. The application of calcareous soil led to increased pH while consequently extractable metal concentrations were decreased in growing media. The pH effect was greater than that of metal complex formation by soluble organic matter (Walker *et al.*, 2004). Application of fresh manure can increase heavy metal solubility, due to the provision of soluble organic compounds which form complexes with the metals (Almås *et al.*, 1999; Shuman, 1999). However, the humic materials, which constitute a major part of the organic matter of compost, can reduce metal solubility by formation of stable metal chelates (Ross, 1994). Differences with respect to effects on metal availability between amendment materials can be related to differences not only in the organic matter, but also in the mineral fractions (pH, salt content) and cation exchange capacity, as well as changes in the redox conditions of the soil (Shuman, 1999; Ross, 1994; Walker *et al.*, 2003).

Metal concentrations and dry weight yield of Agropyron elongatum shoot

The concentrations of metal in plant shoots were given in Tab. 4 and Fig. 3. The shoot Fe, Ni, Cr and Co concentrations were decreased by CS-alone, but this decline tended to increase by FYM. The shoot Zn, Mn, Cu and Pb concentrations increased by CS-alone, but this increment tended to decrease by FYM. The dry weight yield of plant increased CS with FYM more than CS-alone (Fig. 3). The plant grew well and had higher cover in CS with FYM treatments compared to CS-alone (Fig. 4). Farmyard manure is commonly used as tailings amendment because the addition of organic matter can significantly improve the physical characteristics and the nutrient status of mine wastes (Bradshaw and Chadwick,

Tab. 3. The pH, EC values and DTPA-extractable metal concentrations in flotation waste amended with calcareous soil (CS) and farmyard manure (FYM) (as called growing media)[a]

Parameters	$FWCS_{10}$	$FWCS_{20}$	$FWCS_{30}$	$FWCS_{40}$	$FWCS_{10}$		$FWCS_{20}$		$FWCS_{30}$		Significance levels[b]
					FYM_4	FYM_8	FYM_4	FYM_8	FYM_4	FYM_8	
pH	3.53 hi	5.23 e	6.52 b	6.58 a	3.50 i	3.57 h	4.22 g	4.40 f	5.43 d	6.30 c	***
EC (dS m^{-1})	2.83 b	2.57 d	2.39 f	2.46 e	2.70 c	2.56 d	2.20 g	3.14 a	2.70 c	3.13 a	***
Fe (mg kg^{-1})	282.73 d	221.10 f	100.82 i	113.70 h	330.55 b	324.48 c	330.07 b	354.35 a	235.95 e	194.18 g	***
Zn (mg kg^{-1})	1.16 de	0.58 g	0.70 fg	0.89 ef	1.58 bc	3.46 a	1.01 e	1.78 b	1.36 cd	3.36 a	***
Mn (mg kg^{-1})	27.13 d	22.62 ef	13.65 g	16.00 g	25.42 de	37.92 b	34.63 bc	46.59 a	20.36 f	31.21 c	***
Cu (mg kg^{-1})	0.96 g	1.72 d	2.22 c	2.46 b	1.12 f	0.95 g	1.52 e	1.67 d	2.49 b	2.91 a	***
Ni (mg kg^{-1})	6.84 d	5.47 e	3.93 g	4.70 f	9.45 b	11.22 a	9.00 b	7.76 c	4.90 ef	8.10 c	***
Cr (mg kg^{-1})	0.03 c	0.02 c	0.02 c	0.02 c	0.12 a	0.10 b	0.02 c	0.02 c	0.02 c	0.02 c	***
Pb (mg kg^{-1})	0.04 ef	0.08 d	0.13 b	0.15 a	0.04 f	0.05 ef	0.06 e	0.09 d	0.11 c	0.15 a	***
Cd (mg kg^{-1})	0.03 e	0.05 c	0.05 c	0.06 b	0.03 d	0.04 d	0.04 c	0.05 c	0.05 b	0.08 a	***
Co (mg kg^{-1})	0.61 d	0.35 g	0.33 gh	0.46 ef	1.39 b	1.77 a	0.96 c	0.49 e	0.25 h	0.37 fg	***

[a]Values are means (n=4). Values in a row followed by different letters indicate significant differences (p<0.05) between treatments according to a Duncan's multiple range test.
[b]Significance levels: *** p< 0.001

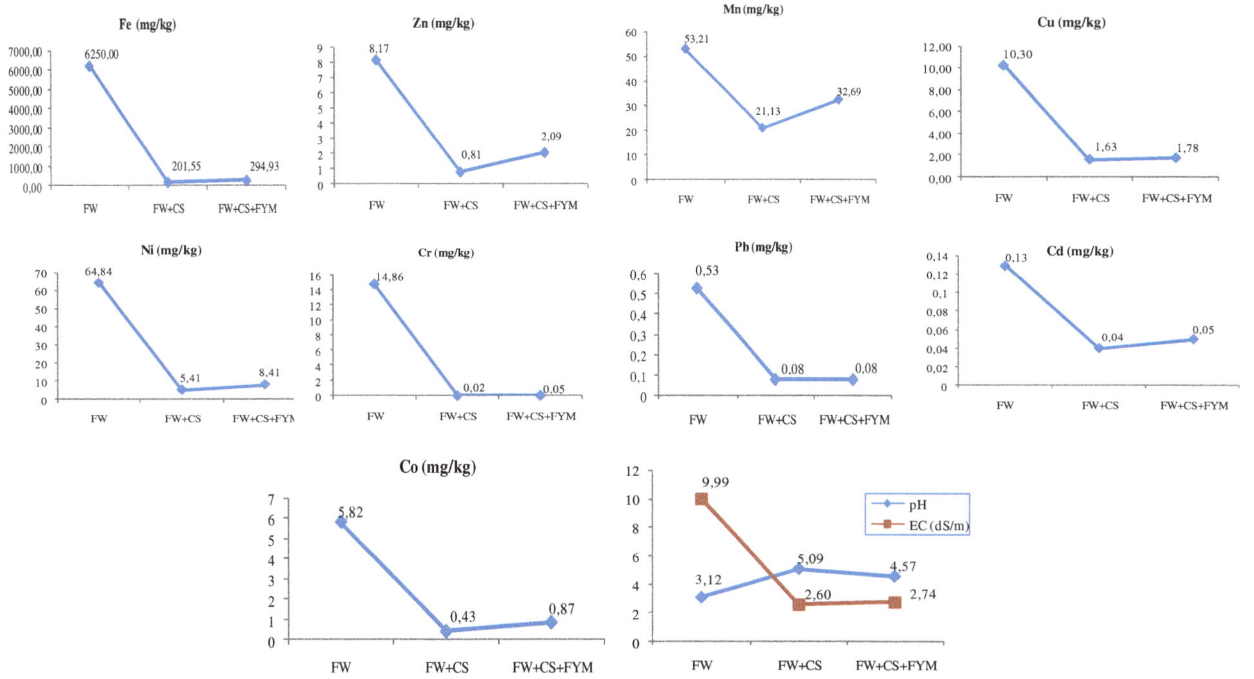

Fig. 2. The values of average pH, EC and DTPA-extract. Fe, Zn, Mn, Cu, Ni, Cr, Pb, Cd, Co concentrations due to treatments of calcareous soil and calcareous soil with farmyard manure in growing media

Tab. 4. Metal concentrations of *Agropyron elongatum* shoot growth in flotation waste amended with calcareous soil (CS) and farmyard manure (FYM) (as called growing media)[a]

Parameters	$FWCS_{10}$	$FWCS_{20}$	$FWCS_{30}$	$FWCS_{40}$	$FWCS_{10}$		$FWCS_{20}$		$FWCS_{30}$		Significance levels[b]
					FYM_4	FYM_8	FYM_4	FYM_8	FYM_4	FYM_8	
Dry weight (g pot^{-1})	25.34 h	29.61 g	35.33 e	32.57 f	31.85 f	40.25 d	40.31 d	46.94 b	41.79 c	48.54 a	***
Fe (mg kg^{-1})	275.80 d	256.13 e	239.32 gh	246.18 f	241.90 g	338.20 a	272.70 d	326.32 b	296.77 c	237.97 h	***
Zn (mg kg^{-1})	60.64 f	75.10 a	76.14 a	71.23 bc	66.89 d	70.06 c	69.27 cd	73.63 ab	62.52 ef	64.07 e	***
Mn (mg kg^{-1})	143.89 c	127.18 d	106.86 f	103.56 f	157.85 b	196.07 a	117.72 e	82.56 h	87.00 g	76.52 i	***
Cu (mg kg^{-1})	1.56 c	1.60 b	1.35 fg	1.64 b	1.32 g	1.48 d	1.46 de	1.37 f	1.74 a	1.43 e	***
Ni (mg kg^{-1})	15.45 c	3.22 d	2.25 de	1.83 e	41.45 a	27.54 b	3.12 d	3.26 d	2.32 de	2.48 de	***
Cr (mg kg^{-1})	1.88 d	1.47 f	1.75 e	1.78 e	1.16 g	2.02 c	1.85 d	2.55 a	2.59 a	2.29 b	***
Pb (mg kg^{-1})	7.82 d	7.76 d	14.81 a	10.71 c	8.51 d	12.61 b	7.99 d	5.50 e	3.21 f	5.43 e	***
Cd (mg kg^{-1})	0.61 f	0.67 e	1.05 b	0.86 c	0.85 c	1.90 a	0.84 c	0.76 d	0.37 h	0.54 g	***
Co (mg kg^{-1})	1.16 c	0.32 de	0.35 d	0.31 e	3.86 a	1.50 b	< LD[c]	< LD	< LD	< LD	***

[a]Values are means (*n*=4). Values in a row followed by different letters indicate significant differences (p<0.05) between treatments according to a Duncan's multiple range test.
[b]Significance levels: *** p< 0.001
[c] Limit of detection (LD<0.1 mg kg^{-1})

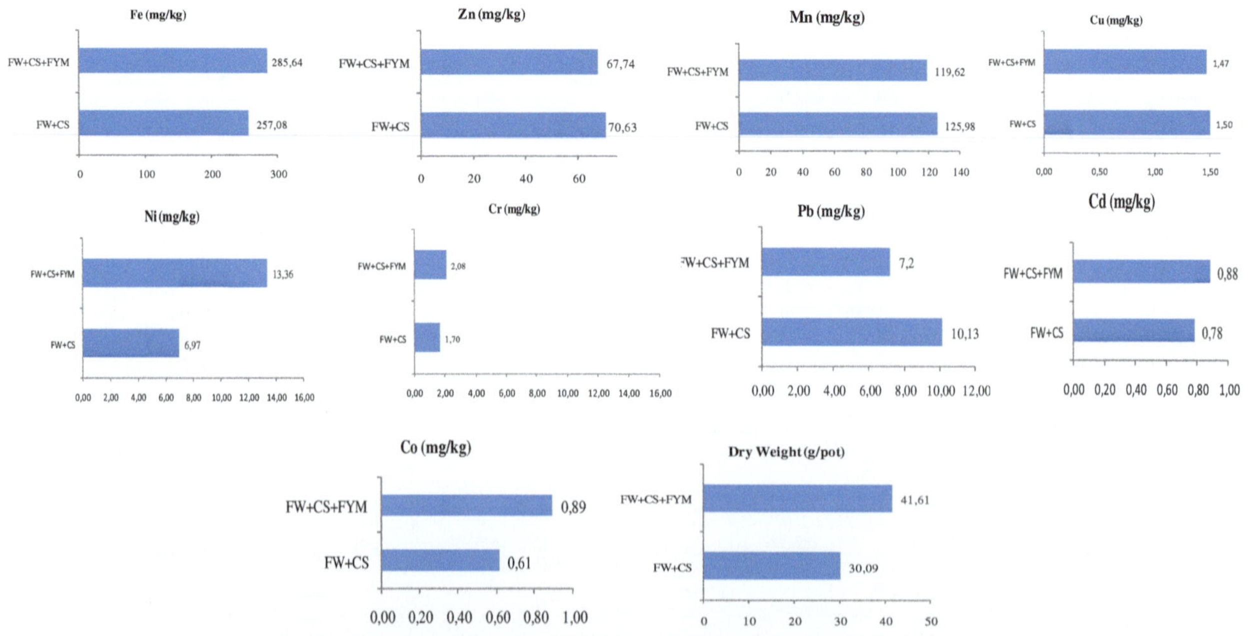

Fig. 3. The values of average Fe, Zn, Mn, Cu, Ni, Cr, Pb, Cd, Co concentrations and dry weight of shoot due to treatments of calcareous soil and calcareous soil with farmyard manure in growing media

Fig. 4. *Agropyron elongatum* growth during the experiment

1980). In addition, both manure compost and sewage sludge contained high amounts of macronutrients (N, P, and K) and organic carbon content, which can compensate for the nutrient deficiencies in Pb/Zn and Cu tailings (Chiu *et al.*, 2006). The effect of the organic amendments on the bioavailability of metals was difficult to observe due to the great variability of total metal concentration and pH. In any case, pH seemed to be a controlling factor in metal availability, to a greater extent than the organic amendments. Liming succeeded in controlling the soil acidification in a soil affected by acid, highly toxic pyritic waste (Clemente *et al.*, 2003).

Conclusion

DTPA-extractable heavy metal concentrations decreased with increasing pH in growing media caused by the applications of CS-alone to flotation waste. However, in applications of CS with FYM, heavy metal concentrations in both growing media and shoot increased, in contrast to increasing pH of the growing media. The higher DTPA-extractable heavy metal concentrations in CS with FYM treatments compared to CS-alone is thought to be the result of chelating effect of farmyard manure. In spite of high heavy metal concentration in growing media and shoot, dry weight yield increased in plant in CS with FYM treatments. This may be due to the improvements in some growing media properties, the presence of inorganic flotation waste, as a result of farmyard manure (organic).

Acknowledgments

The authors would like to express their appreciation to the Scientific Fund of Akdeniz University (Project no: 2008.02.0121.022).

References

Almås A, Singh BR, Salbu B (1999). Mobility of cadmium-109 and zinc-65 in soil influenced by equilibration time, temperature, and organic matter. J Environ Qual 28:1742-1750.

Alvarenga P, Gonçalves AP, Fernandes RM, de Varennes de A, Vallini G, Duarte E, Cunha-Queda AC (2008). Evaluation of composts and liming materials in the phytostabilization of a mine soil using perennial ryegrass. Sci Total Environ 406:43-56.

Basta NT, Gradwohl R, Snethen KL, Schroder JL (2001). Chemical immobilization of lead, zinc, and cadmium in smelter-contaminated soils using biosolids and rock phosphate. J Environ Qual 30:1222-1230.

Bradshaw AD, Chadwick MJ (1980). The restoration of land. Blackwell Scientific, Oxford.

Brown SL, Sprenger M, Maxemchuk A, Compton H (2005). Ecosystem function in alluvial tailings after biosolids and lime application. J Environ Qual 34:1-6.

Chiu KK, Ye ZH, Wong MH (2006). Growth of *Vetiveria zizanioides* and *Phragmities australis* on Pb/Zn and Cu mine tailings amended with manure compost and sewage sludge: A greenhouse study. Bioresource Technology 97:158-170.

Clemente R, Walker DJ, Roig A, Bernal MP (2003). Heavy metal availability in a soil affected by mineral sulphides contamination following the mine spillage at Aznalcóllar (Spain). Biodegradation 14:199-205.

Conesa HM, Faz A, Garcia G, Arnaldos R (2007). Heavy metal contamination in the semiarid area of Cartagena-La Union (SE Spain) and its implications for revegetation. Fresenius Environmental Bulletin 16:1076-1081.

Del MF, Gonzalez V, Garcia I, Sanchez JA, de Haro S, Simon M (2010). Effect of organic and inorganic residues on the solubility of trace elements and phytotoxicity of mining waste. Fresenius Environmental Bulletin 19:2071-2076.

Gutierrez-Gines MJ, Pastor J, Hernandez AJ (2010). Effect of Heavy metals from mine soils on *Avena sativa* L. and education Strategies. Fresenius Environmental Bulletin 19:2083-2086.

Henriquez FS, Fernandez JC (1991). Metal uptake and distrubition in rush (*Juncus conglomerates* L.) plants growing in phrites mine tailings at Lousal, Portugal. Sci Total Environ 102:253-260.

Kacar B (1994). Toprak Analizleri. Bitki ve Toprağın Kimyasal Analizleri. III. Toprak Analizleri. Ankara Üniv. Ziraat Fak. Eğitim Araştırma ve Geliştirme Vakfi Yayınları. No: 3, ss 705, Ankara (in Turkish).

Kacar B, Inal A (2008). Bitki Analizleri. Nobel Yayın No:1241, Fen Bilimleri: 63, 879 pp, Ankara (in Turkish).

Lindsay WL, Norvel WA (1978). Development of a DTPA soil test for zinc, iron, manganese and copper. Soil Sci Amer Jour 42(3):421-428.

Mendez MO, Maier RM (2008) Phytostabilization of mine tailings in arid and semiarid environments-an emerging remediation technology. Environ Health Perspect 116:278-283.

Norland MR, Veith DL (1995). Revegetation of coarse taconite iron ore tailing using municipal waste compost. J Hazard Mater 41:123-134.

Olsen SR, Sommers EL (1982). Phosphorus Soluble in Sodium Bicarbonate, p. 404-430. In: Page AL, Miller RH, Keeney DR (Eds.). Methods of Soil Analysis. Part 2. Chemical and Microbiological Properties. Agronomy Monograph 9, ASA and SSSA, Madison, WI.

Orman Ş, Kaplan M (2007). Effects of calcareous soil and farmyard manure on revegetation of sulphur mine tailings. Fresenius Environmental Bulletin 16:1295-1299.

Pérez-de-Mora A, Burgos P, Madejón E, Cabrera F, Jaeckel P, Schlote M (2006). Microbial community structure and function in a soil contaminated by heavy metals: effects of plant growth and different amendments. Soil Biol Biochem 38:327-341.

Rizzi L, Petruzelli G, Poggio G, Vigna GG (2004). Soil physical changes and plant availability of Zn and Pb in a treatability test of phytostabilization. Chemosphere 57:1039-1046.

Ross SM (1994). Retention, transformation and mobility of toxic metals in soils, p. 63-152. In: Ross SM (Ed.). Toxic Metals in Soil-Plant Systems. John Wiley and Sons Ltd., Chichester, UK.

Roundy BA (1985). Root penetration and shoot elongation of tall wheat grass and basin in relation to salinity. Can J Plant Sci 65:335-343.

Rybicka EH (1996). Impact of mining and metallurgical industries on the environment in Poland. Appl Geochem 11:3-9.

Salomons W (1995). Environmental impact of metals derived from mining activities: processes, predictions, prevention. J Geochem Explor 52:5-23.

Shu WS, Ye ZH, Zhang CY, Wong MH (2005). Natural colonization of plants on five lead/Zinc mine tailings in Southern China. Restoration Ecology 13(1):49-60.

Shuman LM (1999). Organic waste amendments effect on zinc fractions of two soils. J Environ Qual 28:1442-1447.

Simón M, Ortiz I, García I, Fernández J, Dorronsoro C, Aguilar J (1999). Pollution of soils by the toxic spill of a pyrite mine (Aznalcóllar, Spain). Sci Total Environ 242:105-115.

Sutton P, Dick WA (1987). Reclamation of acidic mined lands in humid areas. Adv Agron 41:377-406.

Tordoff GM, Baker AJM, Willis AJ (2000). Current approaches to the revegetation and reclamataion of metalliferous wastes. Chemosphere 41:219-228.

Walker DJ, Clemente R, Roig A, Bernal MP (2003). The effects of soil amendments on heavy metal bioavailability in two contaminated soil Mediterranean soils. Environ Pollut 122:303-312.

Walker DJ, Clemente R, Bernal MP (2004). Contrasting effects of manure and compost on soil pH, heavy metal availability and growth of *Chenopodium album* L. in soil contaminated by pyritic mine waste. Chemosphere 57:215-224.

Williamson A, Johnson MS (1981). Reclamation of metalliferous mine wastes, pp. 185-212. In: Lepp NW (Ed.). Effect of Heavy metal pollution on Plants. Vol. 2. Metals in the Environment. Applied Science Publishers Ltd., Barking.

Williamson NA, Johnson MS, Bradshaw AD (1982). Mine waste rehabilitation: the establishment of vegetation on metal mine waste. Mining Journal Books, London.

Wong JWC, Ip CM, Wong MH (1998) Acid-forming capacity of lead-zinc mine tailings and its implications for mine rehabilitation. Environ Geochem Health 20:149-155.

Wong MH (2003). Ecological restoration of mine degraded soils, with emphasis on metal contaminated soils. Chemosphere 50:775-780.

Yazar Y (2010). Present situation of the Keçiborlu Sulphur Factory Flotation wastes and determination of its environmental effects. Akdeniz University, Science Institute, Department of Soil Science and Plant Nutrition. MSc Thesis, September 2010, 166 p.

Ye ZH, Wong JWC, Wong MH, Lan CY, Baker AJM (1999). Lime and pig manure as ameliorants for revegetation lead/zinc mine tailings: a greenhouse study. Bioresour Technol 69:35-43.

Spatial Distribution of Grapes Sugar Content and its Correlations with Climate Characteristics and Climate Suitability in the Huşi (Romania) Wine Growing Region

Liviu Mihai IRIMIA[1]*, Cristian Valeriu PATRICHE[2], Georgeta Mihaela BUCUR[3], Hervé QUÉNOL[4] , Valeriu V. COTEA[1]

[1]*University of Agricultural Sciences and Veterinary Medicine, Faculty of Horticulture, 3, Mihail Sadoveanu Alley, Iaşi, 700490, Romania; livirimia@uaiasi.ro (*corresponding author); vcotea@uaiasi.ro*

[2]*Romanian Academy, Iaşi Branch, Geography Group, 8, Carol I Street, 700505, Iasi, Romania; pvcristi@yahoo.com*

[3]*University of Agricultural Sciences and Veterinary Medicine, Faculty of Horticulture, 59, Bulevardul Marasti, 011464, Bucharest, Romania; mihaela_g_savu@yahoo.com*

[4]*LETG-Rennes (COSTEL), UMR 6554 CNRS, Université Rennes-2, Rue du Recteur Henry le Moal, 35000, Rennes, France; herve.quenol@uhb.fr*

Abstract

Previous studies have analyzed mainly the spatial distribution of grapes quality parameters at the plot scale, and concluded that grapes sugar content is a variable parameter, with a less predictable spatio-temporal evolution. Our study aimed to identify a pattern of spatial variation of grapes sugar content at a regional scale, under the temperate continental climate of the Huşi (Romania) wine growing region. In order to do this, grapes sugar content was analyzed in relation to climate. The conditioning of grapes sugar content by climatic factors was analyzed based on regression models (Pearson correlation coefficient, coefficient of determination, linear regression). Spatial variation of grapes sugars content was achieved by the regression – kriging approach, using climate suitability as predictor. Spatial variation of obtainable wine types was achieved by classifying the spatial distribution of grapes sugar content, according to DOC wine production regulation for the Husi area. The study revealed that among climatic factors, the actual sunshine duration is the most influential one on grapes sugar content. Mapping grapes sugar content generated a coherent spatial distribution pattern at the vineyard scale, as it significantly correlates with terrain altitude. Based on grapes sugar content spatial variation, two areas with different potential for sugars accumulation into grapes were delineated. The outcomes reveal that grapes sugar content presents significant correlations with climate variables specific to vineyards and with climate suitability for wine production. Knowing these correlations allows revealing patterns that determine the spatial distribution of the grapes sugar content.

Keywords: grapevine, grapes sugar content, spatial distribution, vineyard, wine type

Introduction

Effective vineyard management and obtaining the targeted wine style require thorough knowledge of spatial variation of chemical compounds which generate grapes quality. They are mainly sugars, organic acids, phenolic compounds and aroma compounds, accumulated in berries during their growth and ripening. Among them, sugars are the most important ones (Huglin, 1986), as they determine the alcoholic content, the balance of taste and the final sensory profile of the wine.

When analyzed at a vine level, in known technological and environmental conditions, grapes sugar content is a rather easy to explore parameter, as it is specific to wine grape varieties and it reacts to viticultural practices (Champagnol, 1984; Galet, 1993). However, when analyzed in relation to its spatial variation within the field, it proves to be, a spatially and temporally variable parameter, difficult to analyze and generating contrasting results. Studies conducted up to date on mapping grapes sugar content conclude that "maps from previous years are not useful in the management of harvest quality in the next year" (Tisseyre *et al.*, 2008), or that "soluble solids had a higher level of uncertainty for being predicted in the field" (Santos *et al.*, 2012). In contrast, Bramley (2005), referring to yield quality considers that "its variation appears to be sufficiently temporally stable to justify the identification of 'quality zones' within vineyards", and also that sugar concentrations "are much less variable than total phenolics concentration". Nonetheless, due to the variability of this parameter and the multitude of factors that influence it, no stable pattern of spatial distribution at vineyard scale has been obtained.

Spatial variability of grapes sugar content is generated by lack of uniformity of vineyard' topography, soils and climate (Van Leeuwen, 2010), while its temporal variability is caused by vintage climate (Jackson and Lombard, 1993). What is more, grapes sugar content depends on grapevine variety, yield size (Huglin and Balthazard, 1976), vines age (Galet, 1993) and vineyard training system (Huglin, 1978). Therefore, there is a complex conditioning of this parameter, which requires the analysis of its correlations with a large number of biological, environmental and cultural factors in order to understand the patterns driving its spatial distribution.

The methods used to map the spatial distribution of grapes sugar content within the vineyards are not well defined, they implying yet the difficulty of sampling a large amount of berries for a pertinent depiction of the analyzed area. Taking as a model the technologies involved for in real-time anthocyanins content analysis (Lamb *et al.*, 2003), some harvest attached tools based on refractometry and spectrometry were designed (Wold *et al.* 2001; Workman and Burns 2001; Tisseyre and Taylor, 2005; Sethuramasamyraja *et al.*, 2007) but they are not yet widespread in practice. As an alternative, some authors suggest the use of spatial distribution of the yield size as an indicator, taking into account its correlation with grapes quality indicators (Bramley, 2005), while others consider NDVI (Normalized Difference Vegetation Index) to be a more relevant indicator (Agelet-Fernández *et al.*, 2009).

Unlike previous studies concerning spatial variation of grapes quality parameters at the plot scale (Bramley, 2005; Tisseyre *et al.*, 2008; Trought and Bramley, 2011; Santos *et al.*, 2012), this study analyzes spatial variation of grapes sugar content at the vineyard scale, aiming to reveal a possible model of spatial distribution. So far, grapes sugar content for the 2011 vintage has been analyzed in relation to multiannual averages of the climatic variables specific to the Huşi vineyard, and mapped based on its significant correlation with climatic suitability for wine production generated by them.

Materials and methods

Study area

The study area is the Huşi wine growing region, located in the eastern part of Romania, at 46°67' lat. N and 28°13' long. E, near the town of Huşi (Fig. 1). It has 2139 ha and includes four vineyards: Schit (554.3 ha); Dobrina (570.6 ha); Corni (264.4 ha); and Recea (749.1 ha). The relief is like an amphitheater with an eastern opening and a decreasing altitude from 300 m above sea level (asl.) at the western limit to 40 m asl. at the eastern limit. The Schit and Dobrina vineyards are located above 150 m asl., on slopes with 10 to 30 % inclination, in the higher and hilly western part; the Recea and Corni vineyards are located below 150 m asl., mainly on flat terrains and less inclined slopes, in the lower eastern part of the area. The area is subjected to a temperate continental climate, Dfb according to World Map of the Köppen-Geiger climate classification updated (Kottek *et al.*, 2006).

Sampling the grapes sugar content

The study refers to grapes sugar content (GrSC) of the 2011 vintage. In order to ensure a representative sampling for the study area, berry samples were taken from different 26 locations (Fig. 1), each of them with a surface of about 1 ha. The geographical coordinates of the 26 sampling sites were established by using GPS. For each sampling site, GrSC (g/L) was measured for each of the 'Fetească albă', 'Fetească regală', 'Aligoté' and 'Tămâioasă românească' *V. vinifera* L. wine grape varieties, representing the established wine grape variety assortment. Each wine grape variety was represented by five vines, randomly selected, placed on different rows; from each vine, five berries per cluster were sampled, from four clusters, for a number of 20 berries per vine and 100 berries per variety. To ensure a random sample, the clusters were chosen arbitrarily, without taking into account their size or their location inside the canopy. Also, the berries were sampled from different positions and depths of the cluster. Until they were analyzed in the laboratory, the samples were kept in a freezer, at a temperature of 5 °C. The sugar concentration of grapes was determined in the laboratory, according to OIV protocol (OIV,

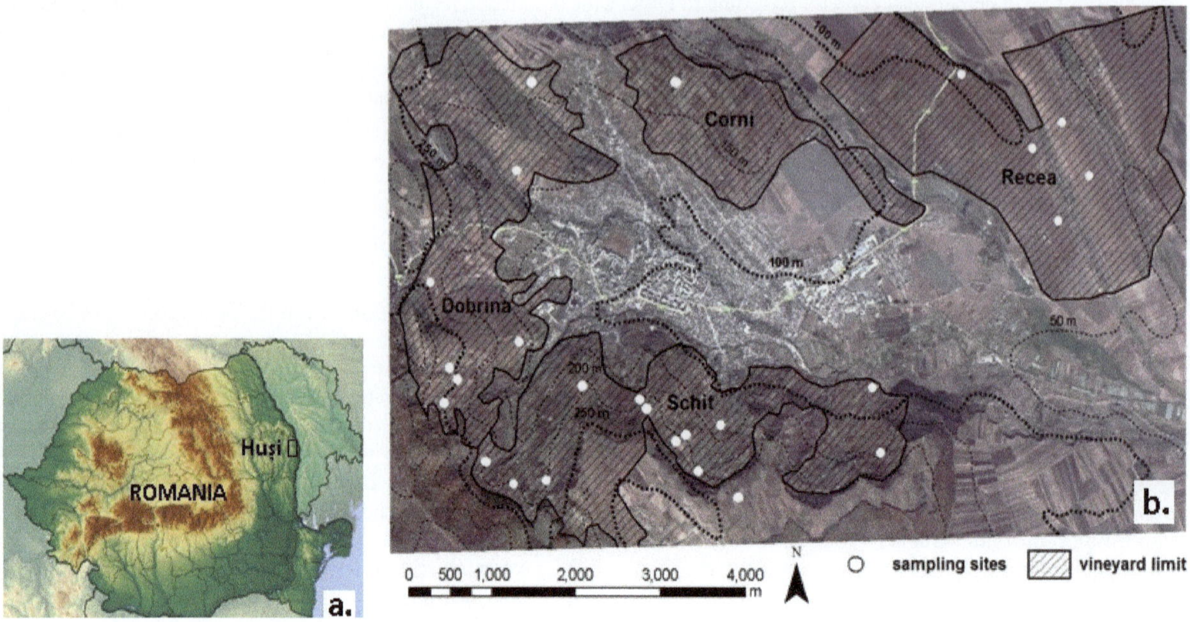

Fig. 1. Location of Huşi wine growing region in Romania (a), its structure and the locations of the 26 sampling sites (b)

1990). To ensure the representativeness of the samples for the sampling sites, the GrSC was expressed as the average of the grapes sugar content for the four wine grape varieties analyzed, thus each of them having equal weight in the berries sample (Table 1). In order to have comparable berries samples, the sampling was carried out in a single day (11 September, 2011), with about 4-5 days before harvesting, from vines with the same training system (Cazenave; rather 30 shoots per vine; 25 to 30 grapes per vine; 3700 vines per ha).

Analysis of climate variables

For each of the 26 sampling sites, there were analyzed the values of nine climate variables, which characterize the climate suitability for different types of wine production in temperate continental climate conditions (Irimia *et al.*, 2014): average annual temperature (AAT, °C); average temperature of the warmest month (TWM, °C); global radiation (GR, kcal/cm^2/Apr 1st to Sept 30); actual sunshine duration (ASD, hours/Apr 1st to Sept 30); precipitation in the growing season (PP, mm/Apr 1st to Sept 30); sum of effective temperatures (Σt_u,°C/Apr 1st to Sept 30); actual heliothermal index (IHa/Apr 1st to Sept 30); bioclimatic index (Ibcv/Apr 1st to Sept 30) and oenoclimate aptitude index (IAOe/Apr 1st to Sept 30) (Table 1). The values of climate variables for each sampling site were extracted from their spatial distributions in the Huşi wine growing region (Irimia *et al.*, 2011), previously established based on multiannual averages (1961-2000) of temperature and precipitation altitudinal gradients implemented in NewLocClim software (FAO, 2003), and on values derived from DEM for the global radiation and sunshine duration (Patriche, 2007).

Analysis of climate suitability for wine production

Climate suitability was assessed in a previous phase of the study based on multiannual averages (1961-2000) of ten climate variables characterizing the Husi wine growing region (Irimia *et al.*, 2014). Each climate variable was mapped, and its values classified and ranked with 5 to 10 points, according to their suitability for wine production. Then the maps were combined in a GIS database, were georeferenced and the average of ranking points was computed at the pixel level (30 x 30 m). The average of ranking points (a.r.p.) expresses climate suitability and varies between 5 and 10 with the following significance: 5 a.r.p.= climate suitability for white table wines, sparkling wines and wines for distillates; 6 a.r.p.= climate suitability for white table wines, sparkling wines, and wines for distillates, as well as for quality white wines in very suitable years, in terms of climate; 7 a.r.p.= climate suitability for quality white wines; 8 a.r.p.= climate suitability primarily for quality white wines and secondarily for red table wines; 9 a.r.p.= climate suitability primarily for quality red wines and secondarily for quality white wines; 10 a.r.p.= climate suitability for quality red wines.

Analysis of the relationships between climate variables and GrSC was performed by computing regression models (Pearson correlation coefficient; coefficient of determination; linear regression) between the GrSC and the multiannual means of the nine climate variables characterizing each of the 26 sampling sites. The results were assessed according to the statistical significance of the linear regression model for a significance level of 0.05. This analysis aimed to highlight the factors driving the GrSC and also to reveal the extent to which they influence this grapes quality parameter.

Analysis of the relationships between climate suitability and GrSC was performed based on the same previously used regression models (Pearson correlation coefficient; coefficient of determination; linear regression) this time computed between the GrSC and the average ranking points (a.r.p.) revealing the climate suitability of each of 26 sampling sites. The results were also assessed according to statistical significance of the linear regression model for a significance level of 0.05. This analysis aimed to reveal the conditioning of the GrSC by local climate suitability.

Spatial distribution of GrSC within the vineyard was achieved based on its significant correlation with local climate suitability revealed by the regression models computed in the previous phase. The spatialization method was the linear regression by using climate suitability as predictor, followed by ordinary kriging of regression residuals. Subsequently, the regression and kriging spatial models were added up in order to achieve the final estimated continuous distribution of GrSC. Interpretation of results is based on: statistical analysis (Min, Max, Range, CV, STD); analysis of the "*spread*" index, respective the difference between the maximum and minimum values expressed as a % of the median value (Bramley, 2005); and analysis of the correlation between the GrSC and the altitude.

Spatial distribution of wine types which can be obtained in accordance with the GrSC for the 2011 vintage was achieved by classifying the spatial distribution of GrSC under the regulations of DOC wine production for the Huşi wine growing region (ONIV). In addition to the requirements regarding the geographical location, wine grape variety, technological parameters of plantation and yield size, this regulation imposes a minimum content of sugar in the berries at the harvest moment, as follows: a GrSC higher than 170 g/L, up to 190 g/L is needed to produce quality wines with geographical denomination (IG); a GrSC higher than 190 g/L up to 196 g/L is required to produce high quality wines with controlled designation of origin (DOC); a GrSC higher than 196 g/L up to 213 g/L is needed to produce controlled designation of origin wines harvested at full maturity (DOC-CMD); a GrSC higher than 213 g/L is needed to produce controlled designation of origin wines late harvest (DOC-CT).

Results

Climatic variables

According to the estimates of the nine climatic variables corresponding to the 26 sampling sites, the Schit and Dobrina vineyards from the higher and hilly western part are characterized as poor in heliothermic resources and wetter, while the Corni and Recea vineyards from the lower and plain part as richer in heliothermal resources and slightly less humid (Table 1). Thermal parameters record minimum values in the Dobrina vineyard (AAT = 9.0 °C; TWM = 20.4 °C; Σtu = 1203.4 °C) and maximum ones in the Recea vineyard (AAT = 10.1°C; TWM = 21.7 °C; Σtu = 1427.1 °C). In a sampling site of the Schit vineyard, thermal parameters record absolute maximum values for the Huşi wine growing region (AAT = 10.2 °C; TWM = 21.8 °C; Σtu = 1449.0 °C), but they are not representative for this

Table 1. The values of climate variables, climate suitability and grapes sugar content (GrSC, g/L) characterizing the 26 sampling sites from the Huşi wine growing region

Sampling sites	Vineyard	Altitude	Aspect	Climate variables* AAT (°C)	TWM (°C)	Σtu (°C)	ASD (hours)	GR (kcal/cm²)	PP (mm)	IAOe	Ibcv	IHa	Climate suitability (a.r.p.) site	Climate suitability (a.r.p.) vineyard	GrSC (g/L) site
1	Schit	60.1	N	10.2	21.8	1449.0	1461.6	87.9	342.2	4702.9	7.8	2.1	9.4		228.9
2	Schit	161.3	NV	9.6	21.2	1333.3	1383.3	83.3	353.9	4447.8	6.8	1.8	8.0		190.0
3	Schit	129.3	NE	9.8	21.4	1369.6	1383.3	85.9	350.2	4503.6	7.0	1.9	8.3		209.8
4	Schit	211.8	NE	9.4	20.8	1276.5	1383.3	79.4	359.7	4359.8	6.5	1.8	7.0		194.6
5	Schit	236.9	N	9.2	20.7	1248.5	1479.0	88.3	362.6	4411.9	6.8	1.8	8.3		217.3
6	Schit	205.0	NV	9.4	20.9	1284.1	1470.3	81.3	358.9	4458.6	6.9	1.9	8.1		223.8
7	Schit	196.1	NV	9.4	20.9	1294.0	1435.5	81.9	357.9	4439.3	6.8	1.9	7.8	8.0	198.0
8	Schit	202.5	V	9.4	20.9	1286.9	1383.3	89.1	358.6	4376.0	6.5	1.8	8.3		211.5
9	Schit	198.8	V	9.4	20.9	1291.0	1287.6	89.5	358.2	4286.7	6.1	1.7	7.7		190.6
10	Schit	219.5	N	9.3	20.8	1267.9	1322.4	80.6	360.6	4285.6	6.2	1.7	6.9		179.7
11	Schit	207.4	SE	9.4	20.9	1281.4	1374.6	92.3	359.2	4358.9	6.5	1.8	8.0		205.7
12	Schit	201.2	NE	9.4	20.9	1288.3	1418.1	85.5	358.5	4413.1	6.7	1.8	8.1		231.8
13	Schit	198.1	NE	9.4	20.9	1291.8	1400.7	84.3	358.1	4401.1	6.6	1.8	8.1		209.3
14	Dobrina	263.7	E	9.1	20.5	1218.8	1365.9	89.0	365.7	4252.5	6.1	1.7	7.4		173.0
15	Dobrina	277.7	E	9.0	20.4	1203.4	1383.3	88.7	367.3	4245.6	6.1	1.7	7.4		195.6
16	Dobrina	276.7	E	9.0	20.4	1204.5	1383.3	88.9	367.2	4247.4	6.1	1.7	7.4		202.8
17	Dobrina	182.8	E	9.5	21.0	1309.0	1400.7	88.6	356.4	4427.6	6.7	1.8	8.5	7.9	214.8
18	Dobrina	215.7	E	9.3	20.8	1272.1	1392.0	91.8	360.2	4361.7	6.5	1.8	8.0		220.5
19	Dobrina	158.5	SV	9.6	21.2	1336.5	1287.6	92.6	353.6	4356.9	6.3	1.7	8.5		209.4
20	Dobrina	161.2	S	9.6	21.2	1333.3	1392.0	90.6	353.9	4456.5	6.8	1.9	8.5		212.1
21	Corni	157.2	S	9.7	21.2	1337.9	1374.6	92.0	353.4	4446.2	6.7	1.8	8.5	8.5	219.0
22	Recea	104.1	SV	9.9	21.5	1398.3	1487.7	90.3	347.3	4651.9	7.6	2.1	9.0		221.4
23	Recea	89.3	E	10.0	21.6	1415.3	1409.4	90.1	345.6	4599.6	7.3	2.0	8.9		220.3
24	Recea	83.4	S	10.0	21.7	1422.1	1470.3	90.7	344.9	4670.8	7.7	2.1	9.4	9.0	217.6
25	Recea	79.1	SV	10.1	21.7	1427.1	1383.3	91.2	344.4	4591.3	7.2	2.0	8.9		213.6
26	Recea	80.2	SV	10.1	21.7	1425.8	1392.0	90.7	344.5	4598.1	7.3	2.0	8.9		197.3

* AAT = average annual temperature (°C); TWM = average temperature of the warmest month (°C); Σt_u = *sum* of effective *temperatures* (°C/Apr 1st to Sept 30); ASD = actual sunshine duration (hours/Apr 1st to Sept 30); GR = global radiation (kcal/cm²/Apr 1st to Sept 30); PP = precipitation in the growing season (mm/Apr 1st to Sept 30); IAOe = oenoclimate aptitude index (Apr 1st to Sept 30); Ibcv = bioclimatic index; IHa = actual heliothermal index (Apr 1st to Sept 30) (Apr 1st to Sept 30); (Irimia *et al.*, 2014).

vineyard, as for all the other 12 sampling sites from this area the thermal parameters are much lower (Table 1).

GR values are the highest in the sampling sites from the Dobrina, Corni and Recea vineyards (90.1 - 92.0 kcal/cm²), which are exposed mainly to south and east, and the lowest in the sampling sites from Schit (80.6-87.9 kcal/cm²), exposed mainly to north (Table 1). ASD varies significantly in the Schit and Dobrina vineyards located on uneven relief and are much homogeneous on the flat terrains of the Recea vineyard; overall, the sampling sites from Recea are characterized by values of about 1400 hours (1383.3-1487.7 hours), while those of the Schit, Dobrina and Corni present very dissimilar values, ranging, in a large interval, from 1287.6 to 1479.0 hours. PP values vary in a narrower range, from 342.2 mm to 367.3 mm, with minimum values in the sampling sites from Recea and maximum ones in those from Dobrina and Schit (Table 1). The bioclimatic indices IAOe, Ibcv and IHa reveal poor heliothermal resources in the sampling sites from Dobrina and Schit, and maximums in the sampling sites from Recea and Corni, thus revealing enhanced heliothermal resources (Table 1).

Climate suitability

The estimates of climate suitability for wine production of the 26 sampling sites vary between 6.9 a.r.p. and 9.4 a.r.p. (Table 1). It is a wide range of variation that shows climatic resources suitable for the production of different types of wine, from sparkling wines to quality red wines. However, the overall averages of the Schit and Dobrina vineyards of the higher zone are of 8.0 a.r.p. and 7.9 a.r.p. respectively, indicating climatic suitability for quality white wines and red table wines; the overall averages of the Corni and Recea vineyards of the lower zone are higher, by 8.6 a.r.p. and 9.0 a.r.p. respectively, revealing the climate suitability for quality red wines production primarily and quality white wines production in secondary.

Correlations between climate variables and GrSC

The analysis of the correlation coefficients between the estimates of climate variables and the GrSC for the 26 sampling sites, reveals the climatic variables driving GrSC, the way they influence GrSC and their representativity in the assessment of the spatial variation of GrSC.

The Pearson correlation coefficient ranging from 0.28 to 0.60 shows the conditioning of GrSC by the local climatic factors (Fig. 2a to Fig. 2i). Climatic conditioning of GrSC is best represented by IHa, Ibcv and IAOe ($r = 0.57 - 0.60$) (Fig. 2a, 2b, 2c). The climatic factor which is most influential in GrSC is ASD which strongly correlates with the GrSC ($r = 0.57$) while the least influential is GR which weakly correlates with GrSC ($r = 0.28$) (Fig. 2g). The temperature, characterized by AAT, TWM and Σtu (Fig. 2d, 2e, 2f), as well as the PP, moderately correlate with GrSC ($r = 0.43$). All correlations are significant for $p < 0.05$, except for the one for the GR.

The coefficient of determination (R^2) reveals that of all climatic factors, the ASD explains best, rather 33 % of the GrSC variance (Fig. 2h), while temperature (AAT, TWM, Σtu) and PP explain only 18 % of the GrSC variance (Fig. 2d, 2e, 2f, 2i).

The simple linear regression shows that, except for PP, the increase of all the other climate variables is associated with the increase of the GrSC (Fig. 2). For the climatic factors, which vary from one year to another, the increase by one unit of the multiannual mean is associated with the increase of GrSC by:

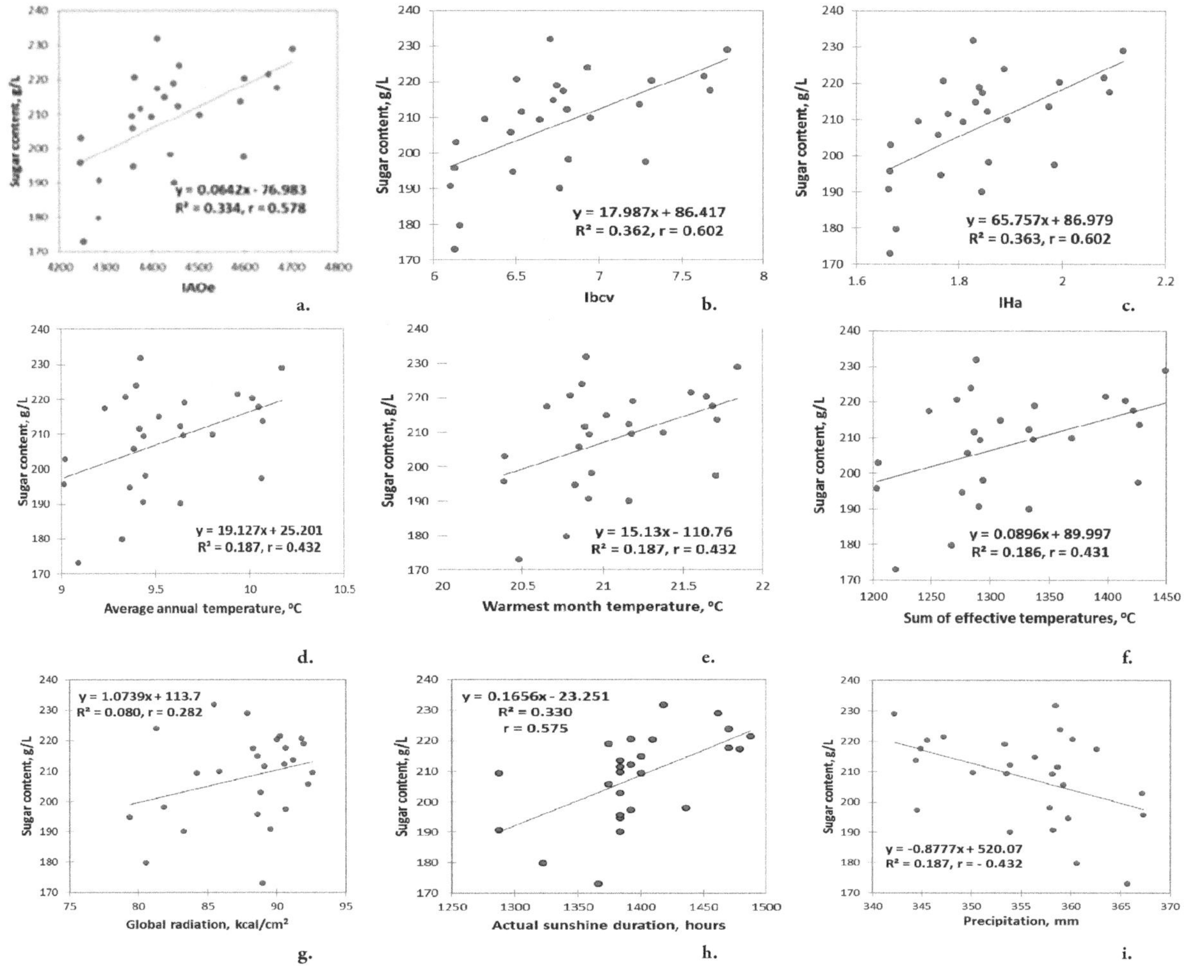

Fig. 2. Regressions between multiannual climate variables and GrSC (g/L) for the 2011 vintage in the Huşi wine growing region

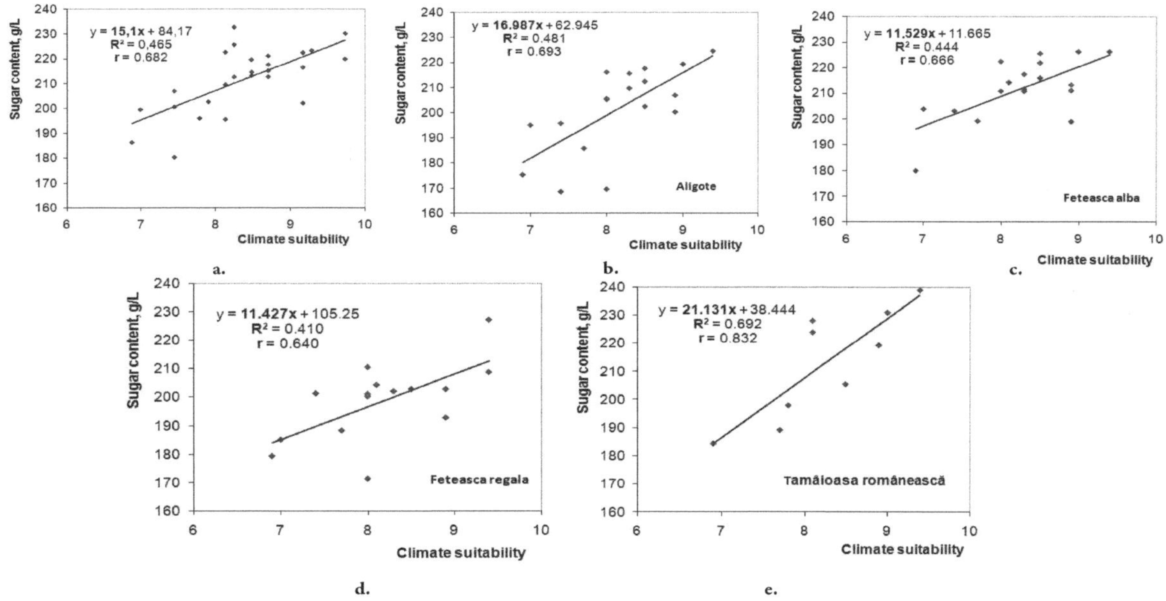

Fig. 3. Regressions between climate suitability for different types of wine production and the GrSC (g/L) for the 2011 vintage in the Huşi wine growing region: a. for the mean of GrSC on 26 sampling sites; b. for 'Aligoté' variety; c. for 'Fetească albă' variety; d. for 'Fetească regală' variety; e. for 'Tămâioasă românească' variety

19.12 g/L in the case of AAT (Fig. 2d); 15.13 g/L in the case of TWM (Fig. 2e); 1.07 g/L in the case of GR (Fig. 2g); and 0.16 g/L in the case of ASD (Fig. 2h). For the PP, the increase by 1 mm of the multiannual mean is associated with the decrease by 0.87 g/L of the GrSC (Fig. 2i).

Correlations between climate suitability and GrSC

Pearson correlation coefficient reveals a strong positive relationship (r = 0.68) between the climate suitability and the GrSC (Fig. 3a); what is more the R^2 shows that climate suitability explains 46.5% of the GrSC variance, and the increase by one unit of the climate suitability/a.r.p. is associated with the increase by 15.1 g/L of the GrSC. These correlations are significant for $p < 0.05$.

When it comes to wine grape variety some differences can be noticed: Pearson correlation coefficient (r) ranges between 0.64 (strong positive correlation) for the 'Fetească regală' variety and 0.83 (very strong positive correlation) for the 'Tămâioasă românească' variety (Fig. 3); the R^2 shows that climate suitability explains 41% of the GrSC variance for the 'Fetească regală' variety up to 69% for the 'Tămâioasă românească' variety. The increase by one unit of the climate suitability is associated with the increase of the GrSC from 11.4 g/L for the 'Fetească regală' variety up to 21.1 g/L for the 'Tămâioasă românească' variety (Fig. 3).

Spatial variation of the GrSC

The map of GrSC for the 2011 vintage shows that it ranges between a maximum of 227.5 g/L in the Recea vineyard and a minimum of 165.3 g/L in the Dobrina and Schit vineyards (Fig. 4). This spatial variation reveals the correlation of GrSC with altitude (r = 0.42; R^2 = 0.18), and the altitudinal decrease of the GrSC, from the lower zone towards the higher zone of the study area. The linear regression between the altitude and the GrSC shows that the increase by one m of the altitude is associated with the decrease by 0.10 g/L of the GrSC (Fig. 5), a difference of about 21 g/L for the 2011 vintage between the lowermost sampling point (60.1 m asl.) and the uppermost sampling point (277.7 m asl.).

Statistical data show that the means of the interpolated GrSC values for the study area vary in a narrow range, between 205.9 g/L for the Schit vineyard and 216.9 g/L for the Recea vineyard (Table 2). The most uniform in terms of GrSC spatial distribution and hence in terms of grapes quality spatial distribution is the Corni vineyard (CV= 1.2), while the less uniform is the Schit vineyard (CV= 4.3). The *spread* index

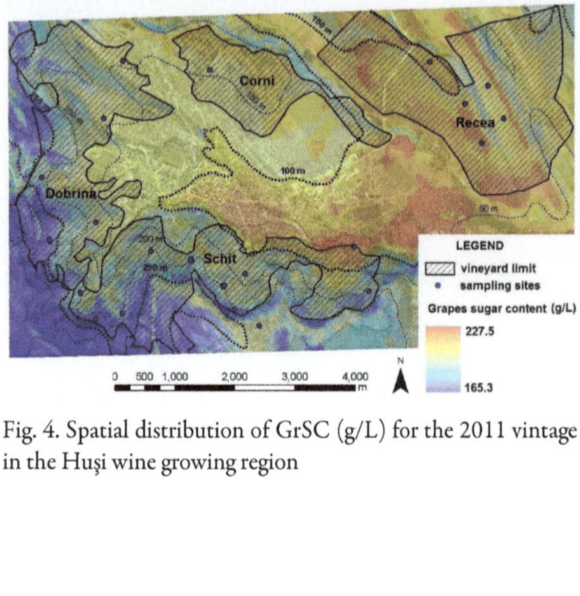

Fig. 4. Spatial distribution of GrSC (g/L) for the 2011 vintage in the Huşi wine growing region

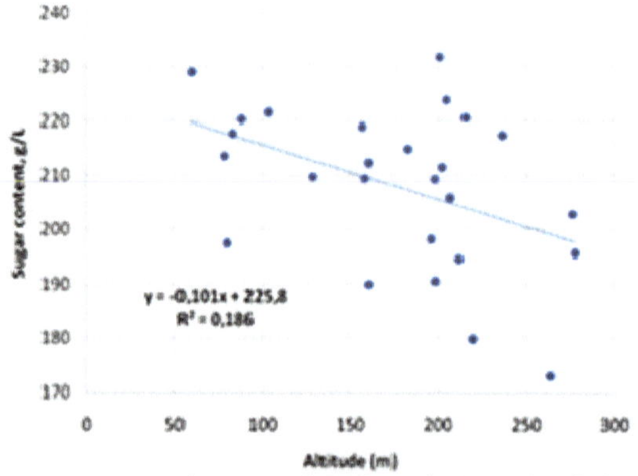

Fig. 5. Regressions between altitude and GrSC (g/L) for the 2011 vintage in the Huşi wine growing region

reveals the same variability: a spread by 7.6 % for the Corni vineyard and by 27.9 % for the Schit vineyard, with an overall average by 16.9 % for the entire study area.

Spatial distribution of wine types which can be obtained depending on spatial distribution of GrSC of 2011 vintage

By classifying the spatial distribution of GrSC of the 2011 vintage according to ranges of GrSC required by the DOC wine production regulation for the Huşi region, it can be noticed that it allows obtaining mainly two types of wine: DOC-CT wine on 52 % of the wine growing region surface and DOC-CMD wine on 41.0 % of the wine growing region surface (Table 3); the GrSC adequate to DOC and IG wines production had insignificant shares, of 3.9 % and 2.8 % respectively of wine growing region area.

The analysis of spatial distribution of GrSC classified according to DOC wine production requirement reveals that the wines types obtainable differentiate according to terrain altitude (Fig. 6): the DOC-CT wine production area extends in the lower zone, up to 150 m asl.; the DOC-CMD wine production area stands between 150-250 m asl.; the DOC and IG wine production zones correspond to high areas at above 250 m asl.

Table 2. Statistics for the spatial distribution of the GrSC within the Huşi wine growing region

Vineyard	Surface (ha)	Sugar content (g/L)						
		Min	Max	Range	Mean	STD	CV	*Spread**
Schit	554.3	168.5	226.1	57.6	205.9	8.8	4.3	27.9
Dobrina	570.6	172.6	214.5	41.9	206.2	7.9	3.8	20.3
Corni	264.4	204.6	220.9	16.3	215.0	2.6	1.2	7.6
Recea	749.7	199.8	225.6	25.7	216.9	3.9	1.8	11.8

* *spread* is the difference between the maximum and minimum values, expressed as a % of the mean value (Bramley, 2005).

Table 3. The structure of wine types which can be obtained in the Huşi wine growing region, depending on spatial distribution of GrSC for the 2011 vintage

GrSC (g/L)	Wine type*	Vineyard								Total surface	
		Schit		Dobrina		Corni		Recea			
		ha	%	ha	%	ha	%	ha	%	ha	%
< 190	IG	35.3	6.3	26.6	4.6	0.0	0.0	0.0	0.0	62.0	2.8
190-196	DOC	39.6	7.1	44.2	7.7	0.0	0.0	0.0	0.0	83.9	3.9
196-213	DOC-CMD	387.8	69.9	356.6	62.4	34.8	13.1	99.3	13.2	878.6	41.0
> 213	DOC-CT	91.4	16.4	143.1	25.0	229.5	86.8	650.3	86.7	1114	52.0
Total		554.3	100	570.6	100	264.4	100	749.7	100	2139	100

* IG = quality wines with geographical denomination; DOC = high quality wines with controlled designation of origin; DOC-CMD = controlled designation of origin wines harvested at full maturity; DOC-CT = controlled designation of origin wines late harvest.

Comparing the spatial distributions of wine types which can be obtained (Fig. 6) and climate suitability specific to the Husi area (Fig. 7), a similitude of two spatial distributions can be noticed. Supported by the strong

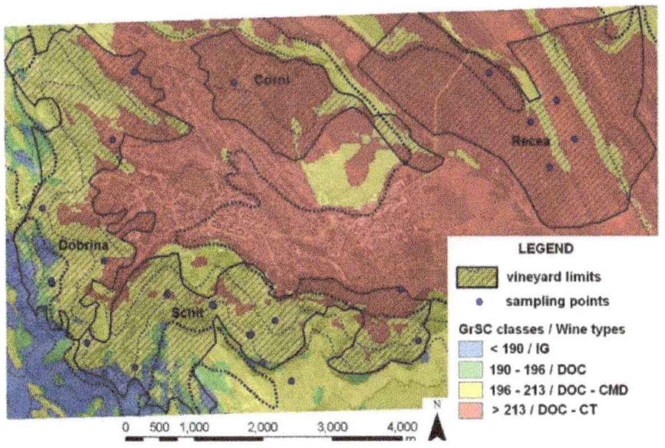

Fig. 6. Spatial distribution of the wine types obtainable within the Husi area according to spatial distribution of the GrSC for the 2011 vintage

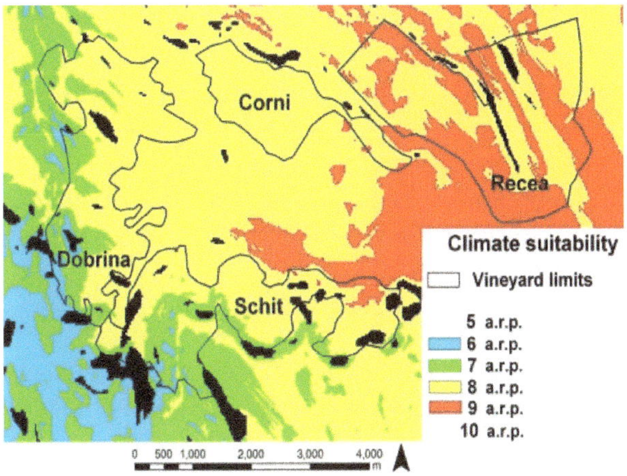

Fig. 7. Map of climate suitability (a.r.p. distribution) within the Husi wine growing region (Irimia *et al.*, 2014)

correlation of GrSC with climate suitability (r = 0.68) this similarity may reveal a pattern driving the spatial distribution of GrSC within the Husi wine growing region.

Discussions

This study analyzed the correlations of the GrSC of the 2011 vintage with the averages of climate variables characterizing the Huşi wine growing region and also with the climate suitability for wine production they generate. Based on regression – kriging analysis, the spatial distribution of GrSC of the 2011 vintage was developed, and by its classification, the map of wine types which GrSC provides was achieved. The two maps show a potential pattern of spatial distribution of GrSC at the vineyard level. This may be a new result given the fact that the research which has been conducted so far (Bramley, 2005; Tisseyre *et al.*, 2008; Santos *et al.*, 2012) has focused mainly on spatial variation of grapes quality parameters at plot scale. Although the analysis at the plot scale provides very accurate results, it reveals only local distribution patterns generated by the soil influence, while at vineyard scale the factor driving the variability is the climate (Van Leeuwen, 2010).

The study revealed strong correlations, statistically significant for $p < 0.05$, between multiannual averages of climate variables which define the climate of the Huşi wine growing region and the GrSC of the 2011 vintage. The ASD, which closely correlates with GrSC (r = 0.57) turned out to be a very influential climatic factor of the GrSC. These results are in accordance with the ones of previous studies (Teodorescu *et al.*, 1987) emphasizing that the years with an exceeding of ASD of 8.9 % are suitable for wine quality, while those with a deficit of ASD of 9.8% are unsuitable for wine quality. Temperature, whose influence was expressed in this study by AAT, TWM and Σtu variables, slightly correlates with GrSC (r = 0.43); although this correlation is weaker compared to the values communicated by Huglin (1978) (r = 0.71-0.83), it indicates that the temperature should be taken into account when analyzing and modelling the spatial distribution of GrSC.

The close correlation (r = 0.68) between the GrSC for the 2011 vintage and climate suitability of the Huşi wine growing region, identified in our research, reveals one of the relations of yield quality dependence which functions at vineyard level. Climate, together with grapevine variety determine differences of grapes composition at regional level (Van Leeuwen, 2010). Climate also determines the type of wine specific to a certain vineyard (Jones *et al.*, 2005). However, the study revealed that the correlation GrSC – climate suitability has a distinct value for each wine grape variety. This requires a customization of the analysis and modelling of the GrSC for each wine grape variety.

Mapping the GrSC based on its correlation with climate suitability (r = 0.68) revealed a coherent pattern of spatial variation at vineyard level. The study showed a significant correlation of the GrSC with altitude (r = 0.42, $p <0.05$) and that under temperate continental climate conditions of the Huşi wine growing region, an increase by 1 m of the altitude, is associated with a decrease by 0.10 g/L of the GrSC. This decrease of the GrSC can be related with the retarding effect of altitudinal increasing on grapes maturation, highlighted by previous studies (Calame *et al.*, 1977). Linearity of the altitudinal distribution of the GrSC is disrupted by the ASD variation that generates local variation of GrSC regardless of the altitude. GrSC variability is characterized by an average of 16.9 % of the *spread* index, value which is comparable with the one provided by other authors (Bramley, 2005).

The map of GrSC allowed revealing, as practical utility, the spatial distribution of wine types that can be obtained within the area. Differentiation of wine types in relation to altitude has been highlighted for other wine growing regions too (Shanmuganathan, 2010). Also, classification of spatial distribution of grapes quality parameters in order to delineate areas of wine quality within the vineyards is already practiced (ASAE, 2007). In our study two areas with different potential for wine production were delineated: an area situated below 150 m asl. where the GrSC records high values which ensure DOC-CT wines obtaining; and an area ranging between 150 and 250 m asl. where the GrSC records lower values which ensure DOC-CMD wines obtaining.

Since vineyards' climate is specific, temporally stable (Jones *et al.*, 2005) and less affected by vintage climate, its influence on grapes quality parameters should generate rather similar consequences from one year to another. Thus, the map of GrSC and also the map of wine types that it provides should present a certain spatio-temporal stability. By knowing the regression coefficients which characterize the relationship between climate variables and GrSC, the impact of their annual variation on wine production zones can be predicted. However, the lack of information on yield quality over several vintages prevents us from considering the study's outcomes as indisputable. Moreover, accurate setting of the spatial distribution of GrSC would require the representation of pedological and biological factors influence. Yet, given the significance of the statistical correlations between climate and GrSC, the map of spatial distribution of the GrSC for the 2011 vintage and the map of obtainable wine types, based on it, can be considered as guidance tools for site specific management within the Huşi wine growing region.

Conclusions

The study provides new information on the factors that determine the spatial variation of grapes sugar content within vineyards. Although it is a spatially and temporally variable parameter, grapes sugar content shows significant correlations with climate variables specific to vineyards and with the climatic suitability for wine production they generate. Knowing these correlations allows revealing patterns that determine spatial distribution of grapes sugar content within the vineyard. In the case of the Huşi wine growing region, based on the significant correlation of GrSC with climate suitability ($r = 0.68$), the map of GrSC for the 2011 vintage and the map of wine types that can be obtained were drawn up. The two maps reveal a coherent pattern of spatial distribution of grapes sugar content at vineyard scale, significantly correlated with terrain altitude. Such maps can be improved by integrating the influence of technological, biological and pedological factors on GrSC and can be used as tools for vineyard site-specific management. Based on grapes sugar content of a single vintage, this study did not allow the assessment of spatio-temporal stability of the obtained pattern of GrSC spatial distribution. By mapping the grapes sugar content for many years and analyzing its correlations with other factors which determine it, a clear pattern of GrSC spatial distribution within the Huşi wine growing region can be achieved.

References

Agelet-Fernández J, Martínez-Casasnovas JA, Arnó J (2009). Selective harvesting zones from remote sensing images and yield data and relation to grape quality parameters for precision viticulture. EFITA Conference 709-715.

ASAE (2007). Interpolation of wine grape quality indicators (Anthocyanin and Brix) and development of differential harvest attachment. ASAE Annual Meeting 071097. doi:10.13031/2013.22971.

Bramley RGV (2005). Understanding variability in winegrape production systems. 2. Within vineyard variation in quality over several vintages. Aust J Grape Wine Res 11(1):33-42.

Calame F, Rochaix M, Simon JL (1977). Observations phénologiques et mesures bioclimatiques dans plusieurs sites viticoles valaisans à différentes altitudes en vue de la délimitation de l'aire viticole. Bulletin de l'OIV 50-559:601-616.

Champagnol F (1984). Eléments de physiologie de la vigne et de viticulture générale. Champagnol F (Ed), Montpellier, 351 p.

FAO - Food and Agriculture Organization of the United Nations (2003). FAOCLIM-2. World-wide Agroclimatic Database v.2.02, FAO/SDRN.

Galet P (1993). Precis de viticulture. Charles Dehan, Montpellier.

Huglin P (1978). Nouveau mode d'évaluation des possibilités héliothermiques d'un milieu viticole. CR Acad Agric Fr 64(13):1117-1126.

Huglin P (1986). Biologie et écologie de la vigne. Payot (Ed), Lausanne, 372 p.

Huglin P, Balthazard J (1976). Données relatives a l'influence du rendement sur le taux des sucres des raisins. Connaiss Vigne Vin 10:175-191.

Irimia L, Patriche CV, Quenol H (2011). GIS applications in viticulture. The spatial distribution of climatic suitability for grapegrowing in Husi wine-growing centre (Romania). Cercet Agron Mold 3(147):75-87.

Irimia L, Patriche CV, Quenol H (2014). Analysis of viticultural potential and deliniation of homogeneous viticultural zones in a temperate climate region of Romania. J Int Sci Vigne Vin 48(3):145-167.

Jackson DI, Lombard PB (1993). Environmental and management practices affecting grape composition and wine quality – a review. Am J Enol Vitic 44(4):409-430.

Jones GV, White MA, Cooper OR, Storchmann K (2005). Climate change and global wine quality. Clim Change 73(3):319-343.

Kottek M, Grieser J, Beck C, Rudolf B, Rubel F (2006). Worldmap of the Köppen-Geiger climate classification updated. Meteorol Z 15:259-263.

Lamb DW, Weedon MM, Bramley RGV (2003). Using remote sensing to map grape phenolics and colour at harvest in a Cabernet Sauvignon vineyard: timing observations against vine phenology and optimising image resolution. Aust J Grape Wine Res 10:46-54.

OIV (1990). Recueil des méthodes internationales d'analyse des vins et des moûts. 368 p. Office International de la Vigne et du Vin. Paris.

ONIV (2015). Specification for the production and marketing of wine with designation of Huşi origin. http://www.oniv.ro/att_view.php?id=433 (in Romanian, accessed in 4 October 2014).

Patriche CV (2007). Aspects concerning the spatialization of radiation balance components. In: Spatial Interpolation for Climate Data, pp. 121-138. The Use of GIS in Climatology and Meteorology. Dobesch H, Dumolard P, Dyras I (Eds), Wiley-ISTE.

Santos AO, Wample RL, Sachidhanantham S, Kaye O (2012). Grape quality mapping for vineyard differential harvesting. Braz Arch Biol Technol 55(2):193-204.

Sethuramasamyraja B, Sachidhanantham S, Yen M, Wample R (2007). Interpolation of wine grape quality indicators (Anthocyanin and Brix) and development of differential harvest attachment. In: Proceedings of the 2007 ASABE Annual Meeting,

Minneapolis, 115-120.

Shanmuganathan S (2010). Viticultural Zoning for the Identification and Characterisation of New Zealand "Terroirs" Using Cartographic Data. Proceedings of GeoCart'2010 and ICA Symposium on Cartography, New Zealand Cartographic Society Inc.

Teodorescu S, Popa AI, Sandu G (1987). Oenoclimate of Romania (in Romanian). Ed. Ştiinţifică şi Enciclopedică, Bucureşti.

Tisseyre B, Mazzoni C, Fonta H (2008). Within-field temporal stability of some parameters in viticulture: Potential toward a site specific management. J Int Sci Vigne Vin 42:27-39.

Tisseyre B, Taylor J (2005). An overview of methodologies and technologies for implementing precision agriculture in viticulture. XII Congresso Brasileiro de Viticultura e Enologia Anais 45-54.

Trought MCT, Bramley RGV (2011). Vineyard variability in Marlborough, New Zealand: Characterizing spatial and temporal changes in fruit composition and juice quality in the vineyard. Australian Journal of Grape and Wine Research 17(1):79-89.

Van Leeuwen C (2010). Terroir: the effect of the physical environment on vine growth, grape ripening and wine sensory attributes, pp. 273-315. In: Reynolds AG (Ed). Managing wine quality. Viticulture and wine quality. Woodhead Publishing, Cambridge.

Wold S, Trygg J, Berglund A, Antti H (2001). Some recent developments in PLS modeling. Chem Intelligent Lab Systems 58:131-150.

Workman Jr J, Burns DA (2001). Comercial NIR instrumentation. In: Burns D, Ciurczak EW (Eds). Handbook of Near-Infrared analysis. pp. 53-70. New York: Marcel Dekker.

In Vitro and *in Planta* Activity of Some Essential Oils against *Venturia inaequalis* (Cooke) G. Winter

Géza NAGY[1]*, Tamás HOCHBAUM[1], Szilvia SÁROSI[2], Márta LADÁNYI[3]

[1]*Corvinus University of Budapest, Faculty of Horticultural Sciences, Department of Plant Pathology, 44. Ménesi Str., H-1118 Budapest, Hungary; geza.nagy@uni-corvinus.hu (*corresponding author); HochbaumT@nebih.gov.hu*
[2]*Corvinus University of Budapest, Faculty of Horticultural Sciences, Department of Medicinal and Aromatic Plants, 29-43 Villányi Str., H-1118 Budapest, Hungary; szilvia.sarosi@uni-corvinus.hu*
[3]*Corvinus University of Budapest, Faculty of Horticultural Sciences, Department of Biometrics and Agricultural Informatics, 29-43 Villányi Str., H-1118 Budapest, Hungary; marta.ladanyi@uni-corvinus.hu*

Abstract

The effect of the essential oils of thyme (*Thymus vulgaris*), cinnamon (*Cinnamomum verum*) and sweet orange (*Citrus sinensis*) on conidial germination of *Venturia inaequalis* was investigated *in vitro*. Moreover, the effect of cinnamon oil was further studied by analyzing its preventive and curative activity as well as its rainfastness in an apple seedling-Venturia pathosystem under growth chamber and greenhouse conditions. The essential oils suppressed the germination of conidia at concentrations of 0.01% and 0.1% effectively. The highest inhibition level was achieved with thyme oil. Cinnamon oil (0.2%) with Silwet Star adjuvant (0.02%) was protective and curative against *Venturia inaequalis* when applied 72 h, 24 h, 1 h before or 24 h, 72 h after inoculation of plants. Effective control was achieved when essential oil was applied by 24 h, 1 h or 24 h pre- or post-inoculation, respectively. The 72 h application was less effective, however, it still showed decent control level. Although the simulated rainfall of light (1 mm), moderate (3 mm) and heavy (5 mm) rain decreased the control efficacy of cinnamon oil, the water amounts did not influence control level. The oil application resulted in a moderate and long lasting control (92% efficacy without rainfall, 65% efficacy with rainfall). The results of this study show that the tested essential oils are appropriate candidates for the development of biofungicides against apple scab.

Keywords: apple scab, conidia, essential oils, preventive and curative activity, rainfastness

Introduction

Apple scab (*Venturia inaequalis* (Cooke) G. Winter) is a widespread and common disease around the world wherever apples are grown (Vaillancourt and Hartman, 2000). The pathogen causes significant loss in the Hungarian apple production. In epidemic years, the number of fungicide applications against the disease often exceeds twenty sprayings in a vegetation period. In recent years, the use of fungicides against apple scab is strictly regulated by EU legislation (eg. Directive 2009/128/EC, Commission regulations No 834/2013, No 1004/2013, No 1138/2013, No 61/2014). In organic production the use of copper fungicides is limited due to soil accumulation as well (Commission Regulation No. 473/2002, Council Regulation (EEC) No. 2092/91). Therefore, the research of alternative solutions to synthetic or copper based fungicides is particularly needed due to the potential risks of the environment pollution and of fungicide resistance. Essential oils of medicinal and aromatic plants are suitable candidates for this purpose. Their antimicrobial and antifungal activity has been known for centuries and it is still an intensively studied field. Since essential oils contain several chemical compounds, the inhibition is based on different biological modes of action. The primary target of essential oils is the fungal plasma membrane. Active compounds cause lesions and reduce ergosterol biosynthesis resulting in severe morphological changes of the hyphal wall (Pinto *et al.*, 2006; Soylu *et al.*, 2010; Tian *et al.*, 2012).

Although there are numerous data available on the research of the effect of plant extracts and essential oils on plant pathogens, fewer results can be found about the effect on *Venturia* species. The plant extracts from *Citrus* sp., *Galega officinalis, Inula viscosa, Quillaja saponaria, Polypodium vulgare* and *Saponaria officinalis* inhibited conidial germination of *Venturia inaequalis* (Cooke) G. Winter effectively and gave good control of apple scab on seedlings in greenhouse studies, however, simulated rainfall decreased the efficacy significantly (Pfeiffer, 2002; Pfeiffer *et al.*, 2004). Plant extracts containing high contents of active ingredients of artemisinin (*Artemisia annua*), chelidonine (*Chelidonium majus*), menthol (*Mentha piperita*), populin (*Populus nigra*), linalool + linalyl acetate (*Salvia sclarea*) and thymol (*Thymus vulgaris*) were tested against the ascospore and conidia germinaton of *Venturia inaequalis* (Cooke) G. Winter. Almost all extracts showed high inhibition of ascospore germination. Artemisinin, thymol and populin extracts effectively inhibited conidial germination as well. The 1% *Populus* extract gave similar control efficacy against apple scab on apple trees as synthetic fungicides in a 2 years field study (Bálint *et al.*, 2014; Theisz *et al.*, 2007). The preventive application of six plant oils (canola, corn, grapeseed, olive, soybean and sunflower) reduced scab infection on apple leaves significantly under controlled condition, however, orchard application of canola and soybean oil with emulsifier did not give better control level than that of the application of emulsifier alone (Northover and Schneider, 1993). The essential oils of *Phellodendron chinense, Russowia sogdiana* and *Zanthoxylum bungeanum* inhibited *in vitro* the mycelial growth of *Venturia pirina* and other fungi effectively. The wide antifungal activity could be attributed to the high contents of monoterpene compounds (Gong *et al.*, 2009, 2011; Tan *et al.*, 2007). Berngtsson *et al.* (2006, 2009) compared the effect of plant extracts and the resistance inducer acibenzolar-S-methyl on conidial germination with their effect on scab infection on apple seedlings. Several extracts inhibited effectively conidial germination, however, showed no or moderate *in vivo* efficacy. They pointed out that besides the direct fungistatic or fungicide effect of the plant extracts, induction of plant resistance might also play an important role in their mode of action.

Although the effect of plant extracts on *Venturia inaequalis* (Cooke) G. Winter has been actively investigated, the effect of essential oils is little-known. This is particularly true to their *in vivo* mode of action, duration of efficacy and rainfastness, which latter highly influence the control efficacy in outdoor conditions.

The effect of the essential oils of thyme (*Thymus vulgaris*) and cinnamon (*Cinnamomum verum*) in different plant – pathogen relationships has been investigated in the Department of Plant Pathology, Corvinus University of Budapest since 2005 (eg. Hochbaum and Nagy, 2013; Horváth *et al.*, 2013; Petróczy *et al.*, 2006).

The objectives of this study were to (i) investigate the effect of thyme, cinnamon and sweet orange oil on the germination of the conidia of *Venturia inaequalis* (Cooke) G. Winter, (ii) reveal *in vivo* mode of action of cinnamon oil against scab on apple seedlings, (iii) assess the length of efficacy of cinnamon oil and (iv) evaluate the effect of rain amount on the efficacy of essential oil treatment.

Materials and methods

Essential oils and adjuvant

The essential oils of cinnamon (*Cinnamomum verum*) (Aromax Inc.), thyme (*Thymus vulgaris*) (Aromax Inc.) and sweet orange (*Citrus sinensis*) (Naturol Ltd.) were selected for *in vitro* and *in vivo* assays. Chemical composition was determined by GC 6890N Gas Chromatograph equipped with 5975 Inert mass selective detector (Tab. 1). According to previous studies (Hochbaum and Nagy, 2013), Silwet Star adjuvant was added in 0.02% concentration to aqueous essential oil dilutions in order to facilitate the dispersion of the oils. The applied concentration did not influence conidium germination. In the wash-off test the adjuvant was not applied.

Tab. 1. Composition (%) of the selected essential oils

Component	RT	LRI	Cinnamon (%)	Thyme (%)	Orange (%)
α-pinene	5.56	938	-	0.67	0.31
camphene	5.95	952	-	1.44	-
sabinene	6.52	976	-	-	0.19
β-myrcene	6.99	995	-	1.47	1.25
α-terpinene	7.79	1018	-	2.15	-
ρ-cymene	8.09	1026	1.77	24.58	-
limonene	8.19	1029	0.86	-	97.64
1,8-cineol	8.38	1034	3.58	6.54	-
γ-terpinene	9.2	1056	-	7.42	-
linalool	10.76	1097	3.65	5.5	0.22
borneol	13.43	1162	-	0.63	-
terpinen-4-ol	13.96	1175	-	0.93	-
α-terpineol	14.55	1189	1.12	-	-
phenylethyl alcohol	14.83	1201	1.64	-	-
thymol	18.81	1290	-	40.64	-
carvacrol	19.2	1300	-	3.6	-
cinnamyl aldehyde	20.8	1351	67.9	-	-
eugenol	21.44	1361	6.83	-	-
geranyl acetate	22.43	1388	-	0.52	-
β-caryophylene	23.68	1420	2.5	3.08	-
methyl trans-cinnamate	24.14	1436	2.09	-	-
ethyl cinnamate	24.2	1440	0.75	-	-
α-humulene	25.07	1454	-	0.32	-
cinnamyl-acetate	26.13	1488	6.62	-	-
caryophylene oxide	30.2	1590	-	0.21	-
benzyl benzoate	35.67	1738	0.33	-	-
Total			99.64	99.7	99.61

Note: RT - Retention time, LRI - Linear retention index relative to C8-C23 n-alkanes on a HP-5 column

Test plants and fungal inoculum

Apple seedlings with 4-6 fully expanded leaves were used for *in vivo* assays. The seedlings were produced from seeds of apple cultivars 'Gala' and 'Jonagold'. Plants were grown in a light-room at 18-24 °C with 12 h photoperiod in vegetable trays and were later transferred into individual pots.

Diseased apple leaves were collected from non-treated orchards and home gardens from different cultivars. Leaves were stored in paper bags in refrigerator (4 °C) until inoculation. Conidia were harvested from the leaves by a small brush into sterile distilled water to obtain conidial suspension of a mixed population. Inoculum concentration was measured by haemocytometer.

In vitro assay of the inhibition of conidial germination

Two-fold dilutions of essential oils dissolved either in water or in 0.1% apple leaf ('Jonagold') brew were prepared. Oil dispersion was facilitated by Silwet Star (0.02%). 60-60 µl of conidium suspensions (7.2 × 10^5 conidia/ml) was measured into the wells of single concave microscope slides. 60-60 µl essential oil suspensions were added to the conidium suspensions. Final concentrations of essential oils were 0.001%, 0.01% and 0.1%. The wells were covered by cover slips to prevent evaporation of the oils. Conidia suspensions were incubated on 20°C in the dark. Germination was assessed microscopically after 96 h by counting 3 × 30 spores per treatment and compared to the germination in water, in apple leaf brew and in 0.02% Silwet Star solution. Conidia were considered germinated when the germ tube was at least twice as long as the one of the conidia. The trial was performed in triplicates. Efficacy of essential oils was calculated by Abbott's formula (Abbott, 1925).

In vivo plant assays

The essential oil of cinnamon was selected for plant assays. According to our earlier *in vitro* studies (Cseh *et al.*, 2014), the essential oil of thyme could be harmful to plants under forced conditions. Orange oil does not show significant *in vivo* inhibition against *Venturia inaequalis* (Cooke) G. Winter in ongoing orchard trials.

Assessment of preventive and curative activity of cinnamon oil

The experiment was carried out in a greenhouse during the winter of 2013-2014. Potted 'Gala' seedlings were placed into plastic boxes. Plants were kept between 10 °C and 27 °C under natural photoperiod supplemented with 12 h artificial light. Relative humidity ranged between 70% and 97%. Artificial inoculation was performed by spraying conidium suspension (4.9-5.4 × 10^5 conidia/ml) onto the upper and lower side of the leaves. Seedlings were sprayed with 0.2% cinnamon oil diluted in 0.02% Silwet Star 72 h, 24 h and 1 h prior to inoculation as protective treatments and 24 h or 72 h after inoculation as curative treatments. Conidium and essential oil suspensions were sprayed onto the plants by hand sprayer until "run-off". After inoculation, relative humidity was maintained above 90% for three days by covering boxes by a plastic lid in order to facilitate conidial germination. Evaluation was carried out on the 14th and 21st days after inoculation by the assessment of 30 leaves per treatment in six blocks according to their infected leaf area. Inhibition of essential oil treatments was compared to inoculated control plants.

Wash-off study of cinnamon oil

The experiment was carried out in a growth room (14-16 °C, 12 h photoperiod, 80-95% relative humidity). Potted 'Jonagold' seedlings were sprayed until "run-off" by a single-nozzled hand sprayer with 0.2% cinnamon oil diluted in distilled water without adjuvant. In order to ensure dilution homogeneity sprayer was being shaken continuously. Treated plants were left to dry for 4 hours (15 °C, 80% RH) before rainfall simulation. To assess the effect of natural rainfall on the rainfastness of the essential oil different quantity of distilled water was sprayed onto the seedlings for simulating light (1 mm), moderate (3 mm) and heavy (5 mm) rainfall. Treated plants were left to dry for 4 hours (15 °C, 80% RH) before artificial inoculation. Untreated, no-water-exposed (0 mm) and water-exposed seedlings were inoculated with conidium suspension (4.9 × 10^5 conidia/ml) in the same manner as described earlier. Evaluation of rainfastness was carried out on the 14th and 21st days by the assessment of 30 leaves per treatment in six blocks according to their infected leaf area.

Statistical analysis

In order to normalize the data and homogenize the variances, in the case of the rate of germinated conidia, arcsin transformation; in the cases of infected leaf area data, ln(x+0.0001) transformation were used. Normality was proved by skewness and kurtosis (Tabachnick and Fidell, 2006). The effect of essential oil treatment on conidial germination was evaluated by univariate ANOVA model having fixed factors as treatments (essential oil types cinnamon, thyme and orange in concentration of 0.001%, 0.01% and 0.1%, Silwet Star and Untreated control) as well as germination media (water, apple leaf brew). The preventive and curative effect of cinnamon oil was evaluated by univariate ANOVA with random block design having fixed factors as treatments (Control, 1 h, 24 h, 72 h or Control, 24 h, 72 h, respectively) as well as the length of time period after inoculation (days 14 and 21) and blocks as different plants. The effect of cinnamon oil under simulated rainfall was evaluated by univariate ANOVA with random block design having fixed factors as treatments (Control, 0 mm rain, 1 mm rain, 3 mm rain and 5 mm rain) as well as time period after inoculation (days 14 and 21) and blocks as different plants. Homogeneity of variances and the normality of the residuals of the ANOVA models run with the transformed data was yet slightly violated, however, considering the homogeneity of sample sizes and the robustness of ANOVA in case the ratio of maximal and minimal variances is under 2, we accepted the results as reliable. In cases of significant factor effects Games-Howell's post hoc test was run.

Results and discussion

Effect on conidial germination

Germination was more intensive in water than in apple leaf brew, however, similar tendency was observed in the inhibition of the germination by the essential oils in both types of applied germination media. In the lowest applied concentration (0.001%), none of the essential oils suppressed germination significantly in comparison with Silwet Star control. Inhibition level was more remarkable in 0.01% concentration. Thyme and sweet orange suppressed germination in apple brew significantly at 94.3% and 82.9% efficacy, respectively. In water, only the inhibition of thyme (96.0% of efficacy) proved to be significant. In 0.1% concentrations, all essential oils suppressed germination in apple brew effectively ($p<0.05$) at 98.9-100.0% efficacy, however, cinnamon oil did not show significant inhibition in water. Both the effects of essential oil treatment and germination media were significant on conidial germination ($F_{tr}(10;44)=46.35$ with $p<0.001$; $F_{med}(1;44)=12.84$ with $p<0.001$, resp.). The interaction was also significant ($F_{tr^*med}(10;44)=2.71$ with $p<0.05$). The treatment effect size was high, meanwhile the media effect size was low (partial $\eta^2=0.91$ and 0.27, resp.). The observed power was above 0.9 in all cases (Fig. 1).

All tested essential oils suppressed conidial germination effectively. The highest inhibition level was achieved with thyme oil. Bálint *et al.* (2014) and Theisz *et al.* (2007) found thyme extracts effective as well. The high contents of phenolic monoterpens, especially thymol (40.6%) and its precursore p-cymene (24.6%), might be responsible for the antifungal effect of thyme oil (Gong *et al.*, 2009), however, the synergistic effect of minor oil components might also contribute to the antifungal activity (Zambonelli *et al.*, 2004). According to Tan *et al.* (2007), the high content of limonene (97.6%) might have caused the good inhibitory effect of orange oil.

In apple leaf brew the germination rates of conidia were lower than in water in most treatment, which could be explained by the accumulation of antifungal substances in old apple leaves (Agrios, 2005).

Preventive and curative activity of cinnamon oil

The preventive and curative applications of 0.02% cinnamon oil decreased the disease severity on inoculated plants significantly compared to inoculated control plants. Oil application 1 h or 24 h before inoculation showed almost complete inhibition (100 to 99% efficacy) by the first while an outstanding inhibition (91 to 92% efficacy) by the second evaluation. Cinnamon oil applied 72 h before inoculation decreased infection level at a rate of 65% (first evaluation) and 51% (second evaluation) efficacy.

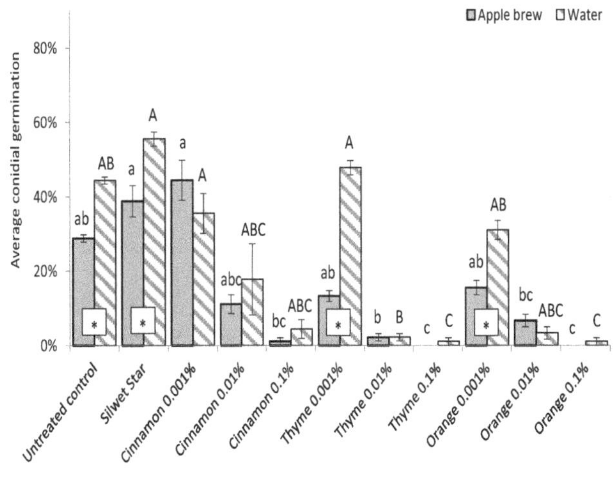

Fig. 1. Mean and standard deviation of conidium germination (%) of *Venturia inaequalis* (Cooke) G. Winter in 0.05% apple brew and in water under different treatments . Different letters are for significantly different groups (apple leaf brew: lower case; water: upper case; p<0.05; Games Howell's test). Asterisk is for significantly different types of media per treatment level.

Curative application of cinnamon showed good disease control even when applied 72 h after inoculation. At the first evaluation, efficacy ranged from 97% (24 h curative application) to 83% (72 h curative application). By the second evaluation, efficacy of cinnamon almost remained the same (96.8%) when applied 24 h after inoculation and decreased by 20% when applied 72 h after inoculation.

Both the effects of preventive and curative application were significant (preventive: $F_{tr}(3;165)=37.03$ with $p<0.001$; curative: $F_{tr}(2;123)=65.13$ with $p<0.001$) while both the effect of time length after inoculation and the interaction were insignificant in both cases (preventive: $F_{time}(1;165)=3.73$ with $p=0.06$; $F_{tr^*time}(3;165)=0.04$ with $p=0.99$, resp.; curative: $F_{time}(1;123)=0.093$ with $p=0.76$; $F_{tr^*time}(2;123)=0.28$ with $p=0.76$, resp.). The treatment effect size was moderate (partial $\eta^2=0.51$ and 0.40 for preventive and curative treatments, respectively) with observed power above 0.99 (Fig. 2).

Foliar application of cinnamon oil inhibited scab infection on apple seedlings effectively under controlled conditions. Interestingly, curative applications gave similar control efficacy as that of the preventive applications. Since curative activity of active substances is based on the inhibition of mycelial growth rather than spore germination, it can be concluded, that the main compounds of the essential oil (cinnamyl aldehyde (67.9%), cinnamyl-acetate (6.6%) and eugenol (6.8%)) effect both spore germination and mycelial growth (Soylu *et al.*, 2010).

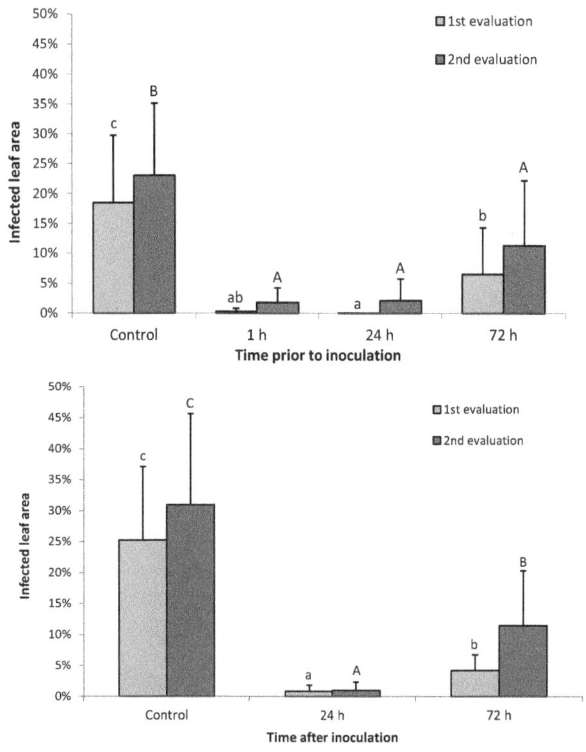

Fig. 2. Mean and standard deviation of infected leaf area (%) of apple leaves affected by preventive (above) and curative (below) treatments of 0.2% cinnamon oil, compared to inoculated untreated control. Different letters are for significantly different groups (1st evaluation: lower case; 2nd evaluation: upper case; p<0.05; Games Howell's test).

According to Velluti *et al.* (2004) cinnamyl aldehyde and eugenol are responsible for the antifungal activity. A strong curative effect of cinnamon oil was observed against Fusarium head blight of winter wheat in previous field experiments as well (Horváth *et al.*, 2013; Nagy *et al.*, 2013). However, Berngtsson *et al.* (2006) found curative application of plant extracts less effective.

The oil components eugenol, cinnamyl aldehyde, cinnamyl-acetate belong to phenylpropanoid molecules, which play an important role in plant defense mechanisms (Bhattacharya *et al.*, 2010; Dixon *et al.*, 2002). Therefore, according to Berngtsson *et al.* (2006, 2009), beside direct antifungal activity cinnamon oil may influence plant defense as well.

The application of cinnamon oil within 24 h pre- or post-inoculation resulted in very effective and long lasting disease control. Decent control level was achieved with its application within 72 h, especially by the curative timing (Fig. 3). This observation emphasizes the importance of disease forecasting for outdoor application of essential oils.

Further research is needed for the determination of the translocation of cinnamon oil within the plant.

Rainfastness of cinnamon oil

Without simulated rainfall cinnamon oil inhibited scab infection on apple seedlings effectively. Control efficacy decreased as a result of rain. However, significant difference in disease severity was not observed among the different rain amounts. The effect of treatment was significant ($F_{treatm}(4;149)=5.96$ with $p<0.001$) that is to say, untreated control revealed significantly different results that those of treated ones. Both the effect of time length after inoculation and the interaction were insignificant ($F_{time}(1;149)=1.32$ with $p=0.25$; $F_{treatm*time}(4;149)=0.48$ with $p=0.75$, resp.). The treatment effect size was low (partial $\eta^2=0.14$) with

Fig. 3. Scab symptoms on the leaves of inoculated control plant (above) and that of curatively (72 h) treated plant by cinnamon oil (below) on the 21th day

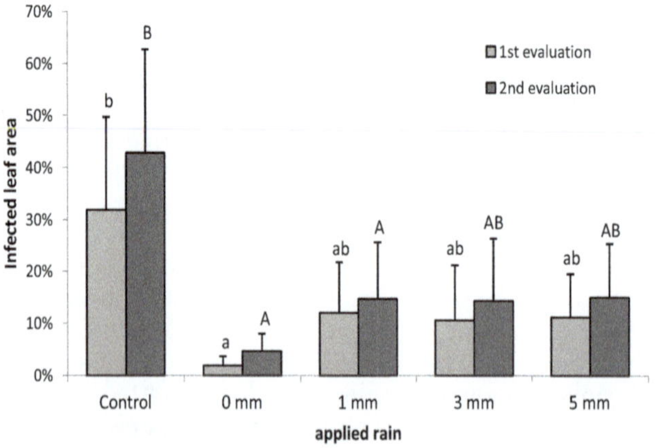

Fig. 4. Mean and standard deviation of infected leaf area (%) of apple leaves affected by cinnamon oil treatment followed by different amount of simulated rainfall, compared to inoculated untreated control. Different letters are for significantly different groups (1st evaluation: lower case; 2nd evaluation: upper case; $p<0.05$; Games Howell's test).

observed power above 0.98 (Fig. 4).

Although control efficacy decreased by simulated rainfall, the water amounts did not influence control level. It can be concluded that part of the active compounds of cinnamon oil remains on the leaf surface due to their lipophilic properties. However, Pfeiffer *et al.* (2004) found that larger amount of rain decreased the efficacy of plant extracts on apple seedlings more considerably.

In spite of simulated rainfall, oil application resulted in moderate (approximately 65% efficacy) and long-lasting disease control in laboratory conditions.

Conclusions

The tested essential oils, especially *Thymus vulgaris* suppress the conidial germination of *Venturia inaequalis* effectively. Cinnamon oil applied with adjuvant has preventive and strong curative effect against apple scab on plants which could be observed within 3 days prior- or post-inoculation. Although the efficacy of cinnamon oil application decreased by simulated rainfall, a moderate control effect was still exhibited even after the plants were exposed to relatively high amount of rain. The results of this study show that the tested essential oils may represent bases for the development of biofungicides against apple scab. However, further field experiments are needed for the investigation of control efficacy in outdoor growing conditions. Moreover, the translocation of essential oils in plant should also be tested in laboratory.

Acknowledgments

This research was supported by the European Union and the State of Hungary, co-financed by the European Social Fund in the framework of TÁMOP 4.2.4. A/1-11-1-

2012-0001 'National Excellence Program'.

References

Abbott WS (1925). A method of computing the effectiveness of an insecticide. J Econ Entomol 18:265-267.

Agrios NG (2005). Plant Pathology, 5th Ed. Elsevier Academic Press, San Diego, p. 207-248.

Bálint J, Nagy Sz, Thiesz R, Nyárádi I-I, Balog A. (2014). Biocontrol strategy to reduce asexual reproduction of apple scab (*Venturia inaequalis*) by using plant extracts. Turk J Agric For 38(1):91-98.

Bengtsson M, Jørgensen HJL, Wulff E, Hockenhull J (2006). Prospecting for organic fungicides and resistance inducers to control scab (*Venturia inaequalis*) in organic apple production. Proceeding from European Joint Organic Congress, 30-31 May, Odense, Denmark, 318-319.

Bengtsson M, Wulff E, Jørgensen HJL, Pham A, Lübeck M, Hockenhull J (2009). Comparative studies on the effects of a yucca extract and acibenzolar-S-methyl (ASM) on inhibition of *Venturia inaequalis* in apple leaves. Eur J Plant Pathol 124(2):187-198.

Bhattacharya A, Sood P, Citovsky V (2010). The roles of plant phenolics in defence and communication during *Agrobacterium* and *Rhizobium* infection. Mol Plant Pathol 11(5):705-719.

Cseh AM, Hochbaum T, Pluhár Zs, Nagy G (2014). Kerti kakukkfű (Thymus vulgaris L.) kemotípusok illóolajának és kivonatainak antifungális és fitotoxikus hatása in vitro körülmények között (in Hungarian). In vitro antifungal and phytotoxic activity of the essential oils of different chemotypes of Thymus vulgaris. Proceedings of the 60th Plant Protection Science Days February 21-22, Budapest, Hungary 74.

Dixon RA, Achnine L, Kota P, Liu CJ, Reddy MSS, Wang L (2002). The phenylpropanoid pathway and plant defence - a genomics perspective. Mol Plant Pathol 3(5):371-390.

Gong YW, Huang YF, Zhou LG, Shi XY, Guo ZJ, Wang MG, Jiang WB (2009). Chemical composition and antifungal activity of the fruit oil of *Zanthoxylum bungeanum* Maxim. (Rutaceae) from China. J Essent Oil Res 21(2):174-178.

Gong Y, Zhou L, Shi X, Ma Z, Guo Z, Wang M, Wang J, Li X (2011) Chemical composition of the fruit essential oil of *Phellodendron chinense* (Rutaceae) from China and its antifungal activity against plant pathogenic fungi. J Essent Oil Res 23(1):108-112.

Hochbaum T, Nagy G (2013). Efficacy of essential oils against the major pathogens and fruit moth of apricot and peach (in Hungarian). Növényvédelem 49(1):8-16.

Horváth A, Kovács B, Nagy G (2013). Application of mint and cinnamon against *Fusarium* disease of winter wheat. Episteme: Czasopismo Naukowo-Kulturalne 18(3):297-304.

Nagy G, Kovács B, Kovács F (2013). Control of *Fusarium* headblight of winter wheat (Application of essential oils in the control of *Fusarium* headblight of winter wheat) (in Hungarian). Biokultúra 24(6):20-22.

Northover J, Schneider KE (1993). Activity of plant oils on diseases caused by *Podosphaera leucotricha*, *Venturia inaequalis*, and *Albugo occidentalis*. Plant Dis. 77(2):152-157.

Petróczy M, Nagy G, Fekete M, Vancsura M, Bánátfy R, Palkovics L (2006). Antifungal activity of essential oils. Acta Microbiol Imm H 53(3):332-333.

Pfeiffer B (2002). Greenhouse experiments on control of *Venturia inaequalis.* First results. Proceedings of the 10th International Conference on Cultivation Technique and Phytopathological Problems in Organic Fruit-Growing and Viticulture 10:81-85.

Pfeiffer B, Alt S, Schulz C, Hein B, Kollar A (2004). Investigations on alternative substances for control of apple scab. Results from conidia germinating tests and experiments with plant extracts. Proceedings of the 11th International Conference on Cultivation Technique and Phytopathological Problems in Organic Fruit-Growing 11:101-107.

Pinto E, Pina-Vaz C, Salgueiro L, Goncalves MJ, Costa-de-Oliveira S, Cavaleiro C, Palmeira A, Rodrigues A, Martinez-de-Oliveira J (2006). Antifungal activity of the essential oil of *Thymus pulegioides* on *Candida*, *Aspergillus* and dermatophyte species. J Med Microbiol 55(10):1367-1373.

Soylu EM, Kurt S, Soylu S (2010). In vitro and in vivo antifungal activities of the essential oils of various plants against tomato grey mould disease agent *Botrytis cinerea*. Int J Food Microbiol 143(3):183-189.

Tabachnick BG, Fidell LS (2006). Using multivariate statistics, 5th Ed. Pearson Education, New York p. 1-1008.

Tan M, Zhou L, Qin M, Li D, Jiang W, Wang Y, Hao X (2007). Chemical composition and antimicrobial activity of the flower oil of *Russowia sogdiana* (Bunge) B. Fedtsch. (Asteraceae) from China. J Essent Oil Res 19(2):197-200.

Thiesz R, Balog A, Ferencz L, Albert J (2007). The effects of plant extracts on apple scab (*Venturia inaequalis* Cooke) under laboratory conditions. Rom Biotech Lett 12(4):3295-3302.

Tian J, Ban X, Zeng H, He J, Chen Y, Wang Y (2012). The mechanism of antifungal action of essential oil from dill (*Anethum graveolens* L.) on *Aspergillus flavus*. PLOS One 7(1):30147.

Vaillancourt LJ, Hartman JR (2000). Apple scab. The Plant Health Instructor DOI: 10.1094/PHI-I-2000-1005-01 Updated 2005.

Velluti A, Marín S, Gonzalez P, Ramos AJ, Sanchis V (2004). Initial screening for inhibitory activity of essential oils on growth of *Fusarium verticillioides*, *F. proliferatum* and *f. graminearum* on maize-based agar media. Food Microbiol 21(6):649-656.

Zambonelli A, D'Aulerio AZ, Severi A, Benvenuti S, Maggi L, Bianchi A (2004). Chemical composition and fungicidal activity of commercial essential oils of *Thymus vulgaris* L. J Essent Oil Res 16(1):69-74.

Effects of Salt and Water Stress on Plant Growth and on Accumulation of Osmolytes and Antioxidant Compounds in Cherry Tomato

Mohamad AL HASSAN[1], Marina MARTÍNEZ FUERTES[1], Francisco José RAMOS SÁNCHEZ[1], Oscar VICENTE[1]*, Monica BOSCAIU[2]

[1]*Universitat Politècnica de València, Instituto de Biología Molecular y Celular de Plantas (UPV-CSIC), CPI, edificio 8E, Camino de Vera s/n, 46022 Valencia, Spain; ovicente@ibmcp.upv.es (*corresponding author)*

[2]*Universitat Politècnica de València, Instituto Agroforestal Mediterráneo (UPV), CPI, edificio 8E, Camino de Vera s/n, 46022 Valencia, Spain*

Abstract

The effects of salt and water stress on growth and several stress markers were investigated in cherry tomato plants. Some growth parameters (stem length and number of leaves) and chlorophyll contents were determined every third day during plant growth, and leaf material was collected after 25 and 33 days of treatment. Both stresses inhibited plant growth; chlorophyll levels, however, decreased only in response to high NaCl concentrations. Proline contents largely increased in leaves of stressed plants, reaching levels high enough to play a major role in cellular osmotic adjustment. Despite reports indicating that tomato does not synthesize glycine betaine, the stress-induced accumulation of this osmolyte was detected in cherry tomato, albeit at lower concentration than that of proline. Therefore, it appears that the plants are able to synthesise glycine betaine as a secondary osmolyte under strong stress conditions. Total sugars levels, on the contrary, decreased in stress-treated plants. Both stress treatments caused secondary oxidative stress in the plants, as indicated by a significant increase in malondialdehyde (MDA) contents. Water stress led to an increase in total phenolics and flavonoid contents and a reduction of carotenoid levels in the leaves; flavonoids also increased under high salinity conditions.

Keywords: proline, glycine betaine, carbohydrates, MDA, carotenoids, phenolics, flavonoids

Introduction

Tomato is grown in a wide range of climatic conditions, from the tropics to subarctic regions, but its optimal cultivation areas are found in warm and rather dry regions, such as Mediterranean countries (Cuartero and Fernández Muñoz, 1999). In these arid or semi-arid zones, water stress and high soil salinity are common environmental factors that can significantly reduce crop yields. One of the most general responses to drought and salinity - as well as to other abiotic stress conditions, also causing cellular dehydration in plants, such as cold, elevated temperatures or exposure to heavy metals - is based on the synthesis and cytoplasmic accumulation of osmolytes, a conserved phenomenon observed in all plants, tolerant as well as sensitive to stress (Munns and Tester, 2008; Parida and Das, 2005; Parvaiz and Satyawati, 2008). Osmolytes are 'compatible solutes', very soluble, low-molecular-weight organic compounds that do not interfere with normal metabolism even when present at high concentrations. While toxic inorganic ions are sequestered in vacuoles, organic osmolytes accumulate predominantly in the cytoplasm, preventing or limiting cellular dehydration (Büssis and Heineke, 1998; Handa *et al.*, 1986; Stewart and Lee, 1974). Reduction of the osmotic potential due to accumulation of osmolytes in response to stress improves the ability of the plant cells to maintain turgor pressure at low water potentials, which is essential for biological processes such as photosynthesis or cell expansion, as well as for maintaining enzymatic activities (Tyree and Jarvis, 1982). Besides their role in osmotic adjustment, osmolytes act as osmoprotective substances, directly stabilising proteins and cell membranes under dehydration conditions. Osmolytes also protect cells from oxidative stress by inactivating 'reactive oxygen species' (ROS) (Hare *et al.*, 1998; Szabados and Savouré, 2010).

The amino acid proline (Pro) and glycine betaine (GB), a quaternary amine, are probably the most common compatible solutes synthesised by plants as a response to abiotic stress (Ashraf and Foolad, 2007; Chen and Murata, 2008; Verbruggen and Hermans, 2008). As for other osmolytes, besides their role in osmoregulation, both compounds can act as 'low-molecular-weight chaperons', contributing to maintain the active conformation of macromolecules in stressed plants, and participate in detoxification of ROS. Moreover, Pro and GB seem to be involved, directly or indirectly, in the regulation of gene expression as signalling molecules, and also serve as metabolites for the cellular storage of carbon and nitrogen during stress, which would be used by the cell once stress has ceased (Munns and Tester, 2008; Szabados and Savouré, 2010). Compatible solutes also include soluble carbohydrates, such as sugars (e.g. sucrose, glucose, fructose or trehalose), sugar alcohols (sorbitol, mannitol, as well as different inositol isomers and derivatives), and the raffinose family of oligosaccharides (Gavaghan *et al.*, 2011; Parida *et al.*, 2002). Although sugars have been shown to act as functional osmolytes in several species, it is not so easy to assess their specific functions in the responses to stress, which can be masked by their multiple additional roles as direct products of photosynthesis, components of the primary metabolism and regulatory molecules (Gil *et al.*, 2013).

A secondary effect of abiotic stresses, including drought and salinity, is the increased generation of 'reactive oxygen species' (ROS), including highly reactive free radicals such as superoxide, singlet oxygen, hydroxyl or perhydroxyl radicals, as well as hydrogen peroxide, molecular oxygen, ozone and other strong oxidant molecules (Apel and Hirt, 2004). ROS are continuously generated by plants as by-products of different metabolic pathways, but under stress their production increases leading to oxidative damage of cellular membranes, proteins, carbohydrates and DNA (Van Breusegem and Dat, 2006). In response to stress, plants activate powerful antioxidant systems, both enzymatic (e.g., superoxide dismutase, catalase, glutathione reductase, several peroxidades) and non-enzymatic (vitamins C and E, carotenoids, flavonoids and other phenolic compounds etc.) (Apel and Hirt, 2004).

Carotenoids are pigments with several functions in plants, besides their direct role in photosynthesis, including their involvement in the mechanisms of oxidative stress tolerance (Gill and Tuteja, 2010). Phenolic compounds also fulfil multiple roles in plants, as structural components of cell walls, participating in the regulation of growth and developmental processes, as well as in the mechanisms of defence against herbivores and pathogens; in addition, they are involved in the responses of plants to practically all types of abiotic stress: UV radiation, extreme temperatures, mineral nutrient imbalance, drought, salinity, heavy metals and herbicides among others (Cheynier *et al.*, 2013; Gould and Lister, 2006). Flavonoids represent the main and most complex subgroup of polyphenols, including more than 9,000 different compounds with a wide array of biological functions (Di Ferdinando *et al.*, 2012; Pollastri and Tattini, 2011; Treutter, 2005, 2006; Winkel-Shirley, 2002). In the last decade, they have been the object of many studies, not only due to the academic interest in elucidating their multiple functions in plants, but also because their alleged beneficial effects for human health, as powerful dietary antioxidants (Wiseman, 2006).

Responses to salinity in tomato, within its salt-tolerance range under natural conditions, have been extensively studied, but only few authors have analysed its responses to higher salt concentrations (Maggio *et al.*, 2004). Subjecting the plants to 'shock treatments' at very high NaCl concentrations, well above those they encounter in the field, may trigger adaptation mechanisms that cannot be observed at low or moderate salinity levels. In addition, there are only a few papers reporting the effects of both salinity and drought on the same plant material (Giannakoula and Ilias, 2013). These two adverse environmental conditions, which will worsen as a consequence of climate change, could seriously affect tomato production in Mediterranean countries in the coming years.

The aims of the present study were to analyse the effects on the growth of tomato plants of water and salt stress treatments, applied under different conditions, including high NaCl concentrations beyond the tolerance threshold, to allow detection of time- and salt concentration- dependent effects. These growth responses were correlated with accumulation of several biochemical stress markers, associated to general mechanisms of abiotic stress tolerance: the main osmolytes (proline, glycine betaine and total soluble sugars), responsible for cellular osmotic adjustment; malondialdehyde (MDA), as a marker of oxidative stress; and some non-enzymatic antioxidants (total carotenoids, phenolics and flavonoids), synthesised in response to secondary oxidative stress. The experiments were carried out in cherry tomato, a variety that has not been extensively studied despite its growing commercial interest.

Materials and methods

Plant material

Cherry tomato (*Solanum lycopersicum* L. var. *cerasiforme*) is clearly differentiated from other tomatoes by the smaller size of its fruits. Plants of this variety are vigorous in their growth and have a higher leaves/fruits ratio; therefore, the fruits receive more photoassimilates, in comparison to larger-size varieties, and have higher sugar and acid contents. Thus, cherry tomatoes are more flavoured and are used mostly for fresh consumption. Seeds of cherry tomato were kindly provided by Pilar Corella, from Rijk Zwaan Ibérica S.A. (Almería, Spain).

Plant growth

The seeds were sown in seed trays containing a mixture of commercial peat and vermiculite (1:1) and were placed in a greenhouse with regulated temperatures ranging between 17 and 23 °C, under a long-day photoperiod (16 h light / 8 h dark). Young plants were transferred to individual pots (11 cm in diameter) on the same substrate 36 days after sowing and were watered for 26 additional days with a standard nutritive solution. Then, salt and water stress treatments started, by watering the plants with increasing NaCl concentrations (150, 300 and 450 mM NaCl in nutritive solution) and by stopping irrigation altogether, respectively; 10 plants were used per treatment. Watering was carried out twice a week by adding 1.5 L of the salt solutions to each tray, which contained 12 standard pots. Control plants were grown in parallel, maintaining the standard irrigation regime with nutritive solution. After starting the treatments, plant height was measured and the number of leaves per plant was counted every three days, to assess the effect of stress on vegetative plant growth. Chlorophyll content was also determined at the same times, using a portable 'Chlorophyll meter' SPAD-502Plus (Konica Minolta).

For each stress treatment, including non-treated controls, leaves in equivalent positions were detached from all plants, 25 and 33 days after starting the treatments, and weighed on a precision balance to calculate the average leaf fresh weight (FW) corresponding to each treatment. Part of the leaf material was stored at -75 °C until further use, and part was dried at 65 °C until constant weight, to measure the average leaf dry weight (DW) for each treatment. Average leaf 'water content' after each treatment, expressed in percentage, was calculated as [(FW – DW) / FW] x 100.

Electric conductivity of the substrate

Soil $EC_{1:5}$ was checked at the beginning and at the end of the treatments. Soil samples were taken from three pots per treatment, air-dried and then passed through a 2-mm sieve.

A soil:water (1:5) suspension was prepared in Milli Q water and stirred for one hour at 600 u/min at 21 °C. Electric conductivity was measured with a Crison Conductivity meter 522 and expressed in dS/m.

Osmolyte contents

Pro contents were determined according to the method of Bates *et al.* (1973) with minor modifications (Vicente *et al.*, 2004). Frozen plant material (100 mg) was ground to a fine powder in a mortar, in the presence of liquid nitrogen. Extraction was carried out with 3% sulfosalicylic acid and cell debris was removed by filtration. One volume of the filtrate was mixed first with one volume of freshly prepared acid ninhydrin (25 mg/mL ninhydrin in 10.44 M acetic acid and 2.4 M phosphoric acid); then one volume of glacial acetic acid was added, the sample was mixed and incubated at 95 °C for 1 h. After stopping the reaction by cooling the samples on ice, they were extracted with two volumes of toluene. The absorbance of the organic phase was determined at 520 nm, using toluene as a blank. A calibration curve was obtained for each assay, using solutions of Pro in 3% sulfosalicylic acid of known, increasing concentrations and subjected to the same treatment as the samples. Pro contents in the plant samples were expressed as 'μmol Pro per gram of dry weight'.

The extraction and quantification of glycine betaine was performed following the method of Grieve and Grattan (1983) with the modifications proposed by Nawaz and Ashraf (2010). Frozen plant material (100 mg) was ground, using a homogenizer (Ultraturrax) in a volume of 2 mL Milli-Q water and centrifuged for 10 minutes at -4 °C and 13,000 rpm. One mL was removed from the supernatant of each extract and mixed with 400 μL HCl 2N. After stirring, 200 μL of each tube was mixed with 80 μL of potassium iodide and brought to 100 mL with Milli-Q water. The tubes were quickly placed in ice and stirred every 20 seconds for 15 min. After 90 min, 800 μL of cold Milli-Q water and 4 mL of 1.2-dichloroethane - kept in the -20 °C freezer until used - were added to each tube. The samples were vortexed for one minute and allowed to settle until the water and the organic phases separated; finally, 1 mL of the lower aqueous phase was aspirated and its absorbance at 265 nm was measured in a spectrophotometer. A standard curve was obtained for each assay, using glycine betaine solutions in Milli-Q water of increasing concentration subjected to the same treatment as the plant samples. Glycine betaine contents in the plant samples were expressed as 'μmol GB per gram of dry weight'.

Total sugars were quantified according to the method described by Dubois *et al.* (1956). Dried material (0.1 g) was crushed and diluted with 3 mL of 80% methanol, left on an agitator for 20-48 h and then mixed with concentrated sulfuric acid and 5% phenol; finally, absorbance of the samples at 490 nm was measured in a spectrophotometer. Glucose solutions were used as standard and the amount of total soluble sugars in each sample was calculated as 'mg. equivalent of glucose per gram of dry weight'.

Malondialdehyde and non-enzymatic antioxidants

Malondialdehyde (MDA), total phenolics and flavonoids contents were measured in the same extracts used for total soluble sugars determination (0.1 g of dried material in 3 mL of 80% methanol). MDA, a product of membrane lipid peroxidation, is considered as an excellent marker of oxidative stress (Del Rio *et al.*, 2005) and is routinely used to assess the degree of oxidative damage induced in plants by different types of stress (e.g. Aghaleh *et al.*, 2009; Li *et al.*, 2010). MDA content was determined by the thiobarbituric acid-reactive-substances (TBARS) assay, modified to correct for the presence of interfering compounds, as described by Hodges *et al.* (1999).

Total phenolic compounds were quantified by reaction with the Folin-Ciocalteu reagent, according to Blainski *et al.* (2013). Absorbance was measured at 765 nm and the results expressed in equivalents of gallic acid (mg eq. GA g^{-1} DW), used as standard. Flavonoid contents were determined following the method described by Zhishen *et al.* (1999); the absorbance was measured at 510 nm, and the amount of flavonoids was expressed in equivalents of catechin (mg eq. C g^{-1} DW), used as standard.

Total carotenoids were measured and quantified following Sims and Gamon (2002): 0.2 g fresh material was crushed and diluted in 80% ice-cold acetone/Tris buffer for 1 hour on a shaker at 4 °C, then centrifuged for 15 min at 3,000 x g. The supernatant was separated and optical density was measured at 663 nm, 646 nm and 470 nm in disposable plastic cuvettes. Total carotenoids values were then converted into μg g^{-1} DW.

Statistic analysis

Data were analysed using the SPSS programme for Windows, v.16.00. Prior to the analysis, the Levene test was applied to check the ANOVA requirements. The significance of differences between treatments was assessed using one-way ANOVA, at 95% confidence level. The post-hoc Tukey test was used to estimate homogeneous groups when more than two samples were compared.

Results

Substrate electric conductivity

The electric conductivity ($EC_{1:5}$) of the plants' substrate was determined before and after the stress treatments (Table 1). In the control pots, not treated with NaCl, the $EC_{1:5}$ value at the end of the experiment was significantly higher than the initial one; this can be easily explained by accumulation of ions present in the nutritive solution. As expected, $EC_{1:5}$ increased even more after watering the plants with NaCl solutions of increasing concentration, in a concentration-dependent manner. The mean $EC_{1:5}$ also increased slightly in the substrate of water stressed plants, as compared to the initial value - most likely due to concentration of ions from the nutritive solution - but the difference was not statistically significant.

Stress-induced inhibition of plant growth

Both salt and water stress negatively affected vegetative plant growth, which was estimated by the increase in plant length and in the number of leaves during the 33 days of treatment (Fig. 1). As compared to the unstressed controls, all NaCl concentrations tested totally blocked growth of tomato plants, with a clear concentration effect: the higher the salt concentration used, the shorter the lag period before growth inhibition was observed

Table 1. Substrate electric conductivity $EC_{1:5}$ (dS/m), measured in soil:water (1:5) extracts at the beginning (day 0) and at the end (day 33) of the salt and water stress treatments

Treatment	(day 0)	0 mM NaCl	150 mM NaCl	300 mM NaCl	450 mM NaCl	Water Stress
$EC_{1:5}$	0.21 ± 0.05a	0.94 ± 0.08b	2.56 ± 0.39c	3.37 ± 0.35d	4.78 ± 0.51e	0.38 ± 0.18a

Different lower case letters indicate significant differences between treatments

Table 2. Effects of increasing salt concentration on MDA and non-enzymatic antioxidant levels in plants after 25 days (sampling 1) and 33 days (sampling 2) of treatment

Parameters	Sampling	Salt treatments (mM NaCl)			
		0	150	300	450
MDA	1	145.07 ± 13.00a	163.45 ± 8.60ab	169.03 ± 10.00ab	214.16 ± 2.90b
(nmol g^{-1}DW)	2	149.30 ± 7.00a	188.24 ± 10.00b	206.02 ± 4.00bc	237.56 ± 20.00c
Total carotenoids	1	648.02 ± 31.00b	613.48 ± 31.00b	546.00 ± 42.00ab	344.37 ± 19.00a
(µg g^{-1} DW)	2	659.53 ± 36.00c	473.01 ± 27.00b	352.98 ± 16.00ab	306.17 ± 18.00a
Total phenolics	1	9.71 ± 0.17a	11.64 ± 1.60ab	13.05 ± 0.90b	14.70 ± 0.83c
(mg eq. GA g^{-1} DW)	2	12.41 ± 0.60a	17.09 ± 0.30b	15.85 ± 3.30b	17.64 ± 2.20b
Total flavonoids	1	10.19 ± 0.30a	11.24 ± 0.80b	12.24 ± 1.10b	12.87 ± 0.50b
(mg eq. C g^{-1} DW)	2	8.74 ± 0.60a	11.03 ± 1.00b	11.64 ± 0.50b	13.74 ± 0.50c

Values shown are means followed by ± SD (n = 5). Different lower case letters in a file indicate significant differences between treatments, for each compound and sampling; Abbreviations: MDA, malondialdehyde; GA, gallic acid; C, catechin; DW, dry weight

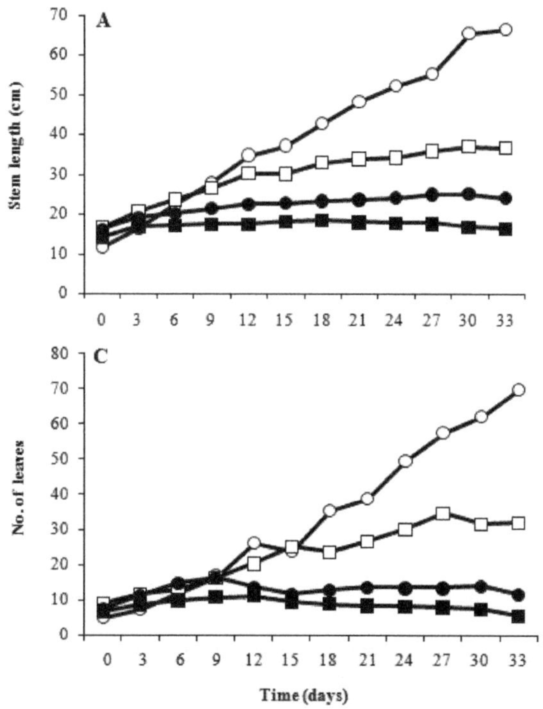

Fig. 1. Mean stem length (A, B) and mean number of leaves (C, D) changes during the salt (A, C) and water (B, D) stress treatments. Control, non-stressed plants (-○-, all panels). 50 mM (-□-), 100 mM (-●-) and 150 mM (-■-) NaCl (panels A, C). Water-stressed plants (-▲-, panels B, D)

(Figs. 1A and C). Water stress also stopped stem growth and the increase in leaf number; this effect was clearly observed 9 days after the last irrigation with nutritive solution, once the substrate had dried (Figs. 1B and D).

Apart from the growth parameters measured continuously, stress-induced inhibition of plant growth was also assessed by determining the mean fresh weight (FW) and dry weight (DW) of the tomato leaves after 25 days (sampling 1) and 33 days (sampling 2) of salt or water stress treatment (Fig. 2). A significant concentration-dependent decrease of FW was observed in salt-treated plants, reaching approximately 75% reduction in those irrigated with 450 mM NaCl, with respect to the non-treated controls. In the presence of this high salt concentration, the plants were already strongly affected after 25 days of treatment and no significant differences were found between the two samplings; at lower salt levels - 150 or 300 mM NaCl - reduction of FW increased with the duration of the treatment (Fig. 2A). Similar qualitative results were obtained when the plants were submitted to water stress, although with stronger differences between the two samplings: after 25 days without irrigation plant FW was reduced by ca. 25%, as average, but by almost 90% after 33 days of treatment (Fig. 2B). The

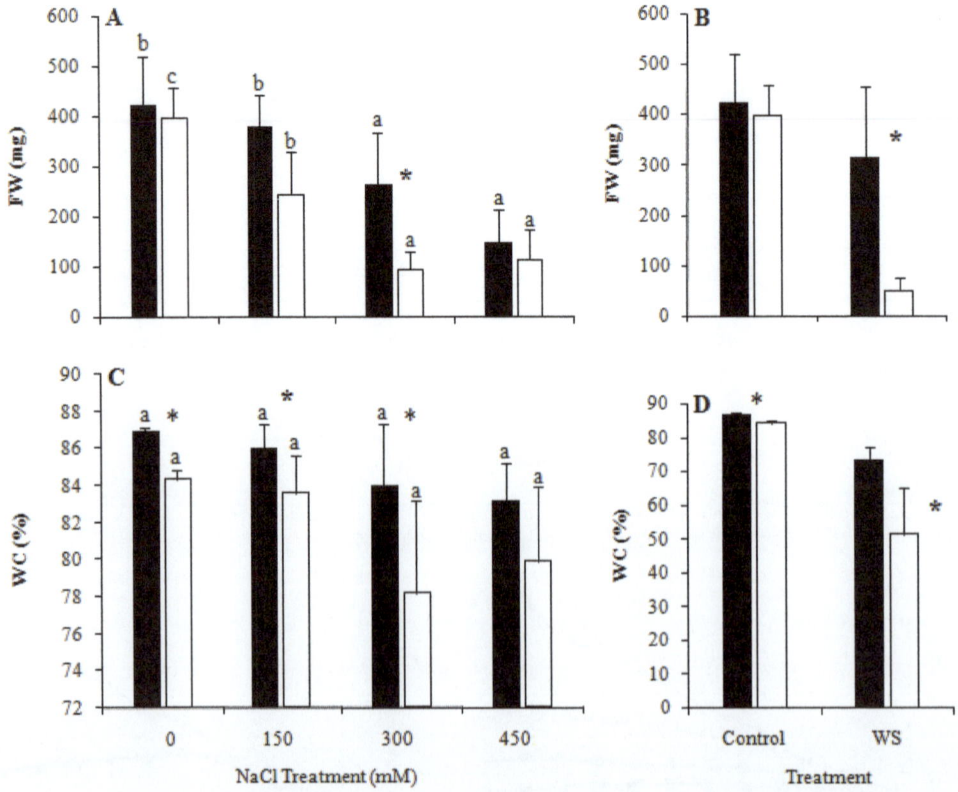

Fig. 2. Average fresh weight (A, B) and relative water content (WC) as percentage of fresh weight (C, D) of tomato leaves after 25 days (sampling 1, black bars) or 33 days (sampling 2, white bars) of salt (A, C) or water (B, D) stress treatments (means ± SD, $n = 3$). Different lower case letters indicate significant differences between treatments, for the same sampling, according to Tukey test ($\alpha = 0.05$); asterisks indicate significant differences between sampling 1 and sampling 2, for each treatment

relative decrease in FW under water stress conditions - as compared with the control plants - was not only a consequence of the inhibition of growth, but also due to a significant loss of water, as shown by the calculated water content of the tomato leaves: it was only 50% of the FW for the plants maintained for 33 days without water, while for those irrigated normally with nutritive solution water content was about 85% of the total leaf weight (Fig. 2D). In the presence of salt, a general, but not statistically significant, decrease of the mean leaf water content was observed with increasing NaCl concentrations, both after 25 and 33 days of treatment, whereas significant differences were detected between the samples collected at different times for each NaCl concentration, up to 300 mM (Fig. 2C).

Chlorophyll content

The chlorophyll content of crop plants is positively correlated with their photosynthetic activity (Gummuluru *et al.*, 1989) and a reduction of chlorophyll level contributes to the inhibition of photosynthesis observed under abiotic stress conditions. A decrease in chlorophyll levels with respect to the controls was indeed detected in plants irrigated with high salt concentrations - 300 and 450 mM NaCl - for two weeks or longer periods (Fig. 3A). On the other hand, no significant differences with non-stressed plants were observed when a moderate - 150 mM - NaCl concentration was used (Fig. 3A) or when the tomato plants were subjected to the water stress treatment (Fig. 3B); this latter observation was consistent with

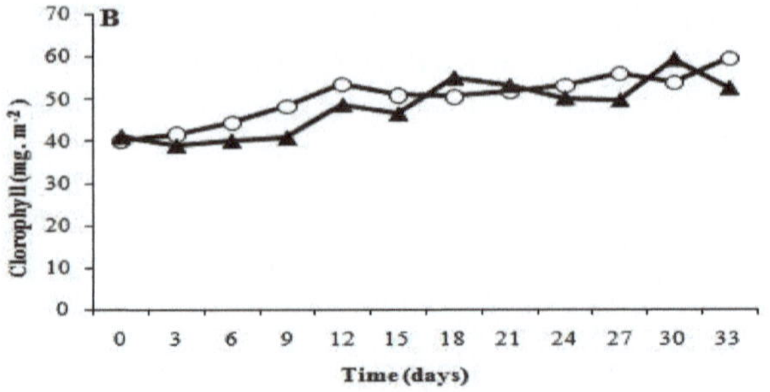

Fig. 3. Mean chlorophyll contents (mg per square meter) measured with a portable chlorophyll meter in leaves of tomato plants during the salt (A) and water (B) stress treatments (n = 10). Control, non-stressed plants (-○-, in both panels). 50 mM (-□-), 100 mM (-●-) and 150 mM (-■-) NaCl (panel A). Water-stressed plants (-▲-, panel B)

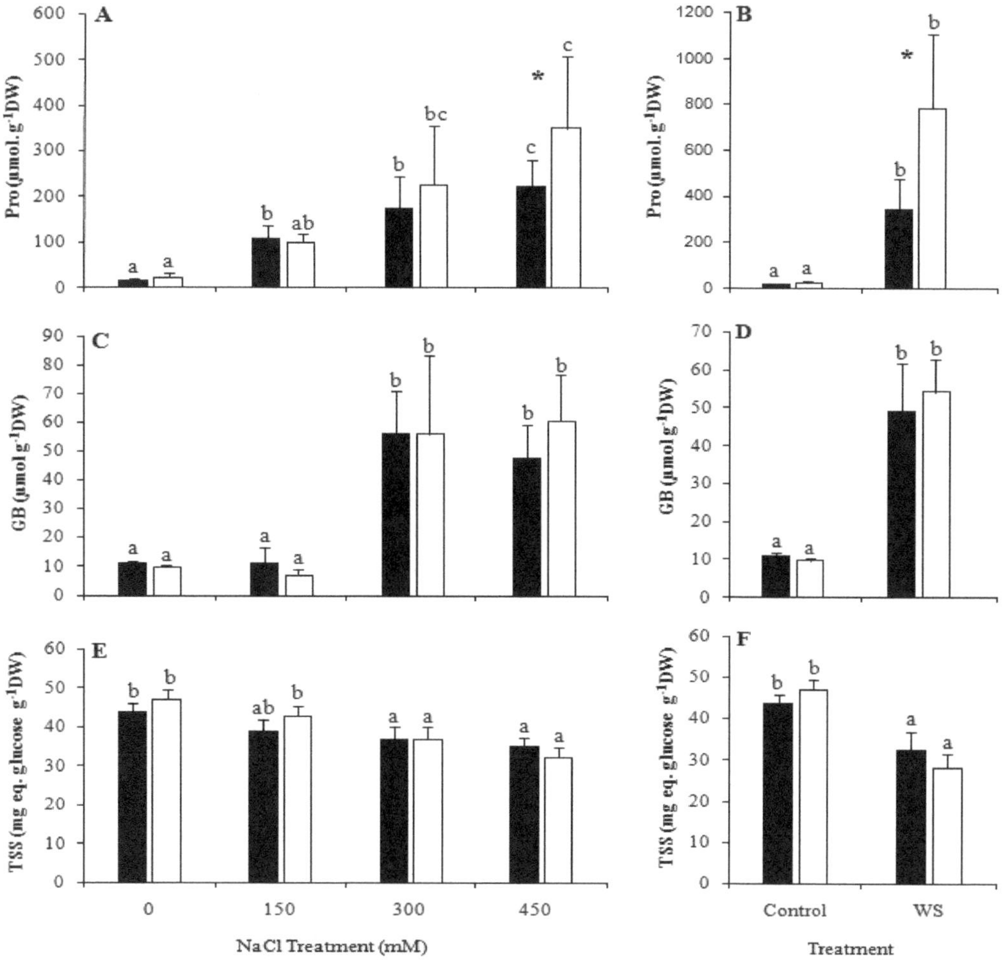

Fig. 4. Proline (Pro, µmol per gram DW) (A, B), glycine betaine (GB, µmol per gram DW) (C, D), and total soluble sugars (TSS, mg equivalent of glucose per gram DW) (E, F) in tomato leaves after 25 days (sampling 1, black bars) or 33 days (sampling 2, white bars) of salt (A, C, E) or water (B, D, F) stress treatments (means ± SD, $n = 3$). Different lower case letters indicate significant differences between treatments, for the same sampling, according to Tukey test ($\alpha = 0.05$); asterisks indicate significant differences between sampling 1 and sampling 2, for each treatment

the visual appearance of the plants, which were clearly wilted, but had not lost their green colour (data not shown).

Proline content

Application of salt treatments significantly increased Pro levels in tomato leaves, in a concentration-dependent manner. Mean Pro values in the presence of 300 mM and 450 mM external NaCl also increased with longer times of treatment although the differences observed between the two samplings of plant material were significant only at 450 mM NaCl. Under the strongest salt stress conditions tested, Pro levels reached about 350 µmol per gram DW, which represented an increase of 16-fold with respect to the non-treated controls (Fig. 4A).

Pro accumulation was more clearly observed in plants kept without irrigation, with increases of about 20-fold and 36-fold after 25 and 33 days, respectively, of water stress (Fig. 4B). Pro contents in the plants subjected to the strongest water stress treatment (33 days without irrigation), in terms of dry weight (ca. 800 µmol g^{-1}) were about double than those determined in plants treated for the same time with 450 mM NaCl (Figs. 4A and B); however, considering the drastic reduction of water content in non-irrigated plants, it can be said that both treatments induced Pro accumulation to similar levels, in absolute terms.

Glycine betaine

Salt and water stress also induced the synthesis and accumulation of GB in tomato (Figs. 4C and D). Yet the absolute values of GB content reached in plants watered with 300 or 450 mM NaCl (50-60 µmol per gram DW) and its relative increase as compared to the control plants (ca. 5-fold) were lower than those of Pro. Moreover, no time-dependent increase in GB levels was detected in this experiment, as the values measured for the two samplings of plant material were not significantly different (Fig. 4C). The same results were obtained for plants subjected to 25 or 33 days of water stress (Fig. 4D).

Total soluble sugars

The total amount of soluble sugars slightly decreased in salt-stressed plants, but significant differences were registered only starting with 300 mM NaCl in both samplings (Fig. 4E). Under water stress conditions the decrease was also significant and more accentuated after a longer drought treatment (Fig. 4F).

Malondialdehyde

MDA is a product of membrane lipid peroxidation and, as mentioned above, is considered a reliable marker of oxidative stress; thus, higher MDA contents should correspond to a higher degree of oxidative stress. As expected, MDA levels increased with all treatments, mostly after longer exposure to stress, so higher values were registered in the second sampling, after 33 days of treatment (Tables 2 and 3). The most affected plants by oxidative stress were those treated with 450 mM NaCl, in which the highest MDA contents were measured (Table 2).

Table 3. Effects of drought stress on MDA and non-enzymatic antioxidant levels in plants after 25 days (sampling 1) and 33 days (sampling 2) of treatment

Parameters	Sampling	Treatments	
		Control	Water stress
MDA	1	145.07 ± 13.00a	173.18 ± 5.20b
(nmol g^{-1}DW)	2	149.30 ± 7.00a	198.35 ± 7.60b
Total carotenoids	1	648.02 ± 31.00b	474.73 ± 11.00a
(μg g^{-1} DW)	2	659.53 ± 36.00b	357.15 ± 3.60a
Total phenolics	1	9.71 ± 0.17a	12.10 ± 1.90b
(mg eq. GA g^{-1} DW)	2	12.41 ± 0.60a	19.48 ± 0.90b
Total flavonoids	1	10.19 ± 0.30b	7.03 ± 0.20a
(mg eq. C g^{-1} DW)	2	8.74 ± 0.60b	5.10 ± 0.50a

Values shown are means followed by ± SD (n = 5). Different lower case letters in a file indicate significant differences between treatments, for each compound and sampling; Abbreviations: MDA, malondialdehyde; GA, gallic acid; C, catechin; DW, dry weight

Total carotenoids, phenolics and flavonoids

Carotenoid levels in leaves decreased in all treatments with respect to the control. In plants treated with 450 mM NaCl this reduction was of about 50%. When comparing the two samplings, lower values were registered in plants exposed for a longer time to stress in all treatments (Tables 2 and 3). Contrary to the pattern of variation of carotenoids, total phenolics increased in plants under stress, especially in those treated with high salt concentrations. Values were significantly higher in plants from the second sampling than in those from the first one in all stress treatments (Tables 2 and 3). A different response to the type of stress was observed in total flavonoid contents, which increased significantly in the presence of NaCl (Table 2) but decreased under drought conditions (Table 3).

Discussion

All our major crops and most wild species are relatively sensitive to environmental stress conditions such as drought or salinity, but tomato is moderately resistant to salt, withstanding a soil EC_{sat} up to 6 dS m^{-1}, so that it can be cultivated in regions exposed to a certain degree of soil salinisation (Maggio *et al.,* 2004). Yet high soil salinity inhibits seed germination and plant growth, and causes reduction in crop yields (Dalton *et al.*, 1997; Romero-Aranda *et al.*, 2002). In this study, young tomato plants have been grown in the presence of increasing concentrations of NaCl, and a concentration- and time- dependent inhibition of vegetative growth has been observed, which was better shown by the reduction of the fresh weight of the plants; this was not unexpected, since inhibition of growth is probably the most general response of plants to stress (Munns, 2002; Munns and Tester, 2008). Despite the fact that high salinity should cause cellular dehydration, in the experimental conditions used here the water content of the plants was reduced very slightly in the presence of salt, suggesting that tomato plants activate relatively efficient mechanisms to cope with the osmotic component of salt stress, and that growth inhibition is mostly due to the ionic toxicity of the salt. Water stress also inhibited growth but, contrary to salinity, in this case a severe dehydration of the plants was observed.

Another difference between salt and water stress treatments refers to changes in chlorophyll contents. In plants treated with high NaCl concentrations, a significant reduction in chlorophyll levels, up to ca. 40% of the non-treated controls, was observed. The decrease in chlorophyll levels in plants affected by salt is due to the inhibition of chlorophyll synthesis, together with the activation of its degradation by the enzyme chlorophyllase (Santos, 2004). Yet, reduction of chlorophyll contents is not the only reason for the inhibition of photosynthesis in the presence of salt, since NaCl also inhibits key enzymes involved in this process, such as Rubisco and PEP carboxylase (Soussi *et al.*, 1998). Although water stress should also reduce chlorophyll levels and photosynthetic activity (Alberte *et al.* 1977; Mafakheri *et al.*, 2010; Sanchez *et al.*, 1983), this was not observed in the present study, probably because the time that the plants were maintained without irrigation was not long enough to detect these effects.

Proline is generally considered as a good indicator of environmental stress in tomato (Clausen, 2005) and there are many reports describing an increase in Pro contents as a response to water or salt stress in this species (Babu *et al.*, 2012; Ghorbanli *et al.*, 2013; Giannakoula and Ilias, 2013; Inal, 2002; Handa *et al.*, 1986; Nahar and Gretzmacher, 2002; Umebese *et al.*, 2009; Yokas *et al.*, 2008), although it should be mentioned that data specific for cherry tomatoes are rather scarce (Maggio *et al.*, 2007; Rosales *et al.*, 2007). What is not so clear is the possible contribution of Pro accumulation to the (relative) resistance of tomato to salinity. When comparing tomato cultivars with differences in their degree of salt tolerance, in some cases the more tolerant cultivars were found to synthesise higher amounts of Pro under stress (Ali *et al.*, 2011), but in others there was no correlation between tolerance and Pro levels (Alian *et al.*, 2000). In the present study, a clear positive correlation between Pro accumulation and the intensity of the applied stress treatments has been established; that is, of Pro levels with the electric conductivity of the substrate - reflecting increasing NaCl concentrations in the nutritive solution - and with the time of exposure to stress. Moreover, Pro reached levels high enough to play a significant role in cellular osmotic adjustment under these stress conditions. Taken together, these data strongly support the notion that Pro is the major physiological osmolyte in cherry tomato, as it has been suggested for other tomato varieties (Nahar and Gretzmacher, 2002).

Usually, a given plant species accumulate preferentially a particular type of osmolyte in response to environmental stress; there have even been attempts to use this preference for one specific type of compatible solute as a taxonomic criterion in wild species (Gorham *et al.*, 1980; Tipirdamaz *et al.*, 2006). In agreement with this idea, and since Pro appears to be the major osmolyte in tomato, it is generally accepted that tomato does not accumulate glycine betaine in natural conditions; in fact, it has been reported that this species lacks two enzymes required for GB biosynthesis (Goel *et al.*, 2011). Nevertheless, exogenous application of GB to tomato plants improves their resistance to drought (Rezaei *et al.*, 2012) and salinity (Chen *et al.*, 2009; Heuer, 2003; Makela *et al.*, 1998). On the other hand, there are some examples of plant species that under severe stress conditions can activate the simultaneous synthesis of several osmolytes to help alleviate the detrimental effects of stress, by contributing to osmotic adjustment and/or acting as 'osmoprotectants', as it has been recently discussed (Tipirdamaz *et al.*, 2006; Gil *et al.*, 2013). This seems to be the case also for tomato, at least for the cherry variety: contrary to the general assumption, in the present study the accumulation of GB, in response to water or salt stress treatments, has been detected in tomato plants. It is true that the maximum levels of GB measured (ca. 50 μmol g^{-1} DW) were about 10-fold lower than those of Pro in the same plants - and also much lower than those measured in taxa that are clear GB accumulators (Boscaiu *et al.*, 2011; Gil *et al.*, 2014). Therefore, GB will have only a modest effect on osmotic balance in stressed tomato plants, but may still contribute significantly to stress resistance due to its putative functions as low-molecular-weight chaperon and/or ROS scavenger. Most published data support the absence of detectable GB in tomato; yet the present results are not the only evidence for accumulation of endogenous GB in response to stress, which has also been reported in plants submitted to chilling, where a threshold level ensures sufficient protection to low temperatures (Park *et al.*, 2006).

The increase of sugars after mild salt and water stress is well-known in tomato fruits, including the cherry variety, and moderate abiotic stress was even recommended as a strategy for improving the quality of tomatoes (Fernández-García *et al.*, 2004; Sgherri *et al.*, 2008; Ullah *et al.*, 1994). Salt stress was found to produce an increase in carbohydrates accumulation, in the form of starch, in the early stages of fruit development, but as soluble sugars in ripe fruits, thus improving the flavour of tomatoes (Yin *et al.*, 2010). There are also several reports of increased amount of sugars in tomato berries under drought (e.g. Nahar and Grezmacher, 2002). However, there are relatively few data published on stress-induced sugar accumulation in vegetative organs in this species, and they generally indicate a reduction of total carbohydrates in leaves (e.g. Amini and Ehsanpour, 2005; Li, 2009), as it has been found in the present study.

In addition to sugars, tomato fruits are rich in several compounds considered as 'health-promoting', such as carotenoids, flavonoids and other phenolics. These secondary metabolites play multiple roles in plants, including scavenging of ROS induced under different stress conditions and causing oxidative stress. A clear symptom of oxidative damage is cell membrane degradation; therefore, MDA - a product of membrane lipid peroxidation - is an excellent marker of oxidative stress (Del Rio *et al.*, 1996). In this study a significant increase of MDA levels in tomato leaves upon salt and water stress treatments of the plants was obseved, in agreement with previous reports showing a salt-induced increase of MDA contents in tomato leaves (Shalata and Thal, 1998), roots (Shalata *et al.*, 2001) and fruits (Rosales *et al.*, 2006).

Mild and moderate saline and water stress produce also an increase in carotenoids levels in tomato fruits. Lycopen, the main carotenoid in tomato, is known as an important natural antioxidant with anti-carcinogenic properties (Krauss *et al.*, 2006). In addition, lycopen is responsible for the red colour of tomatoes, aspect of great economic importance and therefore salt treatments are recommended as an alternative strategy to transgenic crops for obtaining tomatoes with higher amounts of carotenoids (Borghesi *et al.*, 2011). In the present work, carotenoid levels were measured in tomato leaves (not fruits) and a reduction was detected under stress conditions; this is in agreement with previous reports that found a negative correlation between Na^+ and carotenoid contents in tomato leaves (Juan *et al.*, 2005; Tuna *et al.*, 2014).

Salt stress led to a significant increase of phenolics and flavonoids in leaves of plants submitted to the salt treatment and an enhancement of the former in water-stressed plants. There are many publications reporting an increase in the levels of phenolics and flavonoids in tomato fruits in conditions of abiotic stress, which is a topic of direct interest for human health (e.g. Krauss *et al.*, 2006; Ali and Ismail, 2014); in fact, consumption of tomatoes has been recommended to reduce the risk of cardiovascular diseases and cancer (Holiman *et al.*, 1996). Similar studies on leaf material are much scarcer; for example, Sánchez-Rodriguez *et al.* (2011) found an enhancement of some phenolics and flavonoids under moderate water stress (50% of the field capacity) in the more tolerant cultivars of cherry tomatoes, but a reduction in the more sensitive ones, in partial agreement with the data presented here.

Conclusions

Salt and water stress treatments inhibited vegetative growth in cherry tomato, a variety that has not been extensively studied yet, despite its growing commercial interest. In non-irrigated plants, a strong dehydration was partly responsible for reduction of leaf fresh weight, an effect not detected in the presence of NaCl. Both stresses led to the accumulation in the leaves of high levels of Pro, which function as the major osmolyte in cherry tomato, responsible for osmotic adjustment under stress conditions. Glycine betaine also accumulated as a response to salt and water stress - albeit at lower levels than Pro - despite numerous publications reporting its absence in tomato plants; therefore, GB appears to act as a secondary osmolyte that could contribute to stress tolerance in cherry tomato. Both stress treatments caused secondary oxidative stress in the plants, as indicated by a significant increase in malondialdehyde (MDA) contents. The increase in antioxidant phenolic compounds levels in leaves can be considered as part of the response induced to cope with oxidative stress. Contrary to what has been reported for tomato fruits, other metabolites such as total soluble sugars or carotenoids do not increase, but rather decrease in leaves in response to the stress treatments.

References

Aghaleh M, Niknam V, Ebrahimzadeh H, Razavi K (2009). Salt stress effects on growth, pigments, proteins and lipid peroxidation in *Salicornia persica* and *S. europaea*. Biol Plant 53:243-248.

Alberte RS, Thornber JP, Fiscus EL (1977). Water stress effects on the content and organization of chlorophyll in mesophyll and bundle sheath chloroplasts of maize. Plant Physiol 59:351-353.

Ali HEM, Ismail GSM (2014). Tomato fruit quality as influenced by salinity and nitric oxide. Turk J Bot 38:122-129.

Ali SG, Rab A, Khan UA, Nawab K (2011). Enhanced proline synthesis may determine resistance to salt stress in tomato cultivars. Pak J Bot 43:2707-2710.

Alian A, Altman A, Heuer B (2000). Genotypic differences in salinity and water stress resistance tolerance of fresh market tomatoes. Plant Sci 152:59-65.

Amini F, Ehsanpour AA (2005). Soluble proteins, proline, carbohydrates and Na^+/K^+ changes in two tomato (*Lycopersicon esculentum* Mill.) cultivars under *in vitro* salt stress. Am J Biochem Biotechnol 1:204-208.

Appel K, Hirt H (2004). Reactive oxygen species: metabolism, oxidative stress, and signal transduction. Annu Rev Plant Biol 55:373-399.

Ashraf MY, Foolad MR (2007). Roles of glycine betaine and proline in improving plant abiotic stress resistance. Environ Exp Bot 59:206-216.

Babu MA, Singh D, Gothandam KM (2012). The effect of salinity on growth, hormones and mineral elements in leaf and fruit of tomato cultivars PKM1. J Anim Pl Sci 22:159-164.

Bates LS, Waldren RP, Tear LD (1973). Rapid determination of free proline for water-stress studies. Plant Soil 39:205-207.

Blainski A, Lopes GC, de Mello JCP (2013). Application and analysis of the Folin Ciocalteu method for the determination of the total phenolic content from *Limonium brasiliense* L. Molecules 18:6852-6865.

Borghesi E, González-Miret L, Escudero-Gilete ML, Malorgio F, Heredia FJ, Meléndez-Martínez AJ (2011). Effects of salinity stress on carotenoids, anthocyanins and color of diverse tomato genotypes. J Agric Food Chem 59:11676-11682.

Boscaiu M, Tifrea A, Donat P, Mayoral O, Llinares J, Bautista I, Lidón A, Lull C, Vicente O (2011). Seasonal variation of glycine betaine in plants from a littoral salt - marsh in SE Spain. Bull UASVM Horticulture 68:543-544.

Büssis D, Heineke D (1998). Acclimation of potato plants to polyethylene glycol-induced water deficit II. Contents and subcellular distribution of organic solutes. J Exp Bot 49:1361-1370.

Chen S, Gollop N, Heuer B (2009). Proteomic analysis of salt-stressed tomato (*Solanum lycopersicum*) seedlings: effect of genotype and exogenous application of glycinebetaine. J Exp Bot 60:2005-2019.

Chen THH, Murata N (2008). Glycinebetaine: an effective protectant against abiotic stress in plants. Trends Plant Sci 13:499-505.

Cheynier V, Comte G, Davies KM, Lattanzio V, Martens S (2013). Plant phenolics: recent advances on their biosynthesis, genetics, and ecophysiology. Plant Physiol Biochem 72:1-20.

Clausen W (2005). Proline as a measure of stress in tomato plants. Plant Sci 168:241-248.

Cuartero J, Fernández Muñoz J (1999). Tomato and salinity. Sci Hort 78:83-125.

Dalton FN, Maggio A, Piccinni G (1997). Effect of root temperature on plant response functions for tomato: comparison of static and dynamic salinity stress indices. Plant Soil 192:307-319.

Del Rio D, Stewart AJ, Pellegrini N (2005). A review of recent studies on malondialdehyde as toxic molecule and biological marker of oxidative stress. Nutr Metab Cardiovas 15: 316-328.

Di Ferdinando M, Brunetti C, Fini A, Tattini M (2012). Flavonoids as antioxidants in plants under abiotic stresses. In: Ahmad P, Prasad MNV (Eds). Abiotic stress responses in plants: metabolism, productivity and sustainability. Springer, New York pp. 159-179.

Dubois M, Gilles KA, Hamilton JK, Reberd PA, Smith F (1956). Colorimetric method for determination of sugars and related substances. Anal Chem 28:350-356.

Fernández-García N, Martínez V, Cerda A, Carvajal M (2004). Fruit quality of grafted tomato plants grown under saline conditions. J Hort Sci Biotech 79:995-1001.

Gavaghan CL, Li JV, Hadfield ST, Hole S, Nicholson JK, Wilson ID, Howe PWA, Stanley PD, Holmes E (2011). Application of NMR-based metabolomics to the investigation of salt stress in maize (Zea *mays*). Phytochem Anal 22:214-224.

Ghorbanli M, Gafarabad M, Amirkian T, Mamaghani BA (2013). Investigation on proline, total protein, chlorophyll ascorbate and dehydroascorbate changes under drought stress in Akria and Mobil tomato cultivars. Iran J Plant Physiol 3:651-658.

Giannakoula AE, Ilias IF (2013). The effect of water stress and salinity on growth and physiology of tomato. Arch Biol Sci Belgrade 65:611-620.

Gil R, Bautista I, Boscaiu M, Lidón A, Wankhade S, Sánchez H, Llinares J, Vicente O. (2014). Responses of five Mediterranean halophytes to seasonal changes in environmental conditions. AoB Plants, doi:10.1093/aobpla/plu049.

Gil R, Boscaiu M, Lull C, Bautista I, Lidón A, Vicente O (2013). Are soluble carbohydrates ecologically relevant for salt tolerance in halophytes? Funct Plant Biol 40:805-818.

Gill SS, Tuteja N (2010). Reactive oxygen species and antioxidant machinery in abiotic stress tolerance in crop plants. Plant Physiol Biochem 48:909-930.

Goel D, Singh AK, Yadav V, Babbar SB, Murata N, Bansal KC (2011). Transformation of tomato with a bacterial codA gene enhances tolerance to salt and water stresses. J Plant Physiol 168:1286-1294.

Gorham J, Hughes LL, Wyn Jones RG (1980). Chemical composition of salt-marsh plants from Ynys Môn (Anglesey): the concept of physiotypes. Plant Cell Environ 3:309-318.

Gould KS, Lister C (2006). Flavonoid function in plants. In: Andersen ØM, Marham KR (Eds). Flavonoids, chemistry, biochemistry and application, CRC Press, Boca Raton pp 397-442.

Grieve CM, Grattan SR (1983). Rapid assay for determination of water soluble quaternary ammonium compounds. Plant Soil

70:303-307.

Gummuluru S, Hobbs SLA, Jana S (1989). Physiological responses of drought tolerant and drought susceptible durum wheat genotypes. Photo Synthetica 23:479-485.

Handa S, Handa AK, Hasegawa PP, Bressan RA (1986). Proline accumulation and the adaptation of cultured plant cells to water stress. Plant Physiol 80:938-945.

Hare PD, Cress WA, Van Staden J (1998). Dissecting the roles of osmolyte accumulation during stress. Plant Cell Environ 21:535-553.

Heuer B (2003). Influence of exogenous application of proline and glycinebetaine on growth of salt-stressed tomato plants. Plant Sci 165:693-699.

Hodges DM, De Long JM, Forney CF, Robert K, Prange RK (1999). Improving the thiobarbituric acid-reactive-substances assay for estimating lipid peroxidation in plant tissues containing anthocyanin and other interfering compounds. Planta 207:604-611.

Holiman PCH, Hertog MGL, Katan MB (1996). Analysis and health effects of flavonoids. Food Chem 57:43-46.

Inal A (2002). Growth, proline accumulation and ionic relations of tomato (*Lycopersicon esculentum* L.) as influenced by NaCl and Na_2SO_4 salinity. Turk J Bot 26:285-290.

Juan M, Rivero RM, Romero L, Ruiz JM (2005). Evaluation of some nutritional and biochemical indicators in selecting salt-resistant tomato cultivars. Environ Exp Bot 54:193-201.

Krauss S, Schniztler WH, Grassmann J, Woitke M (2006). The influence of different electrical conductivity values in a simplified recirculating soilless system on inner and outer fruit quality characteristics of tomato. J Agric Food Chem 54:441-448.

Li G, Wan SW, Zhou J, Yang ZY, Qin P (2010). Leaf chlorophyll fluorescence, hyperspectral reflectance, pigments content, malondialdehyde and proline accumulation responses of castor bean (*Ricinus communis* L.) seedlings to salt stress levels. Ind Crop Prod 31:13-19.

Li Y (2009). Phsiological responses of tomato seedlings (*Lycopersicon esculentum*) to salt stress. Mod Appl Sci 3:171-176.

Mafakheri A, Siosemardeh A, Bahramnejad B, Struik PC, Sohrabi Y (2010). Effect of drought stress on yield, proline and chlorophyll contents in three chickpea cultivars. Aust J Crop Sci 4:580-585.

Maggio A, De Pascale S, Angelino G, Ruggiero C, Barbieri G (2004). Physiological responses of tomato to saline irrigation in long-term salinized soils. Eur J Agron 21:149-159.

Maggio A, Raimondi G, Martini A, De Pascale S (2007). Salt responses in tomato beyond the salinity tolerance threshold. Env Exp Bot 59:276-282.

Makela P, Jokinen K, Kontturi M, Peltonen-Sainio P, Pehu E, Somersalo S (1998). Foliar application of glycinebetaine novel product from sugar-beet as an approach to increase tomato yield. Ind Crop Prod 7:139-148.

Munns R (2002). Comparative physiology of salt and water stress. Plant Cell Environ 25:239-250.

Munns R, Tester M (2008). Mechanisms of salinity tolerance. Annu Rev Plant Biol 9:651-681.

Nahar K, Gretzmacher R (2002). Effect of water stress on nutrient uptake, yield and quality of tomato (*Lycopersicon esculetum* Mill.) under subtropical conditions. Bodenkultur 53:45-51.

Nawaz K, Ashraf M (2010). Exogenous application of glycinebetaine modulates activities of antioxidants in maize plants subjected to salt stress. J Agron Crop Sci 196:28-37.

Parida A, Das AB, Das P (2002). NaCl stress causes changes in photosynthetic pigments, proteins and other metabolic components in the leaves of a true mangrove, *Bruguiera parviflora*, in hydroponic cultures. J Plant Biol 45:28-36.

Parida A, Das AB, Sanada Y, Mohanty P (2004). Effects of salinity on biochemical components of the mangrove *Aegiceras corniculatum*. Aquat Bot 80:77-87.

Parida AK, Das AB (2005). Salt tolerance and salinity effect on plants: a review. Ecotoxic Environ Safe 603:324-349.

Park EJ, Jeknic Z, Chen THH (2006). Exogenous application of glycinebetaine increases chilling tolerance in tomato plants. Plant Cell Physiol 47:706-714.

Parvaiz A, Satyawati S (2008). Salt stress and phyto-biochemical responses of plants: a review. Plant Soil Environ 53:89-99.

Pollastri S, Tattini M (2011). Flavonols: old compounds for old roles. Ann Bot 108:1225-1233.

Rezaei MA, Jokar J, Ghorbanli M, Kaviani B, Kharabian-Masouleh A (2012). Morpho-physiological improving effects of exogenous glycine betaine on tomato (*Lycopersicum esculentum* Mill.) cv. PS under drought stress conditions. Plant Omics 5:79-86.

Romero-Aranda R, Soria T, Cuartero JJ (2002). Greenhouse mist improves yield of tomato plants grown under saline conditions. J Am Soc Hortic Sci 127:644-648.

Rosales MA, Rios JJ, Castellano R, López-Carrión AI, Romero L, Ruiz JM (2007). Proline metabolism in cherry tomato exocarp in relation to temperature and solar radiation. J Hort Sci Biotech 82:739-744.

Rosales MA, Ruiz JM, Hernández J, Soriano T, Castilla N, Romero L (2006). Antioxidant content and ascorbate metabolism in cherry tomato exocarp in relation to temperature and solar radiation. J Sci Food Agric 86:1545-1551.

Sanchez RA, Hall AJ, Trapani N, Cohen de Hunau R (1983). Effects of water stress on the chlorophyll content, nitrogen level and photosynthesis of leaves of two maize genotypes. Photosynth Res 4:35-47.

Sánchez-Rodríguez E, Moreno DA, Ferreres F, Rubio-Wilhelmi MM, Ruiz JM (2011). Differential responses of five cherry tomato varieties to water stress: Changes on phenolic metabolites and related enzymes. Phytochemistry 72:723-729.

Santos CV (2004). Regulation of chlorophyll biosynthesis and degradation by salt stress in sunflower leaves. Sci Hort 103:93-99.

Sgherri C, Kadlecová Z, Pardossi A, Navari-Izzo F, Izzo R (2008). Irrigation with diluted seawater improves the nutritional value of cherry tomatoes. J Agri Food Chem 56:3391-3397.

Shalata A, Mittova V, Volokita M, Guy M, Tal M (2001). Responses of the cultivated tomato and its wild-tolerant relative *Lycopersicon pennellii* to salt-dependent oxidative stress: The root antioxidative system. Physiol Plant 112:487-494.

Shalata A, Tal M (1998). The effect of salt stress on lipid peroxidation and antioxidants in the leaf of the cultivated tomato and its wild salt-tolerant relative *Lycopersicon pennellii.* Physiol Plant 104:169-174.

Sims DA, Gamon JA (2002). Relationships between leaf pigment content and spectral reflectance across a wide range of species, leaf structures and developmental stages. Remote Sens Environ 81:337-354.

Slimestad R, Fossen T, Verheul TJ (2008). The flavonoids of tomatoes. J Agric Food Chem 56:2436-2441.

Soussi M, Ocaña A, Lluch C (1998). Effect of salt stress on growth, photosynthesis and nitrogen fixation in chick-pea (*Cicer arietinum* L.) cultivars. J Exp Bot 49:1339-1347.

Stewart GR, Lee JA (1974). The role of proline accumulation in halophytes. Planta 120:279-289.

Szabados L, Savouré A (2010). Proline: a multifunctional amino acid. Trends Plant Sci 15:89-97.

Tipirdamaz R, Gagneul D, Duhaze C, Ainouche A, Monnier C, Ozkum D, Larher F (2006). Clustering of halophytes from an inland salt marsh in Turkey according to their ability to accumulate sodium and nitrogenous osmolytes. Environ Exp Bot 57:139-153.

Treutter D (2005). Significance of flavonoids in plant resistance and enhancement of their biosynthesis. Plant Biol 581-591.

Treutter D (2006). Significance of flavonoids in plant resistance: a review. Environ Chem Lett 4:147-157.

Tuna AT (2014). Influence of foliarly applied different triazole compounds on growth, nutrition, and antioxidant enzyme activities in tomato (*Solanum lycopersicum* L.) under salt stress. Austral J Crop Sci 8:71-79.

Tyree MT, Jarvis PG (1982). Water in tissue and cells. In: Lange OL, Nobel PS, Osmond OB, Ziegler M (Eds). Encyclopedia of Plant Physiology Vol 2. Springer, Berlin pp 35-77.

Ullah SM, Gerzabeck MH, Sonja G (1994). Effect of seawater and soil-salinity on ion uptake, yield and quality of tomato (fruit). Bodenkultur 45:227-237.

Umebese CE, Olatimilehin TO, Ogunsusi TA (2009). Salicylic acid protects nitrate reductase activity, growth and proline in amaranth and tomato plants during water deficit. Am J Agric Biol Sci 4:224-229.

Van Breusegem F, Dat JF (2006). Reactive oxygen species in plant cell death. Plant Physiol 141:384-390.

Verbruggen N, Hermans C (2008). Proline accumulation in plants: a review. Amino Acids 35:753-759.

Vicente O, Boscaiu M, Naranjo MA, Estrelles E, Belles JM, Soriano P (2004). Responses to salt stress in the halophyte *Plantago crassifolia* (Plantaginaceae). J Arid Environ 58:463-481.

Winkel-Shirley B (2002). Biosynthesis of flavonoids and effect of stress. Curr Opin Plant Biol 5:218-223.

Wiseman H (2006). Isoflavonoids and human health. In: Andersen ØM, Marham KR (Eds). Flavonoids, chemistry, biochemistry and application. CRC Press, Boca Raton pp 397-442.

Yin Y-G, Kobayashi Y, SanukiA, Kondo S, Fukuda N, Ezura H, Sugaya S, Matsukura C (2010). Salinity induces carbohydrate accumulation and sugar-regulated starch biosynthetic genes in tomato (*Solanum lycopersicum* L. cv. 'Micro-Tom') fruits in an ABA and osmotic stress-independent manner. J Exp Bot 61:563-574.

Yokaş I, Tuna AL, Bürün B, Altunlub H, Altan F, Kaya C (2008). Responses of the tomato (*Lycopersicon esculentum* Mill.) plants to exposure to different salt forms and rates. Turk J Agric For 32:319-329.

Zhishen J, Mengcheng T, Jianming W (1999). The determination of flavonoid contents in mulberry and their scavenging effects on superoxide radicals. Food Chem 64:555-559.

34

Molecular Cloning and Expression Analysis of a MADS-Box Gene (*GbMADS2*) from *Ginkgo biloba*

Xiaohui WANG[1], Junhuan CHENG[1], Feng XU[1*], Xingxiang LI[2], Weiwei ZHANG[1], Yongling LIAO[3], Shuiyuan CHENG[4], Xianbin LI[5]

[1]*College of Horticulture and Gardening, Yangtze University, Jingzhou, 434023, China; 529179525@qq.com; 374334033@qq.com; xufengl98@126.com (*corresponding author); wwzhangchn@163.com*
[2]*Medical School, Yangtze University, Jingzhou, 434023, China; lxx@yangtzeu.edu.cn*
[3]*College of Forestry, Nanjing Forest University, Nanjing, 210037, China; liaoyongling@yeah.net*
[4]*Wuhan Polytechnic University, Wuhan 430023, China; s_y_cheng@sina.com*
[5]*Jingzhou Soil and Fertilizer Station, Jingzhou 434025, China; 676593359@qq.com*

Abstract

As a kind of transcription factors gene family, MADS-box genes play an important role in plant development processes. To find genes involved in the floral transition of *Ginkgo biloba*, a MADS-box gene, designated as *GbMADS2,* was cloned from *G. biloba* based on EST sequences by RT-PCR. Sequence analysis results showed that the cDNA sequence of *GbMADS2* contained a 663 bp length ORF encoding 221 amino acids protein, which displayed typical structure of plant MADS-box protein including MADS, I, and K domains and C terminus. The sequence of GbMADS2 protein was highly homologous to those of MADS-box proteins from other plant species with the highest homologous to AGAMOUS (CyAG) from *Cycas revoluta*. The phylogenetic tree analysis revealed that *GbMADS2* belonged to *AGAMOUS* clade genes. Real-time PCR analysis indicated that expression levels of *GbMADS2* gene in female and male flower were significantly higher than those in root, stem, and leaves, and that *GbMADS2* expression level increased along with time of flower development. The spatial and time-course expression profile of *GbMADS2* implied that *GbMADS2* might be involved in development of reproductive organs. The isolation and expression analysis of *GbMADS2* provided basis for further studying the molecular mechanism of flower development in *G. biloba*.

Keywords: bioinformatics analysis, *GbMADS2*, spatial expression pattern, time-course expression pattern

Introduction

Plant flower formation and development is a complex process that is controlled by environmental and endogenous signals (Brambilla and Fornara, 2013), with four main pathways: autonomous, photoperiod, vernalization, and gibberellin-mediated (Bernier and Périlleux, 2005). Many genes involved in flower development act in concert with physiological signals to establish complex gene regulatory networks that regulate flowering and floral organ development. Many of these genes belong to the MADS-box gene family, which is a large transcription factor family with diverse roles, especially floral development (Hemming and Trevaskis, 2011). Also, MADS genes are great significant for studying the system evolution of plant groups because of their universality and highly conservatism in the plant kingdom (Purugganan *et al.*, 1995). In flowering plants, MADS-box genes control diverse developmental processes in organs ranging from roots to flowers and fruit development (Ng and Yanofsky, 2001). In regard to biological metabolism, researcher have found an important function of MADS which as transcription factors regulate the biosynthetic pathways of secondary metabolites flavonoids (Lalusin *et al.*, 2006). Moreover, MADS-box genes involve in plant growth and development process, such as synthesis of gibberellin (Jackson, 2009), the fruits ripening (Ireland *et al.*, 2013), plant dormancy (Saito *et al.*, 2013), the root elongation (Zhang and Forde, 1998), callus differentiation (Fornara *et al.*, 2004), shoot development (Lee *et al.*, 2008).

Ginkgo biloba, the only surviving genus and species of the Ginkgoaceae family, is a dioecious tree, native of China. Resistant to city smoke and industrial fumes, it has become a favorite ornamental tree in parks and streets of numerous towns all over the world. Moreover, *G. biloba* accumulates, mainly in its leaves, two types of compounds of pharmaceutical importance, that are terpenes (ginkgolides and bilobalide) and flavonoids. The leaf extracts are widely used for the treatment of peripheral or cerebral circulatory disorders, especially in geriatric practice (Van Beek and Montoro, 2009). *G. biloba* is a kind of Mesozoic precious plants with a special long juvenile phase, general 15 to 20 years to bear fruit after planting. As a unique relict species belonging to a plant lineage known as Ginkgophyta, *G. biloba* has many unique flowering habits compared with other plants (Chen *et al.*, 2008). Therefore, it is important theoretical and practical significance to study floral organ development and fruit formation on *G. biloba*. However, up to now, little literature reported isolation and function of the genes involved in *G. biloba* flower organ development. To explore the molecular mechanism of *G. biloba* flower organ development, this study isolated a MADS-box gene, designated as *GbMADS2*, from *G. biloba* and analyzed the gene expression patterns in different *G. biloba* organs. This work aimed to provide the theoretical basis for further studies on the molecular mechanism of this gene participating in flowering in *G. biloba*.

Materials and methods

Plant materials

15-year-old grafts of *G. biloba* were grown in a greenhouse at Yangtze University, China. To test the spatial expression pattern of *GbMADS2*, the roots, leaves, fruits, stems, male and female flowers of *G. biloba* grafts were collected. For measurement of the time-course expression patter of *GbMADS2* in ginkgo flowers, the male and female flower samples on 2, 4, 6, 8, and 10 days after flower bud sprouting. All the samples were collected and immediately frozen in liquid nitrogen, and kept at -80 °C prior to total RNA extraction.

Cloning of the full-length cDNA of GbMADS2

Total RNA was extracted from female flowers using the CTAB method (Cai *et al.*, 2007). The quality and concentration of the RNA was all determined by agarose gel electrophoresis and spectrophotometer analysis. The primers MADSFP (5′-TGCTCTAAATTCCAACAGAAAAG-3′) and MADSRP (5′-GTAATATATCCCGCCCATAAACT-3′) were designed and synthesized (Shanghai Sangon, China) based on the EST sequence of *G. biloba*. A 712 bp-length fragment was obtained using the one-step RT-PCR kit (Dalian TaKaRa, China) under the following conditions: 50°C for 30 min and 94°C for 3 min, followed by 35 cycles of amplification at 94 °C for 1 min, 51 °C for 1 min, and 72 °C for 1 min. The PCR product was purified, cloned into the pMD18-T vector (Dalian TaKaRa, China), and then sequenced. Subsequent BLAST results confirmed that the amplified product was the sequence of the MADS gene.

Bioinformatics analysis and molecular evolution analyses

The obtained nucleotide sequence and deduced amino acid sequence were compared through database search using BLAST program (NCBI, http://www.ncbi.nlm.nih.gov). Multiple sequence alignment was performed with the software Vector NTI suit 10.0 program. The protein sequences of GbMADS2 and other MADSs were obtained from GenBank and aligned with the software Vector NTI suit 10.0 program. Phylogenetic tree was constructed using CLUSTAL W 1.83 and MEGA 4.0. The reliability of the tree was measured by bootstrap analysis with 100 replicates.

Quantitative Real Time-PCR analysis of GbMADS2

To investigate expression pattern of *GbMADS2* gene in different tissues, total RNA of the roots, leaves, fruits, stems, male and female flowers of *G. biloba* were isolated. First-strand cDNA synthesis was carried out in triplicate for each sample according to the instructions of the manufacturer (Real-Time PCR Reagent Kit, Dalian TaKaRa, China). Quantitative Real-time PCR (qRT-PCR) was performed using a Perkin-Elmer 7000 thermal cycler with SYBR Premix Ex Taq™ II Kit (Dalian TaKaRa, China) according to the protocol of the manufacturer. Reactions were performed in triplicate using 2 μL of Master Mix, 0.5 M of each primer, 2 μL of diluted cDNA template, and nuclease free water to a final volume of 20 μL. The PCR reaction conditions were 95 °C for 3 min, and then 30 cycles of 94 °C for 1 min, 60 °C for 30 s, and 72 °C for 30 s, with a final extension at 72 °C for 3 min. Fluorescence was measured at the end of each annealing step. Raw data were analyzed with Light Cycler software, and expression level was normalized to *G. biloba* house-keeping gene *18S* (*Gb18S*, GenBank accession no. D16448) to minimize the variation in the cDNA template levels. The primers for *GbMADS2* gene (MADS2-S: 5′-GTGGCTCTCATCGTCTTCTC-3′; MADS2-A: 5′-ATTTGCTGTCTCAGTTTTCC-3′) and *Gb18S* gene (18S-S: 5′-ATAACAATACTGGGCTCATCG-3′; 18S-A: 5′-TTCGCAGTGGTTCGTCTTTC-3′) were designed using the Sequence Detection System software. Real-time PCR data were technically replicated with error bars, representing mean ± SE (n = 3).

Results

Cloning and characterization of GbMADS2

The amplified products of RT-PCR were separated on 1% agarose gel stained with ethidium bromide. The result showed that the PCR products size was about 700 bp (Fig. 1), which had a same size with the expected MADS-box gene sequence. The recyclable PCR products were sequenced, and showed that the cDNA sequence of PCR products contained 712 bp. The results of BlastN analysis on NCBI showed that this cDNA sequence had high similarity with that of other MADS genes. The nucleotide sequence of *GbMADS2* was 84% and 80% identical to MADS-box gene from *Cycas edentata* and *Rafflesia cantleyi*, and 82%, 82%, and 81% identical to *AGAMOUS* gene from *Macrozamia spiralis Ceratozamia mexicana*, *Zamia integrifolia*, and 74% identical to *AGAMOUS-like* gene from *Magnolia denudata*, respectively (Table 1), indicating the gene we isolated

Table 1. Nucleotide sequence of *GbMADS2* similarity to the *MADS-box* genes from other plant species

Species	Gene ID	GenBank Accession No.	Homology
Cycas edentata	*MADS-box*	AF492455	84%
Macrozamia spiralis	*AG*	HM596862	82%
Ceratozamia mexicana	*AG*	HM596861	82%
Zamia integrifolia	*AG*	HM596866	81%
Rafflesia cantleyi	*MADS-box*	KF730033	80%
Magnolia denudata	*AGAMOUS-like*	JQ326231	74%

is a member of MADS-box gene family. Therefore, this gene was designated as GbMADS2 (GenBank Accession No. KP260628). As shown in Fig. 2, the nucleotide sequence of *GbMADS2* gene contained a 663 bp length open reading frame (ORF), encoding a predicted protein sequence of 221 amino acid residues. Furthermore, the homologous sequence of MADS genes among different species showed that *GbMADS2* gene might keep a strong conserved role during the molecular evolution (Smaczniak *et al.*, 2012).

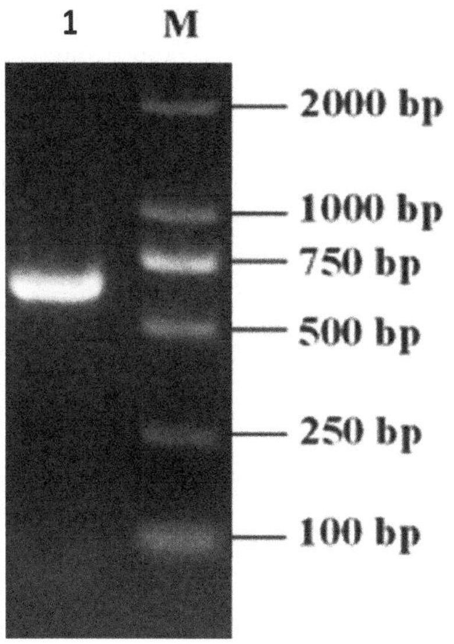

Fig. 1. Elecrtophoresis pattern of amplified product of *GbMADS2* from *Ginkgo biloba.* 1 - Amplified products of GbMADS2 cDNA ; M - DNA Marker DL2000

```
1     TGCTCTAAATTCCAACAGAAAAGATGGGCCGTGGGAAGATTGAGATGAAGAGGATTGAGA
1                             M  G  R  G  K  I  E  M  K  R  I  E
61    ACACTACAAACCGACAGGTCACTTTCTGCAAGCGTCGAAATGGTCTGTTGAAGAAAGCGT
13    N  T  T  N  R  Q  V  T  F  C  K  R  R  N  G  L  L  K  K  A
121   ATGAATTATCCGTGCTTTGTGATGCGGAAGTGGCTCTCATCGTCTTCTCCAGTCGCGGGA
33    Y  E  L  S  V  L  C  D  A  E  V  A  L  I  V  F  S  S  R  G
181   GACTCTATGAGTTTGCTAATAACAGTGTGAAGAGAACAATCGACAGATACAAGAAGACTT
53    R  L  Y  E  F  A  N  N  S  V  K  R  T  I  D  R  Y  K  K  T
241   GCGCTGACAACTCTCACGGTGGAGCCATTTCAGAGTGTAATTCTCAGTACTGGCAACAGG
73    C  A  D  N  S  H  G  G  A  I  S  E  C  N  S  Q  Y  W  Q  Q
301   AGGCAGGAAAACTGAGACAGCAAATTGATATTCTGCAAAATGCAAATAGACACTTGATGG
93    E  A  G  K  L  R  Q  Q  I  D  I  L  Q  N  A  N  R  H  L  M
361   GGGACGCGCTTACATCTTTAAGTGTAAAGGAGCTTAAGCAGCTAGAAATTCGACTTGAGA
113   G  D  A  L  T  S  L  S  V  K  E  L  K  Q  L  E  I  R  L  E
421   GGGGCATTAGCAGGGTTCGATCAAAGAAGAATGAAATGTTGCTTGAGGAGATAGAGATTA
133   R  G  I  S  R  V  R  S  K  K  N  E  M  L  L  E  E  I  E  I
481   TGCAAAGAAGGGAACACATATTACTGGCGGAGAACCAGTTTCTTCGCACTAAGATAGCTG
153   M  Q  R  R  E  H  I  L  L  A  E  N  Q  F  L  R  T  K  I  A
541   AATGTGAAAGCAGCCAAAATGCAAATATGCTGCCGGGTCCCGAATTCGATGCATTGCCTG
173   E  C  E  S  S  Q  N  A  N  M  L  P  G  P  E  F  D  A  L  P
601   GATTTGATTCTCGACATTTCCTACATGCAAGTATAATGGATGCTCACCATTATGCACAAC
193   G  F  D  S  R  H  F  L  H  A  S  I  M  D  A  H  H  Y  A  Q
661   AGGATCAAACGGCCCTTCAACTTGGATGAAGTTTATGGGCGGATATATTAC
213   Q  D  Q  T  A  L  Q  L  G *
```

Fig. 2. The nucleotide acid sequence and deduced amino acid sequence of GbMADS2. The underline for specific primers MADSFP and MADSRP

Characterization of the deduced GbMADS2 protein

The deduced GbMADS2 had high similarities with other plant MADS-box proteins by using DNAMAN 6.0. It was found that GbMADS2 exhibited 91% identity with AGAMOUS (CyAG) from *Cycas revolute*, 83.7% identity with DAL2 from *Picea abies* and 61.4% identity to AGL11 from *Arabidopsis thaliana,* respectively (Fig. 3). BlastP analysis showed that GbMADS2 possessed typical MADS-box protein structure of MIKC type (type II) plants.

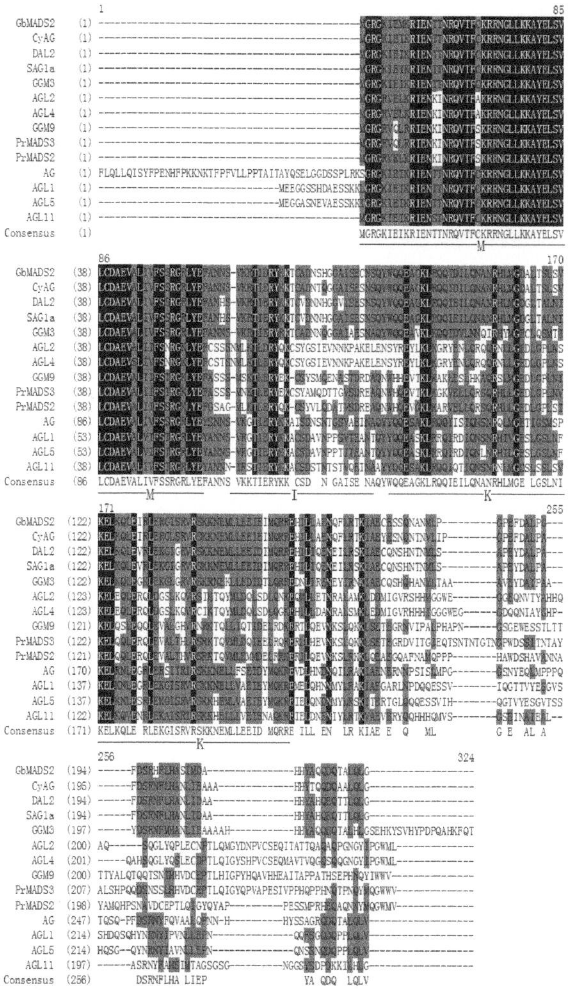

Fig. 3. The homologue comparison of GbMADS2 amino acid sequence with other MADS-box protein. The completely identical amino acids are indicated with white foreground and black background. The conserved amino acids are indicated with black foreground and grey background. Non-similar amino acids are indicated with black foreground and white background. The specie and protein name and GenBank accession number are as following: *Ginkgo biloba*: GbMADS2 (AKE50811); *Arabidopsis thaliana*: AG (CAB78898), AGL1 (M55550), AGL5 (M55553), AGL2 (M55551), AGL4 (M55552) and AGL11 (U20182); *Cycas revolute*: CyAG (AF492455); *Picea abies*: DAL2 (X79280); *Picea mariana*: SAG1a (U69482); *Gnetum gnemon*: GGM3 (AJ132209) and GGM9 (AJ132215); *Pinus radiata*: PrMADS2 (U42400) and PrMADS3 (U76726)

According to the modular domain structure of type II MADS protein, GbMADS2 could be divided into four conservative domains (Fig. 3), including MADS domain at 1-57 amino acids, K domain at 91-156 amino acids, I domain containing 33 amino acids, and a C terminus. However, GbMADS2 did not show high homology with the sequence of N-terminus extension of AG proteins of *Arabidopsis thaliana*, but similar to the AG subfamily of reported MADS-box protein from other plants. Compared the different domains similarities of GbMADS2 with CyAG, DAL2 and AGL11, respectively, they had the highest homology in MADS domain. It was found that GbMADS2 exhibited 94.74%, 100%, and 100% similarities in MADS domain as well as 8.5%, 86.4% and 69.7% similarities in K domain of CyAG, DAL2 and AGL11, respectively (Fig. 3).

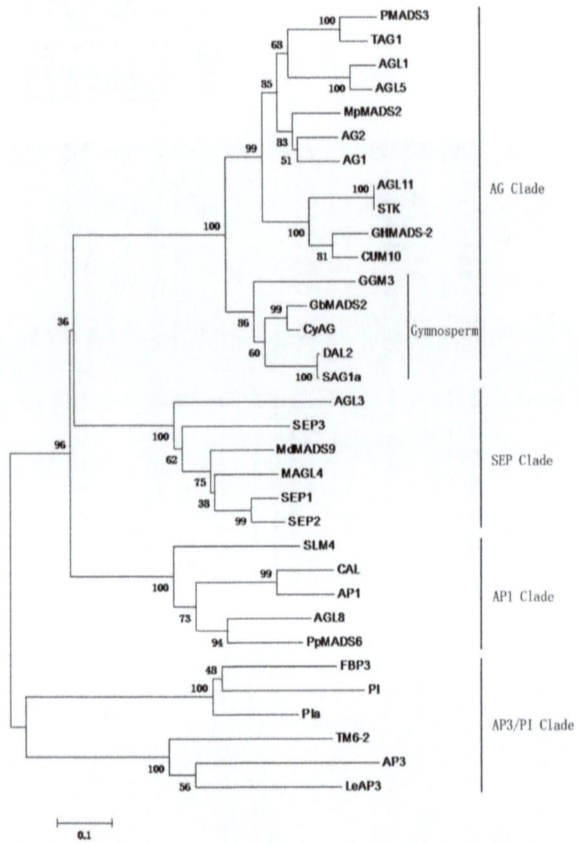

Fig. 4. Phylogenetic tree analysis of protein encoded by MADS-box genes. The specie and protein name and GenBank accession number are as following: *Arabidopsis thaliana*: STK (NP_001078364), CAL (NP_564243), AP1 (CAA78909), AGL8 (Q42429), AGL3 (P29383), SEP1(P29382), SEP2 (P29384), SEP3 (O22456), PI (NP_197524) and AP3 (NP_191002); *Petunia hybrida*: PMADS3 (BAB79434) and FBP3 (CAA50549); *Lycopersicon esculentum*: TAG1 (Q40168) and LeAP3 (AAC42583); *Eschscholzia californica*: AG2 (AAZ53206) and AG1 (AAZ53205); *Magnolia praecocissima*: MpMADS2 (BAB70737); *Gossypium hirsutum*: GHMADS-2 (AAN15183); *Cucumis sativus*: CUM10 (AAC08529); *Prunus persica*: PpMADS6 (AAU29514); *Malus domestica*: MdMADS9 (CAA04920); *Populus tremuloides*: MAGL4 (AAL08423); *Lotus corniculatus* var. *japonicus*: PIa (AAX13299); *Carica papaya*: TM6-2 (ABQ51322); *Silene latifolia*: SLM4 (CAA56658)

Molecular evolution analysis

To investigate the evolutionary relationships in GbMADS2 and MADSs from other plant species, we choose some typical MADS-box genes from GenBank to analyze molecular evolution of GbMADS2 on the basis of the four subfamilies (AP1, AP3/PI, SEP and AG) of flowers homologous alien genes of MADS-box gene. A phylogenetic tree was constructed by using NJ method of MEGA software. As shown in Fig. 4, GbMADS2 belonged to the subfamily of AG genes in the branch of AP1, AP3/PI, SEP, AG, and had the closest relationship with CyAG of *C. revolute*. Taken into account together with high sequence homologies of GbMADS2 with other plant AG protein, our data suggested that GbMADS2 belonged to AG subfamily of MADS-box gene families. These results suggested that GbMADS2 shared a common evolutionary with other plant AG proteins based on conserved structure and sequence characteristics, such as amino acid homologies and conserved motifs, respectively.

Expression analysis of GbMADS2

To investigate the expression of *GbMADS2* gene in different tissues of *G. biloba*, qRT-PCR was carried out using gene-specific primers and cDNAs from different ginkgo tissues as templates. As shown in Fig. 5, GbMADS2 was strongly expressed in male and female flowers, but negligible amounts of its transcripts were detected in roots, stems, and leaves vegetative tissues of *G. biloba*. For further examine the role of *GbMADS2* during flower development, we measured the time-course expression pattern of *GbMADS2* in flowers.

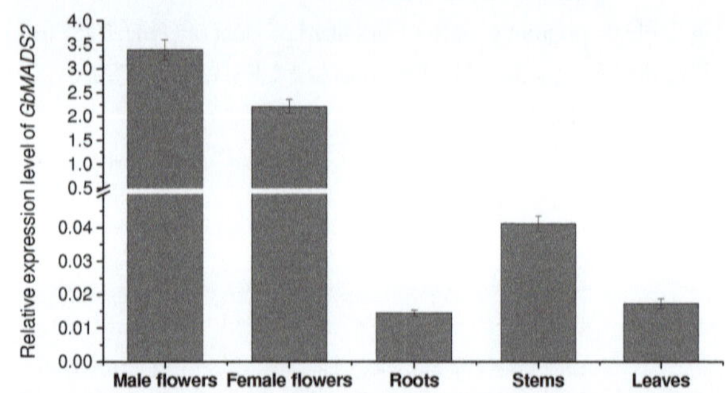

Fig. 5. Expression level of *GbMADS2* in different tissue of *Ginkgo biloba*

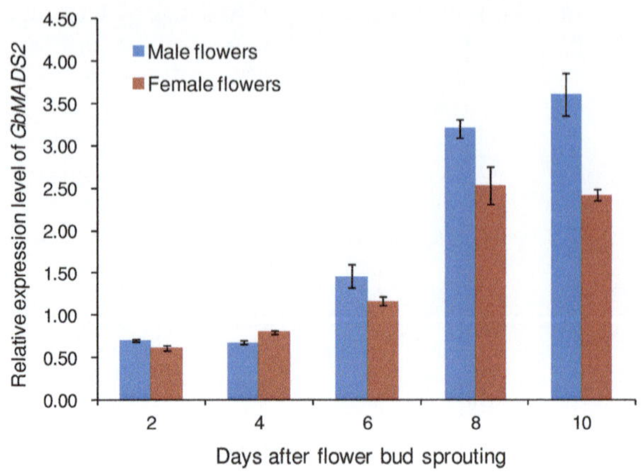

Fig. 6. Time-course expression profile of GbMADS2 during flower development in *Ginkgo biloba*

The results showed that the transcript level of *GbMADS2* significantly increased along with the development of male and female flowers (Fig. 6), implying that *GbMADS2* might be involved in flower development.

Discussions

MADS-box genes have been divided in two large groups, termed type I and type II genes (Alvarez-Buylla *et al.,* 2000). Type I and type II MADS-box genes were generated by a gene duplication that occurred before the divergence of plants, animals and fungi more than a billion years ago (Becker *et al.,* 2000). In plants, the type I MADS-box genes function are still not clear, type II MADS-box genes which include all plant MADS-box genes for which expression patterns or mutant phenotypes are known (De Bodt *et al.,* 2003; Kaufmann *et al.,* 2005). The type I MADS-box genes only contain MADS domain-box structure (Nam *et al.,* 2004), while Type II genes comprise MADS-box, I, K and C structure with different degree of conservative, and called MIKC-type MADS-box (Ma *et al.,* 1991; Münster *et al.,* 1997; Nam *et al.,* 2004; Kaufmann *et al.,* 2005). The sequence analysis results revealed that GbMADS2 contained M, I, and K domains and C terminus, which are typical MADS-box structure of plants. Therefore, *GbMADS2* belongs to MIKC type MADS-box gene and might be related to floral organ formation and development (Coen and Meyerowitz 1991; Heijmans *et al.,* 2012).

Most genomes of plants exist in varying quantity of MADS-box genes duplication. Therefore, it is possible for providing the raw materials for the evolution of the organism new form by gene duplication and function diversification. MADS-box genes form a multiple gene families at many genes duplication of the evolution process, especially in angiosperms (Aswath and Kim, 2005). By phylogeny reconstruction, almost all of the angiosperm MADS-box genes can be subdivided into 12 major gene clades (Becker and Theißen, 2003). Phylogenetic tree showed that gymnosperms and angiosperms had seven same subfamilies of MADS domain, that was to say the existing seed plants have at least seven different MIKC type genes: AG, AGL2, AGL6, DEF/GLO, GGM13, STMADS11 and TM3 gene subfamily in their common ancestor (Smaczniak *et al.,* 2012). Most of MIKC type MADS genes are involved in the development of sporophyte reproductive organs, such as the ovule (Busi *et al.,* 2003). The phylogenetic tree analysis revealed that GbMADS2 gene belonged to AG subfamily, implying that the function of GbMADS2 is probably similar to AG clade genes.

Real-time PCR results revealed that expression level of GbMADS2 in reproductive was significantly higher than that in vegetative tissues, similar to the *Arabidopsis* and cotton AG genes (Mizukami *et al.,* 1992; Liu *et al.,* 2010), indicating GbMADS2 might play a role in the development of reproductive tissues in *G. biloba*. Similar to the time-course expression profile of *OsMADS1* gene in rice (Jeon *et al.,* 2000), we also found the expression level of *GbMADS2* exhibited an increased tendency during flower development, further confirming *GbMADS2* contribute to flower development in *G. biloba*. Taken together, the sequence analysis and the expression pattern data supports the conclusion that *GbMADS2* probably represents a function orthologue of the AG gene family, implicating that the gene may participate in flower development.

Conclusions

In summary, isolation and characterization of *GbMADS2* gene constitute a start point for our studies of flower development in *G. biloba*. We have also recently isolated several MADS-box genes of *G. biloba* and characterized their expression (manuscript in submission). Experiments are underway to further characterize the possible specific role of each gene in *G. biloba* that will enable us to infer a flower development model of this plant and define ways to enhance its productivity exploiting molecular genetic techniques.

Acknowledgements

This work was supported by the Youth Talent Foundation of Hubei Province (2013CFA039), the Key Scientific Research Program of Chinese Ministry of Education (212112 and 210137), International Science and Technology Cooperation Project of Hubei Province (2013BHE029 and 2013BHE039), and the Graduate Innovative Project of Jiangsu Province (2013CXZZ13-D551).

References

Alvarez-Buylla ER, Liljegren SJ, Pelaz S, Gold SE, Burgeff C, Ditta GS, Francisco Vergara-Silva F, Yanofsky MF (2000). MADS-box gene evolution beyond flowers: expression in pollen, endosperm, guard cells, roots and trichomes. Plant J 24(4):457-466.

Aswath CR, Kim SH (2005). Another story of MADS - box genes - their potential in plant biotechnology. Plant Growth Regul 46(2):177-188.

Becker A, Winter KU, Meyer B, Saedler H, Theißen G (2000). MADS-box gene diversity in seed plants 300 million years ago. Mol Biol Evol 17(10):1425-1434.

Becker A, Theißen G (2003). The major clades of MADS-box genes and their role in the development and evolution of flowering plants. Mol Phylogenet Evol 29(3):464-489.

Bernier G, Périlleux C (2005). A physiological overview of the genetics of flowering time control. Plant Biotechnol J 3(1):3-16.

Brambilla V, Fornara F (2013). Molecular control of flowering in response to day length in rice. J Integr Plant Biol 55(5):410-418.

Busi MV, Bustamante C, D'Angelo C, Hidalgo-Cuevas M, Boggio SB, Valle EM, Zabaleta E (2003). MADS-box genes expressed during tomato seed and fruit development. Plant Mol Biol 52(4):801-815.

Cai R, Xu F, Chen L, Cheng S (2007). Modification of total RNA isolation method from different *Ginkgo biloba* organs. Biotechnology 17(4):38-41. (In Chinese with English abstract)

Chen L, Cai R, Xu F, Cheng S (2008). Plants flowering gene regulation and *Ginkgo* juvenile phase genetic improvement research progress. Henan Agric Sci 2:13-16 (In Chinese with English abstract).

Coen ES, Meyerowitz EM (1991). The war of the whorls: genetic interactions controlling flower development. Nature 353(6339):31-37.

De Bodt S, Raes J, Florquin K, Rombauts S, Rouze P, Theiben G, Van de Peer Y (2003). Genome-wide structural annotation and evolutionary analysis of the type I MADS-box genes in plants. Mol Evol 56(5):573-586.

Fornara F, Parenicová L, Falasca G, Pelucchi N, Masiero S, Ciannamea S, Lopez-Dee Z, Altamura MM, Colombo L, Kater MM (2004). Functional characterization of OsMADS18, a member of the AP1/SQUA subfamily of MADS box genes. Plant Physiol 135(4):2207-2219.

Heijmans K, Morel P, Vandenbussche M (2012). MADS-box genes and floral development: The dark side. J Exp Bot 63(15):5397-5404.

Hemming MN, Trevaskis B (2011). Make hay when the sun shines: The role of MADS-box genes in temperature dependant seasonal flowering responses. Plant Sci 180(3):447-453.

Ireland HS, Yao JL, Tomes S, Sutherland PW, Nieuwenhuizen N, Gunaseelan K, Winz RA, David KM, Schaffer RJ (2013). Apple *SEPALLATA1/2-like* genes control fruit flesh development and ripening. Plant J 73(6):1044-1056.

Jackson SD (2009). Plant responses to photoperiod. New Phytol 181(3):517-531.

Jeon JS, Jang S, Lee S, Nam J, Kim C, Lee SH, Chung YY, Kim SR, Lee YH, Cho YG, An G (2000). *"leafy hull sterile1"* is a homeotic mutation in a rice MADS box gene affecting rice flower development. Plant Cell 12(6):871-884.

Kaufmann K, Melzer R, Theißen G (2005). MIKC-type MADS-domain proteins: structural modularity, protein interactions and network evolution in land plants. Genes 347(2):183-198.

Lalusin, AG, Nishita K, Kim SH, Ohta M, Fujimura, T. (2006). A new MADS-box gene (*IbMADS10*) from sweet potato (*Ipomoea batatas* (L.) Lam) is involved in the accumulation of anthocyanin. Mol Genet Genomics 275(1):44-54.

Lee S, Woo YM, Ryu S (2008). Further characterization of a rice AGL12 group MADS-box gene, *OsMADS26*. Plant Physiol 147(1):1156-168.

Liu X, Zuo K, Xu J, Li Y, Zhang F, Yao H, Wang Y, Chen Y, Qiu C, Sun X, Tang K (2010). Functional analysis of *GbAGL1*, a D-lineage gene from cotton (*Gossypium barbadense*). J Exp Bot 61(4):1193-1203.

Ma H, Yanofsky MF, Meyerowitz EM (1991). *AGL1-AGL6*, an Arabidopsis gene family with similarity to floral homeotic and transcription factor genes. Genes Dev 5(3):484-495.

Mizukami Y, Ma H (1992). Ectopic expression of the floral homeotic gene *AGAMOUS* in transgenic Arabidopsis plants alters floral organ identity. Cell 71(1):119-131.

Münster T, Pahnke J, Di Rosa A, Kim JT, Martin W, Saedler H, Theissen G (1997). Floral homeotic genes were recruited from homologous MADS-box genes preexisting in the common ancestor of ferns and seed plants. Proc Natl Acad Sci USA 94(6):2415-2420.

Nam J, Kim J, Lee S, An G, Ma H, Nei M (2004). Type I MADS-box genes have experienced faster birth-and-death evolution than type II MADS-box genes in angiosperms. Proc Natl Acad Sci USA 101(7):1910-1915.

Ng M, Yanofsky MF (2001). Function and evolution of the plant MADS-box gene family. Nature Rev Genet 2(3):186-195.

Purugganan MD, Rounsley SD, Schmidt RJ, Yanofsky MF (1995). Molecular evolution of flower development: diversification of the plant MADS-box regulatory gene family. Genetics 140(1):345-356.

Saito T, Bai S, Ito A, Sakamoto D, Saito T, Ubi BE, Imai T, Moriguchi T (2013). Expression and genomic structure of the dormancy-associated MADS box genes *MADS13* in Japanese pears (*Pyrus pyrifolia* Nakai) that differ in their chilling requirement for endodormancy release. Tree Physiol 33(6):654-667.

Smaczniak C, Immink RG, Angenent GC, Kaufmann K (2012). Developmental and evolutionary diversity of plant MADS-domain factors: insights from recent studies. Development 139(17):3081-3098.

Van Beek T, Montoro P (2009). Chemical analysis and quality control of *Ginkgo biloba* leaves, extracts, and phytopharmaceuticals. J Chromatogr A 1216(11):2002-2032.

Zhang H, Forde BG (1998). An *Arabidopsis* MADS box gene that controls nutrient-induced changes in root architecture. Science 279:407-409.

Effect of Planting Systems and Growing Media on Earliness, Yield and Quality of Strawberry Cultivation under Soilless Culture

Nafiye ADAK[1]*, Hamide GUBBUK[2]

[1]*Akdeniz University, Technical Sciences Vocational School, Environmental Protection and Control Program, 07058, Antalya, Turkey; nafiye@akdeniz.edu.tr (*corresponding author)*
[2]*Akdeniz University, Faculty of Agriculture, Department of Horticulture, 07058, Antalya, Turkey; gubbuk@akdeniz.edu.tr*

Abstract

Soilless growing is becoming an attractive option because of the contamination of agricultural land due to the use of excessive fertilisers and insecticides to get higher productivity per unit area and the yield advantage with soilless culture getting more yield per unit area than the conventional growing. Effects of two planting systems and various growing media on earliness, yield and quality (colour L, C, h°, the rupture of calyx, firmness, total soluble solid, acidity and vitamin C) in strawberry (*Fragaria* × *ananassa* Duch.) cultivation under soilless culture was studied. Two different planting seasons i.e., summer and fall were chosen and growing media was composed of peat (Pt), perlite (Pr), cocopeat (Co), volcanic tuff (Vlc) and their combinations. Results showed that fall planting system was favourable in terms of earliness but summer planting gave better results in terms of total yield. Moreover, the first harvest of summer planting was February, whereas the first harvest of the fall planting was in December. Early yield in the fall planting comprised 36.0% of the total yield. The Co and the Co+Vlc media were more favourable than other media with regard to yield, earliness and fruit quality. In addition vegetation period was found from January to May and it was observed that as the total soluble solids values increased in fruits, acidity and vitamin C values decreased.

Keywords: *Fragaria* × *ananassa* Duch., 'Camarosa' cultivar, substrate, quality

Introduction

Turkey is among the leading countries cultivating strawberry, after the USA and Mexico, with an annual production of 353 173 tonnes (FAO, 2012). In Turkey, strawberry cultivation is carried out under a wide range of ecological conditions, such as protected cultivation in coastal strips using short-day cultivars, or open-field cultivation in upland with neutral cultivars. Applications to increase the earliness and yield have been tested and applied, especially in the Mediterranean coastline (Adak *et al.*, 2003). In Turkey, the seedling types used include bare root, plug and frigo (cold stored) and the planting systems used are summer and early fall planting. In soil cultivation, plant spacing is adjusted according to the seedling type and the soil structure on a rate of approximately 50 000-70 000 plant ha^{-1}.

Soilless culture methods, used mostly for products such as tomatoes and peppers, have also been initiated be used for strawberry plants as well. The number of plants per unit area has increased especially via soilless culture applications; thus, the yield per unit area has also increased (Paraskevopoulou-Paroussis *et al.*, 1995; Ozeker *et al.*, 1999; Paranjpe *et al.*, 2003a; Paranjpe *et al.*, 2008). Fruit quality has also increased due to the fact that this type of cultivation allows for controlled cultivation. Substrate selection is important for soilless cultivation system planning along with the cultivation system (Favaro and Marano, 2003). In this regard, Paranjpe *et al.* (2003b) stated that pine shaving is heavily used in Florida besides substrates such as perlite and peat. Linardakis and Manios (1991) have stated that the highest yield with vertical systems is attained from perlite (80%) + peat (20%); and Lopez-Medina *et al.* (2004) from coconut coir. Lieten (2004) suggested that rockwool has adverse effects on the vegetative development of strawberries and their root developments, whereas coconut coir is well suited for use with strawberries. In Turkey and the rest of the world, consumption of peat resources, decrease in perlite reserves, and waste problem of rockwool etc. have directed people towards the use of renewable materials that can be easily obtained by local resources at cheap prices. Hence, alternative substrates should be tried and compared to peat, perlite and rockwool, which are used widely both in the world and in Turkey (Barkham, 1993; Robertson, 1993; Frolking *et al.*, 2001; García Gómez *et al.*, 2002).

The objective of this study was to examine the soilless strawberry cultivation possibilities in Turkey and under Mediterranean ecological conditions and to determine the effects of the most suited planting systems and growing media on earliness, yield and quality.

Materials and methods

This study was conducted under greenhouse conditions of the private sector research area (36° 50' 37" N; 30° 50' 31" E) which is a high, unheated and modern greenhouse (10 m width x 40 m length x 6.5 m height) in 2012-2013. 'Camarosa' (*Fragaria* × *ananassa* Duch.) strawberry cultivar, respectively summer and fall planting systems were used. Frigo seedlings were used for summer planting and the planting was carried out on June 15th, 2012. In contrast, plug type seedlings were used for the fall planting and the planting was carried out on September 15th, 2012. The following growing media were used:

1. Coconut coir (100%) (Co); 2. Peat (100%) (Pt); 3. Coconut coir+Volcanic tuff (50% + 50%) (Co+Vlc); 4. Peat+Volcanic tuff (50% + 50%) (Pt+Vlc); 5. Peat + Perlite (50% + 50%) (Pt+Pr).

The physical properties of the growing media are as follows: weight per unit volume was 0.18 g cm^{-3} for peat, 0.13 g cm^{-3} for coconut coir, and 0.66 g cm^{-3} for volcanic tuff; the total porosity was 93% for peat, 66% for perlite, 91% for coconut coir, and 71% for volcanic tuff; the air capacity was 33% for peat, 53% for perlite, 35% for coconut coir, and 55% for volcanic tuff; the water holding capacity was 65% for peat, 38% for perlite; 62% for coconut coir, and 35% for volcanic tuff.

This experiment was laid out under greenhouse conditions and the highest number of seedlings per unit area was used as much as possible. Punched pots having 75 cm in length, 25 cm in width and 25 cm in depth and white in colour were used and were placed on benches designed with a width of 25 cm, length of 13.5 m and height of 75 cm which was placed to have 1% slope to ensure free drainage. Among the benches, 75 cm distances were left in order to perform the cultural practices. This trial was carried out at two different times with plug and frigo seedlings, with 10 seedlings planted in each pot (13.33 seedling m^{-2}).

Dosatron injection system equipped with time adjustment irrigation and fertilisation properties was used and the plant nutrient solution was prepared in accordance with Lieten (2004). pH of the solution was set to 6.0 throughout the cultivation season and the Electrical Conductivity (EC) values were between 1.5-1.8 mS cm^{-1} depending on the plant growing cycle (Lieten, 2008; Cantliffe *et al.*, 2008).

Drainage channels were placed underneath the pots during the experiment and optimum irrigation was attained by measuring the drainage amount obtained from this channel in one day. Accordingly, fertigation was calculated depending on drainage amounts which were 30% for summer months and 20% for winter months (Cantliffe *et al.*, 2008; Montesano *et al.*, 2010; Gul *et al.*, 2011). Fertigations of 266.66 ml/plant/day (10 irrigations per day) from the end of July to the end of September, 213.28 ml/plant/day (8 irrigations per day) from the end of September to the beginning of March and 239.94 ml/plant/day (9 irrigations per day) from the beginning of March to the end of May were carried out.

In order to ensure pollination, bumble bees were placed starting from the beginning of November and these hives were replaced every 6 weeks. Chemical control methods were sprayed against diseases and pests such as red spiders, aphids, leaf worms and powdery mildew disease throughout the production period.

During the 2012-2013 season, crown diameter measurements via digital calliper (at the widest section of the crown) and leaf number measurements were carried out on plants in December and in February. Leaf nutrient concentrations were evaluated on 15 young leaves (the 15 most recently matured leaves) which were selected randomly from each parcel during March. Nitrogen content of the leaves was determined using the modified Kjeldahl method; phophorous content using the vanado molibdo phosphoric yellow method, whereas the potassium, calcium, magnesium concentrations were measured by using Atomic Absorption Spectrophotometry (Kacar, 1972). First harvesting dates along with early and total yield values were determined. Early yield was determined in December and in January, whereas total yield was in December and in June. Yield values were calculated by dividing the total weight of harvested fruits from one parcel during the vegetation period by the number of plants. Colour, calyx removal, fruit firmness, total soluble solid (TSS), titrable acid and vitamin C values were also determined. These fruit analyses were carried out on 15 (fifteen) randomly selected fruits from each parcel in each harvest. Fruit skin colours were measured via brand chromometer in terms of "L", "a" and "b" values, which were expressed in terms of C (chroma) and h° (hue). C and h° values were determined according to the formulae given below:

$$C: \sqrt{(a^2+b^2};\qquad h^o: \tan^{-1}(b/a).$$

The h° value expresses the redness and yellowness of the colour mathematically: as the hue value decreases the colour approaches red and as it increases the colour moves away from red. In contrast, the chroma value mathematically represents the brightness and opacity of the colour and higher chroma values indicate that the colour is brighter. Calyx removal and firmness in fruits was determined via digital penetrometer. TSS content of the fruit sap was measured using a hand refractometer and titratable acidity was determined by titrating the samples obtained from the fruit sap with NaOH (0.1 N) until they reached pH 8.2. Citric acid content was then determined as given below:

% Acid=Citric acid constant (0.007) x Used NaOH x NaOH factor x 100.

Vitamin C (L-Ascorbic Acid) amount was calculated via spectrophotometric method and was expressed as mg ascorbic acid 100 ml^{-1} fresh fruit according to Pearson and Churchill (1970).

Studies were planned according to the experimental design named randomised parcels in factorial order with four replicates of 20 plants per replicate. LSD test was used to compare the means. Statistical analyses were performed using SAS and MSTAT softwares.

Results

Crown diameter, leaf number

Effect of planting systems on crown diameter was determined to be statistically significant, whereas its effect on leaf number was not significant (Table 1). Effect of the growing media on crown diameter and leaf number was statistically significant. Highest crown diameter and leaf number values were observed in the Co and the Co+Vlc media. The PS x GM interaction was not found to be statistically significant (Table 1).

Leaf nutrient concentrations

Effects of planting systems and growing media on leaf N, P,

K, Ca and Mg concentrations were statistically insignificant (Table 2). Even though nutrient concentration changed based on the planting systems and growing media, following ranges were measured during march measurements for N 2.78% and 2.85%; P 0.63% and 0.68%; K 3.00% and 3.22%; Ca between 1.65% and 1.68%; and Mg 0.48% and 0.50%. Regarding interaction between PS x GM, insignificant relationships were fixed.

Table 1. Effect of various planting systems and growing media on crown diameter and leaf number in strawberry

Factors	Crown diameter (mm)	Leaves plant^{-1} (no)
Summer	37.72 b*	22.73
Fall	40.20 a	21.33
LSD %5	2.120	NS
Co	43.49 a	23.25 a
Pt	36.67 b	21.17 ab
Co+Vlc	44.07 a	23.42 a
Pt+Vlc	36.05 b	20.67 b
Pt+Pr	34.54 b	21.67 ab
LSD %5	4.738	2.461
** PS x GM	NS	NS

**PS (Planting systems) x GM (Growing media)
*Values in the same column that are followed by different letters are significantly different using LSD comparison test.

Table 2. Effect of various planting systems and growing media on the N, P, K, Ca and Mg content in strawberry leaves

Factors	N (%)	P (%)	K (%)	Ca (%)	Mg (%)
Summer	2.85	0.64	3.11	1.68	0.49
Fall	2.81	0.65	3.10	1.67	0.49
LSD %5	NS	NS	NS	NS	NS
Co	2.85	0.68	3.14	1.67	0.49
Pt	2.83	0.65	3.22	1.67	0.50
Co+Vlc	2.82	0.63	3.00	1.65	0.48
Pt+Vlc	2.87	0.63	3.08	1.68	0.50
Pt+Pr	2.78	0.64	3.07	1.68	0.48
LSD %5	NS	NS	NS	NS	NS
PS x GM	NS	NS	NS	NS	NS

Earliness and total yield

For the "frigo" (cold stored) seedlings planted on 17th July, 2012 during the 2012-2013 growing period, first flowering came out on 10th January, 2013; whereas the first harvesting was done on 3rd of February, 2013 (summer plantation). For the plug seedlings planted on 15th September, 2012, first flowering took place on 30th October, 2012 whereas first harvesting occurred on 7th December, 2012 (fall plantation) (Table 3).

Earliness was not determined for the summer plantings with frigo seedlings (Table 4) whereas an early yield of 264.67 g plant^{-1} was obtained during the fall plantings with plug seedlings (from December to the end of February). Summer planting was determined to be more advantageous in comparison to fall planting in terms of total yield. In fact, a total yield of 998.67 g plant^{-1} was obtained during the summer planting and a total yield of 725.33 g plant^{-1} was obtained for the fall planting.

Whereas the effects of growing media on early yield were not determined to be statistically significant, their effects on total yield were determined to be statistically significant. Highest total yield values were determined in the Co+Vlc and the Co media, while the lowest yields were determined in the Pt media (Table 4).

Fruit skin color (L, C, h°)

Effects of planting systems and growing media on fruit skin colour L, C, h° were not determined to be statistically significant (Table 5). Even though it differs according to the planting systems and growing media, value of L varied between 29.20 and 31.64; value of C varied between 37.72 and 39.62; and the value of h° varied between 24.65 and 28.20.

Calyx removal, fruit firmness

Effects of planting systems on calyx removal and fruit firmness were statistically insignificant (Table 6). Even though no statistically significant effect of growing media on calyx removal was determined, the growing media had an important effect on fruit firmness Highest value of fruit firmness was 0.80 kg cm^{-2} in the Co+Vlc medium. The interaction between PS x GM were also not statistically significant.

Total soluble solid, Titrable acid and Vitamin C content

Effects of planting systems on TSS, titratable acid and Vitamin C were not determined to be statistically significant (Table 7). In contrast, the effect of growing media on TSS was not statistically significant and the highest TSS amount was detected in mixture media, such as the Co+Vlc, Pt+Vlc and the Pt+Pr. In terms of acid and Vitamin C, no difference was determined among the applications and the acid amount varied between 1.10% and 1.12%; whereas the Vitamin C amount varied among 55.35% and 57.21%.

Table 3. First flowering and first harvesting dates of strawberry for planting systems in the 2012-2013 growing period

Planting System	Planting Date	First Flowering Date	First Harvesting Date
Summer	17.07.2012	10.01.2013	3.02.2013
Fall	15.09.2012	30.10.2012	7.12.2012

Table 4. Effect of various planting systems and growing media on early and total yield of strawberry

Factors	Early yield (g plant^{-1})	Total yield (g plant^{-1})
Summer	-	998.67 a
Fall	264.67	725.33 b
LSD %5		30.88
Co	276.67	865.00 ab
Pt	263.33	836.67 b
Co+Vlc	283.33	903.33 a
Pt+Vlc	240.00	860.00 ab
Pt+Pr	260.00	845.00 b
LSD %5	NS	48.824
PS x GM	NS	NS

Table 5. Effect of planting systems and growing media on fruit skin colour (L, C, h°) of strawberry fruits

Factors	Fruit Skin Colour		
	L	C	h°
Summer	29.60	39.16	26.15
Fall	31.64	38.25	26.57
LSD %5	NS	NS	NS
Co	31.45	38.87	24.65
Pt	30.93	39.62	28.19
Co+Vlc	30.74	39.55	24.82
Pt+Vlc	30.79	37.76	26.04
Pt+Pr	29.20	37.72	28.20
LSD %5	NS	NS	NS
PS x GM	NS	NS	NS

Table 6. Effect of planting systems and growing media on the calyx removal of strawberry fruit and fruit firmness

Factors	Calyx Removal (kg)	Fruit Firmness (kg cm^{-2})
Summer	0.87	0.70
Fall	0.85	0.75
LSD %5	NS	NS
Co	0.87	0.78 ab
Pt	0.85	0.73 ab
Co+Vlc	0.87	0.80 a
Pt+Vlc	0.86	0.70 ab
Pt+Pr	0.85	0.67 b
LSD %5	NS	0.11
PS x GM	NS	NS

Table 7. Effect of various planting systems and growing media on TSS, acidity and Vitamin C in strawberry fruits

Factors	TSS (%)	Acid (%)	Vitamin C (%)
Summer	7.57	1.10	55.68
Fall	7.60	1.12	56.64
LSD %5	NS	NS	NS
Co	7.51 b	1.11	56.80
Pt	7.48 b	1.12	57.21
Co+Vlc	7.69 a	1.12	55.72
Pt+Vlc	7.62 a	1.10	55.35
Pt+Pr	7.65 a	1.11	55.71
LSD %5	0.0958	NS	NS
PS x GM	NS	NS	NS

Table 8. Monthly variation in TSS, Acidity and Vitamin C content of strawberry fruits during the 2012-2013 production season

Factors	TSS (%)	Acid (%)	Vitamin C (%)
January	6.33 e	1.28 a	65.55 a
February	7.13 d	1.22 b	63.56 a
March	7.28 c	1.09 c	57.43 b
April	8.14 b	1.04 d	51.16 c
May	9.06 a	0.92 e	43.10 d
LSD %5	0.0958	0.0371	2.7474

Effects of months on the TSS, acid and Vitamin C values of fruits during the 2012-2013 experimental period are shown in Table 8. As can be seen in this table, the effects of time on TSS, acid and Vitamin C were found be statistically significant. As a matter of fact, the TSS value increases as acidity and Vitamin C values decrease between January and May. The TSS value varied between 6.33% and 9.06%; the acidity values varied among 1.28% and 0.92%; and the Vitamin C values among 65.55% and 43.10%.

Discussion

Our results showed that earliness, yield and quality changed according to planting system. Different results have been also obtained in the previous studies. Tropea (1980) stated that the total yield obtained from frigo seedlings was higher but later in comparison to those of plug seedlings, whereas Pipattanawong *et al.* (1995) stated that the vegetative development was very strong for frigo seedlings. Similar results have been obtained in the current study as well. In the current study, the early yield for fall planting of plug seedlings was determined to be higher and the total yield was determined to be lower than those of frigo seedlings. The first harvesting dates varied from researcher to researcher. Takeda (1999a) reported the first harvest date for soilless strawberry culture systems in November-December; Adak *et al.* (2003) reported the first flowerings for conventional fall planting in the beginning of November and the first harvest in the middle of December; Paranjpe *et al.* (2003c) stated that the first harvesting was made at the end of November for the protected hanging system; and Vasilakakis *et al.* (1999) reported that the first harvesting was made at the end of December-beginning of January for the cultivation system in perlite medium with plug seedlings.

Yield per plant values also differed for studies carried out in various media by different researchers. Linardakis and Manios (1991) stated that the highest yield was obtained in peat +perlite medium with 250 g plant^{-1}. Takeda (1999a) stated that the highest yield via nutrient film technique method for the 'Camarosa' type reached to 1.2 kg plant^{-1}; Takeda (1999b) reported a yield of 700 g plant^{-1} in the planting of the 'Chandler' type with plug seedlings. Takeda and Hokanson (2003) stated that the variation was between 0.6 kg plant^{-1} and 1.2 kg plant^{-1}, even though the yield per plant values changed according to the type of plant in the soilless culture studies that they carried out using frigo transplants; Cantliffe *et al.* (2008) stated that they acquired a yield per plant value of 250 g for plug transplants grown in coconut coir and pine shaving medium; and Hotchmunt *et al.* (2008) stated that the highest yield of 725 g plant^{-1} was obtained from the 'Camarosa' type and in peat+perlite medium and that the early yield varied between 93 and 107 g plant^{-1} as a result of the studies carried out using plug seedlings. However, studies regarding coconut coir applications are very limited and the few studies that have been carried out were determined to be advantageous in terms of yield (Lopez-Medina *et al.*, 2004; Lieten, 2008). Navarro *et al.* (2010) stated that total fruit yield per plant decreased significantly with increasing salinity in perlite. Carmona *et al.* (2012) stated that the grapevine marc and grapevine stalk substrates had no limiting characteristics for its use as a medium for the cultivation of ornamental plants in container, and could replace conventional substrates, such as peat and coconut fiber.

Research carried out for the pomological analyses on fruits obtained via soilless culture have also differed among researchers. Vasilakakis *et al.* (1999) have stated that cultivation varies between 4.5% and 9.5% and titrable acid content varies between 0.8% and 1.8% for strawberry fruits in column culture. Tagliavini *et al.* (2005) stated that the TSS content varies between 6.2% and 4.7%, and the titrable acid content varies between 10.3 and 7.3 meq/l malic acid/100 g and Eltez and Tüzel (2007) reported that the TSS content varies between 7.625% and 5.825%, whereas the pH value varies between 4 and 3.98 and the vitamin C amount varies between 21.370 and 27.870 mg/100 ml. In addition, Paranjpe *et al.* (2004) stated that as the temperature increases the quality of fruit decreases (firmness, TSS, acidity) throughout the growing season and Cantliffe *et al.* (2008) stated that TSS amount did not vary from medium to medium.

In this study, earliest harvest was obtained in the fall planting of plug transplants in the beginning of December. No difference was determined between media in terms of yield and quality. The Co and the Co+Vlc media were determined to be more advantageous than the other media in terms of total yield per plant and all of the fruit quality criteria analysed. Thus, these results were found to be mostly in agreement with the aforementioned results.

Conclusions

In this study, which examines soilless strawberry cultivation possibilities in Turkey in Mediterranean ecological conditions via protected soilless culture, various effects of planting systems and growing media on earliness and yield have been found. The first harvest was obtained in the beginning of February for the summer planting of frigo transplants, whereas for the all planting of plug transplants, the first harvest was obtained in the beginning of December. Hence, the early yield determined in December and January was only obtained for the fall plantings. Summer planting of frigo transplants seems to be more advantageous in terms of total yield. A yield increase of almost 27% was obtained for the summer planting in comparison with that of the fall planting. In terms of growing media, the Co and the Co+Vlc media have been determined to be more advantageous in terms of yield, earliness and all of the other examined criteria. Also, subject to the advancement of vegetation (from January to May), acidity and vitamin C values decreased as the TSS value increased.

Acknowledgements

This paper was supported by the Scientific Research Projects Administration Unit of Akdeniz University. The authors thanks to Dr. Tomas Ayala-Silva, Agricultural Research Service, United States Department of Horticulture (USDA-ARS, Florida/U.S.); Dr. Anil Khar, ARS Senior Scientist, Maharashtya, India; Emeritus Prof. Richard A. Weismiller, University of Maryland, U.S.A, for their critical review of this manuscript.

References

Adak N, Gübbük H, Pekmezci M (2003). The growing possibilities of some strawberry cultivars grown under protected cultivation in Antalya conditions (in Turkish). Türkiye IV. Ulusal Bahçe Bitkileri Kongresi Antalya 313-315.

Barkham JP (1993). For peat's sake: conservation or exploitation? Biodiv Conserv 2:556-566.

Cantliffe DJ, Castellanos JZ, Paranjpe AV (2008). Yield and quality of greenhouse grown strawberries as affected by nitrogen level in coco coir and pine bark media. Proc Fla State Hort Soc 120:157-161.

Carmona E, Moreno MT, Aviles M, Ordovas J (2012). Composting of wine industry wastes and their use as a substrate for growing soilless ornamental plants. Spanish Journal of Agricultural Research 10(2):482-491.

Eltez RZ, Tüzel Y (2007). The effects of different soilless culture techniques in "A-shape" system on the yield and quality of greenhouse strawberry production (in Turkish). Ege Üniversitesi Ziraat Fakültesi Dergisi 44(1):15-27.

FAO (2012). Statistical database. *http://faostat.fao.org/site/339/default.aspx* (accessed in August 9, 2014).

Favaro JC, Marano RP (2003). Alterations in the physical and physico-chemical properties of a substrate based on composted sawdust and perlite with polycyclic tomato crops. Spanish Journal of Agricultural Research 1(3):105-109.

Frolking S, Roulet NT, Moore TR, Richard PJH, Lavoie M, Muller SD (2001). Modeling northern peatland decomposition and peat accumulation. Ecosystems 4:479-498.

Garcia Gomez A, Bernal MP, Roig A (2002). Growth of ornamental plants in two composts prepared from agroindustrial wastes. Bioresource Technology 83(2):81-87.

Gul A, Tuzel Y, Tuzel H, Irget M, Kidoglu F, Tepecik M (2011). Effects of nutrition and irrigation on sweet pepper production in volcanic tuff. Spanish Journal of Agricultural Research 9(1):221-229.

Kacar BN (1972). Plant Analysis (in Turkish). Ankara Üniversitesi Ziraat Fakültesi Yayınları Ankara 453:644-645.

Lieten F (2004). Substrates as an alternative to methyl bromide for strawberry fruit production in Northern Europe in both protected and field production. Proceedings of International Conference on Alternatives to Methyl Bromide 41-46.

Linardakis DK, Manios BI (1991). Hydroponic culture of strawberries in plastic greenhouse in a vertical system. Acta Hort 287:317-326.

Lopez Medina J, Peralbo A, Flores F (2004). Closed soilless growing system: a sustainable solution for strawberry crop in Huelva (Spain). Acta Hort 213-215.

Montesano F, Parente A, Santamaria P (2010). Closed cycle subirrigation with low concentration nutrient solution can be used for soilless tomato production in saline conditions. Scientia Horticulturae 124(3):338-344.

Navarro JM, Garrido C, Flores P, Martinez V (2010). The effect of salinity on yield and fruit quality of pepper grown in perlite. Spanish Journal of Agricultural Research 8(1):142-150.

Ozeker E, Eltez RZ, Tuzel Y, Gul A, Onal K, Tanrisever A (1999). Investigations on the effect of different growing media on the yield and quality of strawberries grown in vertical bags. Acta Hort 491:409-412.

Paranjpe AV, Cantliffe DJ, Lamb EM, Stoffella PJ, Powell CA (2003a). Winter strawberry production in greenhouses using soilless substrates: an alternative to methyl bromide soil fumigation. Proc Fla State Hort Soc 116:98-105.

Paranjpe AV, Cantliffe DJ, Rondon S, Chandler CK, Brecht JK, Brecht EJ, Cordasco K (2003b). Trends in fruit yield and quality, susceptibility to powdery mildew (*Sphaerotheca macularis*), and aphid (*Aphis gossypii*) infestation for seven strawberry cultivars grown without pesticides in a passively ventilated greenhouse using pinebark as soilless substrate. Proc Fla State Hort Soc 116:63-72.

Paranjpe AV, Cantliffe DJ, Lamb EM, Stoffella PJ, Powell CA (2003c). Increasing winter strawberry production in north-central Florida using passive ventilated greenhouses and high plant densities. Acta Hort 626:269-276.

Paranjpe A, Cantliffe DJ, Lamb EM, Stoffella PJ (2004). Economic feasibility of production strawberries in a passively ventilated greenhouse in North-Central Florida. Proc Fla State Hort Soc 117:27-37.

Paranjpe AV, Cantliffe DJ, Stoffella PJ, Lamb EM, Powell CA (2008). Relation of plant density to fruit yield of 'Sweet Charlie' strawberry grown in a pine bark soilless medium in a high-roof passively ventilated greenhouse. Scientia Horticulturae 115:117-123.

Paraskevopoulou-Paroussi G, Grafiadellis M, Paroussis E (1995). Precocity, plant productivity and fruit quality of strawberry plants grown in soil and soilless culture. Acta Hort 408:109-118.

Pearson D, Churchill AA (1970). The chemical analyses of foods. Gloucester Place 104:233.

Pipattanawong N, Fujishige N, Yamane K, Ogata R (1995). Growth and development of four day-neutral strawberries under hydroponic system with or without chilling. Journal of the Japanese Society for Horticultural Science 64(1):95-102.

Robertson RA (1993). Peat, horticulture and environment. Biodiv Conserv 2(5):541-547.

Tagliavini M, Baldi E, Lucchi P, Antonelli M, Sorrenti G, Baruzzi G, Faedi W (2005). Dynamics of nutrients uptake by strawberry plants (*Fragaria x ananassa* Dutch.) grown in soil and soilless culture. Europ J Agronomy 23(1):15-25.

Takeda F (1999a). Strawberry production in soilless culture systems. Acta Hort 481:289-295.

Takeda F (1999b). Out-of-season greenhouse strawberry production in soilless substrate. Adv Strawberry Res 18:4-15.

Takeda F, Hokanson SC (2003). Strawberry fruit and plug plant production in the greenhouse. Acta Hort 626:283-285.

Tropea M (1980). The control of strawberry plant nutrition in the sack culture. Proceedings Fifth International Congress on Soilless Culture 477-484.

Vasilakakis M, Alexandridis A, El Fadl S, Anagnostou K (1999). Effect of substrate (new or used perlite), plant orientation on the coloumn and irrigation frequency on yield efficiency of strawberry plants (cv. Selva) and fruit quality. Cahiers Options Mediterraneennes 31:357-363.

Morphological and Molecular Characterization and Relationships of Turkish Local Eggplant Heirlooms

Hatice Filiz BOYACI*, Volkan TOPCU, Akin TEPE, Isilay Karasahin YILDIRIM, Mehmet OTEN, Aytekin AKTAS

*Bati Akdeniz Agricultural Res. Inst., Dept. of Vegetable Crops and Ornamentals, 07100 Antalya, Turkey; filiz_boyaci@yahoo.com (*corresponding author)*

Abstract

A total of 38 eggplant genotypes, of which 32 were heirloom accessions collected from different regions of Burdur province five were different local genotypes from other provinces, and one was a cultivar, were used as reference in this study. The phylogenetic relationships among these heirlooms were evaluated using 40 morphologic descriptors and five randomly amplified polymorphic RAPD markers. The horizontal dendrograms were created by using UPGMA with both morphologic and molecular data. Burdur heirloom accessions showed high genetic diversity based on morphological and molecular data. The genetic similarity rates ranged from 0.29 to 0.91 according to the morphological data, and ranged from 0.84 to 0.98 according to the molecular data. Molecular data generated by RAPD method, compared to morphological data, were insufficient to reveal genetic diversity. Therefore, in order to confirm genetic variations, studies based on other molecular methods are necessary. The regional genetic populations include a wide eggplant genetic diversity which can be good source for the breeding studies performed in the future.

Keywords: accession, breeding, diversity, genetic variation, phylogenetic relationship, RAPD, UPGMA

Introduction

Eggplant, which was called as the king of vegetables (Daunay and Janick, 2007), is an indigenous plant of India (Weese and Bohs, 2010). It has been well known since BC III and cultivated for 1,500 years in Asia (Kashyap *et al.*, 2003). It is cultivated as a perennial in tropical areas, while it is cultivated as annual in subtropical areas (Kowalska, 2008). Eggplant is placed in *Solanum* genus and includes wide genotypic and phenotypic variation (Fukuoka *et al.*, 2010). Eggplant is thought to have been developed from the wild ancestor, *Solanum insanum* has small, round, green, thick-skinned and bitter taste fruits (Barchi *et al.*, 2010). Cultivation *S. insanum* of had been performed in China, India and Thailand (Daunay *et al.*, 2001). Large fruiting eggplants were cultured in India in early time, and small fruiting was cultured at IV century in China and at IX century in Africa (Sekera *et al.*, 2007). First cultivated eggplants were described as high tall plants, with big spines on the calyx, small, bitter fruiting and with high seed content in fruits (Swarup, 1995). Mutation, natural pollination and hybridization, together with selection gave rise to genetic diversity, as well as in decreasing of prickles and bitterness at fruit, changing of fruit shape, size and color (Frary *et al.*, 2007). Genetic diversity accumulated and many different heirlooms emerged in countries where it was cultivated (Prohens *et al.*, 2003). Entrance of the eggplant to Turkey was carried out by the silk-road. The genetic diversity accumulated in producing areas and by trade of eggplant throughout centuries in Anatolia (Janick, 2001). Eggplant cultivation was done in open field until the second half of the 1970s in Turkey and then cultivation under protected cultivation started. The eggplant cultivation in greenhouse was begun with local varieties. However, the growers preferred hybrid F1 varieties which have the cylindrical and dark purple or black colored fruiting and using of them has become commonly in a short period (Ekiz and Boyaci, 2001). Also, the increase of use F1 hybrids in the open field cultivation was observed in recent years. Steadily decrease was seen in the cultivation of local varieties compared to the hybrids (Cericola *et al.*, 2013). The genetic diversity is low among the genotypes which have dark purple-black fruits (Muñoz-Falcón *et al.*, 2009). In recent years, one of the important problems faced by eggplant breeding programs, as well as in other species, is a narrowing of the genetic base. To create variations, time-consuming and expensive methods are needed, including mutation breeding, interspecific hybridization and biotechnological approaches. The genetic variation contained by heirlooms among is seen in the previous studies (Demir *et al.*, 2010; Muñoz-Falcón *et al.*, 2008, 2009; Prohens *et al.*, 2003, 2008, 2011). Therefore, collection and characterization of genetic resources is required for the improvement of new varieties. In this respect, molecular characterization is more reliable then morphological characterization (Li *et al.*, 2010).

Here we investigate the genetic relationships between Turkish heirlooms using morphological and molecular data is presented here.

Materials and methods

Morphological and molecular studies were conducted at Bati Akdeniz Agricultural Research Institute, Antalya, Turkey in 2010-2013.

In total 38 materials were used in the study of these. 32 were local heirlooms collected from Burdur province during survey studies. Materials origin, providing method and place/company of origin are given in Table 1. Long purple commercial variety (YRL 68) and heirlooms originating in other province (YRL 1, YRL 3, YRL 6, YRL 57, and YRL 79) were used as reference cultivars for classification of Burdur province heirlooms.

Five RAPD primers OPH-02, OPL-04, OPB-07, OPO-10, OPL-16 (10 mer) reported as highly polymorphic in previously studies (*Demir et al.*, 2010; Nunome *et al.*, 2001) were selected to detect polymorphisms and identify genetic relationship of the heirlooms.

Survey

The survey studies were performed at nine different locations in Burdur province in September both in 2010 and 2011. In total 32 materials were collected from different farms. The locations of collected eggplants in Burdur province were represented on the map (Fig. 1) and their geographic coordinates were defined (Table 2).

Morphological observations

For each accession, the seeds were sown in seedling treys containing peat moss. Seedlings in 4-5 leaves stages were planted in glasshouse. Twenty plants were planted for each genotype. Morphological observations were performed according to 40 descriptors chosen among the International Board for Plant Genetic Resources Institute (IBPGRI), International Union for the Protection of New Varieties of

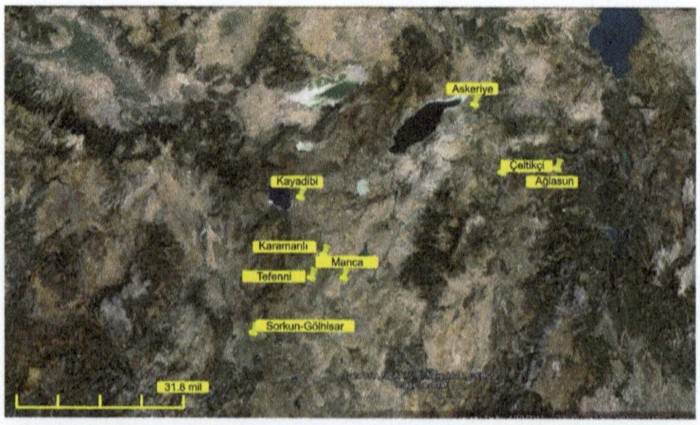

Fig. 1. Geographic map of district and village in Burdur province of Turkey including collection places

Table 1. Origin, providing method and place/company of origin of the materials used in the study

No	Name of the materials	Origin	Providing method	Place/company of origin
1	YRL 1	Antalya-Kumluca District	Agricultural district offices	Grower
2	YRL 3	Antalya-Kumluca District	Agricultural district offices	Grower
3	YRL 6	Antalya-Kumluca District	Agricultural district offices	Grower
4	YRL 7	Burdur-Aglasun /Cine Village	Survey in the field	Grower
5	YRL 8	Burdur-Aglasun /Cine Village	Survey in the field	Grower
6	YRL 9	Burdur-Aglasun /Cine Village	Survey in the field	Grower
7	YRL 12	Burdur/Askeriye Village	Survey in the field	Grower
8	YRL 14	Burdur-Aglasun /Cine Village	Survey in the field	Grower
9	YRL 15	Burdur-Aglasun /Cine Village	Survey in the field	Grower
10	YRL 19	Burdur-Aglasun /Cine Village	Survey in the field	Grower
11	YRL 20	Burdur-Aglasun /Cine Village	Survey in the field	Grower
12	YRL 24	Burdur-Celtikci/Tekke Village	Survey in the field	Grower
13	YRL 26	Burdur-Celtikci/Tekke Village	Survey in the field	Grower
14	YRL 27	Burdur-Celtikci/Tekke Village	Survey in the field	Grower
15	YRL 28	Burdur-Celtikci/Tekke Village	Survey in the field	Grower
16	YRL 30	Burdur-Celtikci/Tekke Village	Survey in the field	Grower
17	YRL 34	Burdur-Celtikci/Tekke Village	Survey in the field	Grower
18	YRL 35	Burdur-Celtikci/Tekke Village	Survey in the field	Grower
19	YRL 36	Burdur-Celtikci/Tekke Village	Survey in the field	Grower
20	YRL 43	Burdur-Celtikci/Tekke Village	Survey in the field	Grower
21	YRL 44	Burdur-Celtikci/Tekke Village	Survey in the field	Grower
22	YRL 45	Burdur-Celtikci/Tekke Village	Survey in the field	Grower
23	YRL 46	Burdur-Askeriye Village	Survey in the field	Grower
24	YRL 47	Burdur-Askeriye Village	Survey in the field	Grower
25	YRL 49	Burdur-Askeriye Village	Survey in the field	Grower
26	YRL 50	Burdur-Askeriye Village	Survey in the field	Grower
27	YRL 51	Burdur-Askeriye Village	Survey in the field	Grower
28	YRL 52	Burdur-Askeriye Village	Survey in the field	Grower
29	YRL 57	Mugla-Fethiye/Günesli Village	Survey in the field	Grower
30	YRL 58	Burdur-Karamanlı/Manca Village	Survey in the field	Grower
31	YRL 59	Burdur-Karamanlı/Manca Village	Survey in the field	Grower
32	YRL 61	Burdur-Gölhisar/Sorkum Village	Survey in the field	Grower
33	YRL 62	Burdur-Yesilova/Kayadibi Village	Survey in the field	Grower
34	YRL 64	Burdur-Tefenni District	Survey in the field	Grower
35	YRL 65	Burdur-Tefenni District	Survey in the field	Grower
36	YRL 68	Long purple	Purchased	Company
37	YRL 75	Burdur City Centrum	Agricultural district offices	Grower
38	YRL 79	Antalya-Akseki/Uzumdere Village	Survey in the field	Grower

Table 2. Geographical coordinates of local heirloom eggplant collected in Burdur province of Turkey

Location	Geographical coordinates
Aglasun /Cine Village	37°35'0.07"N/30°38'1.34"E
Celtikci Tekke Village	37°34'36.86"N/30°26'40.11"E
Burdur/Askeriye Village	37°45'41.88"N/30°21'15.99"E
Karamanlı-Manca Village	37°18'18.74"N/29°53'19.89"E
Karamanlı-Centrum	37°22'2.35"N/29°49'19.71"E
Gölhisar-Sorkum Village	37° 9'54.49"N/29°34'39.38"E
Yesilova-Kayadibi Village	37°31'19.56"N/29°44'35.03"E
Tefenni-Centrum	37°18'27.39"N/29°46'40.57"E

Plants (UPOV) plant feature criteria and some of the criteria for the breeders. Descriptors include plant, leaf, flower and fruit traits observations and measurements (Table 3). Skin color of the each eggplant specimens was measured with a portable tristimulus reflectance colorimeter Minolta CR-400 Chroma Meter (Konica Minolta Sensing, Inc., Osaka, Japan), and the parameters were expressed in CIE L*a*b* system, where L* is lightness (brightness-darkness) ranged from 0 to 100 units, a* is light intensity in red (+) or green (–) spectrum, b* is intensity in yellow (+) or blue (–) spectrum. Chroma (C: $\sqrt{(a^2+b^2)}$) measures color saturation or intensity and the hue angle ($h° = \tan^{-1}(b^*/a^*)$ determines the red, yellow, green, blue, purple, or intermediate colors between adjacent pairs of these basic colors. The L*, a*, and b*, *C* and *h* ° values obtained from six samples of each eggplant accession. Every record represents the average of three readings which were taken from the equatorial region, spaced equidistantly.

DNA extractions and PCR analysis

DNA extractions from young leaves were performed according to a modified Doyle and Doyle (1990) method by using CTAB protocol (Mutlu *et al.*, 2008). RAPD analysis was performed according to Demir *et al.* (2010). The amplification reactions were released containing 20 ng DNA, 0.5 unite *Taq* polymerase, 2.5 μL 10X buffer, 3.5 μL 25 mM $MgCl_2$, 2 μL 2.5 mM dNTPs, 2 μL RAPD primer for RAPD-PCR. DNA was amplified in a thermal cycler. It was programmed for an initial 5 min denaturation step at 94 °C, followed by 35 cycles of a 30 s denaturation step at 94 °C, 1 min annealing at 35 °C, 45 s extension step at 72 °C, followed by a final 8 min extension step at 72 °C.

Statistical analysis

Genetic similarity was analyzed by the UPGMA (Unweighted pair-group method, arithmetic average) clustering procedure using the software NTSYS (Numerical Taxonomy Multivariate

Table 3. Descriptors used for characterization and evaluation of eggplant accessions used in the study

Traits	Description
Plant habit	Score range (1=Open, 3=Bushy, 5=Semi open)
Plant height	Score range (1=Long, 3=Intermediate, 5=Short)
Stem thickness	Score range (1=Thick, 3=Intermediate, 5=Thin)
Stem hairiness	Score range (1=Dense, 3=Intermediate, 5=Tenuous)
Stem color	Score range (1=Grayish, 3=Green, 5=Green-purple, 7=Grayish-green-purple, 9=Grayish-green, 11= Grayish-purple, 13=Purple)
Shoot tip color	Score range (1=Grayish, 3=Green, 5=Green-purple, 7=Grayish-green-purple, 9=Grayish-green, 11=Grayish-purple, 13=Purple)
Length of internodes	Score range (1=Long, 3=Intermediate, 5=Short)
Leaf color	Score range (1=Light green, 3=Green, 5= Dark green)
Leaf size	Score range (1=Large, 3=Intermediate, 5=Small)
Leaf hairiness	Score range (1=Dense, 3=Intermediate, 5=Tenuous)
Presence of spine on petiole	Score range (1=Many, 3=Intermediate, 5=Few, 7=Absent)
Bud size	Score range (1=Large, 3=Intermediate, 5=Small)
Bud hairiness	Score range (1=Dense, 3=Intermediate, 5=Tenuous, 7=Absent)
Presence of spine on bud	Score range (1=Many, 3=Intermediate, 5=Few, 7=Absent)
Flower color	Score range (1=Light purple, 3=Purple, 5= Dark purple)
Flower size	Score range (1=Large, 3=Intermediate, 5=Small)
Calyx size	Score range (1=Large, 3=Intermediate, 5=Small)
Fruit shape	Score range (1=Long, 3=Intermediate, 5=Short, 7=Ovoid, 9=Pear shaped)
Dominated fruit color	Score range (1=White, 3= Green, 5=Purple, 7=Black)
Range of dominated fruit color	Score range (1=Regular, 3=Mottled, 5=Stripe, 7=Mealy)
Fruit stalk length	Score range (1=Long, 3=Intermediate, 5=Short)
Presence of spine on fruit stalk	Score range (1=Few, 3=Intermediate, 5=Many, 7=Absent)
Fruit calyx prickles	Score range (1=Few (1-5 pricks), 3=Mid (6-20 pricks), 5=Many (more than 20)
Fruit brightness	Score range (1=Bright, 3=Matt)
Fruit end shape	Score range (1=Flat, 3= Pointed, 5=Round)
Fruit curvature	Score range (1=Present, 3=Absent)
Fruit end button size	Score range (1=Large, 3= Intermediate, 5=Small)
Fruit length (cm)	The average measurement of ten fruits
Fruit diameter (cm)	The average measurement of ten fruits
Average fruit weight (g)	The average measurement of ten fruits
Presence of groove on fruit	Score range (1=Present, 3=Absent)
Fruit flesh firmness	Score range (1=Tightly, 3= Floppy, 5=Spongy)
Fruit flesh color	Score range (1=Greenish, 3=Greenish-cream, 5=White, 7=White- cream, 9=Greenish-white, 11=Cream)
Presence of hole in fruit	Score range (1=Present, 3=Absent)
Degree of fruit curvature	Score range (1=Slight, 3= Mid, 5=Much)
Soluble solids	The average measurement of five fruits juice samples
The length of fruit coated by calyx	Score range (1=Less than 20%, 3= between 20-70%, 5=More than 70%)
Tendency to parthenocarpy	Score range (1=Present, 3=Absent)
Presence of seed in fruit	Score range (1=Few, 3=Intermediate, 5=Many)
Seed maturity	Score range (1=Immature, 3=Mature)

Analysis System) pc 2.2 version (Rohlf, 1998), using morphological and molecular data. For morphological data, each genotype was characterized with description number (Table 3). For molecular analysis data, each genotype was identified for each primer based on the presence (1) and absence (0) of bands. In order to show the variations STANDARDIZATION module was used. Correlation matrix adapted to SIMINT module was used to determine of correlation coefficient. The dendrograms were drawn using the clustering method UPGMA via the SHAN module. The cophenetic correlation coefficient was calculated with Mantel method to evaluate the efficiency of clustering.

Results

A high morphological diversity was observed among eggplant genotypes characterized by quantitative and qualitative descriptors. Some of the phenotypic observation data related with important variable traits (calyx prickles, fruit size, weight, shape and curvature, skin color, groove etc.) are reported in Table 4. The fruits calyx of all genotypes had prickles. However, nearly 50% of them had lower numbers of prickles. Most of the genotypes had between 20-70% of their fruit length covered by the calyx. The degree of fruit curvature in 50% of the genotypes was slight. Nearly 30% percent of the genotypes had fruits with grooved appearance. Dominant fruit color was mostly purple and distribution of this color was 50% regular, 22% mealy, 18% mottled, and 2% striped according to the genotypes. Fruit flesh color was generally white and hole in the fruit was mostly absent. The lowest total soluble solid content (brix value) was 2.8. The highest brix value (6.1) was determined in YRL 15 which was collected from Cine village The average diameter of the fruits ranged from 32.67 to 73.22 mm. Fruit length and fruit weight average ranged between 10.9 and 23.3 cm, 0.100 and 0.235 kg, respectively.

Skin color characteristics of the eggplant heirlooms are shown in Table 5. Fruit color varied greatly among different genotypes. Regarding the skin Lightness (L*), higher values were obtained from genotypes which were mostly, cream, yellowish, orange-yellowish or goldenred in color (L* > 70). When L* values were ranged from 48 to 58 together with h^c values ranged from 80 to 112, it is observed that these genotypes' skin color were generally greenish-yellow, chartreuse or green in color. Regarding the b* values, very low values near or below zero mostly indicated that the skin color was cyan (if $h°$ > 285); blue (if $h°$ > 240); purple (if $h°$ > 285); magenta (if $h°$ > 350). It is observed from the research, that fruits which had lower $h°$ values (1.05 < $h°$ < 12.60) were found to be red in color. Chroma values were varied from 3.82 to 24.26.

Totally 40 basic morphological descriptors were used for to determine the phylogenetic relationships among the Burdur local heirlooms. The Eigen value was 84. A 2-way Mantel test (Mantel, 1967) method was performed. Approximate Mantel t- statistic test were t = 10.0925, p = 1.0000. The matrix correlation (r) was 0.72. The similarity rates according to the coefficient similarity of genotypes ranged between 0.29 and 0.91.

Two major groups were revealed using the dendrogram generated by the UPGMA method using morphological data (Fig. 2). First group (Group A) was consisted of YRL 75, YRL 65, YRL 61 and YRL 59. These genotypes showed low genetic similarity with reference genotypes and cultivar (Long purple). Second group was divided into three subgroups. Group B consist of YRL 19 which showed high genetic similarity with reference cultivar YRL 68 (Long purple). The highest genetic similarity was observed in Group D between genotypes YRL 47 and YRL 51. The fruits of eggplant Burdur heirlooms belonging to the Group C are shown in Fig. 3.

Table 4. Some of the phenotypic descriptors related important variable traits of eggplant

Genotype	Fruit calyx prickles	Fruit length covered by the calyx	Degree of fruit curvature	Fruit grooved	Dominated fruit color	Distribution of dominant color	Fruit flesh firmness	Fruit flesh color	Seed content	Seed maturity	Hole in the fruit	Brix	Fruit diameter (mm)	Fruit lenght (cm)	Fruit weight (kg)
YRL 1	few	between 20-70%	slight	absent	green	mottled	floppy	greenish-cream	intermediate	mature	absent	5.2	34.80	19.8	0.116
YRL 3	few	between 20-70%	slight	present	purple	striped	tightly	white	many	mature	present	4.2	49.79	14.2	0.140
YRL 6	mid	between 20-70%	slight	absent	green	mottled	floppy	greenish-cream	intermediate	immature	absent	4.2	36.60	23.3	0.140
YRL 7	mid	less than 20%	mid-	absent	black	regular	floppy	cream	intermediate	mature	absent	5.1	40.33	18.7	0.133
YRL 8	mid	less than 20%	slight	present	purple	regular	tightly	greenish-cream	intermediate	mature	present	4.7	46.86	15.7	0.133
YRL 9	few	between 20-70%	slight	absent	black	regular	tightly	greenish-cream	intermediate	mature	absent	3.8	45.38	18.5	0.138
YRL 12	few	less than 20%	slight	absent	purple	regular	spongy	white	intermediate	mature	present	2.8	40.98	20.5	0.125
YRL 14	mid	between 20-70%	none	absent	black	regular	tightly	greenish-cream	few	mature	present	4.3	43.60	17.2	0.133
YRL 15	mid	between 20-70%	slight	present	black	regular	tightly	greenish-cream	few	immature	present	6.1	35.07	18.8	0.110
YRL 19	few	between 20-70%	slight	absent	black	regular	tightly	greenish-cream	intermediate	mature	absent	4.1	47.32	17.5	0.146
YRL 20	mid	between 20-70%	slight	absent	black	regular	spongy	greenish-cream	many	immature	present	5.2	42.17	20.6	0.160
YRL 24	few	less than 20%	slight	absent	purple	regular	spongy	greenish-cream	few	immature	present	5.2	38.96	18.3	0.100
YRL 26	few	between 20-70%	mid-	absent	green	mealy	floppy	white	intermediate	mature	present	3.2	45.73	16.7	0.143
YRL 27	mid	between 20-70%	slight	absent	purple	mealy	floppy	white	intermediate	mature	present	4.2	58.13	18.0	0.213
YRL 28	few	less than 20%	mid-	absent	black	regular	floppy	white	intermediate	mature	present	3.4	53.13	16.3	0.177
YRL 30	mid	less than 20%	mid-	absent	purple	mealy	spongy	white	intermediate	immature	present	3.1	47.43	14.5	0.133
YRL 34	mid	between 20-70%	slight	present	purple	mealy	spongy	white	few	mature	present	3.1	54.11	22.1	0.232
YRL 35	mid	between 20-70%	mid-	absent	purple	mealy	spongy	white	few	mature	present	3.0	53.66	17.5	0.207
YRL 36	mid	between 20-70%	mid-	absent	purple	mealy	floppy	white	intermediate	mature	present	3.2	52.05	14.8	0.163
YRL 43	mid	between 20-70%	none	present	purple	mottled	spongy	white	intermediate	immature	present	3.2	57.87	15.1	0.187
YRL 44	mid	between 20-70%	slight	absent	purple	mealy	spongy	white	intermediate	mature	present	3.4	51.29	15.2	0.168
YRL 45	mid	between 20-70%	slight	present	purple	mealy	tightly	white	few	mature	present	3.9	50.50	14.6	0.173
YRL 46	mid	less than 20%	mid-	present	purple	regular	floppy	white	intermediate	mature	absent	4.1	42.79	16.4	0.125
YRL 47	few	less than 20%	mid-	present	purple	regular	floppy	white	intermediate	immature	absent	4.1	35.05	18.0	0.127
YRL 49	few	less than 20%	mid-	absent	purple	regular	tightly	white	intermediate	immature	present	3.8	44.98	19.7	0.153
YRL 50	mid	less than 20%	mid-	absent	purple	regular	tightly	white	intermediate	mature	present	3.1	43.15	19.2	0.143
YRL 51	few	less than 20%	mid-	present	purple	regular	floppy	white	intermediate	mature	absent	3.1	44.82	17.4	0.147
YRL 52	few	less than 20%	mid-	absent	purple	regular	tightly	white	intermediate	mature	present	2.9	42.73	17.5	0.133
YRL 57	few	between 20-70%	mid-	absent	green	mottled	tightly	greenish-cream	intermediate	immature	present	5.1	32.67	19.1	0.100
YRL 58	few	less than 20%	mid-	absent	purple	regular	tightly	white	few	mature	present	4.2	47.03	18.5	0.148
YRL 59	mid	between 20-70%	mid-	absent	purple	mottled	floppy	white	intermediate	mature	present	4.7	51.40	13.0	0.155
YRL 61	mid	between 20-70%	none	present	purple	regular	tightly	white	intermediate	mature	present	4.1	64.76	10.9	0.183
YRL 62	few	between 20-70%	slight	absent	black	regular	tightly	greenish-cream	few	immature	present	4.9	46.09	16.3	0.143
YRL 64	few	less than 20%	slight	absent	purple	regular	floppy	white	intermediate	mature	present	3.8	51.86	22.9	0.228
YRL 65	mid	less than 20%	slight	present	purple	mottled	tightly	white	intermediate	mature	present	3.8	73.22	11.6	0.235
YRL 68	mid	between 20-70%	slight	present	black	regular	tightly	greenish-cream	medium	immature	present	3.1	50.76	14.2	0.150
YRL 75	mid	between 20-70%	none	present	black	regular	tightly	greenish-cream	intermediate	mature	present	4.2	65.97	11.9	0.137
YRL 79	few	between 20-70%	slight	present	green	mottled	floppy	greenish-cream	intermediate	immature	present	4.2	47.71	13.6	0.122

Table 5. Skin color characteristics of eggplant

Genotype	L*	a*	b*	C	h°	color
YRL 1	48.33 ± 6.38	-0.67 ± 4.80	18.38 ± 4.04	19.04 ± 3.88	88.77 ± 15.99[X]	chartreuse
YRL 3	73.36 ± 4.06	9.05 ± 2.52	6.31 ± 1.09	11.27 ± 1.52	36.39 ± 11.67	yellowish-orange
YRL 6	51.54 ± 8.50	-6.18 ± 5.43	22.47 ± 6.03	23.75 ± 6.71	101.46 ± 13.81	chartreuse
YRL 7	26.21 ± 0.66	6.03 ± 1.76	0.36 ± 0.51	6.05 ± 1.78	62.44 ± 131.52	yellowish
YRL 8	31.83 ± 3.32	11.44 ± 2.19	2.69 ± 1.07	11.78 ± 2.31	12.60 ± 4.48	redish
YRL 9	29.28 ± 3.64	10.22 ± 3.40	1.25 ± 1.45	10.38 ± 3.48	5.85 ± 6.76	redish
YRL 12	31.71 ± 3.20	21.64 ± 2.52	-0.98 ± 0.75	21.67 ± 2.54	357.56 ± 1.74	magenta
YRL 14	25.83 ± 0.99	4.97 ± 0.79	0.36 ± 0.88	5.03 ± 0.93	182.80 ± 175.41	cyan
YRL 15	25.23 ± 0.22	4.07 ± 1.31	0.002 ± 0.26	4.08 ± 1.31	299.07 ± 131.60	purple
YRL 19	25.48 ± 0.51	3.81 ± 0.78	-0.03 ± 0.22	3.82 ± 0.79	238.99 ± 167.40	blue
YRL 20	25.82 ± 0.92	4.84 ± 1.15	0.003 ± 0.28	4.85 ± 1.15	299.43 ± 131.95	purple
YRL 24	29.02 ± 3.24	17.85 ± 4.63	-0.42 ± 0.54	17.86 ± 4.65	243.94 ± 146.42	blue
YRL 26	55.22 ± 5.86	-7.43 ± 2.45	17.56 ± 3.60	19.10 ± 4.20	112.31 ± 3.60	greenish
YRL 27	68.54 ± 5.60	8.83 ± 2.37	5.71 ± 1.76	10.88 ± 1.00	34.29 ± 14.89	orange
YRL 28	29.89 ± 2.86	9.70 ± 0.84	1.57 ± 1.36	9.93 ± 0.72	9.29 ± 8.33	red
YRL 30	66.80 ± 2.88	11.23 ± 1.66	6.30 ± 1.17	12.99 ± 1.04	29.80 ± 7.71	orange
YRL 34	77.73 ± 3.23	4.64 ± 1.69	9.76 ± 1.00	10.98 ± 0.16	64.53 ± 10.30	cream-yellowish
YRL 35	75.07 ± 2.53	6.01 ± 1.29	8.28 ± 1.20	10.38 ± 0.33	53.91 ± 9.74	yellow
YRL 36	72.34 ± 4.46	8.41 ± 2.74	6.96 ± 1.55	11.30 ± 1.24	41.16 ± 14.66	orange-yellow
YRL 43	56.55 ± 9.09	19.35 ± 5.14	2.09 ± 2.76	19.81 ± 4.51	68.93 ± 128.77	yellow
YRL 44	72.50 ± 8.32	7.51 ± 4.42	7.49 ± 2.90	11.58 ± 2.54	48.10 ± 21.20	goldenrod
YRL 45	76.29 ± 4.16	5.00 ± 2.26	10.97 ± 2.10	12.40 ± 1.05	64.49 ± 13.53	yellow
YRL 46	37.41 ± 5.48	22.90 ± 1.68	-0.49 ± 1.07	22.94 ± 1.68	298.96 ± 131.62	purple
YRL 47	32.80 ± 4.44	20.59 ± 3.19	-0.50 ± 1.36	20.64 ± 3.23	357.03 ± 2.04	magenta
YRL 49	32.04 ± 3.54	23.02 ± 2.99	-0.51 ± 0.47	23.04 ± 3.00	240.48 ± 166.70	blue
YRL 50	36.77 ± 6.11	22.17 ± 2.18	-0.07 ± 1.29	22.20 ± 2.15	179.90 ± 177.00	cyan
YRL 51	37.55 ± 4.73	21.85 ± 3.29	-0.89 ± 0.93	21.88 ± 3.30	285.88 ± 142.39	purple
YRL 52	35.51 ± 2.61	23.76 ± 1.51	-1.37 ± 0.48	23.81 ± 1.49	356.67 ± 1.24	magenta
YRL 57	44.44 ± 8.19	4.04 ± 6.62	9.70 ± 5.12	13.35 ± 1.43	63.87 ± 38.34	yellowish
YRL 58	37.80 ± 5.11	23.66 ± 2.46	-1.61 ± 0.66	23.72 ± 2.43	356.05 ± 1.90	magenta
YRL 59	56.43 ± 6.66	18.22 ± 4.99	1.86 ± 2.45	18.55 ± 4.72	80.11 ± 137.94	greenish-yellow
YRL 61	51.83 ± 4.36	14.86 ± 2.14	0.41 ± 3.95	15.45 ± 1.56	242.80 ± 156.44	blue
YRL 62	31.03 ± 5.72	10.50 ± 2.59	2.53 ± 2.32	10.96 ± 2.90	12.19 ± 9.68	redish
YRL 64	37.51 ± 4.79	24.23 ± 0.59	0.02 ± 1.29	24.26 ± 0.61	285.99 ± 142.37	purple
YRL 65	73.19 ± 4.48	8.17 ± 2.89	7.90 ± 0.91	11.66 ± 1.61	45.71 ± 12.66	orange-yellow
YRL 68	35.81 ± 5.99	9.95 ± 2.41	6.17 ± 3.81	12.01 ± 3.63	27.50 ± 15.13	dark-orange
YRL 75	25.39 ± 0.72	6.03 ± 1.01	0.12 ± 0.14	6.03 ± 1.01	1.05 ± 1.04	redish
YRL 79	57.34 ± 3.69	0.12 ± 2.26	19.50 ± 2.14	19.64 ± 2.08	88.98 ± 6.91	chartreuse

[X]: Means represent three 10-fruit samples ± SD.

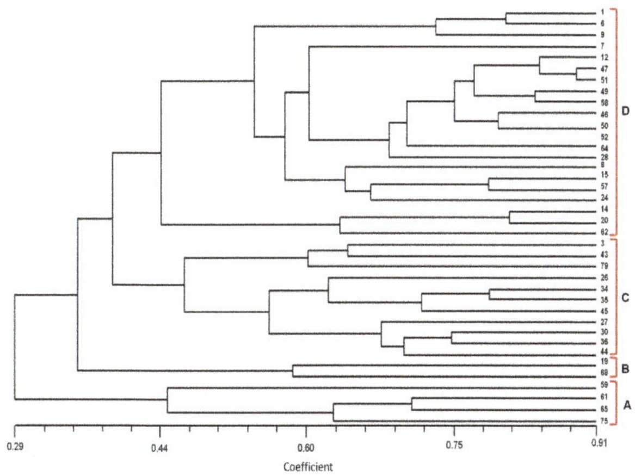

Fig. 2. UPGMA dendrogram showing phylogenetic relationships of local Burdur eggplant heirlooms together with reference cultivars using morphological data

Fig. 3. The fruits of Burdur eggplant heirlooms belonging to the Group C

A total of 65 amplified RAPD bands were generated. Twenty nine bands were polymorphic and the mean percentage of polymorphism was 44.61%. OPO-10 primer produced the maximum numbers of bands (18). Although the OPB-07 primer produced the minimal number of bands (8), it revealed a 100% polymorphism (Table 6). The OPH-2, OPB-07, OPO-10 and OPL-16 primers' PCR products and their band patterns are shown in Fig. 4a-d.

Table 6. Primer code, sequence, GC content, number of polymorphic bands and percent polymorphism of each RAPD primer used

Primer code	Nucleotide sequence	No. of amplified bands	No. of polymorphic bands	Polymorphic bands (%)
OPH-02	5'-TCG GAC GTG A-3'	14	7	50
OPL-04	5'-GAC TGC ACA C-3'	16	4	25
OPB-07	5'-GGT GAC GCA G-3'	8	8	100
OPO-10	5'-TCA GAG CGC C-3'	18	5	27.77
OPL-16	5'-AGG TTG CAG G-3'	9	5	55.55
TOTAL		**65**	**29**	**44.61**

The phylogenetic relationships among 38 genotypes were evaluated using 29 polymorphic loci of the genomic DNA generated using randomly amplified polymorphic DNA (RAPD) technique. The approximate Mantel t-test statistic were t = 8.9263, p = 1.0000. The matrix correlation (r) was found to be 0.64. The similarity rates according to the coefficient similarity of genotypes ranged between 0.84 and 0.98.

Using molecular data two major groups emerged in the dendrogram generated with UPGMA method (Fig. 5). First group branched into two subgroups and were showed in brackets as Group A and Group B. Group A was consisted of YRL 15 and YRL 27. Group E included most of the genotypes had high similarities with reference cultivar. The highest genetic similarity was observed in Group D with 98% percentage similarity index between genotypes YRL 6 and YRL 44. YRL 1 and YRL 27 were detected as the most distantly genotypes from each other in the cluster. Fruits of Burdur eggplant heirlooms YRL 6 and YRL 44, which showed highest genetic similarity with 98% percentage are shown in Fig. 6. Also, the fruits of Burdur heirlooms which grouped together in cluster E in the UPGMA dendrogram using molecular data and reference culture are shown in Fig. 7.

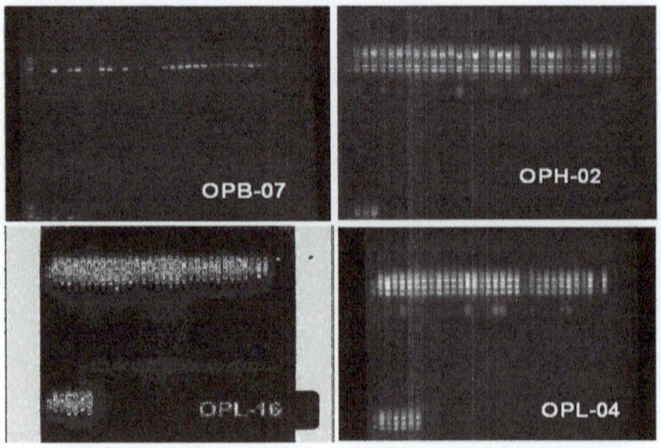

Fig. 4. PCR products of Burdur eggplant heirlooms generated using RAPD primers a:OPB-07, b:OPH-02, c:OPO-016, d:OPL-04

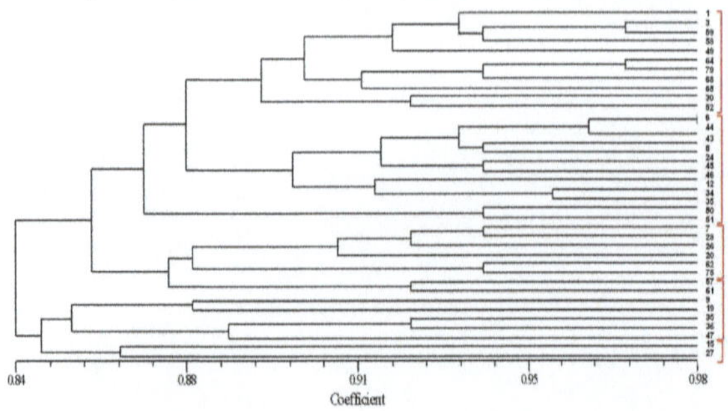

Fig. 5. UPGMA dendrogram showing phylogenetic relationships of local Burdur eggplant heirlooms together with reference cultivars using molecular data

Fig. 6. Fruits of Burdur eggplant heirlooms YRL 6 and YRL 44, which showed highest genetic similarity with 98% percentage

Fig. 7. Fruits of eggplant Burdur heirlooms with reference cultivar (YRL 68) situated in GROUP E generated by UPGMA using molecular data

Discussions

Eggplant has a wide genetic diversity in the regions where it is cultivated, although they are not native to the region (Muñoz-Falcón *et al.*, 2008). In spite of the fact that Turkey is not the place origin of eggplant, wide genetic diversity has been reported in Turkey (Demir *et al.*, 2010; Tümbilen *et al.*, 2011a, 2011b). It is clearly evident in this study data that Burdur, which is a small geographical region, had a rich genetic diversity. Also, a wide genetic variability was determined in both Spain and Jordan local genotypes (Prohens *et al.*, 2003). Local genotypes can contribute to enhancing the gene pool used in breeding studies and to help increase heterosis (Muñoz-Falcón *et al.*, 2009). In recent years, some factors like cultivation of commercial varieties instead of heirlooms, construction of buildings on agricultural land, and innovation in cultivation methods have led to erosion of plant genetic resources (Cericola *et al.*, 2013). Therefore, there is a need collecting and identification of local heirlooms before they disappear (Muñoz-Falcón *et al.*, 2008).

Some characters that contributed to genetic diversity were as flowering dates, the number of seeds per fruits, fruit features, and the growth pattern of plants. These features are controlled by several genes in eggplant (Frary and Doğanlar, 2003). *Solanum melongena* accessions could characterize these descriptors like bigger and flabby fruits, less flowers/inflorescence, few fruits/plant and higher acidity etc. compared to the wild relatives (Polignano *et al.*, 2010). Consistent with previously works, a higher diversity for most morphological descriptors was recorded in the collection of Burdur local heirlooms identified in this study. Fruit color can be cream, green, red, reddish-purple, dark purple or black, and some varieties produce fruit which is where the genetic variation necessary for future varietal improvement and for addressing future breeding challenges will be found.

Molecular markers linked with agronomic traits are useful tools for marker assisted selection and mapping candidate genes studies in breeding programs (Nunome *et al.*, 2009;

Wang *et al.*, 2010). Some of the RAPD markers used were determined to have relationship with coloring of stem, calyx and fruiting in eggplant (Frary *et al.*, 2003). This provides a great advantage for identifying features affecting by ecological conditions (Biswas *et al.*, 2009; Nunome *et al.*, *2001).* Relationships among eggplant materials have been studied by molecular studies for use by eggplant breeders (Furini and Wunder, 2004). As mentioned above the RAPD markers revealed as a potential useful tool for determination of genetic diversity. It was found that RAPD analysis in eggplant with four primers had been adequate to identify of genetic diversity (Biswas *et al.*, 2009). Also, RAPD markers were more effective than ISSR for revealing genetic diversification as reported by Ali *et al.* (2011). *Tiwari et al.* (2009) previously reported that even two of the 29 RAPD primers were found to be sufficient for identification of local cultivars. All RAPD markers used in the study have produced polymorphic bands as expected. However, if fruit criteria are taken into consideration for distinguishing in the dendrogram and grouping, it is not enough for the breeders. More informative DNA markers can be used to provide better progress in eggplant breeding studies (Stàgel *et al.*, 2008). Simple Sequence Repeats (SSRs) methods were found more successful for distinguishing closely related eggplant cultivars (Hurtado *et al.*, 2012; Muñoz-Falcón *et al.*, 2009; Prohens *et al.*, 2008). The approach of using SSR markers instead of RAPD markers can provide better results in the phylogenetic relationships studies. Muñoz-Falcón *et al.* (2009) reported that if the morphological and molecular data are considered together, they can be provide sufficient and useful information for the breeders. Similarly to the findings of other studies, we suggest that the molecular evidences need to be supplemented by morphological data to validate the phylogenetic relationships among the genotypes. It is very important to note that the genetic variations can't be detected by only showing a dendrograms generated by molecular data.

Conclusions

The local populations are of great importance for the breeders so that they adapted well to their cultivated areas. There is a need collecting and identification of these heirlooms before integrated to the breeding programs. The aim of the study was to investigate the diversity among heirlooms cultivated in Burdur province. A high genetic diversity was determined widely among them. These materials can be of a potential value for the breeders.

Acknowledgements

This research was financially supported by grants from the General Directorate of Agricultural Research and Policy, Republic of Turkey Ministry of Food, Agriculture and Livestock under BBSS-10-12 project number.

References

Ali Z, Xu ZL, Zhang D Y, He XL, Bahadur S, Yi JX (2011). Molecular diversity analysis of eggplant (*Solanum melongena)* genetic resources. Genetics and Molecular Research 10(2):1141-1155.

Barchi L, Lanteri S, Portis E, Stàgel A, Valè G, Toppino L, Rotino GL (2010). Segregation distortion and linkage analysis in eggplant (*Solanum melongena* L.). Genome 53(10):805-815.

Biswas MS, Akhond MAY, Al-Amin A, Khatun M, Kabir MR (2009). Genetic relationship among ten promising eggplant varieties using RAPD markers. Plant Tissue Culture and Biotechnology 19(2):119-126.

Cericola F, Portis E, Toppino L, Barchi L, Acciarri N, Ciriaci T, Sala T, Rotino GL, Lanteri S (2013). The population structure and diversity of eggplant from Asia and the Mediterranean basin. PloS ONE 8(9):DOI:10.1371/journal.pone.0073702.

Daunay MC, Lester NR, Gebhardt C, Hennart W, Jahn M (2001). Genetic resources of eggplant (*Solanum melongena* L.) and allied species: a new challenge for molecular genetics and eggplant breeders. In: Van Den Berg RG, Barendse GW, Mariani C (Eds). Solanaceae V. Nijmegen University, Press Nijgemen, The Netherlands, pp. 251-274.

Daunay MC, Janick J (2007). History and Iconography of Eggplant. Chronica Horticulturae 47(3):16-22.

Demir K, Bakır M, Sarıkamış G, Acunalp S (2010). Genetic diversity of eggplant (*Solanum melongena*) germplasm from Turkey assessed by SSR and RAPD markers. Genetics and Molecular Research 9(3):1568-1576.

Ekiz H, Boyacı HF (2001). Pepper and eggplant varieties in greenhouses on the coast of Mediterranean in Antalya. XIth Meeting on Genetics and Breeding of Capsicum and Eggplant, 9-12 April, Antalya, 241-245.

Frary A, Doganlar S (2003). Comparative genetics of crop plant domestication and evolution. Turkish Journal of Agriculture and Forestry 27:59-69.

Frary A, Doganlar S, Daunay MC, Tanksley SD (2003). QTL analysis of morphological traits in eggplant and implications for conservation of gene function during evolution of Solanaceous species. Theoretical and Applied Genetics 107(2):359-370.

Frary A, Doganlar S, Daunay MC (2007). Eggplant, p. 287-314. In: Genome Mapping and Molecular Breeding in Plants. Kole C (Ed). Springer, Acad Press.

Fukuoka H, Yamaguchi H, Nunome T, Negoro S, Miyatake K, Ohyama A (2010). Accumulation, functional annotation, and comparative analysis of expressed sequence tags in eggplant (*Solanum melongena* L.), the third pole of the genus *Solanum* species after tomato and potato. Gene 450(1-2):76-84.

Furini A, Wunder J (2004). Analysis of eggplant (*Solanum melongena* L.) related germplasm: morphological and AFLP data contribute to phylogenetic interpretations and germplasm utilization. Theoretical and Applied Genetics 108(2):197-208.

Hurtado M, Vilanova S, Plazas M, Gramazio P, Fonseka HH, Fonseka R, Prohens J (2012). Diversity and relationships of eggplants from three geographically distant secondary centers of diversity. PLoS ONE 7(7):DOI:10.1371/journal.pone.0041748.

Janick J (2001). Asian crops in North America. HortTechnology 11:510-513.

Kashyap V, Kumar SV, Collonnier C, Fusari F, Haicour R, Rotino

GL, Sihachakr D, Rajam RM (2003). Biotechnology of eggplant. Scientia Horticulturae 97:1-25.

Kowalska G (2008). Flowering biology of eggplant and procedures intensifying fruit-set. Acta Scientiarum Polonorum, Hortorum Cultus 7(4):63-76.

Li H, Chen H, Zhuang T, Chen J (2010). Analysis of genetic variation in eggplant and related *Solanum* species using sequence-related amplified polymorphism markers. Scientia Horticulturae 125(1):19-24.

Mantel NA (1967). The detection of disease clustering and a generalized regression approach. Cancer research 27(2 Part 1):209-220.

Muñoz-Falcón JE, Prohens J, Vilanova S, Nuez F (2008). Characterization, diversity, and relationships of the Spanish striped (*Listada*) eggplants: a model for the enhancement and protection of local heirlooms. Euphytica 164(2):*405-419.*

Muñoz-Falcón JE, Prohens J, Vilanova S, Ribas F, Castro A, Nuez F (2009). Distinguishing a protected geographical indication vegetable (Almagro eggplant) from closely related varieties with selected morphological traits and molecular markers. Journal of the Science of Food and Agriculture 89(2):320-328.

Muñoz-Falcón JE, Prohens J, Vilanova S, Nuez F (2009). Diversity in commercial varieties and landraces of black eggplants and implications for broadening the breeders' gene pool. Annals of Applied Biology 154(3):453-465.

Mutlu N, Boyaci FH, Göçmen M, Abak K (2008). Development of SRAP, SRAP-RGA, RAPD and SCAR markers linked with a *Fusarium* wilt resistance gene in eggplant. Theoretical and Applied Genetics 117(8):1303-1312.

Nunome *T*, Ishiguro *K*, Yoshida *T*, Hirai *M (2001).* Mapping of fruit shape and color development traits in eggplant (*Solanum melongena* L.) based on RAPD and AFLP markers. Breeding Science 51(1):19-26.

Nunome T, Negoro S, Kono I, Kanamori H, Miyatake K, Yamaguchi H, Ohyama A, Fukuoka H (2009). Development of SSR markers derived from SSR-enriched genomic library of eggplant (*Solanum melongena* L.). Theoretical and Applied Genetics 119(6):1143-1153.

Polignano G, Uggenti P, Bisignano V, Della Gatta C (2010). Genetic divergence analysis in eggplant (*Solanum melongena* L.) and allied species. Genetic Resources and Crop Evolution 57(2):171-181.

Prohens J, Valcarcel JV, Fernandez de Cordova P, Nuez F (2003). Characterization and typification on Spanish eggplant landraces. Capsicum Eggplant News 22:135-138.

Prohens J, Muñoz-Falcón JE, Vilanova S, Nuez F (2008). Use of molecular markers for the enhancement of local varieties of vegetables for Protected Designations of Origin and Geographical Indications. Bulletin of University of Agricultural Sciences and Veterinary Medicine Cluj-Napoca. Horticulture 65(1):16-20.

Prohens J, Muñoz-Falcón JE, Vilanova S, Nuez F (2011). Comparison of Morphological, AFLP and SSR Markers for the Protection of Eggplant Germplasm. Acta Hort 898:123-131.

Reed DH, Frankham R (2001). How closely correlated are molecular and quantitative measures of genetic variation? a meta-analysis. Evolution 55(6):1095-1103.

Rohlf FJ (1998). NTSYS-PC: Numerical taxonomy and multivariate analysis system. Release 2.20j. Exeter Software, Setauket, N.Y.

Sekara A, Cebula S, Kunicki E (2007). Cultivated eggplants - origin, breeding objectives and genetic resources, a review. Folia Horticulturae 19(1):97-114.

Stàgel A, Portis E, Toppino L, Rotino GL, Lanteri S (2008). Gene-based microsatellite development for mapping and phylogeny studies in eggplant. BMC Genomics 9(1):357.

Swarup V (1995, March). Genetic resources and breeding of aubergine (*Solanum melongena* L.). In I International Symposium on Solanacea for Fresh Market 412 pp. 71-79.

Tiwari SK, Karihaloo JL, Nowsheen H, Gaikwad AB (2009). Molecular characterization of brinjal (*Solanum melongena* L.) cultivars using RAPD and ISSR markers. *Journal of Plant Biochemistry and Biot*echnology 18(2):189-195.

Tümbilen Y, Frary A, Daunay MC, Doganlar S (2011). Application of EST-SSRs to examine genetic diversity in eggplant and its close relatives. Turkish Journal of Biology 35(2):125-136.

Tümbilen Y, Frary A, Mutlu S, Doganlar S (2011). Genetic diversity in Turkish eggplant (*Solanum melongena*) varieties as determined by morphological and molecular analyses. International Research Journal of Biotechnology 2:16-25.

Wang Q, Zhao F, Sun Q, Yang A (2010). Genetic diversity of eggplant revealed by SSR markers. 4th International Conference on Bioinformatics and Biomedical Engineering June 18-20, 2010 Chengdu, China.

Weese TL, Bohs L (2010). Eggplant origins: out of Africa, into the Orient. Taxon 59(1):49-56.

Permissions

All chapters in this book were first published in NBHAC, by University of Agricultural Sciences and Veterinary Medicine, Cluj-Napoca; hereby published with permission under the Creative Commons Attribution License or equivalent. Every chapter published in this book has been scrutinized by our experts. Their significance has been extensively debated. The topics covered herein carry significant findings which will fuel the growth of the discipline. They may even be implemented as practical applications or may be referred to as a beginning point for another development.

The contributors of this book come from diverse backgrounds, making this book a truly international effort. This book will bring forth new frontiers with its revolutionizing research information and detailed analysis of the nascent developments around the world.

We would like to thank all the contributing authors for lending their expertise to make the book truly unique. They have played a crucial role in the development of this book. Without their invaluable contributions this book wouldn't have been possible. They have made vital efforts to compile up to date information on the varied aspects of this subject to make this book a valuable addition to the collection of many professionals and students.

This book was conceptualized with the vision of imparting up-to-date information and advanced data in this field. To ensure the same, a matchless editorial board was set up. Every individual on the board went through rigorous rounds of assessment to prove their worth. After which they invested a large part of their time researching and compiling the most relevant data for our readers.

The editorial board has been involved in producing this book since its inception. They have spent rigorous hours researching and exploring the diverse topics which have resulted in the successful publishing of this book. They have passed on their knowledge of decades through this book. To expedite this challenging task, the publisher supported the team at every step. A small team of assistant editors was also appointed to further simplify the editing procedure and attain best results for the readers.

Apart from the editorial board, the designing team has also invested a significant amount of their time in understanding the subject and creating the most relevant covers. They scrutinized every image to scout for the most suitable representation of the subject and create an appropriate cover for the book.

The publishing team has been an ardent support to the editorial, designing and production team. Their endless efforts to recruit the best for this project, has resulted in the accomplishment of this book. They are a veteran in the field of academics and their pool of knowledge is as vast as their experience in printing. Their expertise and guidance has proved useful at every step. Their uncompromising quality standards have made this book an exceptional effort. Their encouragement from time to time has been an inspiration for everyone.

The publisher and the editorial board hope that this book will prove to be a valuable piece of knowledge for researchers, students, practitioners and scholars across the globe.

List of Contributors

Xiao-Dong CAI
Yangtze University, College of Horticulture and Gardening, Jingzhou, Hubei, P.R. China

Gui-Yuan WANG
Yangtze University, College of Horticulture and Gardening, Jingzhou, Hubei, P.R. China

Wen-Juan CAO
Yangtze University, College of Horticulture and Gardening, Jingzhou, Hubei, P.R. China

Péter PÓTI
Szent István University, Institute of Animal Husbandry, Páter Karoly 1, H-2100 Gödöllő, Hungary

Ferenc PAJOR
Szent István University, Institute of Animal Husbandry, Páter Karoly 1, H-2100 Gödöllő, Hungary

Ákos BODNÁR
Szent István University, Institute of Animal Husbandry, Páter Karoly 1, H-2100 Gödöllő, Hungary

Károly PENKSZA
Szent István University, Institute of Botany and Plant Physiology, Páter Karoly 1, H-2100 Gödöllő, Hungary

János TŐZSÉR
Szent István University, Institute of Animal Husbandry, Páter Karoly 1, H-2100 Gödöllő, Hungary

Csaba GYURICZA
Szent István University, Institute of Crop Production, Páter Karoly 1, H-2100 Gödöllő, Hungary

Arezu MOVLUDI
University of Mohaghegh Ardabili, Faculty of Agriculture, Department of Agronomy, Daneshgah Street, Ardabil, Iran

Ali EBADI
University of Mohaghegh Ardabili, Faculty of Agriculture, Department of Agronomy, Daneshgah Street, Ardabil, Iran

Soodabeh JAHANBAKHSH
University of Mohaghegh Ardabili, Faculty of Agriculture, Department of Agronomy, Daneshgah Street, Ardabil, Iran

M. DAVARI
University of Mohaghegh Ardabili, Faculty of Agriculture, Department of Plant Protection, Daneshgah Street, Ardabil, Iran

Ghasem PARMOON
University of Mohaghegh Ardabili, Faculty of Agriculture, Department of Agronomy, Daneshgah Street, Ardabil, Iran

Marija MARKOVIĆ
Belgrade University, Faculty of Forestry, Ulica kneza Višeslava, 1, 11000 Belgrad, Serbia

Mihailo GRBIĆ
Belgrade University, Faculty of Forestry, Ulica kneza Višeslava, 1, 11000 Belgrad, Serbia

Matilda DJUKIĆ
Belgrade University, Faculty of Forestry, Ulica kneza Višeslava, 1, 11000 Belgrad, Serbia

Zhanwu GAO
Key Laboratory of Vegetation Ecology of Ministry of Education, Institute of Grassland Science, Northeast Normal University, Changchun, 130024, China
Department of Geography, Baicheng Normal College, Baicheng, 137000, China

Jiayu HAN
Key Laboratory of Vegetation Ecology of Ministry of Education, Institute of Grassland Science, Northeast Normal University, Changchun, 130024, China

Chunsheng MU
Key Laboratory of Vegetation Ecology of Ministry of Education, Institute of Grassland Science, Northeast Normal University, Changchun, 130024, China

Jixiang LIN
Alkali Soil Natural Environmental Science Center, Northeast Forestry University, Key Laboratory of Saline-Alkali Vegetation Ecology Restoration in Oil Field, Ministry of Education, Harbin, 150040, China

Xiaoyu LI
Key Laboratory of Wetland Ecology and Environment, Northeast Institute of Geography and Agroecology, Chinese Academy of Science, Changchun 130012, China

Lidong LIN
Dongtou Fisheries Science and Technology Research Institute, Dongtou, 325700, China

Shengnan SUN
Key Laboratory of Vegetation Ecology of Ministry of Education, Institute of Grassland Science, Northeast Normal University, Changchun, 130024, China

Vasilica STAN
University of Agronomic Sciences and Veterinary Medicine, 59 Marasti Bd., Sector 1, 011464, Bucharest, Romania

Gina FÎNTÎNERU
University of Agronomic Sciences and Veterinary Medicine, 59 Marasti Bd., Sector 1, 011464, Bucharest, Romania

Mircea MIHALACHE
University of Agronomic Sciences and Veterinary Medicine, 59 Marasti Bd., Sector 1, 011464, Bucharest, Romania

Santiago VILANOVA
Instituto de Conservación y Mejora de la Agrodiversidad Valenciana, Universitat Politècnica de València, Camino de Vera 14, 46022 Valencia, Spain

Maria HURTADO
Instituto de Conservación y Mejora de la Agrodiversidad Valenciana, Universitat Politècnica de València, Camino de Vera 14, 46022 Valencia, Spain

Adriana CARDONA
Instituto de Conservación y Mejora de la Agrodiversidad Valenciana, Universitat Politècnica de València, Camino de Vera 14, 46022 Valencia, Spain

Mariola PLAZAS
Instituto de Conservación y Mejora de la Agrodiversidad Valenciana, Universitat Politècnica de València, Camino de Vera 14, 46022 Valencia, Spain

Pietro GRAMAZIO
Instituto de Conservación y Mejora de la Agrodiversidad Valenciana, Universitat Politècnica de València, Camino de Vera 14, 46022 Valencia, Spain

Francisco J. HERRAIZ
Instituto de Conservación y Mejora de la Agrodiversidad Valenciana, Universitat Politècnica de València, Camino de Vera 14, 46022 Valencia, Spain

Isabel ANDÚJAR
Instituto de Conservación y Mejora de la Agrodiversidad Valenciana, Universitat Politècnica de València, Camino de Vera 14, 46022 Valencia, Spain

Jaime PROHENS
Instituto de Conservación y Mejora de la Agrodiversidad Valenciana, Universitat Politècnica de València, Camino de Vera 14, 46022 Valencia, Spain

Geza BUJDOSO
Corvinus University of Budapest, Department of Pomology, 29-43, Villányi út, Budapest, Hungary
National Agricultural Research and Innovation Centre, Fruitculture Research Institute, 2, Park utca, Budapest, Hungary

Gyorgy VEGVARI
Corvinus University of Budapest, Department of Soil Science and Water Management, 29-43, Villányi út, Budapest, Hungary

Veronika HAJNAL
Corvinus University of Budapest, Department of Pomology, 29-43, Villányi út, Budapest, Hungary

Gitta FICZEK
Corvinus University of Budapest, Department of Pomology, 29-43, Villányi út, Budapest, Hungary

Magdolna TOTH
Corvinus University of Budapest, Department of Pomology, 29-43, Villányi út, Budapest, Hungary

Chun-Yan LIU
Yangtze University, College of Horticulture and Gardening, No 88 Jingmi Raod, Jingzhou, Hubei 434025, China

Yong-Ming HUANG
Yangtze University, College of Horticulture and Gardening, No 88 Jingmi Raod, Jingzhou, Hubei 434025, China

Ying-Ning ZO
Yangtze University, College of Horticulture and Gardening, No 88 Jingmi Raod, Jingzhou, Hubei 434025, China

Qiang-Sheng WU
Yangtze University, College of Horticulture and Gardening, No 88 Jingmi Raod, Jingzhou, Hubei 434025, China

Ozlem ALAN
Ege University, Odemis Vocational Training School, 35760, Izmir, Turkey

Gulcan KINACI
Eskisehir Osmangazi University, Faculty of Agriculture, Department of Field Crops, 26160, Eskisehir, Turkey

Engin KINACI
Eskisehir Osmangazi University, Faculty of Agriculture, Department of Field Crops, 26160, Eskisehir, Turkey

Zekiye BUDAK BASCIFTCI
Eskisehir Osmangazi University, Faculty of Agriculture, Department of Field Crops, 26160, Eskisehir, Turkey

Kenan SONMEZ
Eskisehir Osmangazi University, Faculty of Agriculture, Department of Horticulture, 26160, Eskisehir, Turkey

Yasemin EVRENOSOGLU
Eskisehir Osmangazi University, Faculty of Agriculture, Department of Horticulture, 26160, Eskisehir, Turkey

Imren KUTLU
Eskisehir Osmangazi University, Faculty of Agriculture, 26160, Eskisehir, Turkey

Md. Rezwan MOLLA
Plant Genetic Resources Centre, Bangladesh Agricultural Research Institute, Gazipur-1701, Bangladesh

M. Rawshan ALI
Molecular Breeding Laboratory, Plant breeding Division, Bangladesh Agricultural Research Institute, Gazipur-1701, Bangladesh

Mirza HASANUZZAMAN
Department of Agronomy, Faculty of Agriculture, Sher-e-Bangla Agricultural University, Dhaka-1207, Bangladesh

Mahamud Hossain AL-MAMUN
Olericulture Division, Horticultural Research Centre, Bangladesh Agricultural Research Institute, Gazipur-1701, Bangladesh

Asgar AHMED
Molecular Breeding Laboratory, Plant breeding Division, Bangladesh Agricultural Research Institute, Gazipur-1701, Bangladesh

M. A. N. NAZIM-UD-DOWLA
Plant Genetic Resources Centre, Bangladesh Agricultural Research Institute, Gazipur-1701, Bangladesh

Md. Motiar ROHMAN
Plant Genetic Resources Centre, Bangladesh Agricultural Research Institute, Gazipur-1701, Bangladesh

Edyta PACZOS-GRZEDA
University of Life Sciences in Lublin, Institute of Plant Genetics, Breeding and Biotechnology, Akademicka Str 15, 20-950 Lublin, Poland

Piotr Tomasz BEDNAREK
Plant Breeding and Acclimatization Institute - NRI, Radzikow, 05-870 Blonie, Poland

Aneta KOROLUK
University of Life Sciences in Lublin, Institute of Plant Genetics, Breeding and Biotechnology, Akademicka Str 15, 20-950 Lublin, Poland

Zygmunt NITA
Strzelce Plant Breeding Company, Glowna Str 20, 99-307 Strzelce, Poland

Zofia BANASZAK
DANKO Plant Breeding Company, Choryn 27, 64-000 Koscian, Poland

Andrzej BICHONSKI
Malopolska Plant Breeding Company, Polanowice, 32-090 Slomniki, Poland

Marek CHMIEL
Malopolska Plant Breeding Company, Polanowice, 32-090 Slomniki, Poland

Agnieszka GRADZIELEWSKA
University of Life Sciences in Lublin, Institute of Plant Genetics, Breeding and Biotechnology, Akademicka Str 15, 20-950 Lublin, Poland

Katarzyna NOWACZYK
DANKO Plant Breeding Company, Choryn 27, 64-000 Koscian, Poland

Aleksandra SZOLKOWSKA
DANKO Plant Breeding Company, Choryn 27, 64-000 Koscian, Poland

Krystyna WERWINSKA
Strzelce Plant Breeding Company, Glowna Str 20, 99-307 Strzelce, Poland

Patrycja WIECZOREK
University of Life Sciences in Lublin, Institute of Plant Genetics, Breeding and Biotechnology, Akademicka Str 15, 20-950 Lublin, Poland

Saeid ABU-ROMMAN
Department of Biotechnology Al-Balqa' Applied University, Al-Salt, 19117, Jordan

Mazen ATEYYAT
Department of Plant Production and Protection. Faculty of Agricultural Technology, Al-Balqa' Applied University, Al-Salt, 19117, Jordan

Xiao-Bu CAI
Department of Resources and Environment, Tibet Agricultural and Animal Husbandry College, Tibet, 860000, China

Yue-Lin PENG
Department of Plant Science, Tibet Agricultural and Animal Husbandry College, Tibet, 860000, China

Min-Na YANG
Department of Plant Science, Tibet Agricultural and Animal Husbandry College, Tibet, 860000, China

Tao ZHANG
Institute of Grassland Science, Northeast Normal University, Key Laboratory of Vegetation Ecology, Ministry of Education, Changchun, 130024, China

Qiang ZHANG
Institute of Geographic Sciences and Natural Resources Research, Chinese Academy of Sciences, Beijing 100101, China

Hakoomat ALI
Bahauddin Zakariya University, Department of Agronomy, Faculty of Agricultural Sciences and Technology, Multan, 60800 Pakistan

Zuhair HASNAIN
Bahauddin Zakariya University, Department of Agronomy, Faculty of Agricultural Sciences and Technology, Multan, 60800 Pakistan

Ahmad N. SHAHZAD
Bahauddin Zakariya University, Department of Agronomy, Faculty of Agricultural Sciences and Technology, Multan, 60800 Pakistan

Naeem SARWAR
Bahauddin Zakariya University, Department of Agronomy, Faculty of Agricultural Sciences and Technology, Multan, 60800 Pakistan

Muhammad K. QURESHI
Bahauddin Zakariya University, Department of Plant Breeding and Genetics, Faculty of Agricultural Sciences and Technology, Multan, 60800 Pakistan

Shazia KHALIQ
Bahauddin Zakariya University, Department of Environmental Sciences, Multan, 60800 Pakistan

Muhammad F. QAYYUM
Bahauddin Zakariya University, Department of Soil Science, Multan, 60800 Pakistan

Veronika HAJNAL
Corvinus University of Budapest, Department of Pomology, 29-43 Villányi út, Budapest, Hungary

Zarif OMID
Corvinus University of Budapest, Department of Pomology, 29-43 Villányi út, Budapest, Hungary

Márta LADÁNYI
Corvinus University of Budapest, Department of Mathematics and Informatics, 29-43 Villányi út, Budapest, Hungary

Magdolna TÓTH
Corvinus University of Budapest, Department of Pomology, 29-43 Villányiút, Budapest, Hungary

László SZALAY
Corvinus University of Budapest, Department of Pomology, 29-43 Villányi út, Budapest, Hungary

Meral INCESU, Berken CIMEN
Cukurova University, Faculty of Agriculture, Department of Horticulture, 01330, Adana, Turkey

Turgut YESILOGLU
Cukurova University, Faculty of Agriculture, Department of Horticulture, 01330, Adana, Turkey

Bilge YILMAZ
Cukurova University, Faculty of Agriculture, Department of Horticulture, 01330, Adana, Turkey

Roxana Maria MADJAR
University of Agronomic Sciences and Veterinary Medicine, Faculty of Agriculture, 59 Marasti Blvd., District 1, 011464, Bucharest, Romania

Gina VASILE SCAEȚEANU
University of Agronomic Sciences and Veterinary Medicine, Faculty of Agriculture, 59 Marasti Blvd., District 1, 011464, Bucharest, Romania

Mircea MIHALACHE
University of Agronomic Sciences and Veterinary Medicine, Faculty of Agriculture, 59 Marasti Blvd., District 1, 011464, Bucharest, Romania

Catalina CALIN
Petroleum-Gas University of Ploiesti, Faculty of Petroleum Refining and Petrochemistry, 39 Blvd. Bucuresti, 100680, Ploiesti, Romania

Valentin Sebastian DAN
University of Agricultural Sciences and Veterinary Medicine, 3-5 Manastur St., Cluj-Napoca, Romania

Adrian PETICILA
University of Agronomic Sciences and Veterinary Medicine, Faculty of Horticulture, 59 Marasti Blvd., District 1, 011464, Bucharest, Romania

Fatih SEN
Ege University, Faculty of Agriculture Department of Horticulture 35100 Bornova, Izmir, Turkey

Rustu Efe OKSAR
Ege University, Faculty of Agriculture Department of Horticulture 35100 Bornova, Izmir, Turkey

Mina GOLKARIAN
Ege University, Faculty of Agriculture Department of Horticulture 35100 Bornova, Izmir, Turkey

Sevde YALDIZ
Ege University, Faculty of Agriculture Department of Horticulture 35100 Bornova, Izmir, Turkey

Guillermo NIÑO-MEDINA
Centro de Investigación en Alimentación y Desarrollo (CIAD) A.C., Unidad Culiacán, Carretera Culiacán a El Dorado, km 5.5, C.P. 80110, Culiacán, Sinaloa, Mexico
Universidad Autónoma de Nuevo León, Facultad de Agronomía, Francisco Villa S/N, C.P. 66050, Col. Ex-Hacienda El Canadá, Escobedo, Nuevo León, Mexico

Dolores MUY-RANGEL
Centro de Investigación en Alimentación y Desarrollo (CIAD) A.C., Unidad Culiacán, Carretera Culiacán a El Dorado, km 5.5, C.P. 80110, Culiacán, Sinaloa, Mexico

Alfonso GARDEA-BÉJAR
Centro de Investigación en Alimentación y Desarrollo, (CIAD) A.C., Unidad Hermosillo, Carretera a La Victoria km 0.6, C.P. 83304, Hermosillo, Sonora, Mexico

Gustavo GONZÁLEZ-AGUILAR
Centro de Investigación en Alimentación y Desarrollo, (CIAD) A.C., Unidad Hermosillo, Carretera a La Victoria km 0.6, C.P. 83304, Hermosillo, Sonora, Mexico

Basilio HEREDIA
Centro de Investigación en Alimentación y Desarrollo (CIAD) A.C., Unidad Culiacán, Carretera Culiacán a El Dorado, km 5.5, C.P. 80110, Culiacán, Sinaloa, Mexico

Manuel BÁEZ-SAÑUDO
Centro de Investigación en Alimentación y Desarrollo (CIAD) A.C., Unidad Culiacán, Carretera Culiacán a El Dorado, km 5.5, C.P. 80110, Culiacán, Sinaloa, Mexico

Jorge SILLER-CEPEDA
Centro de Investigación en Alimentación y Desarrollo (CIAD) A.C., Unidad Culiacán, Carretera Culiacán a El Dorado, km 5.5, C.P. 80110, Culiacán, Sinaloa, Mexico

Rosabel VÉLEZ DE LA ROCHA
Centro de Investigación en Alimentación y Desarrollo (CIAD) A.C., Unidad Culiacán, Carretera Culiacán a El Dorado, km 5.5, C.P. 80110, Culiacán, Sinaloa, Mexico

Okan OZKAYA
Cukurova University, Agriculture Faculty, Horticulture Department, Balcali, Adana, Turkey

Songül ÇÖMLEKÇIOGLU
Cukurova University, Agriculture Faculty, Horticulture Department, Balcali, Adana, Turkey

Hatice DEMIRCIOGLU
Cukurova University, Agriculture Faculty, Horticulture Department, Balcali, Adana, Turkey

Qiang-Sheng WU
Yangtze University, College of Horticulture and Gardening/Institute of Root Biology, Jingzhou, Hubei 434025, People's Republic of China

Qiu-Dan NI
Yangtze University, College of Horticulture and Gardening/Institute of Root Biology, Jingzhou, Hubei 434025, People's Republic of China

Yan-Cong QUE
Yangtze University, College of Horticulture and Gardening/Institute of Root Biology, Jingzhou, Hubei 434025, People's Republic of China

Wei HUANG
Yangtze University, College of Horticulture and Gardening/Institute of Root Biology, Jingzhou, Hubei 434025, People's Republic of China

Leo SABATINO
University of Palermo, Department of Agricultural and Forest Sciences, Viale delle Scienze, 90128, Palermo, Italy

Fabio D'ANNA
University of Palermo, Department of Agricultural and Forest Sciences, Viale delle Scienze, 90128, Palermo, Italy

Giovanni IAPICHINO
University of Palermo, Department of Agricultural and Forest Sciences, Viale delle Scienze, 90128, Palermo, Italy

Abdullah KARASU
Uludag University, Mustafakemalpasa Vocational School, Department of Plant Production, 16500 Bursa, Turkey

Hayrettin KU CU
Uludag University, Mustafakemalpasa Vocational School, Department of Plant Production, 16500 Bursa, Turkey

Mehmet ÖZ
Uludag University, Mustafakemalpasa Vocational School, Department of Plant Production, 16500 Bursa, Turkey

Gamze BAYRAM
Uludag University, Faculty of Agriculture, Department of Field Crops, 16059 Bursa, Turkey

Sina COSMULESCU
University of Craiova, Agriculture & Horticulture Faculty, Department of Horticulture & Food Science, A.I.Cuza Street, 13, Craiova, Romania

Ion TRANDAFIR
University of Craiova, Sciences Faculty, Department of Chemistry, Calea Bucuresti Street, 107, Craiova, Romania

Violeta NOUR
University of Craiova, Agriculture & Horticulture Faculty, Department of Horticulture & Food Science, A.I.Cuza Street, 13, Craiova, Romania

Mira IONICA
University of Craiova, Agriculture & Horticulture Faculty, Department of Horticulture & Food Science, A.I.Cuza Street, 13, Craiova, Romania

Felicia TUTULESCU
University of Craiova, Agriculture & Horticulture Faculty, Department of Horticulture & Food Science, A.I.Cuza Street, 13, Craiova, Romania

Marcel PARVU
"Babes-Bolyai" University, Faculty of Biology and Geology, Department of Biology, 42 Republicii Street, 400015 Cluj-Napoca, Romania

Laurian VLASE
"Iuliu Hatieganu" University of Medicine and Pharmacy, Faculty of Pharmacy, Department of Pharmaceutical Technology and Biopharmaceutics, 12 Ion Creanga Street, 400010 Cluj-Napoca, Romania

Alina E. PARVU
"Iuliu Hatieganu" University of Medicine and Pharmacy, Faculty of Medicine, Department of Pathophysiology, 3 Victor Babes Street, 400012 Cluj-Napoca, Romania

Oana ROSCA-CASIAN
"Babes-Bolyai" University, "A. Borza" Botanical Garden, 42 Republicii Street, 400015 Cluj-Napoca, Romania

Ana-Maria GHELDIU
"Iuliu Hatieganu" University of Medicine and Pharmacy, Faculty of Pharmacy, Department of Pharmaceutical Technology and Biopharmaceutics, 12 Ion Creanga Street, 400010 Cluj-Napoca, Romania

Ovidiu PARVU
"Babes-Bolyai" University, Faculty of Mathematics and Computer Science, 1 Kogalniceanu Street, 400084 Cluj-Napoca, Romania

Florin PACURAR
University of Agricultural Sciences and Veterinary Medicine Cluj-Napoca, Faculty of Agriculture, Calea Mana tur 3-5, 400372 Cluj-Napoca, Romania

Ioan ROTAR
University of Agricultural Sciences and Veterinary Medicine Cluj-Napoca, Faculty of Agriculture, Calea Mana tur 3-5, 400372 Cluj-Napoca, Romania

Albert REIF
University of Freiburg, Faculty of Environment and Natural Resources, Tennenbacherstr. 4, 79085 Freiburg, Germany

Roxana VIDICAN
University of Agricultural Sciences and Veterinary Medicine Cluj-Napoca, Faculty of Agriculture, Calea Mana tur 3-5, 400372 Cluj-Napoca, Romania

Vlad STOIAN
University of Agricultural Sciences and Veterinary Medicine Cluj-Napoca, Faculty of Agriculture, Calea Mana tur 3-5, 400372 Cluj-Napoca, Romania

Stefanie M. GÄRTNER
University of Freiburg, Faculty of Environment and Natural Resources, Tennenbacherstr. 4, 79085 Freiburg, Germany

Robert B. ALLEN
Landcare Research, P.O. Box 69040, Lincoln 7640, New Zealand

Bulent KOSE
Ondokuz Mayis University, Faculty of Agriculture, Department of Horticulture, 55139 Atakum, Samsun, Turkey

Hossam Saad EL-BELTAGI
Cairo University, Biochemistry Department, Faculty of Agriculture, Gamma st, Giza, Cairo, Egypt

Osama K. AHMED
Cairo University, Biochemistry Department, Faculty of Agriculture, Gamma st, Giza, Cairo, Egypt

Adel E. HEGAZY
University of Sadat City, Genetic Engineering and Biotechnology Institute (GEBRI), Plant Biotechnology Department, Egypt

Huseyin KALKAN
Akdeniz University, Faculty of Agriculture, Department of Soil Science and Plant Nutrition, Antalya, Turkey

Mustafa KAPLAN
Akdeniz University, Faculty of Agriculture, Department of Soil Science and Plant Nutrition, Antalya, Turkey

Şule ORMAN
Akdeniz University, Faculty of Agriculture, Department of Soil Science and Plant Nutrition, Antalya, Turkey

Liviu Mihai IRIMIA
University of Agricultural Sciences and Veterinary Medicine, Faculty of Horticulture, 3, Mihail Sadoveanu Alley, Iasi, 700490, Romania

Cristian Valeriu PATRICHE
Romanian Academy, Iasi Branch, Geography Group, 8, Carol I Street, 700505, Iasi, Romania

Georgeta Mihaela BUCUR
University of Agricultural Sciences and Veterinary Medicine, Faculty of Horticulture, 59, Bulevardul Marasti, 011464, Bucharest, Romania

Herve QUENOL
LETG-Rennes (COSTEL), UMR 6554 CNRS, Université Rennes-2, Rue du Recteur Henry le Moal, 35000, Rennes, France

Valeriu V. COTEA
University of Agricultural Sciences and Veterinary Medicine, Faculty of Horticulture, 3, Mihail Sadoveanu Alley, Iasi, 700490, Romania

Géza NAGY
Corvinus University of Budapest, Faculty of Horticultural Sciences, Department of Plant Pathology, 44. Ménesi Str., H-1118 Budapest, Hungary

Tamás HOCHBAUM
Corvinus University of Budapest, Faculty of Horticultural Sciences, Department of Plant Pathology, 44. Ménesi Str., H-1118 Budapest, Hungary

Szilvia SÁROSI
Corvinus University of Budapest, Faculty of Horticultural Sciences, Department of Medicinal and Aromatic Plants, 29-43 Villányi Str., H-1118 Budapest, Hungary

Márta LADÁNYI
Corvinus University of Budapest, Faculty of Horticultural Sciences, Department of Biometrics and Agricultural Informatics, 29-43 Villányi Str.,H-1118 Budapest, Hungary

Mohamad AL HASSAN
Universitat Politècnica de València, Instituto de Biología Molecular y Celular de Plantas (UPV-CSIC), CPI, edificio 8E, Camino de Vera s/n, 46022 Valencia, Spain

Marina MARTÍNEZ FUERTES
Universitat Politècnica de València, Instituto de Biología Molecular y Celular de Plantas (UPV-CSIC), CPI, edificio 8E, Camino de Vera s/n, 46022 Valencia, Spain

Francisco José RAMOS SÁNCHEZ
Universitat Politècnica de València, Instituto de Biología Molecular y Celular de Plantas (UPV-CSIC), CPI, edificio 8E, Camino de Vera s/n, 46022 Valencia, Spain

Oscar VICENTE
Universitat Politècnica de València, Instituto de Biología Molecular y Celular de Plantas (UPV-CSIC), CPI, edificio 8E, Camino de Vera s/n, 46022 Valencia, Spain

Monica BOSCAIU
Universitat Politècnica de València, Instituto Agroforestal Mediterráneo (UPV), CPI, edificio 8E, Camino de Vera s/n, 46022 Valencia, Spain

Xiaohui WANG
College of Horticulture and Gardening, Yangtze University, Jingzhou, 434023, China

Junhuan CHENG
College of Horticulture and Gardening, Yangtze University, Jingzhou, 434023, China

Feng XU
College of Horticulture and Gardening, Yangtze University, Jingzhou, 434023, China

Xingxiang LI
Medical School, Yangtze University, Jingzhou, 434023, China

Weiwei ZHANG
College of Horticulture and Gardening, Yangtze University, Jingzhou, 434023, China

Yongling LIAO
College of Forestry, Nanjing Forest University, Nanjing, 210037, China

Shuiyuan CHENG
Wuhan Polytechnic University, Wuhan 430023, China

Xianbin LI
Jingzhou Soil and Fertilizer Station, Jingzhou 434025, China

Nafiye ADAK
Akdeniz University, Technical Sciences Vocational School, Environmental Protection and Control Program, 07058, Antalya, Turkey

Hamide GUBBUK
Akdeniz University, Faculty of Agriculture, Department of Horticulture, 07058, Antalya, Turkey

Hatice Filiz BOYACI
Bati Akdeniz Agricultural Res. Inst., Dept. of Vegetable Crops and Ornamentals, 07100 Antalya, Turkey

Volkan TOPCU
Bati Akdeniz Agricultural Res. Inst., Dept. of Vegetable Crops and Ornamentals, 07100 Antalya, Turkey

Akin TEPE
Bati Akdeniz Agricultural Res. Inst., Dept. of Vegetable Crops and Ornamentals, 07100 Antalya, Turkey

Isilay Karasahin YILDIRIM
Bati Akdeniz Agricultural Res. Inst., Dept. of Vegetable Crops and Ornamentals, 07100 Antalya, Turkey

Mehmet OTEN
Bati Akdeniz Agricultural Res. Inst., Dept. of Vegetable Crops and Ornamentals, 07100 Antalya, Turkey

Aytekin AKTAS
Bati Akdeniz Agricultural Res. Inst., Dept. of Vegetable Crops and Ornamentals, 07100 Antalya, Turkey